Applications of Nanotechnology in Microbiology

Ramesh S. Chaughule • Anushree S. Lokur
Editors

Applications of Nanotechnology in Microbiology

Springer

Editors
Ramesh S. Chaughule
Department of Chemistry
Ramnarain Ruia Autonomous College
Mumbai, India

Anushree S. Lokur
Department of Microbiology
Ramnarain Ruia Autonomous College
Mumbai, India

ISBN 978-3-031-49932-6 ISBN 978-3-031-49933-3 (eBook)
https://doi.org/10.1007/978-3-031-49933-3

This Springer imprint is published by the registered company Springer Nature Switzerland AG
The registered company address is: Gewerbestrasse 11, 6330 Cham, Switzerland

Paper in this product is recyclable.

Foreword 1

I am delighted to compose this foreword for this book that professors Ramesh Chaughule and Anushree Lokur have expertly curated, both of whom are distinguished contributors to this volume as well. The amalgamation of microbiology and nanotechnology has exhibited remarkable success in offering pioneering solutions to safeguard human health and the environment. When nanotechnology is seamlessly integrated with microbiology, it has showcased its efficacy across diverse domains, including medicine, pharmaceuticals, industry, agriculture, and environmental applications. This integration has effectively surmounted the challenges posed by microbiological limitations.

This book stands as a compendium of the most recent scientific advancements, encompassing fundamental applications of microbial nanotechnology across various sectors aimed at combating bacterial pathogens, exploiting nanotechniques within microbiology, and showcasing innovative strides in the domain of medical microbiology.

The book encompasses a wide array of chapters contributed by leading experts in the exciting realm of nanomaterial application within microbiology, covering topics such as microbial biosynthesis of nanomaterials, research in food sciences, vaccine development, biofilm investigation, microfluidic device utilization, drug delivery mechanisms, diagnostic tool innovations, and environmental and wastewater remediation strategies.

This collection of timely and indispensable articles furnishes invaluable information that will prove beneficial to an array of disciplines. Research scientists in academic institutions and industry, students, clinicians, and technologists will all find immense value in this book.

The field of nanotechnology and microbiology is rapidly evolving, and this book serves as a timely resource to bring us up to speed with the latest developments and trends in the field. I extend my heartfelt congratulations to Profs. Chaughule and

Lokur, as well as the publisher for their significant contributions to advancing knowledge in this dynamic area of research.

Charlotte Drake Cardeza Professor of Medicine, Ulhas P. Naik
Director, Cardeza Center for Hemostasis, Thrombosis, and Vascular Biology,
Associate Director, Cardeza Foundation for Hematologic Research, Sidney Kimmel Medical College, Jefferson (Philadelphia University + Thomas Jefferson University)
Philadelphia, PA, USA
October 9, 2023

Foreword 2

In science and technology, a few frontiers hold as much promise and intrigue as the convergence of nanotechnology and microbiology. The book you hold in your hands, "Applications of Nanotechnology in Microbiology," is a comprehensive exploration of this dynamic field, authored by leading experts who have dedicated their research to unveiling the profound impacts that nanoscale materials can have on the world of microbiology.

At the heart of this book are 17 chapters that span a broad spectrum of topics, showcasing the versatile nature of nanotechnology in microbial studies and its far-reaching applications. The authors take us on a captivating journey through the fascinating world of microbiology, offering insights into cutting-edge research that harnesses the potential of nanomaterials to address some of the most pressing challenges in healthcare, biotechnology, and environmental science.

The book opens with a chapter on the "Biosynthesis of Metal Nanoparticles Using bacterial Metabolites and Their applications," which demonstrates the ingenious use of microbial resources to create functional nanomaterials. Following this, "Insights on Microbes Mediated Greener Synthesis of Nanoparticles" delves into this exciting approach's ecological advantages and challenges.

Readers will also find chapters exploring the essential role of nanotechnology in combating infectious diseases, focusing on "Use of Nanomaterials as an Antimicrobial and Antiviral Regimen" and "Vaccine Nanotechnology for the Prevention of Infectious Diseases." These chapters provide a glimpse into groundbreaking research efforts that aim to revolutionize disease prevention and treatment.

The book further extends its reach into the domain of food microbiology, highlighting the impact of nanotechnology on food safety, supply, and pathogen detection. Chapters like "Application of Nanotechnology in Food Microbiology" and "Nano-material Based Sensing Platforms for Food Borne Pathogen Detection"" provide valuable insights into how nanomaterials can enhance the quality and safety of the food we consume.

As the reader progresses through the chapters, the potential of nanotechnology to address a broad range of microbial challenges becomes apparent. From controlling dental microbial biofilms to developing advanced antifungal nanomaterials and exploring the use of nanosilver diamine fluoride in dental caries treatment, this book offers a comprehensive view of how nanotechnology is changing the landscape of microbial control and treatment.

The integration of nanotechnology with cutting-edge research tools is also explored in chapters such as "Next Generation Sequencing and Solid State Nanopores," which delve into the use of nanotechnology to enhance genomic analysis. In addition, "Emerging Microfluidics Devices for Microbial Studies" sheds light on the promising role of microfluidic devices in advancing our understanding of microbes.

Wastewater remediation, a critical global challenge, receives attention in the chapter "Insights of Nano-biotechnology as Bio-adsorbents for Wastewater Remediation," highlighting the sustainable potential of nanomaterials in addressing environmental concerns.

"Applications of Nanotechnology in Microbiology" is an invaluable resource for researchers, academics, students, and anyone with a curious mind and a passion for understanding the revolutionary potential of the microscopic world.

Warmest congratulations to Dr Ramesh S. Chaughule and Dr Anushree Lokur for editing and publishing this remarkable book. Your tireless efforts and dedication have resulted in a comprehensive and invaluable resource that sheds light on the transformative intersection of nanotechnology and microbiology. I have no doubt that it will inspire countless individuals to delve into the exciting world of nanotechnology in microbiology.

Visiting Professor, Indian Institute of Technology Bombay, Mumbai, India
Former Director, Agharkar Research Institute, Pune, India
10 October 2023

Kishore M. Paknikar

Preface

Nanotechnology is mainly concerned for the creation of functional materials, devices, and systems through control of matter in 1–100 nm length scale exploiting physical, chemical, and biological properties. Microbiology is the scientific study of living microorganisms, such as bacteria, viruses, fungi, algae, and protozoa. Microbiology encompasses various aspects of these microorganisms, including their structure, function, genetics, ecology, and interactions with other living organisms. Thus their behavior, evolution, and the complex interactions with their surrounding are studied.

The combination of microbiology and nanotechnology has proven to be highly successful in offering innovative solutions for safeguarding human health and the environment. The wealth of information derived from these fields is immense, and despite being a relatively new interdisciplinary science, it has already yielded significant benefits in human, environmental, and ecological sciences. Nanotechnology, when integrated with microbiology, has demonstrated its effectiveness in diverse domains such as medicine, pharmaceuticals, industry, agriculture, and environmental applications, effectively addressing existing limitations posed by microbiological challenges. "Microbial nanotechnology" is a compilation of the latest scientific advancements in the fundamental application of microbial nanotechnology across various sectors to combat bacterial pathogens, applications of nanotechniques in microbiology, and innovative advances in the area of medical microbiology. Applications of microbial nanotechnology are currently emerging with new areas being explored.

We are pleased to introduce this fourth book published by Springer Nature "Applications of Nanotechnology in Microbiology" that serves as a highlight of the extensive subject areas in the domain of microbiology to aspiring and working scientists and showcase its potential applications in various fields, including food, pharmacology, medicine, water, plants, environmental remediation, etc. The book covers primarily six sections. The first section covers microbial biosynthesis of nanomaterials. The second section covers vaccines research. The third section is on foods. The fourth one is for biofilm research. The fifth one is for microfluidic devices, drug delivery, diagnostic tools, and applications; and the sixth one is for environmental and wastewater remediation.

1. Biosynthesis of nanomaterials by microorganisms is a recently attracting interest as a new, exciting approach toward the development of "greener" nanomanufacturing compared to traditional chemical and physical approaches. The chapter by Preethi Kathirvel et al. highlights the significance of combining nanotechnology and microbiology in the biosynthesis of metal nanoparticles using bacterial metabolites.

   The biogenic approach offers an eco-friendly and cost-effective alternative to chemical and physical methods. The unique characteristics of microbiologically synthesized metal nanoparticles open up a wide range of applications in sectors such as medicine, cosmetics, food, agriculture, textiles, and biotechnology industries, promising a sustainable and innovative future. The chapter by Gita Singh and Chandra highlights the potential of microbes in synthesizing nanoparticles as a greener alternative to traditional chemical methods. Microbial synthesis offers advantages such as non-toxicity, cost-effectiveness, and scalability for commercial applications. The chapter discusses different microbial agents, synthesis conditions, and mechanisms involved, along with applications and potential challenges for the future advancement of this environmentally friendly approach to nanoparticle synthesis.
2. The NMs have gained significant attention in the field of antimicrobial and antiviral research, driven by the escalating concern of antibiotic resistance. These materials, whether synthetic, semi-synthetic, or natural, offer promising therapeutic applications as drug delivery vehicles and broad-spectrum inhibitors against microbial and viral diseases, benefiting from their large surface area to volume ratio. The chapter by Chakraborty et al. focuses on the potential therapeutic use of nanoparticles as antibacterial, antifungal, and antiviral agents, discussing their mechanisms of action, advantages, drawbacks, and possible side effects. The significance of infectious diseases and the rise of antibiotic-resistant superbugs underscore the importance of vaccines as an effective public health measure.

   The success of lipid nanoparticle-based mRNA vaccines during the COVID-19 pandemic showcases the potential of nanotechnology in vaccine development. Biologically derived nanoparticles with native pathogen-associated molecular patterns can reduce the need for synthetic adjuvants and possess beneficial qualities like biodegradability and biocompatibility, making them scalable for large-scale production. The chapter by Sontakke et al. discusses various materials, processes, administration approaches, and future prospects of vaccine nanotechnology for preventing infectious diseases.

   The chapter by Bharti et al. focuses on comparing free and immobilized antimicrobial nanoparticles, exploring their applications in various fields, and addressing concerns related to toxicity and environmental impact. Understanding the differences, potential uses, risks, and challenges associated with these nanomaterials, this will help inform decision-making for scientists, policymakers, and businesses looking to apply antimicrobial nanotechnology effectively and responsibly in medicine, water purification, food packaging, and consumer goods.

3. Nanotechnology has emerged as a valuable tool in the field of food microbiology, offering significant contributions to food science and technology. Chapter by Guha et al. explains this through the use of various organic, inorganic, and bio-nanomaterials, nanotechnologies that can detect and inhibit the growth of food-spoilage microorganisms on food surfaces, extend shelf-life, and enhance the survival of probiotics under challenging conditions.

   Nanoparticle-based biosensors have also been developed to detect food-borne pathogens and hazardous substances in food, ensuring the production of contamination-free and safe food products, thus having implications on public health. The recent discovery of the CRISPR-Cas9 genome editing platform has significant potential in plant breeding and plant-microbe interactions. The chapter by Bharathan and Turarbekova discusses the historical context of selective domestication of crop plants and farm animals, which has evolved with the help of advanced biological tools. It also highlights the potential role of nanomaterials in disease suppression to ensure global food security, suggesting that nanotechnology can play a crucial role in advancing the food supply chain.

   Traditional methods for pathogen detection are time-consuming and may yield false negative results. Nanotechnology-mediated biosensors offer reliable, cost-effective, rapid, and sensitive diagnostic tools, utilizing novel bio-receptors and nanomaterial-based transducers to enhance selectivity and sensitivity for early pathogen detection in complex food matrices. The chapter by Pathania et al. explores the significance of nanomaterial-based sensing platforms for detecting foodborne pathogens.

4. Traditional treatment options for biofilm eradication are limited, but nanotechnology-based approaches offer promising possibilities. The use of nanoparticle-based systems as active therapeutic agents to target specific microbial and biofilm features, as well as their potential as drug delivery vehicles, is being studied to address the challenges of controlling and treating dental microbial biofilms effectively. The chapter by Shetty et al. highlights the significance of dental microbial biofilms and their association with various dental disorders.

   Various types of nanoparticles and nano-systems have shown potential in combating *C. albicans* biofilms, making nanotechnology a promising new avenue for preventing and treating fungal infections. Nanotechnology offers a promising therapeutic approach to improve the stability, solubility, and bioabsorption of antifungal agents, as well as targeted delivery to biofilm sites to enhance their efficacy and reduce side effects. The chapter by Shelar and others discusses the challenges of antimicrobial resistance and biofilm formation, specifically focusing on *Candida albicans* infections.

   Silver diamine fluoride (SDF) has gained popularity as an effective material for arresting dental caries. The use of nano-scaled SDF is particularly beneficial for young patients who may have difficulty tolerating traditional dental treatments, reducing the need for general anesthesia or sedation. The chapter by Aksoy et al. explores the mechanisms of action, significance of nanoparticles, synthesis pathways, potential adverse effects, and clinical protocols associated with SDF and nano SDF in dental caries management.

5. Microfluidics devices allow for fluid manipulation at a small scale, enabling various translational research applications. Nanotechnologies integrated with microfluidics contribute to reduced sample consumption, lower costs, faster analysis, increased sensitivity, and automation, making them promising tools for advancing microbial studies. The chapter by Khachane et al. highlights the significance of microfluidics technologies in microbial studies.

   Nanocarriers such as liposomes, solid-lipid nanoparticles, polymer-based nanoparticles, and metal-based nanoparticles are explored to enhance drug delivery efficacy. These nanosystems enable targeted drug delivery, improve bioavailability, and overcome barriers to effectively deliver microbe-based formulations at the desired site of action, addressing challenges associated with traditional drug delivery methods. The chapter by Shirsat et al. highlights the significance of nanotechnology in formulating microorganism-based drug delivery systems.

   Nanotechnology-based electrochemical biosensors offer advantages in rapid, sensitive, and cost-effective detection of viral biomarkers, making them valuable tools for point-of-care testing. The chapter by Cetinkaya et al. discusses the importance of detecting viral diseases and the challenges they pose to public health. It explores recent developments in electrochemical biosensors using nanomaterials, while also highlighting the advantages and disadvantages of point-of-care testing compared to traditional diagnostic methods.

   Next-generation sequencing (NGS) and nanotechnology-based nanopore sequencing have revolutionized DNA sequencing by providing details at the single-molecule level, enhancing accuracy and speed. Molecular dynamics (MD) simulations are explored to understand the effects of nanopore material properties and geometry on sequencing results. The chapter by Mallakmir et al. discusses the rapid evolution of genome sequencing and its wide applications in life science research. Also it highlights the growing list of NGS applications in various fields.

6. Environmental remediation specifically focuses on the environmental applications of microbial nanotechnology. It covers recent advancements and emerging trends in environmental remediation using microbial systems and nanotechnologies. MDC (microbial desalination cell) integrates microbial fuel cell (MFC) and electrodialysis for wastewater treatment, desalination, and renewable energy production. The chapter by Khiran and others highlights the potential of seawater as a source of freshwater through this newly developed technology MDC and the use of microorganisms, and algae involved in the anode chambers can successfully help to overcome challenges associated with MFC.

   Nanobiotechnology offers a promising solution through the use of bioadsorbents, which are nanoparticles with tunable structural properties and high surface area. These bioadsorbents demonstrate efficient performance in removing emerging contaminants, such as heavy metals, drugs, and microbes from wastewater, making them a potential tool for wastewater treatment. The chapter by Naaz et al. discusses the role of biosynthesized nanoparticles in developing bioadsorbents and their structural properties and applications, along with future prospects in nanobiotechnology for wastewater remediation.

As this interdisciplinary science continues to evolve, further innovations are expected to enhance human well-being, environmental sustainability, and ecological balance, provided adequate safety measures and regulations are implemented.

The editors wish to thank all the distinguished and expert contributors for their enthusiastic participation in this endeavor. We are confident that the book will serve as a valuable guide for researchers and students of medicine, pharmacy, environmental science, materials chemistry, food science, and microbiology. Dr. Chaughule wishes to thank his family members and staff of Ramnarain Ruia autonomous college for all the support. Dr. Anushree Lokur wishes to thank the staff of Microbiology department, administration, and the family for full support.

Last but not the least, the editors sincerely thank the staff of the Springer Nature for their support from time to time.

Mumbai, India Ramesh S. Chaughule
Mumbai, India Anushree S. Lokur

Contents

1 **Biosynthesis of Metal Nanoparticles Using Bacterial Metabolites and Their Applications** 1
Preethi Kathirvel, Gayathri Devi Kandasamy, and Mouliraj Palanisamy

2 **Insights on Microbes-Mediated Greener Synthesis of Nanoparticles: Advantages and Challenges** 41
Gita Singh and Sudeshna Chandra

3 **Use of Nanomaterials as an Antimicrobial and Antiviral Regimen** 59
Ashok Chakraborty, Anil Diwan, and Jayant Tatake

4 **Vaccine Nanotechnology for the Prevention of Infectious Diseases** 77
Sumit Sontakke, Poonam Sharma, Rebecca Worsham, Kishan Kadur, Chandana Bhargavi, and Olugbenga Morebise

5 **Exploring the Application, Safety, and Challenges of Free Versus Immobilized Antimicrobial Nanomaterials** 97
Sharda Bharti, Pooja Nag, Kapil Sadani, Soumyo Mukherji, and Suparna Mukherji

6 **Application of Nanotechnology in Food Microbiology: Implication on Public Health** 135
Smita Guha, Ashok Chakraborty, and Debjit Chakraborty

7 **Application of CRISPR Technology and Nanomaterials to Advance Food Supply** 157
Narayanaswamy Bharathan and Zhibek Turarbekova

8 Nanomaterial-Based Sensing Platforms for Food-Borne Pathogen Detection 193
Preeti K. Pathania, Navjeet Kaur Lotey, Dev Kashyap, and Priyanka Sabherwal

9 Dental Microbial Biofilms: Control and Treatment Through Nanotechnology Approaches 229
Heeresh Shetty, Varsha Shukla, Anushree S. Lokur, and Ramesh S. Chaughule

10 Recent Advances in Antifungal Nanomaterials for Combating Biofilm Infection Caused by *Candida albicans* 271
Amruta Shelar, Pravin N. Didwal, and Rajendra Patil

11 The Effect of Nano Silver Diamine Fluoride in Arresting Dental Caries 291
Merve Aksoy, Cenkhan Bal, Mohamed Brahmi, and Nurgul K. Bakirhan

12 Emerging Microfluidics Devices for Microbial Studies 309
Saurabh Khachane, Vaibhav Lagad, Rutuja Vikhe, and Saurabh Kumar

13 Nanotechnology Approaches for Microbe-Based Formulations and Drug Delivery 333
Vaishali A. Shirsat, Achyut Chalodiya, Rutuja Kadam, and Divya Jaiswal

14 Nanotechnology-Based Electrochemical Diagnostic Tools for the Detection of Viral Diseases: Advantages and Disadvantages 363
Ahmet Cetinkaya, S. Irem Kaya, Banu Kaskatepe, Nurgul K. Bakirhan, and Sibel A. Ozkan

15 Next-Generation Sequencing and Solid-State Nanopores 387
Snehal Mahajan-Mallakmir, Sanket Mahajan, Anushree S. Lokur, and Ramesh S. Chaughule

16 An Overview of Anode in Microbial Fuel Cell: Current Challenges and Opportunities 407
S. Khiran, K. Vignaprasad, Manickam Matheswaran, and Samsudeen Naina Mohamed

17 Insights of Nanobiotechnology as Bio-adsorbents for Wastewater Remediation 421
Romana Naaz, Sameer Ahmad, and Weqar Ahmad Siddiqi

Index 439

Chapter 1
Biosynthesis of Metal Nanoparticles Using Bacterial Metabolites and Their Applications

Preethi Kathirvel, Gayathri Devi Kandasamy, and Mouliraj Palanisamy

1.1 Introduction

Nanotechnology is an emerging branch of science that deals with excising the structure of matter between 1 and 100 nanometres on the atomic, molecular, and supramolecular levels for the development of desired properties and functions and diverse applications. This rapidly expanding field has allowed for the design and development of multifunctional nanoparticles for various applications such as consumer goods, nanomedicine, nanoelectronics, biomaterials, and energy generation [1]. There has been enormous development in the field of nanotechnology in terms of nanoparticle synthesis pathways, their characterization techniques, mode of action, and applications. Physical and chemical methods of nanoparticle synthesis have proven to be less time-consuming, effective, and efficient. However, increased production of metal nanoparticles and oxides by chemical and physical methods is showing ecotoxicological effects when released into the environment. Hence, 'Green synthesis' or 'Biosynthesis' of nanoparticles using microorganisms (bacteria, fungi, and viruses), yeasts, and algae is gaining considerable attention [3]. In the last few decades, nanoscience has attracted the attention of the scientific community worldwide for the sustainable production of various nanoparticles (NPs), using innovative techniques, which find applications in various industries such as pharmaceutical, medical diagnostics, disease treatment, and also in energy, electronics, chemical, and agriculture and space industries [1]. Green nanotechnology is defined as the development of clean technologies with the goal of minimizing potential environmental and human health risks associated with manufacturing and encouraging the replacement of existing products with new nanomaterial-based products that are significantly more eco-friendly throughout their lifecycle. The disciplines of

P. Kathirvel (✉) · G. D. Kandasamy · M. Palanisamy
Department of Microbial Biotechnology, Bharathiar University,
Coimbatore, Tamil Nadu, India

R. S. Chaughule, A. S. Lokur (eds.), *Applications of Nanotechnology in Microbiology*, https://doi.org/10.1007/978-3-031-49933-3_1

microbiology and nanotechnology individually have strengthened the field of science and technology by delivering novel solutions for human well-being and maintaining environmental and ecological balance [3]. Yet, owing to instances where the frequent and inappropriate use of drugs (antibiotics) has led to the emergence of multidrug resistance in microorganisms, and the nanoparticle delivery is affecting the food cycle, it is urgently required to develop interdisciplinary research practices combining nanotechnology and microbiology (Microbial Nanotechnology) to provide innovative solutions for the human health, and environmental and ecological damage [4]. Biosynthesis of nanoparticles and the simultaneous use of microorganisms make the use of nanotechnology more sustainable and environmentally friendly. The green synthesis of nanoparticles from fungal, and bacterial enzymes, and pigments can be a potential solution. They act as reductive agents for the metal complex salt and generate metallic nanoparticles [5]. A deeper understanding of biosynthetic pathways, along with the potential provided by genetic engineering, are driving research towards the breakthrough development of microbial-based nanoparticle synthesis for future scaling-up and industrial exploitation of these promising 'nano-factories' [6]. This book chapter mainly emphasizes the utilization of various bacterial metabolites for metal nanoparticle synthesis and understanding the possible mechanisms involved in the fabrication of metal nanoscale particles. The present chapter will discuss the various applications of microbial approaches in metal nanoparticle synthesis, as depicted in Fig. 1.1.

1.2 Nanoparticles (NPs)

A nanoparticle, ultrafine particle, or nanomaterials is often described as a matter particle with a dimension of 1–100 nanometres (nm). The term is also applied to bigger particles with sizes of up to 500 nm, as well as fibres and tubes with sizes of <100 nm in only two directions. Metal particles <1 nm in size are commonly referred to as atom clusters. Because their smaller size drives very different physical or chemical properties, such as colloidal properties and ultrafast optical effects or electric properties, NPs are usually distinguished from microparticles (1–1000 m), "fine particles" (sized between 100 and 2500 nm), and "coarse particles" (sized between 2500 and 10,000 nm). However, in general for most applications, <100 nm is deemed to be effective for applications due to easier penetration and similar sizes to biomolecules. The smaller size of nanomaterials provides myriad research opportunities for biologists. Owing to their dimensions matching the scale of biomolecules, nanomaterials can interact with complex biological systems in unique ways [2]. NPs show unique and considerably changed chemical, physical, and biological properties compared to the bulk of the same chemical composition, due to their high surface-to-volume ratio [7]. Some of the major groups of nanoparticles are described in Fig. 1.2.

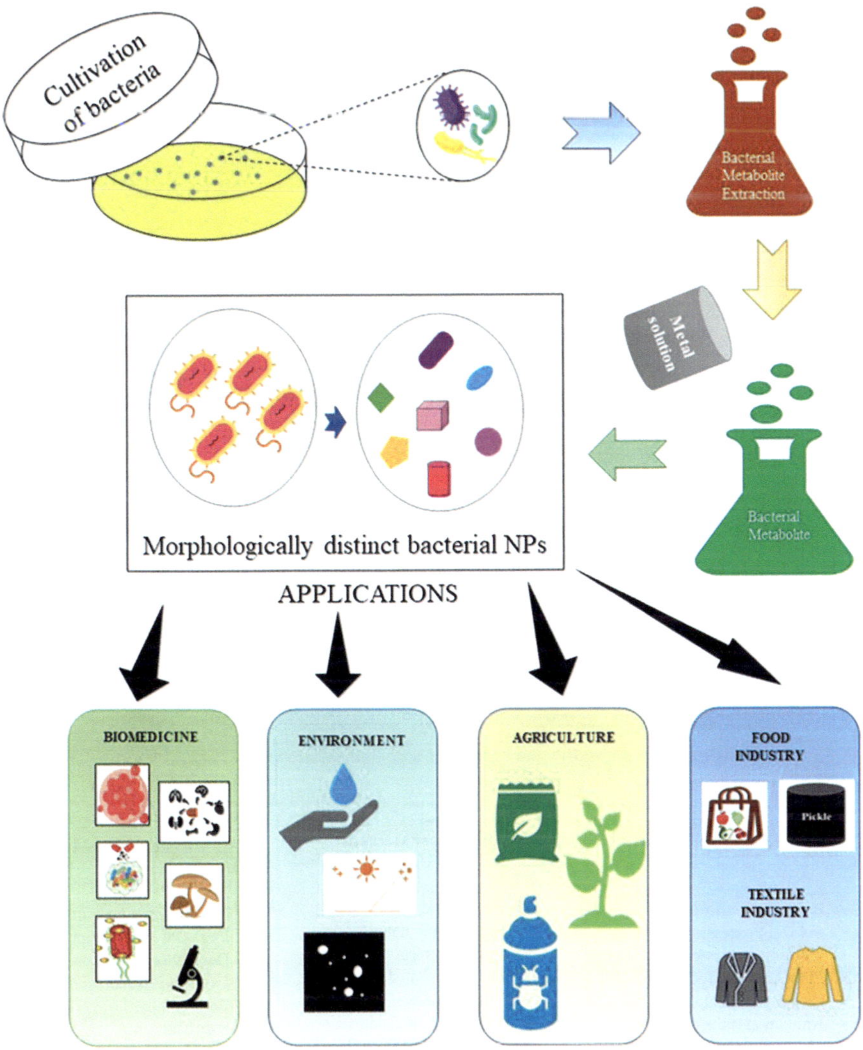

Fig. 1.1 Graphical abstract

1.3 Synthesis of Nanoparticles

Various methods can be employed for the synthesis of NPs, but all these methods are broadly classified into two main classes i.e. (1) Bottom-up approach and (2) Top-down approach [10] as shown in Fig. 1.3. These two approaches further divide into various subclasses to synthesize nanoparticles, namely physical, chemical, and biologically based on the reaction condition, operation and adopted protocols. The bottom-up method depends on the nanoparticles' chemical and biological synthesis, while top-down approaches generally refer to the physical or chemical route [2].

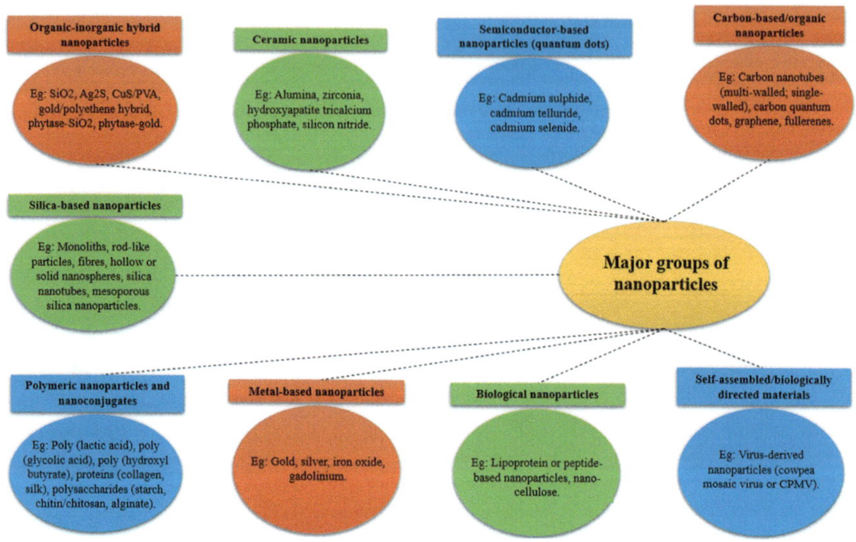

Fig. 1.2 Major groups of nanoparticles/nanomaterials in the application

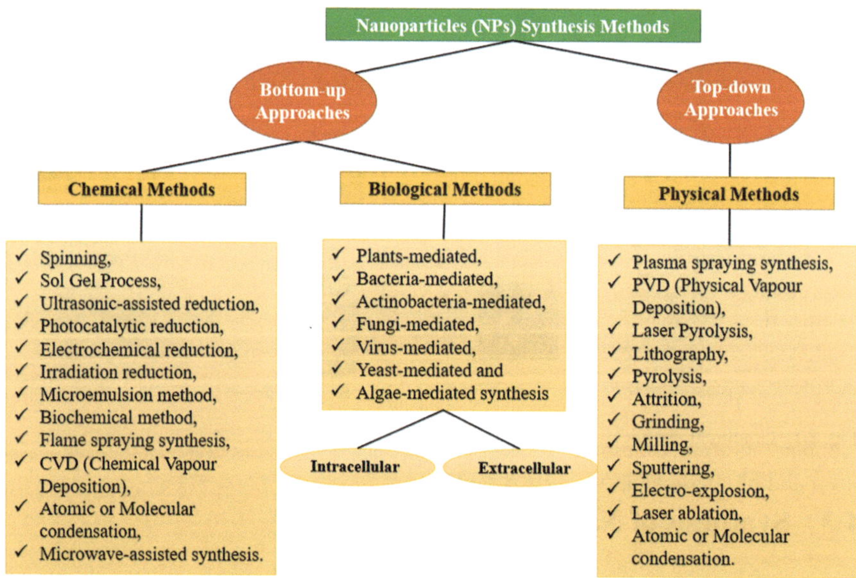

Fig. 1.3 Different approaches and methods for the synthesis of nanoparticles

1.3.1 *Top-Down Approach*

In this method, a destructive approach is employed. Starting from the bulk molecule, which decomposed into smaller molecules and then these molecules are converted into suitable NPs. Examples of top-down approach are grinding, physical vapour deposition (PVD), and other decomposition techniques [10]. The top-down approach involves the mechanical or physical method of reducing the size by gradually breaking down the bulk materials into the nanoscale structure [2].

1.3.2 *Bottom-Up Approach*

Bottom-up approach is employed in contrast to top-down approaches in which NPs are created from relatively simpler substances; thus, this approach is also known as the building-up approach. Sedimentation, chemical vapour decomposition (CVD), and reduction procedures are examples of this case [10]. The bottom-up strategy is primarily focused on the chemical and biological assembly of nanoscale atoms or molecules into molecular structures.

1.3.2.1 Biological Methods

Biological methods are used in the synthesis of metal and metal oxide nanoparticles of desirable size and morphology as they enhance the properties of nanoparticles in a greener route [9]. The biosynthesis of metal nanoparticles by biogenic sources such as plants and microbes are currently under exploitation. Green synthesis approaches such as biological methods provide a sustainable, economical, and less harsh nanoparticle synthesis method compared to physical and chemical methods. Moreover, biological synthesis offers control over the size and shape of required applications. This is now well-known that many microbes can produce inorganic materials either intracellular or extracellularly. Microorganisms such as bacteria, actinomycetes, fungi, yeasts, viruses, and algae are being explored as reducing or stabilizing agents to synthesize metal nanoparticles such as gold, silver, cadmium, copper, platinum, titanium, palladium, and zinc, which find uses in numerous industrial and biomedical application [2, 11].

1.4 Various Metals Used for Nanoparticles Synthesis

Metal nanoparticles have received a lot of attention due to their distinct characteristics such as catalytic activity, magnetic and optical properties, and antibacterial and anticancer effects. The most important property of metal and metal-oxide nanoparticles is their large surface area to volume ratio, which boosts their interaction with other molecules [8]. During the last decades, research interest in metal nanoparticles and their production has increased significantly because of their innovative applications in various sectors such as agriculture, food technology, environment, and textile industries (Table 1.1). The arrival of greener methods of synthesis has extended the areas of applications to pharmaceutical, food, and biomedical sectors because of their biocompatibility nature [9]. Nanoparticles can be produced using different metals, such as silver (Ag), gold (Au), copper (Cu) and copper oxide (CuO), zinc (Zn) and zinc oxide (ZnO), iron (Fe_2O_3), palladium (Pd), nickel oxide (NiO), magnesium oxide (MgO), selenium (Se), platinum (Pt), and titanium dioxide (TiO_2) [12].

1.5 Bacterial Metabolites

An important area of research in nanotechnology deals with the synthesis of nanoparticles of different morphologies, sizes, and mono-dispersity to improve their properties. In this regard, there is a growing need to develop reliable, non-toxic, clean, eco-friendly, and green experimental protocols for the synthesis of NPs. To achieve this objective, researchers are started to use natural processes such as the use of enzymes, microbial enzymes, vitamins, polysaccharides, biodegradable polymers, microorganisms, and biological systems for the synthesis of NPs [7]. One approach that shows the immense potential is based on the biological synthesis of metal nanoparticles using bacterial metabolites. Intra- and extracellular microbial enzymes and secondary metabolites secreted by microorganisms play a key role in the reduction of metal ions into their respective nanoparticles [12]. Researchers have started to employ biomass or cell extracts of bacteria for synthesize NPs. Bacteria are considered a potential bio-factory for the synthesis of metal NPs such as gold, silver, platinum, palladium, titanium, cadmium, and so forth [7] (Table 1.1).

1.5.1 Bacterial Pigments

The microbial pigments are used for various industrial applications due to their rapid growth on low-cost medium and can produce multicolour shades. Also, they are eco-friendly and non-hazardous compared to synthetic dyes. Recently, quite a lot of methods have been established that include the use of biological materials to synthesize NPs. Notable among them are bacterial pigments (Table 1.2), although have received scanty attention compared to other biological materials [57].

Table 1.1 Bacteria-mediated synthesis of nanoparticles

Bacterial source	Size (nm)	Applications	Reference
Silver nanoparticles			
Bacillus cereus	10–30	Antibacterial activity	[13]
Deinococcus radiodurans	Variable	Antibacterial, antibiofilm, and anticancer activity	[14]
Arthrobacter sp.	9–72	Antimicrobial activity	[15]
Bacillus brevis	41–62	Antibacterial activity	[16]
Oscillatoria limnetica	3.30–17.97	Antimicrobial and anticancer activity	[17]
Klebsiella pneumonia	26.84–44.42	Antimicrobial activity	[25]
Escherichia coli	5–50	Antimicrobial activity	[26]
Exiguobacterium aurantiacumm	5–50	Antimicrobial activity	[26]
Brevundimonas diminuta	5–50	Antimicrobial activity	[26]
Alcaligenes faecalis	30–50	Antimicrobial and antibiofilm activity	[27]
Streptacidiphilus durhamensis	8–48	Antibacterial and anticancer activity	[28]
Bacillus brevis NCIM 2533	41–68	Antibacterial activity	[16]
Streptomyces xinghaiensis OF1	5–20	Antimicrobial activity and synergistic effect with antibiotics	[29]
Nocardiopsis flavascens RD30	5 and 50	Cytotoxicity	[30]
Pseudomonas aeruginosa	35–60	Antibacterial activity	[31]
Streptomyces parvulus	1–40	Antimicrobial activity	[32]
Bacillus sp. *AZ1*	9–32	Antimicrobial activity	[33]
Gordonia amicalis	5–25	Antioxidant scavenging activity	[34]
Ochrobactrum rhizosphaerae	10	Antimicrobial activity	[35]
Streptomyces rochei MHM13	22–85	Antimicrobial activity and anticancer activity	[36]
Sinomonas mesophila	4–50	Antimicrobial activity	[37]
Gold nanoparticles			
Vibrio alginolyticus	100–150	Anticancer and antioxidant activity	[38]
Paracoccus haeundaensis BC74171	20.93 ± 3.46	Antioxidant activity and antiproliferative effect	[39]
Micrococcus yunnanensis	53.8	Antibacterial and anticancer activity	[40]
Mycobacterium sp BRS2A-AR2	5–55	Anticancer activity	[41]
Caldicellulosiruptor changbaiensis	<20	Antibacterial and antibiofilm activity	[42]
Bacillus subtilis	20–25	Degradation of methylene blue	[18]
Streptomyces griseoruber	5–50	Catalytic degradation of methylene blue	[43]

(continued)

Table 1.1 (continued)

Bacterial source	Size (nm)	Applications	Reference
Lactobacillus kimchicus DCY51	5–30	Antioxidant activity	[44]
Gordonia amarae	15–40	Application in rapid sensing of copper ions	[45]
Gordonia amicalis	5–25	Antioxidant scavenging activity	[34]
Copper nanoparticles			
Marinomonas, Rhodococcus, Pseudomonas, Brevundimonas, Bacillus.	Variable	Antibacterial and antifungal activity	[20]
Halomonas elongate	Variable	Antimicrobial activity	[21]
Streptomyces capillispiralis Ca-1	3.6–59	Antimicrobial activity	[46]
Marine endophytic actinomycetes	Variable	Antibacterial activity	[47]
Palladium nanoparticles			
Shewanella loihica PV-4	2–7	Degradation of methyl orange dye	[19]
Platinum nanoparticles			
Jeotgalicoccus coquinae ZC15	5.74	Antioxidant and antibacterial activity	[48]
Kocuria rosea MN23	5.85		
Pseudomonas kunmingensis ADR19	3.95		
Psychrobacter faecalis FZC6	2.49		
Pseudomonas putida KT2440	8.06		
Sporosarcina psychrophila KC19	4.24		
Vibrio fischeri NRRL B-11177	3.84		
Shewanella oneidensis MR-1	3–40	Catalytic reduction of nitrophenol and azo dyes	[49]
Streptomyces sp.	20–50	Anticancer activity	[50]
Shewanella oneidensis MR-1	61.03	Catalytic reduction of 4-nitrophenol	[51]
Zinc-oxide nanoparticles			
Aeromonas hydrophila	57.72	Antibacterial and antifungal activity	[22]
Lactobacillus sporogens	145.7	Antibacterial activity	[23]
Serratia ureilytica	Variable	Antimicrobial activity	[24]
Halomonas elongata IBRC-M 10214	18.11 ± 8.93	Antimicrobial activity	[52]
Bacillus megaterium (NCIM2326)	45 ~ 95	Antimicrobial activity	[53]
Lactobacillus paracasei LB3	1179 ± 137	Antimicrobial activity	[54]
Staphylococcus aureus	10 ~ 50	Antimicrobial activity	[55]
Streptomyces sp.	20 ~ 50	Antimicrobial activity	[56]

Table 1.2 Synthesis of nanoparticles by pigments

Pigments	Bacterial source	Type of nanoparticle	Biological activity	Reference
Actinorhodin	*Streptomyces coelicolor*	Silver NPs	Antibacterial activity	[9]
Pink coloured pigment	*Streptomyces* sp. NS-05	Silver NPs	Antibacterial activity	[58]
Violacein	*Janthinobacterium lividum*	Silver NPs	Antibacterial activity	[59]
		Titanium dioxide NPs	Anti-stain and self-cleaning properties	
Pyocyanin	*P. aeruginosa* PA6	Silver NPs	Antibacterial activity	[60]
Phycocyanin	*Nostoc linckia*	Silver NPs	Antibacterial activity	[61]
Flexirubin	*Chryseobacterium artocarpi* CECT 8497	Silver NPs	Anticancer activity	[62]
Prodigiosin	*Serratia marcescens*	Silver and gold NPs	Antiparasitic activity	[63]
Violacein	*Chromobacterium violaceum*			
Phycocyanin	*Limnothix* sp. 37–2-1	Silver NPs	Antibacterial activity	[64]

1.5.2 Bacterial Polysaccharides and Exopolysaccharides (EPSs)

Bacterial exopolysaccharides are produced extracellularly and play important roles in surface adherence and cell–cell communication. EPSs act as reducing and capping agents (stabilization) to produce nanoparticles. Therefore, the EPSs are used as an alternative for the microbiological production of numerous metal nanoparticles. Bacterial EPSs mainly comprise carbohydrates and non-carbohydrate components, which are responsible for their anionic nature. These components tend to increase the lipophilicity of the bacterial EPSs and directly influence their interaction with cations such as metal ions (Ag^+, Au^+, and ZnO^+) during the synthesis of NPs. Metal ions in contact with EPS are chelated, then reduced and stabilized via electrostatic bonds by various functional groups. For example, oxidation of -OH groups to form C=O groups and oxidation of -CHO groups to form -COOH groups play an important role during metal nanoparticle synthesis. The polymeric structure of EPSs creates a network by -H bonding in which nanoparticles stabilize with subsequent prevention of their agglomeration and precipitation [65]. Various types of functional groups are present in EPS of both gram-positive and gram-negative bacteria and act as reducing and stabilizing agents for the purpose of synthesizing NPs with the capping and chelating processes [66]. This helps in the regulation of various characteristics of NPs, such as size, shape, and particle dispersion [67]. Dextran is one of the predominant components of EPS that helps in the synthesis of graphene nanoparticles. Chemically, dextran is a complex branched glucan comprising glucose residues that remain interlinked with α-(1,6) glycosidic linkages and are mainly produced by certain groups of lactic acid bacteria like *Leuconostoc mesenteroides* and *Streptococcus mutans* [65, 68]. Curdlan is another type of exopolysaccharide, which is a water-insoluble polymeric substance comprising (1, 3)-β-D-glucan repeated units that are joined by β-(1, 3)-glycosidic bonds and are produced predominantly by *Agrobacterium* sp., *Alcaligenes faecalis*, and *Rhizobium* sp. It can be carboxylated or oxidized to curdlan derivatives for the purpose of synthesizing and stabilizing NPs [65, 69]. Levan is another bacterial EPS obtained from the bacteria *Acetobacter xylinum* NCIM 2526 that has been used for the synthesis of silver and gold nanoparticles that have catalytic activity, as tested by the reduction of 4-nitrophenol and methylene blue. In this study, levan acted as a stabilizer or capping agent, and reducing agent during the chemical reduction method and the thermal reduction method, respectively [70, 71]. Raveendran et al., demonstrated the synthesis of extremophilic bacterial sulphated polysaccharide mauran-based nanoparticles to address various applications in the field of bio-nanotechnology. It was found that the mauran nanoparticles are highly stable for at least 8 weeks in the solution and can be used for drug encapsulation and effective delivery under different pH conditions [72]. Bankura et al., synthesized size-controlled Ag NPs using the aqueous solution of dextran, which acted as both a reductant and a stabilizer [73]. Leung et al., synthesized Ag NPs with the help of carboxymethylated curdlan. The persistence of the negatively charged groups (hydroxyl and carboxyl groups) resulted in a better reduction of the silver ions. In another study, it was revealed that succinoglycan bacteria produce structurally known exopolysaccharides that were used for the synthesis of silver nanoparticles [74]. Bacterial cellulose nanofibers have been shown to act as templates for the synthesis of gold NPs to produce nanocomposites formed by gold NP nanofibers in aqueous suspension. These nanocomposites have been used to immobilize the horseradish peroxidase enzyme to make a biosensor to detect H_2O_2 [75]. It is a polymeric substance produced by a diazotrophic bacteria, *Sinorhizobium meliloti,* that helped in the reduction of the metal by inducing oxidation of the aldehyde group to a carboxyl group with the help of nucleophilic insertion [76]. Sphingans (gellan) have been used as reducing and stabilizing agents to synthesize gold NPs for drug delivery formulations that inhibit the human glioma cell lines LN-18 and LN-229 [77].

1.6 Biosynthesis of Metal Nanoparticles Using Bacterial Metabolites

Bacteria is one of the most important biogenic sources for producing nanoparticles in various aspects. Figure 1.4 depicts the methodology of nanoparticle synthesis using bacteria and their metabolites. One of its main advantages is the secretion of intracellular and extracellular enzymes that work to reduce metal ions and thus

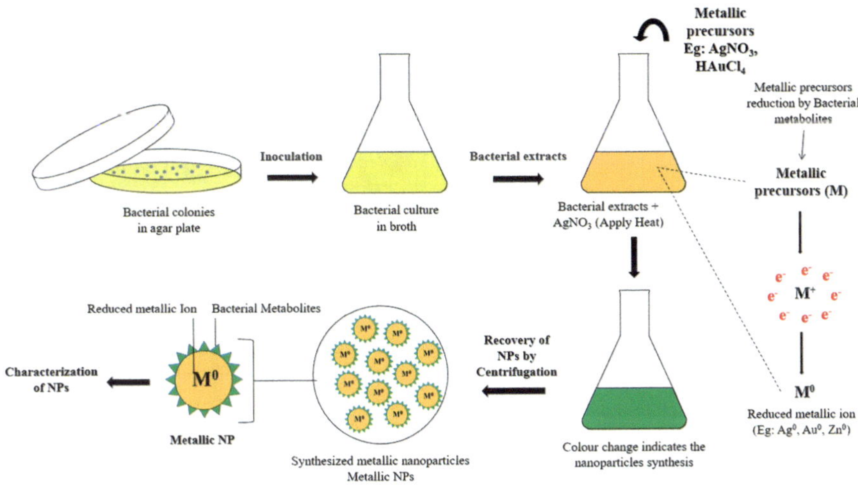

Fig. 1.4 Biosynthesis of nanoparticles using bacterial metabolites

produce nanoparticles. Because of their rapid growth, simplicity of culture, and preservation in vitro, their use in nanotechnology is rather inexpensive. And also, nanoparticle production can be controlled by manipulating several conditions, including temperature, pH, the concentration of metal ions, and reaction time, making them ideal for use in biosynthesis to produce nanoparticles [25].

1.7 Applications of Bacteria-Derived Nanoparticles

The amalgamation of nanotechnology and microbiology holds expanded applications in various fields, including waste valorization, degradation of pollutants, food-borne pathogen detection, shelf-life extension of food products, biosynthesis of medically important nanoparticles, drug delivery, nanofertilizers, smart textiles, nanofibers, etc. Some of the prevalent agricultural, biomedical, textile, environmental, and food technological applications of bacteria-mediated nanoparticles are depicted in Fig. 1.5 as well as in Table 1.3 and summarized in detail as follows:

1.7.1 Applications in Biomedicine

The application of biofabricated nanoparticles in biomedical research opens new possibilities. NPs have been extensively used in biomedicine owing to their small size, more surface area, non-toxicity, controlled biocompatible dimensions, and superparamagnetic properties. The most notable application in the medical field is

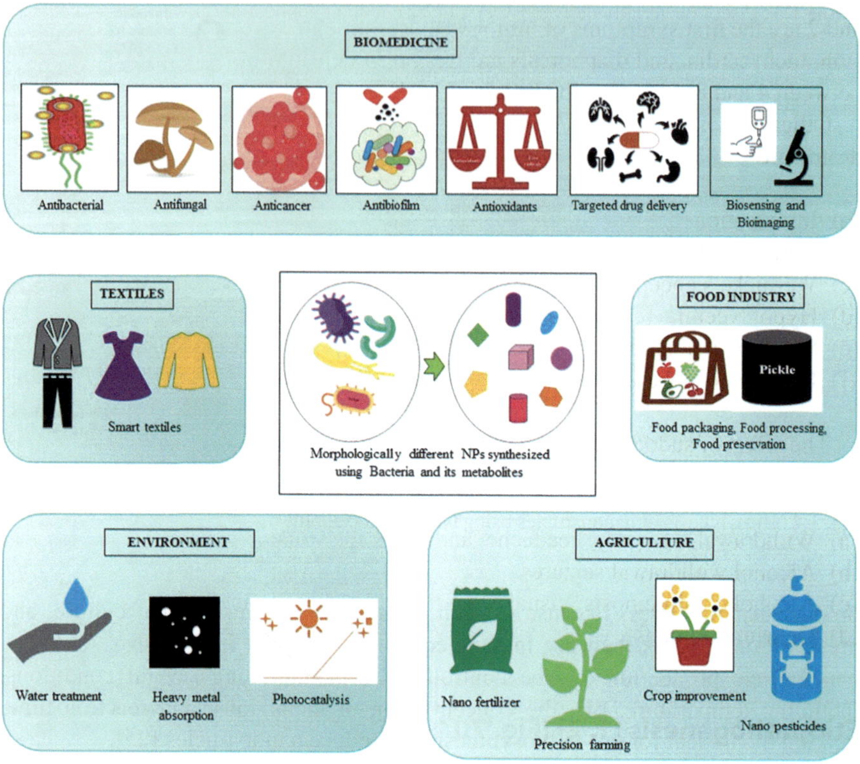

Fig. 1.5 Multifaceted applications of nanoparticles synthesized using bacteria and its metabolites

their antimicrobial, anti-larvicidal, and antioxidant therapies [78]. In general, several biofabricated nanoparticles, including gold, silver, copper, zinc, iron, and titanium could provide an innovative solution in biomedical and pharmaceutical fields as antimicrobials, antioxidants, antibiofilm, antiageing, anticancer, anti-inflammatory, anti-larvicidal, and imaging or diagnostic agents [2]. Bacteria-derived nanoparticles can be used to create materials that function like bacteria for a vast array of applications, such as drug delivery for cancer cells, imaging, magnetic nanomarkers to identify various pathogens, and immunological regulation. This section focuses on bacteria-derived nanoparticles as viable biomaterials, introducing many bacterial species with excellent potential and feasible fabrication techniques.

1.7.1.1 Antimicrobial Agents

The emergence of the multidrug resistance (MDR) trait among pathogenic bacteria has spurred the search for new antimicrobial nanoparticles. A key benefit of biogenic synthesis is the pre-existing presence of naturally occurring stabilizing or capping agents, such as polysaccharides or proteins, on the surface of the nanoparticle at the

Table 1.3 List of bacterially derived nanoparticles and their applications in various domains

Sl. No.	Bacterial source	Type of nanoparticle	Applications	Reference
1.	*Paraclostridium benzoelyticum*	Zinc oxide	Antibacterial, antidiabetic, anti-inflammatory	[80]
2.	*Pseudomonas aeruginosa*	Cadmium sulphide	Removal of cadmium	[159]
3.	*Proteus vulgaris*	Iron oxide	Antibacterial, antioxidant, cytotoxic activity	[97]
4.	*Leuconostoc mesenteroides*	Silver	Cotton fabric impregnation	[120]
5.	*Bacillus niacini*	Zirconium	Antibiofilm activity	[160]
6.	*Enterobacter aerogenes*	Zinc phosphate	Antibacterial, anticorrosion activity	[161]
7.	*Cuprividus* sp.	Silver	Antibacterial, antibiofilm activity	[162]
8.	*Lactobacillus casei*	Selenium	Antioxidant, anticancer	[163]
9.	*Lactobacillus acidophilus*	Silver	Antimicrobial, antioxidant, antibiofilm agent	[105]
10.	*Lactobacillus paracasei*	Selenium	Antifungal	[164]
11.	*Bacillus thuringiensis*	Silver	Insecticidal activity	[149]
12.	*Pseudomonas silesiensis*	Copper	Anticancer, antimicrobial	[165]
13.	*Bacillus subtilis*, *Bacillus amyloliquefaciens*	Silver	Antibacterial and larvicidal activity	[166]
14.	*Shigella flexneri*	Copper	Reduces cadmium stress in plants	[142]
15.	*Bacillus licheniformis*	Silver	Antimicrobial and antiviral activity	[167]
16.	*Pseudomonas stutzeri*	Selenium	Anticancer activity	[168]
17.	*Bacillus siamensis*	Silver	Plant growth promotion	[169]
18.	*Pseudomonas fluorescens*	Copper	Biocontrol agent	[170]
19.	*Lysinibacillus* sp.	Tellurium	Bioremediation	[171]
20.	*Bacillus licheniformis*	Silver	Photocatalytic dye degradation, antibacterial activity, cytotoxicity	[172]
21.	*Serratia marcescens*, *Burkholderia cepacia*	Silver	Biocontrol agent	[173]
22.	*Bacillus cereus*	Zinc oxide	Nanopesticides	[147]
23.	*Lactobacillus pentosus*, *Lactobacillus crustorum*, *Lactobacillus spicheri*	Silver	Antimicrobial food packaging	[174]
24.	*Marinospirillum alkaliphilum*	Silver	Antimicrobial, textile effluent treatment	[175]

(continued)

Table 1.3 (continued)

Sl. No.	Bacterial source	Type of nanoparticle	Applications	Reference
25.	*Escherichia* sp.	Copper	Dye detoxification, textile effluent treatment	[154]
26.	*Leuconostoc lactis*	Silver	Textile dye degradation	[176]
27.	*Lactobacillus paracasei*	Silver	Antimicrobial, antioxidant, antibiofilm activity	[177]
28.	*Idiomarina* sp.	Lead sulphide	Bioimaging	[113]
29.	*Myxobacteria virescens*	Silver	Food packaging	[129]
30.	*Enterobacter* sp.	Magnesium oxide	Nanofertilizers	[141]
31.	*Bacillus cereus*	Silver	Removal of heavy metals	[152]

time of synthesis, which drastically reduces the amount of post-production processing. The research team of Saeed (2020) [26] worked on the biological approach for the fabrication of silver nanoparticles (AgNPs) using the secondary metabolites of *E. coli, Brevundimonas diminuta,* and *Exiguobacterium aurantiacumm*. The antimicrobial activity of the biofabricated nanoparticles has been investigated against multidrug-resistant pathogens and methicillin-resistant *Staphylococcus aureus*. Results show that the AgNPs synthesized using bacterial metabolites exhibit great potential as antimicrobial agents. Because of their antibacterial properties, AgNPs of bacterial origin are widely employed in wound dressings, catheters, and numerous domestic supplies [79]. ZnONPs bioprinted with the bacterium *Paraclostridium benzoelyticum*5610 biomass were found to possess antidiabetic, antibacterial, and anti-inflammatory activities by performing several in vitro and in vivo experiments [80]. It is exciting to highlight that nanoparticles fabricated utilizing bacterial extracts are effective at killing other microbial species and increase the effectiveness of current medications to combat antimicrobial resistance traits. In addition, the biogenic nanoparticles synthesized using the cellular extracts of *Brevibacterium casei* were explored as an anticoagulant in human plasma [81], and antimicrobial carbon dots fabricated using the extracts of *Lactobacillus acidophilus* showed promising antimicrobial activity [82]. The study conducted by Yusof and co-workers, [83] demonstrated the biofabrication of AgNPs employing the cell biomass of *Lactobacillus plantarum,* which exhibited considerable dose-dependent antibacterial activity against both gram-positive and gram-negative bacteria, indicating its potential application as an antibacterial agent. In addition, the effectiveness of NPs against bacteria can be improved if they are used in combination with conventional antibiotics. For example, Singh et al., 2015 investigated the synergetic action of bioinspired AgNPs using *Brevibacterium frigoritolerans* in combination with commercially available antibiotics and reported that all the antibiotics exhibited significant antimicrobial efficacy against disease-causing pathogens [84]. The fabrication of AgNPs using the endophytic bacterium *Bacillus zanthoxyli* had strong antibacterial activity against *Salmonella typhi enterica, Staphylococcus aureus, Pseudomonas aeruginosa, E. coli,* and *Bacillus subtilis* [85]. Bioinspired AgNPs using a novel endophytic bacterium, *Rothia endophytica,* might have unique therapeutic significance as an anticandidal agent or a vital element in anticandidal medications [86]. These findings expanded endophyte biotechnology's potential applications in the field of nanomedicine. It is intriguing to observe from the above examples that nanoparticles biosynthesized using bacteria and their metabolites are effective at killing other microbial species and improve the effectiveness of current medications to combat antimicrobial resistance traits.

1.7.1.2 Antioxidant Activity

Antioxidants are chemical molecules produced as a result of the body's response to external stimuli and environmental stress that help reduce cell damage by neutralizing free radicals. Almost all organisms contain indigenous antioxidant defence and repair systems to guard against oxidative damage; however, these systems are frequently insufficient to entirely prevent the damage. As a result, using antioxidant supplements to minimize oxidative damage to the human body is advised [87]. When compared to the traditional methods of antioxidant supplementation, nanoparticles have numerous benefits, including enhanced bioavailability, an eco-friendly nature, selective antioxidant supplementation, and controlled release at the target site. Antioxidant compounds can be shielded against degradation by employing nanotechnology-based delivery systems that increase bioavailability and have physicochemical drug-like features [88]. According to some recent studies, the biological synthesis of nanoparticles using bacteria and their metabolites such as silver, gold, platinum, and selenium has notable antioxidant activity. Strong antioxidant activity was demonstrated for biologically derived AgNPs utilizing endophytic *Pantoea anthophila,* which may be the result of functional group adsorption from the endophytic extract on the surface of AgNPs. Bioinspired AgNPs are used as powerful antioxidants or a valuable component of antioxidant formulations in the pharmaceutical and biomedical industries due to their promising radical scavenging efficacy [89]. Endophytic bacterium-mediated synthesis of silver nanoparticles exhibited significant free radical scavenging activity, indicating that bacteria-derived NPs have the potential to be developed as promising antioxidant agents [90]. On the Vero normal cell line, *Pseudomonas indica*-mediated AgNPs displayed remarkable antioxidant effects without cytotoxicity as well as antifungal efficacy against *M. racemosus, R. microspores,* and *S. racemosum* [91]. Stabilized biologically synthesized selenium nanoparticles (SeNPs) by *Pantoea agglomerans* demonstrated the highest antioxidant activity [92], whereas SeNPs produced via a biosynthetic approach using probiotic *Bacillus subtilis* exhibited remarkable antioxidant activity in terms of DPPH and ABTS scavenging action [93]. The marine bacterium

Lysinibacillus odysseyi-mediated biosynthesis of AuNPs was found to possess multifarious activities, including antibacterials, antioxidants, and DNA cleavage [94]. In addition, the antioxidant activity of the nanoparticles fabricated using the bacteria was thought to be mostly determined by the physicochemical properties of the synthesized nanoparticles, namely the size and zeta potential of the nanoparticles. This is evident from the results of Eramabadi et al., 2020, [48] in which the tiniest platinum nanoparticles (PtNPs) produced by gram-negative *Psychrobacter faecalis* had higher antioxidant activity, whereas the most negatively charged PtNPs produced by gram-positive *Jeotgalicoccus coquinae* demonstrated superior antioxidant activity. The primary cause of the enhanced antioxidant activity in the above examples is mainly due to the bioactive substances present on the surface of biogenically synthesized nanoparticles. Enhanced antioxidant activity of the biogenically synthesized nanoparticles is due to the presence of bioactive compounds on their surface.

1.7.1.3 Anticancer Agents

Cancer is an intricate disease with a wide range of presentation, progression, and prognosis. It is well-known that a complex confluence of hereditary and environmental variables results in cancer, a multifactorial disease. Many anticancer medications are unable to effectively exert their pharmacological effects and reach their target location in sufficient concentrations without causing irreparable unintended harm to healthy tissues and cells. By overcoming biological obstacles to deliver therapeutic drugs directly, nanotechnology offers a plethora of alternatives for the treatment of cancer. Metal nanoparticles' distinctive physical and chemical properties, such as their diverse optical properties, high surface-to-volume ratio, ability to be surface-personalized, and simplicity of production, present novel therapeutic possibilities for cancer [95]. Researchers tested the anticancer activity of several biogenerated nanoparticles made by bacteria against cancerous cells. Patil and his co-workers [39] synthesized gold nanoparticles (AuNPs) using the novel marine bacterium *Paracoccus haeundaensis* BC74171T and evaluated their antioxidant and antiproliferative activity against non-cancerous (HaCaT, HEK293) and cancerous cell lines (AGS, A549). From the study, it has been proven that biogenic synthesis of AuNPs using bacteria is a facile method since they are non-toxic to human cells and have a dose-dependent antiproliferative effect on cancer cells, indicating their applications in biomedicine. The findings of Nandhini et al., 2021 [96] suggest that spherical AuNPs synthesized using *Enterococcus* sp. inhibit the proliferation of HepG2 cells via intracellular ROS-mediated apoptosis. Decreased expression of proliferating cell nuclear antigen protein may be the possible reason behind the antiproliferative effect of AuNPs against hepatocellular carcinoma. Majeed et al., 2021 [97] concluded that iron oxide nanoparticles (IONPs) of bacterial origin (*Proteus vulgaris*) showed a good cytotoxic effect against brain cancer cells as well as preventing cell migration and cell–cell interaction. The bacteria-derived AuNPs emphasize being biocompatible and cost-effective with cytotoxic and anti-inflammatory effects against colon cancer [38]. In another study, biologically synthesized AgNPs derived from *Bacillus* sp. were tested in vitro against MCF7 breast cancer cells and found to exhibit superior anticancer activity via induction of apoptotic mechanisms [98]. Venil et al., 2016 [62] synthesized AgNPs from the flexirubin pigment extracted from *Chryseobacterium artocarpi and* investigated its anticancer activity on cancer cell lines. The result demonstrated that biogenerated AgNPs were found to inhibit 99% of human breast cancer cell lines (MCF-7). AgNPs bioderived from marine *E. coli* exhibited considerable cytotoxic influence against human breast cancer cell lines and lung cancer cell lines (A549), thereby implying that they have a great deal of potential to be powerful anticancer drugs [99]. These findings confirmed the potential of biogenic nanoparticles as an effective anticancer agent.

1.7.1.4 Antibiofilm Agents

Antibiotics have been used frequently to treat infections caused by pathogens, which has resulted in the emergence of resistance that can lead to the formation of biofilms on medical equipment. The goal of eliminating bacterial biofilm infections is currently being achieved with the use of promising nanotechnology methods. Recent advances in nanotechnology guarantee the prevention of infections involving drug-resistant biofilms [100]. Bimetallic gold-silver nanoparticles derived from *Shewanella oneidensis* appear to be a promising nano-antibiotic for combating bacterial resistance in well-established bacterial biofilms [101]. According to Khaleghi et al., 2019 [102], the AgNPs biosynthesized using *Bacillus thuringiensis* not only inhibited the growth of biofilm but also had the ability to remove pre-formed biofilm. The results of Jayabalan et al., 2019 [103] proved that microbially synthesized ZnONPs using *Pseudomonas putida* exhibited a high degree of biofilm detachment property against the tested microbial biofilms (*Enterococcus faecalis* and *Bacillus thuringiensis*). Nontoxic selenium nanoparticles were biofabricated extracellularly using the probiotic strain *Lactobacillus acidophilus* and evaluated for their antibiofilm activity against five different resistant and sensitive bacterial isolates. The results showed that the SeNPs exhibited a notable dose-dependent increase in the degradation of pre-formed biofilm and were also found to be more effective against *S. aureus, P. aeruginosa,* and *E. coli* as compared to *K. pneumonia* and *B. subtilis* [104]. In addition, the biogenic silver nanoparticles fabricated using *Lactobacillus acidophilus* exhibited significant biofilm inhibition against multi-drug resistant enteroaggregative *E. coli* [105].

1.7.1.5 Targeted Drug Delivery

Nanoparticles can deliver drugs and bioactive agents in a time-controlled or site-specific manner. Pharmaceutical nanotechnology focuses on drug formulation in biocompatible nanoforms for improved drug delivery. NPs improve drug efficiency and safety by improving bioavailability, providing targeted drug distribution, increasing drug stability, controlled drug release, and extending drug action in the target tissue [106]. NPs can deliver a variety of medications to body parts such as the lungs, spleen, brain, liver, lymphatic system, artery walls, or systemic circulation for both short- and long-term dosages. The combination of various polymers can be used to alter the rates of degradation and medication release. Genes, proteins, biological macromolecules, vaccines, hydrophilic pharmaceuticals, and hydrophobic medications are all capable of being delivered by nanoparticles [107]. Cancer is a group of diseases characterized by abnormal tissue growth that can lead to the formation of tumours that can spread into other tissues and cause significant effects in the patient, with severities and complications potentially leading to fatal outcomes [108]. Conventional cancer treatment procedures (surgery, radiotherapy, and chemotherapy) have their own risks, such as the non-specificity and toxicity of the drugs, which attack both healthy and cancerous cells. Furthermore, tailored therapy administration to the damaged organ and early detection of this debilitating disease are still in their infancy. As a result, it is critical to discover alternate means of diagnosing and treating this disease. It has been claimed that nanomedicines can be successfully used for tumour diagnostics and treatment through targeted drug delivery [1]. According to the literature, numerous types of nanoparticles biofabricated by employing bacteria and their metabolites have been employed against various cancer cell lines. El-Naggar et al., 2018 [109] tested the in vitro anticancer efficiency of AgNPs biofabricated using phycoerythrin extracted from *Nostoc carneum* and found stronger cytotoxicity against MCF-7 breast cancer cell lines than human lung fibroblast (WI38) and human amnion (WISH) cell lines. Drug consolidation into nanocarriers can protect a drug from deterioration, reduce drug dosage, and speed up the curing process with controlled drug release [110].

1.7.1.6 Bioimaging and Biosensors

Biological imaging has drawn more attention in recent years from both healthcare and research institutions. Numerous imaging techniques have been established to either comprehend biological processes in living cells, tissues, and organs or to identify and measure human disease in order to produce improved therapy methodologies [111]. Imaging biomolecules is required to explore biomolecular interactions in vivo and in vitro, as well as to understand the function and process of the biomolecules, along with appropriate labelling. In addition to positron emission tomography, computed tomography, single-photon emission computed tomography, magnetic resonance imaging, and ultrasonography, nanoparticle-based contrast agents are used in bioimaging [112]. The study of Srivastava et al., 2017 [113] demonstrated the benign synthesis of fluorescent lead sulphide nanoparticles (PbS_2NPs) by *Idiomarina* sp. and its bioimaging applications. The biogenic fluorescent nanoparticles were found to be noncytotoxic, and the fluorescence of the NPs turned out to be sustained in an array of microenvironments. HeLa cells internalized PbS_2NPs, which were then uniformly dispersed throughout the cells' cytoplasm without penetrating their nuclei. As a result, these nanoparticles are adaptable fluorophores that may be exploited for myriad applications in analysis and imaging. Raveendran et al., 2013 [72] reported the green synthesis of extremophilic bacterial sulphated polysaccharide-based nanoparticles and confirmed its use in cellular imaging techniques. Along with chitosan (CH), mauran (MR), a sulphated exopolysaccharide derived from the halophilic bacteria *Halomonas maura* was used to synthesize nanoparticles. Anionic MR and cationic CH were simply complexed with polyelectrolytes to generate MR/CH nanoparticles. Fluorescein isothiocyanate (FITC) tagged MR/CH nanoparticles demonstrated an outstanding method of tracking system and cell imaging that can be established more easily than other incompatible and hazardous imaging techniques. The application of dye-tagged MR/CH nanoparticles for a safe and nontoxic mode of live cellular imaging was demonstrated by the possibility that the free sulphated groups present in the nanoparticles could compete with host proteoglycan cell receptors in binding various antigenic determinants during disease mechanisms.

Theranostic effectiveness can be increased by using analytical devices, namely biosensors that assess physiological signals or identify analytes [114]. In the realm of medical sciences, improving biosensor performance is essential for promoting human health and extending human life [112]. For diagnostic purposes, particular biomarkers in biological samples have been identified via nanomaterial-based biosensors. Nanomaterials have been used as transducer materials in the construction of biosensors because of their exceptional properties, which have enhanced performance and analytical signals. The fabrication of environmentally friendly biosensors depends heavily on green nanomaterials produced utilizing microorganisms [115]. For example, as reported in the literature by Luo et al., 2014 [116] a sensitive electrochemical aptamer-based biosensor was developed to detect toxin A of *Clostridium difficile* using biogenically synthesized gold nanoparticles by *Bacillus stearothermophilus.*

1.7.2 Applications in Textiles

The spread of microbe-borne illnesses in healthcare settings, such as hospitals, is a serious issue that endangers human health. Textiles are regarded as one of the primary causes of infection transmission between patients and healthy people. As a result, it is critical to design medical textiles capable of reducing the emergence of microbial infection by introducing pathogenic bacterial resistance to medical textiles. Through inclusion into textiles during manufacturing, nanoparticles provide excellent methods for achieving this notion [117]. In this regard, eco-friendly nanoparticles are currently being employed for the production of antimicrobial fabrics. The textile antibacterial guard is quite intriguing and useful to human health. Antimicrobial compounds such as chitosan, hemp, metal fibres, N-halamine, Cu_2O, Ag, TiO_2, etc. have been introduced into fabrics. Silver is the most basic antibacterial nanoparticle utilized on textile surfaces. It functions as a doping antibacterial agent and demonstrates exceptional antimicrobial activity without altering its mechanical qualities [118]. Biogenically synthesized AgNPs were integrated into several textile fabrics, including mill-scoured polyester fabric, wool/polyester blend, and coloured polyester, to create antibacterial and UV-protective textiles. The antimicrobial activity of AgNPs-treated textile fabrics against the tested microorganisms (*Escherichia coli, Pseudomonas aeruginosa, Staphylococcus aureus, Salmonella typhi*, and *Candida albicans*) was excellent, and it remained stable after five repeated washing cycles. Furthermore, the UV blocking efficiencies of AgNPs finished fabrics were unaffected by washing, confirming the AgNPs' robust attachment to the textile fabrics [119]. Silver nanoparticles synthesized from the dextran produced by *Leuconostac mesenteroides* dextrasucrase were used to impregnate cotton fabric in order to create a textile nanocomposite material with high antibacterial activity. The control dextran-treated cotton fabric exhibited no antibacterial activity in this investigation, while the cotton textiles loaded with AgNPs showed a considerable microbial decrease, showing that the synthesized AgNPs offer enough antimicrobial activity [120]. Green synthesized nanoparticles coated over commercial products have recently piqued the interest of researchers due to the possibility of enhanced properties such as water resistance, antibacterial, mechanical, high hardness, and optical, making them technologically enticing for various applications in the paint, electrical, paper, textile, and cosmetic industries. Tharani et al., 2020 [121] developed an eco-friendly simple bio-inspired approach for the synthesis of chitosan-AgNPs (CS-AgNPs) employing the extracellular biomass of *L. reuteri* as a capping and reducing agent. Cotton fabrics infused with CS-AgNPs demonstrated a possible zone of inhibition against the *B. subtilis* pathogen. As a result of the findings, the synthesized CS-AgNPs can be employed as a possible antibacterial nanomaterial in the textile industry to reduce the growth of bacterial colonies. Because of the numerous advantages such as lower cost, stability, self-cleaning ability, whiteness, mechanical strength, and UV-blocking properties, much research emphasis is currently being focused on the antimicrobial properties of ZnONPs and their exploitation as antimicrobial agents for textile finishing against a broad spectrum of pathogens. Dhandapani et al., 2014 [46] reported the production of antimicrobial textile materials containing ZnONPs using a biochemical precipitation method mediated by ureolytic bacteria. Wet interfacial contact experiments on ZnONPs loaded cotton fabric demonstrate significant killing effectiveness against *Staphylococcus aureus* and *Escherichia coli* due to persistent interfacial contact between the nanocrystals and bacterial species on the cotton fabric. The nanoparticles incorporated into the cotton fabric function as a barrier, slowing bacterial multiplication and attachment. When biogenically synthesized ZnONPs were coated on cotton fabric, they provided resistance against gram-positive bacteria, self-cleaning ability, and UV-blocking capacity against malachite green dye [122]. In this regard, different types of nanoparticles have been biogenically synthesized utilizing bacteria and combined with textiles. Furthermore, biosensing nanoparticles may be added to the fabrics using similar procedures to those used to embed antibacterial nanoparticles in order to monitor illness conditions, which would benefit patients. Thus, smart textiles with antibacterial efficacy and illness monitoring capability will be useful in the future of medical textile industries for controlling bacteria-mediated infection and monitoring disease conditions in patients [123]. Illegal waste disposal from textile companies containing recalcitrant pollutants is a global issue that is exacerbated in underdeveloped countries. Ahmed et al., 2020 used an eco-friendly method to synthesize AgNPs using *Bacillus marisflavi* TEZ7 and used them as photocatalysts to degrade not only synthetic dyes but also actual textile effluents, followed by phytotoxicity testing and molecule identification [124].

1.7.3 Applications in Foods

Due to the global expansion of the human population, one of the most pressing concerns for poor and emerging countries is how to consume food sustainably and safely. Food safety issues are evolving at a mysterious rate in the twenty-first century. Nanotechnology is a burgeoning revolution with enormous promise in fields ranging from mechanics to medicine, including the food business. The frequent use of chemical pollutants and pesticides in food has a direct impact on its nutritional quality. In particular, from a microbiological food safety standpoint, nanotechnology has proven to boost food safety due to its substantial role at every stage of the food chain, including food processing, preservation, packaging, storage, and quality monitoring. Packing materials are vital for food product protection because they prevent spoilage due to physicochemical or biological mechanisms and preserve overall quality throughout storage and handling. In this regard, nanoparticles play a vital role in food safety and preservation, as well as the creation of sensors. Furthermore, the use of NPs in nano-packaging components has been proven to increase storage product stability, long-distance packing, and the prevention of food spoilage [125].

Metal and metal oxide NPs have been shown to possess exceptionally effective antibacterial and antioxidant characteristics, as well as nanobiosensors for tracking and monitoring food quality. Nanoparticles are employed as carriers to deliver enzymes, anti-browning agents, antioxidants, and other bioactive components, extending the shelf life even after the packaging has been opened. Most of the nanoparticles used in food packaging have antimicrobial capabilities, functioning as carriers for antimicrobial peptides and offering protection against microbial deterioration [126]. Metallic nanoparticles are beneficial for developing antimicrobial goods that have the potential to extend the shelf life of food by preventing microbial development, and as a result, they have sparked an abundance of interest. They interact with various microbial cells, destroy them, and have the ability to limit the formation of biofilms [127]. Because of their ability to encapsulate active compounds and improve functionality, various nanoparticles such as zinc oxide, titanium dioxide, magnesium oxide, silver nanoparticles, fullerene derivatives, zerovalent iron, and carbon nanotubes have demonstrated antibacterial activity. Food ingredients are also nano-packaged using nanoparticles. This use of nanotechnology provides solutions for food packaging by modifying the penetration activities of foils, improving chemical, mechanical, and microbiological obstacle effects, and boosting heat resistance. Metallic nanoparticles, such as AgNPs, are utilized not only as antibacterial preservatives but also play beneficial functions in food preservation [128].

Silver nanoparticles are crucial particles used over a long period of time for material packaging [128]. Bhople et al., 2016 reported the fabrication of AgNPs utilizing *Myxobacteria virescens* and their influence on the shelf life of apples. The study covers the impregnation of AgNPs onto butter paper using three distinct methods: the direct production of AgNPs on paper, the glass rod method (with binder), and the glass rod method (without binder), with the antibacterial action against pathogenic bacteria being evaluated. Among these methods for coating AgNPs on butter paper, paper with the direct synthesis of AgNPs on it demonstrated significant antibacterial potential against *Klebsiella pneumoniae, Escherichia coli, Pseudomonas aeruginosa, Staphylococcus aureus*, and *Salmonella typhi* and was thus used for apple wrapping to increase shelf life. The results revealed that the antimicrobial paper containing AgNPS extends the shelf life of apples by up to 15 days [129]. Due to their significant advantages over conventional polymers, biomolecules, including exopolysaccharides (EPS) produced by probiotic bacteria, are appropriate for the fabrication of nanoparticles with potent biological activity. This is evident from the findings of Rajoka et al., 2020 where the silver nanoparticles bioinspired using EPS produced by *Lactobacillus brevis* isolated from Chinese koumiss possess numerous applications in the food and agricultural industries [130]. Nanosensors have recently been created for a variety of applications in fields such as agriculture, healthcare, the environment, food, and defence. The introduction of nanosensors alerts consumers to food deterioration or contamination by detecting pesticides, toxins, and microbiological contamination in food items based on colour and flavour formation [126]. These sensors are used in conjunction with plastic packaging to monitor physical factors such as temperature, pH, and colour of the packed food and to detect chemical species such as gases emitted from spoiled food in order to ensure quality and safety. Nanosensors are also being developed to detect microbiological pollutants in packaging plants as well as on the surfaces of food products. Food testing employing nanosensors will be less expensive than shipping samples to laboratories for examination [131]. Because of their conjugation with diverse nanostructures such as nanoparticles, nanorods, carbon nanotubes, nanofibers, nano biofilms, quantum dots, and nanowires, these devices have a high surface-to-volume ratio with exceptional electric and optical characteristics. When functionalized with specific antibodies, nanoparticles can help in the detection of pesticides, specific toxins, and viruses, all of which aid in the surveillance of food quality. Furthermore, one of the most efficient nanobiosensors, the innovative low-cost nanoluminescent spray, emits visible light, enabling simple identification of pathogen strains in various food products. This spray is composed of several magnetic nanoparticles that react with pathogens found in food and generates a visible colour that can be easily spotted [128].

1.7.4 Applications in Agriculture

Agriculture is often regarded as the backbone of developing countries, with more than 60% of the population relying on it for their sustenance. Productivity, economic viability, environmental effects, and social well-being are crucial indicators for agricultural systems. At the same time, the agricultural industry faces several issues, such as inefficient resource utilization, climate change, and excessive use of chemical fertilizer. The indiscriminate use of various pesticides, herbicides, chemical fertilizers, and other agrochemical products to increase crop output has a negative impact on soil fertility, biodiversity, and ultimately ecosystems [132]. At this alarming juncture, interest in nanotechnology has increased in the agricultural sector in contrast to conventional agricultural practices. Nanotechnology is a weapon against elements that impair plant health. Farmers prefer to utilize nanoparticles over other conventional methods because of their small size, long shelf life, excellent efficiency, and ease of handling [133]. The main issue regarding employing nanotechnology tools in the agricultural sector includes specific applications like nanoherbicides, nanofertilizers, and nanopesticides for tracking nutrients and product levels to enhance productivity without water and soil contamination and to provide safety from pests, insects, and infectious diseases [134]. This can be done by enhancing the efficacy of drugs at lower dosages by utilizing biogenic nanoparticles [132]. The accurate, effective, and need-based usage of nanotechnology has enormous benefits for supporting sustainable agriculture with increased food and nutritional security as well as alleviating environmental concerns with the precise use of agrochemicals and water [135]. The improved bioavailability, reactivity, and bioactivity of NPs as well as their effects on surfaces and adhesion, all play a role in their use in agriculture. Although there is ample opportunity for using nanotechnology in agriculture, the field is still in its early stages [136]. Some of the precise and emerging applications of biogenic nanoparticles synthesized using bacteria in agriculture are described below.

Nanomaterials, characterized by a particle size of <100 nm, can either be nutrients (micro or macro) themselves, or serve as carriers or additives for other nutrients (e.g. by combining with minerals). They can also be generated by encapsulating nutrients into the nanomaterials [137]. According to several reports, only less than half of the nutrients sprayed in typical bulk fertilizers are actually utilized by plants. Contrarily, nanofertilizers are seen as essential to precision farming because they supply nutrients in a site-specific and controlled manner, thereby increasing nutrient use effectiveness and minimizing wastage [136]. By lowering nitrogen loss from leaching, emissions, and long-term assimilation by soil microbes, they have the potential to have a significant influence on energy, the environment, and the economy. Economic systems based on nanofertilizers are going to establish the best circumstances for a cleaner environment, increased productivity, and better resource management. Opening the door for a thriving agro-economy [138]. An additional advantage regarding the usage of nanofertilizers is the requirement in a smaller amount to have the same impact as conventional fertilizers. Utilizing lower quantities made it easier to apply and helped cut down on transportation expenses [136].

Saqib et al., 2022 reported the synthesis of ZnONPs utilizing endophytic *Enterobacter hormaechei* and tested their use as a biofertilizer in foliar sprays at various dosages. The results revealed a considerable improvement in the growth of the rice plant and higher levels of proteins, chlorophyll, and carotenoids [139]. Similarly, Bettencourt et al., 2020 suggested using endophytic *Paenibacillus polymyxa* fermented supernatant containing auxin chemicals as a stabilizing agent to generate mono- and bimetallic iron and manganese oxide nanoparticles [140]. The biogenic iron and manganese nanoparticles synthesized using bacteria had a substantial impact on plant growth, particularly in terms of root growth, germination rates, and fresh weight in maize plantlets, and thus can be employed as micronutrient nanofertilizers. In order to develop a sustainable agricultural system biogenic nanoparticle made with *Enterobacter* sp. may operate as a stabilizing agent to boost the effectiveness of beneficial biological entities in plants. According to the findings of Ahmed et al., 2021, biogenic MgONPs greatly increased biomass chlorophyll content, and plant height, and activated the natural antioxidant defences. Additionally, it was found that biogenic MgONPs dramatically decreased ROS accumulation by impeding the transfer of arsenic from the soil to the rice plant. The unique physicochemical characteristics of NPs, including their huge surface area, tiny size, and capping molecules, determine the precise effects of green MgONPs on arsenic translocation. The study concluded that biogenic MgONPs might be employed to create an effective nanofertilizer for long-term rice cultivation in soils contaminated with metals [141].

Nanotechnology has proven to be an effective method for creating nanofertilizers and other nanoscale formulations with prospective implications in the field of agriculture, including increasing yields and reducing soil pollution. Diverse research has documented the possible uses of NPs as heavy metal chelators to stop their acropetal transfer in plants cultivated in metal-contaminated circumstances, which lends credence to this theory. Noman et al., 2020a explored the biogenic production of CuNPs using metal-resistant *Shigella flexneri* SNT22 as a green nanofactory and eco-friendly technique and its application to reduce cadmium (Cd) stress in wheat plants [142]. The outcomes showed that by reducing Cd inflow and improving nutrient uptake, biogenic CuNPs boosted the biomass and length of wheat plants. Additionally, biogenic CuNPs altered the cellular ionic equilibrium, thereby reducing Cd toxicity. The immobilization of Cd onto the wide surface areas of CuNPs as a result of the capping of various protein molecules may be responsible for the increase in biomass of wheat plants as well as the decreased acropetal Cd translocation. As a result, a green CuNPs-based strategy may be the most significant option for remote approaches to reducing Cd stress in plants. Noman et al., 2021 investigated the effects of different amounts of biogenic CuNPs synthesized using *Klebsiella pneumoniae* NST2 on the morpho-physiochemical characteristics of maize plants grown in saline environments [143]. By scavenging ROS and increasing the activity of antioxidant machinery, the biogenic CuNPs also protect maize plants from oxidative damage. This application modified the ionic homeostasis in plant systems, namely by reducing the uptake of $Na+$ and $Cl-$ ions, to further boost growth as well as fresh and dry biomass production in salinized maize plants. In conclusion, bioengineered nanoformulations may be used as eco-friendly and sustainable substitutes for the conventional methods routinely employed to regulate the detrimental effects of salt on plants. By scavenging ROS and increasing the activity of antioxidative machinery, the naturally produced CuNPs also protect maize plants from oxidative damage. This application modified the ionic homeostasis in plant systems, namely by reducing the uptake of Na + and Cl-ions, to further boost growth as well as fresh and dry biomass production in salinized maize plants. In conclusion, bioengineered nanoformulations may be used as environmentally friendly and sustainable substitutes for the conventional methods routinely employed to regulate the detrimental effects of salt on plants.

The use of nanoscale materials as carriers for active components or chemicals in the field of pesticides is more effective in terms of application and dosage requirements. There is a necessity for alternatives because farming procedures currently employ conventional methods like integrated pest management, which are insufficient. In this regard, nanotechnology satisfies the need for an environmentally benign alternative to manage insect infestations. The ability of microorganisms to produce NPs that are hostile to pests and insects serves as a tool in the nanotechnological technique for insect pest management [144]. One of the most challenging areas of the pesticide industry is the nanoformulation of conventional pesticides with polymers or metal nanoparticles. They must provide a variety of benefits, including increased effectiveness and longevity, biodegradability, improved wettability and dispersibility, a relatively small quantity of active components, a lack of toxicity, and practical pesticidal capabilities. By manipulating the outer shell of the nanoscale, nanoencapsulation of pesticides is advantageous for a controlled and gradual release of the active component, which releases low doses over a prolonged time period and avoids excess run-off of undesirable pesticides [145]. Nanopesticides will minimize the rate of application because the amount of product that is truly effective is at least 10–15 times lower than that administered with the conventional formulation. As a result, a considerably smaller amount is needed to have much better and more sustained control [146]. Biogenic ZnONPs synthesized using native *Bacillus cereus* RNT6 demonstrated significant antimicrobial activity against *Burkholderia glumae* and *Burkholderia gladioli*. Additionally, ultrastructure analyses of bacterial phytopathogens that were treated with ZnONPs showed a significant level of structural damage and a wide range of cell wall morphology. These findings demonstrated that green ZnONPs can be used to formulate nanopesticides and have the propensity to protect rice plants as bactericidal agents [147]. Similar results were obtained when biogenic zirconium oxide nanoparticles (ZrONPs) generated employing *Enterobacter* sp. were found to have potent in vitro antifungal activity against *Pestalotiopsis versicolor*. ZrONPs were adsorbed on the *P. versicolor* cell membrane, disrupting the hyphal structure and causing it to shrink. They also caused damage to the pathogen's cell walls, chromatin, mitochondria, and ribosomes. As a result, the findings of the study showed that biogenic ZrONPs can be used to formulate fungicides and have a significant potential to protect bayberry plants as antifungal agents [148]. The effectiveness of the metal nanoparticles synthesized using bacteria against plant infections and insect pests has been established by previous investigations. Sayed et al., 2017 observed that the AgNPs synthesized using *Bacillus thuringiensis* was found to be significantly virulent and exhibit potent insecticidal activity against the larvae of cabbage looper, *Trichoplusiani* and black cutworm, *Agrotisipsilon.* As a result, nanoparticles synthesized using bacteria could be used for producing new biopesticide formulations [149].

1.7.5 Applications in Environment

Nanotechnological advancement has led to the development of technologies that could replace or improve current pollution monitoring strategies, prevention, and treatment of hazardous materials contaminating water, soil, air, and sediment [150]. Researchers have demonstrated the effectiveness of bacteria-based NPs on environmental issues, and a few of these studies are listed below.

One of the most sensitive environmental issues worldwide is heavy metal poisoning in wastewater. Therefore, there is a need to find novel and efficient techniques for their removal from water and wastewater because of their substantial health and environmental problems. Numerous studies have reported the use of nanoparticles for the adsorption of heavy metals due to the ease with which their surface functionality can be altered and their high surface area to volume ratio for improved efficiency and adsorption capacity [151]. *Bacillus cereus*-based biogenically synthesized AgNPs were stabilized at the alumina surface and checked for the maximum removal of heavy metals from the pharmaceutical effluent. According to the findings, at an ideal pH, the alumina-supported zerovalent silver nanoparticles (Al-ZVAgNPs) eradicated 98.13% and 98.76% chromium and lead, respectively, from the pharmaceutical industrial effluent. Additionally, the supported nanoparticles are more stabilized at the alumina surface and are easier to separate from the cleaned pharmaceutical effluent, enabling them to be reused. As a result, it was determined that this technique may be used to remediate contaminants from wastewater as well as decontaminate heavy metals from industrial effluent [152]. Similar to this, it was shown that biogenic manganese oxide nanoparticles synthesized utilizing the manganese-oxidizing bacteria *Bacillus mycoides* and *Bacillus subtilis* were successful in removing heavy metals (Zn, Ni, Hg, and Co). According to the study, the nanoparticle types Mn_2O_3 (Bixbyte) and Mn_3O_4 (Hausmannite) can generate a crystalline peak that is highly effective at removing several metals, particularly zinc, cobalt, and mercury, with nickel as the exception. Consequently, the finding demonstrates that a green strategy might potentially remove hazardous metals and could be used as a promising technique to remediate wastewater containing multi-metals [153].

Organic contaminants such as pharmaceutical waste, textile dyes, and pesticides, exist frequently in wastewater, which is particularly detrimental to terrestrial and aquatic ecosystems. They can create substantial bioaccumulation in the environment, which will have a lengthy half-life and the smallest trace can cause biological variation. They are poisonous because they are extremely polar and persistent in environmental degradation processes. As a result, there is a need for a new generation of ecologically friendly wastewater treatment facilities that can reduce the negative effects on the environment and human health. The heterogeneous photocatalysis with nanoparticle technology appears to be an appealing pre-treatment option among the new generations of wastewater treatment to improve the degradation of organic contaminants and the biodegradability of wastewater to undergo further downstream treatments [151]. A number of biogenic nanoparticles generated utilizing bacteria have been used to clean up environmental contaminants. *Escherichia* sp.-derived copper nanoparticles have been proven to be effective in treating textile wastewater. Furthermore, the azo dye degradation experiments show that the biosynthesized CuONPs decoloured 97.07%, 90.55%, 88.42%, and 83.61% of Congo red, malachite green, direct blue-1, reactive black-5 dyes, respectively, after 5 h of exposure to direct sunlight [154]. Similarly, methylene blue and Congo red dye were decolourized using cell-associated gold and silver nanoparticles derived from *Pseudomonas lipolytica* and reduced with sodium borohydride ($NaBH_4$). AuNPs act as a conduit for the passage of electrons from BH_4^- (donor) to the methylene blue dye. AuNPs have shown greater catalytic efficiency in comparison with Congo-red decolourization. As a result, *Pseudomonas lipolytica*-derived nanoparticles stabilized by peptides successfully decolourized two hazardous dyes [155]. Since 4-nitrophenol is one of the pollutants emitted into the environment during the manufacturing of pesticides, insecticides, and herbicides, catalytic reduction of 4-nitrophenol is essential. As 4-aminophenol has numerous uses as a corrosion inhibitor, anti-corrosion lubricant, etc., its conversion from 4-nitrophenol raises significant environmental and economic concerns [156]. Using $NaBH_4$, the biosynthesized AgNPs from *Bacillus amyloliquefaciens* completely reduced 4-NP to 4-AP after 15 min, demonstrating a potent chemocatalytic effect. The biosynthesized AgNPs were used as catalysts to breakdown the kinetic barrier for electron transfer from donor (BH_4^-) to acceptor 4-NP [157]. Abbas et al., 2022 reported the immobilization of biogenically synthesized ZnONPs utilizing *Bacillus cereus* on calcium alginate beads for the degradation of methylene blue under solar irradiation. The findings showed that the immobilized biogenic ZnONPs have a high capacity to degrade MB dye up to 150 mg/L concentration. A reliable and environmentally beneficial method for the large-scale treatment of textile effluents was also made feasible by the immobilized nanomaterial, which considerably improved numerous water quality metrics of real textile effluent [158].

1.8 Conclusion

Nanoparticles have demonstrated a great deal of utilization for scientists and researchers in various sectors, including biomedicine, food technology, textiles, agriculture, and the environment. The green synthesis technique presents a non-toxic, safe, and sustainable alternative for the fabrication of nanoparticles contrary to the physical and chemical approaches. Though developing effective ways for the synthesis of these nanoparticles remains a challenge, the use of microbes is an ideal choice to deal with the creation of cost-effective and eco-friendly nanoparticles. Since recent studies show that the fabrication of nanoparticles using bacteria offers several advantages, this study focuses on the bacteria-mediated synthesis of nanoparticles and their myriad applications in various domains. Future research should be focused on the large-scale synthesis of nanoparticles using bacteria because it involves the use of more secure, sustainable reducing and capping agents as well as the avoidance of any hazardous, expensive, and toxic chemicals for the synthesis and stabilization processes.

References

1. Koul, B., Poonia, A. K., Yadav, D., & Jin, J. O. (2021). Microbe-mediated biosynthesis of nanoparticles: Applications and future prospects. *Biomolecules, 11*(6), 886. https://doi.org/10.3390/biom11060886
2. Ghosh, S., Ahmad, R., Zeyaullah, M., & Khare, S. K. (2021). Microbial nano-factories: Synthesis and biomedical applications. *Frontiers in Chemistry, 9*, 626834. https://doi.org/10.3389/fchem.2021.626834
3. Ball, A. S., Patil, S., & Soni, S. (2019). Introduction into nanotechnology and microbiology. *Methods in Microbiology, 46*, 1–18. https://doi.org/10.1016/bs.mim.2019.04.003
4. Donia, A., Al-Madboly, L., & Bokhari, H. (2021). Microbial nanotechnology. *Importance & Applications of Nanotechnology, 6*, 10–15.
5. Mandeep, & Shukla, P. (2020). Microbial nanotechnology for bioremediation of industrial wastewater. *Frontiers in Microbiology*, 11. https://doi.org/10.3389/fmicb.2020.590631
6. Grasso, G., Zane, D., & Dragone, R. (2019). Microbial nanotechnology: Challenges and prospects for green biocatalytic synthesis of nanoscale materials for sensoristic and biomedical applications. *Nanomaterials, 10*(1), 11. https://doi.org/10.3390/nano10010011
7. Iravani, S. (2014). Bacteria in nanoparticle synthesis: Current status and future prospects. *International Scholarly Research Notices, 2014*, 1–18. https://doi.org/10.1155/2014/359316
8. Gahlawat, G., & Choudhury, A. R. (2019). A review on the biosynthesis of metal and metal salt nanoparticles by microbes. *RSC Advances, 9*(23), 12944–12967. https://doi.org/10.1039/C8RA10483B
9. Adelere, I., & Lateef, A. (2016). A novel approach to the green synthesis of metallic nanoparticles: The use of agro-wastes, enzymes, and pigments. *Nanotechnology Reviews, 5*, 567–587. https://doi.org/10.1515/ntrev-2016-0024
10. Khan, I., Saeed, K., & Khan, I. (2019). Nanoparticles: Properties, applications and toxicities. *Arabian Journal of Chemistry, 12*(7), 908–931. https://doi.org/10.1016/j.arabjc.2017.05.011
11. Iravani, S. (2011). Green synthesis of metal nanoparticles using plants. *Green Chemistry, 13*(10), 2638–2650. https://doi.org/10.1039/C1GC15386B
12. Messaoudi, O., Bendahou, M., Messaoudi, O., & Bendahou, M. (2020). *Biological synthesis of nanoparticles using endophytic microorganisms: Current development*. Intech Open. https://doi.org/10.5772/intechopen.93734
13. Prakash, A., Sharma, S., Ahmad, N., Ghosh, A., & Sinha, P. (2011). Synthesis of Agnps by Bacillus cereus bacteria and their antimicrobial potential. *Journal of Biomaterials and Nanobiotechnology, 2*(2), 155–161. https://doi.org/10.4236/jbnb.2011.22020
14. Kulkarni, R. R., Shaiwale, N. S., Deobagkar, D. N., & Deobagkar, D. D. (2015). Synthesis and extracellular accumulation of silver nanoparticles by employing radiation-resistant Deinococcus Radiodurans, their characterization, and determination of bioactivity. *International Journal of Nanomedicine, 10*, 963–974. https://doi.org/10.2147/IJN.S72888

15. Yumei, L., Yamei, L., Qiang, L., & Jie, B. (2017). Rapid biosynthesis of silver nanoparticles based on flocculation and reduction of an exopolysaccharide from *Arthrobacter* Sp. B4: Its antimicrobial activity and phytotoxicity. *Journal of Nanomaterials, 2017*, e9703614. https://doi.org/10.1155/2017/9703614
16. Saravanan, M., Barik, S. K., Mubarak Ali, D., Prakash, P., & Pugazhendhi, A. (2018). Synthesis of silver nanoparticles from bacillus brevis (NCIM 2533) and their antibacterial activity against pathogenic bacteria. *Microbial Pathogenesis, 116*, 221–226. https://doi.org/10.1016/j.micpath.2018.01.038
17. Hamouda, R. A., Hussein, M. H., Abo-Elmagd, R. A., & Bawazir, S. S. (2019). Synthesis and biological characterization of silver nanoparticles derived from the cyanobacterium Oscillatoria Limnetica. *Scientific Reports, 9*(1), 1–17. https://doi.org/10.1038/s41598-019-49444-y
18. Srinath, B. S., Namratha, K., & Byrappa, K. (2018). Eco-friendly synthesis of gold nanoparticles by bacillus subtilis and their environmental applications. *Advanced Science Letters, 24*(8), 5942–5946. https://doi.org/10.1166/asl.2018.12224
19. Ahmed, E., Kalathil, S., Shi, L., Alharbi, O., & Wang, P. (2018). Synthesis of ultra-small platinum, palladium and gold nanoparticles by Shewanella Loihica PV-4 electrochemically active biofilms and their enhanced catalytic activities. *Journal of Saudi Chemical Society, 22*(8), 919–929. https://doi.org/10.1016/j.jscs.2018.02.002
20. John, M. S., Nagoth, J. A., Zannotti, M., Giovannetti, R., Mancini, A., Ramasamy, K. P., Miceli, C., & Pucciarelli, S. (2021). Biogenic synthesis of copper nanoparticles using bacterial strains isolated from an antarctic consortium associated to a psychrophilic marine ciliate: Characterization and potential application as antimicrobial agents. *Marine Drugs, 19*(5), 263. https://doi.org/10.3390/md19050263
21. Rad, M., Taran, M., & Alavi, M. (2018). Effect of incubation time, CuSO4 and glucose concentrations on biosynthesis of copper oxide (CuO) nanoparticles with rectangular shape and antibacterial activity: Taguchi method approach. *Nano Biomedicine and Engineering, 10*(1), 25–33. https://doi.org/10.5101/nbe.v10i1.p25-33
22. Jayaseelan, C., Rahuman, A. A., Kirthi, A. V., Marimuthu, S., Santhoshkumar, T., Bagavan, A., Gaurav, K., Karthik, L., & Rao, K. V. B. (2012). Novel microbial route to synthesize ZnO nanoparticles using Aeromonas Hydrophila and their activity against pathogenic bacteria and fungi. *Spectrochimica Acta. Part A, Molecular and Biomolecular Spectroscopy, 90*, 78–84. https://doi.org/10.1016/j.saa.2012.01.006
23. Mishra, M., Paliwal, J. S., Singh, S. K., Selvarajan, E., Subathradevi, C., & Mohanasrinivasan, V. (2013). Studies on the inhibitory activity of biologically synthesized and characterized zinc oxide nanoparticles using lactobacillus Sporogens against staphylococcus aureus. *Journal of Pure and Applied Microbiology, 7*(2), 1263–1268.
24. Dhandapani, P., Siddarth, A. S., Kamalasekaran, S., Maruthamuthu, S., & Rajagopal, G. (2014). Bio-approach: Ureolytic bacteria mediated synthesis of ZnO nanocrystals on cotton fabric and evaluation of their antibacterial properties. *Carbohydrate Polymers, 103*, 448–455. https://doi.org/10.1016/j.carbpol.2013.12.074
25. Saleh, M. N., & Alwan, S. K. (2020). Bio-synthesis of silver nanoparticles from bacteria Klebsiella pneumonia: Their characterization and antibacterial studies. *Journal of Physics Conference Series, 1664*(1), 012115. https://doi.org/10.1088/1742-6596/1664/1/012115
26. Saeed, S., Iqbal, A., & Ashraf, M. A. (2020). Bacterial-mediated synthesis of silver nanoparticles and their significant effect against pathogens. *Environmental Science and Pollution Research, 27*, 37347–37356. https://doi.org/10.1007/s11356-020-07610-0
27. Divya, M., Kiran, G. S., Hassan, S., & Selvin, J. (2019). Biogenic synthesis and effect of silver nanoparticles (AgNPs) to combat catheter-related urinary tract infections. *Biocatalysis and Agricultural Biotechnology, 18*, 101037. https://doi.org/10.1016/j.bcab.2019.101037
28. Buszewski, B., Railean-Plugaru, V., Pomastowski, P., Rafińska, K., Szultka-Mlynska, M., Golinska, P., Wypij, M., Laskowski, D., & Dahm, H. (2018). Antimicrobial activity of biosilver nanoparticles produced by a novel Streptacidiphilus Durhamensis strain. *Journal of Microbiology, Immunology, and Infection, 51*(1), 45–54. https://doi.org/10.1016/j.jmii.2016.03.002

29. Wypij, M., Czarnecka, J., Świecimska, M., Dahm, H., Rai, M., & Golinska, P. (2018). Synthesis, characterization and evaluation of antimicrobial and cytotoxic activities of biogenic silver nanoparticles synthesized from Streptomyces Xinghaiensis OF1 strain. *World Journal of Microbiology and Biotechnology, 34*(2), 23. https://doi.org/10.1007/s11274-017-2406-3
30. Ranjani, A., Gopinath, P. M., Ananth, S., Narchonai, G., Santhanam, P., Thajuddin, N., & Dhanasekaran, D. (2018). Multidimensional dose–response toxicity exploration of silver nanoparticles from nocardiopsis flavascens RD30. *Applied Nanoscience, 8*(4), 699–713. https://doi.org/10.1007/s13204-018-0824-7
31. Naik, M. M., Prabhu, M. S., Samant, S. N., Naik, P. M., & Shirodkar, S. (2017). Synergistic action of silver nanoparticles synthesized from silver resistant estuarine pseudomonas aeruginosa strain SN5 with antibiotics against antibiotic resistant bacterial human pathogens. *Thalassas International Journal of Marine Science, 33*(1), 73–80. https://doi.org/10.1007/s41208-017-0023-4
32. Silva-Vinhote, N. M., Caballero, N. E. D., Silva, T. de A, Quelemes, P. V., AraÃojo, A. R. de, Moraes, A. C. M. de, CÃ¢mara, A. L. dos S, Longo, J. P. F., Azevedo, R. B., Silva, D. A. da, Leite, J. R. de S. de A, & Teixeira, M. F. S. (2017). Extracellular biogenic synthesis of silver nanoparticles by actinomycetes from amazonic biome and its antimicrobial efficiency. *African Journal of Biotechnology, 16*(43), 2072–2082. https://doi.org/10.5897/AJB2017.16148
33. Deljou, A., & Goudarzi, S. (2016). Green extracellular synthesis of the silver nanoparticles using thermophilic bacillus Sp. AZ1 and its antimicrobial activity against several human pathogenetic bacteria. *Iranian Journal of Biotechnology, 14*(2), 25–32. https://doi.org/10.15171/ijb.1259
34. Sowani, H., Mohite, P., Munot, H., Shouche, Y., Bapat, T., Kumar, A. R., Kulkarni, M., & Zinjarde, S. (2016). Green synthesis of gold and silver nanoparticles by an actinomycete gordonia amicalis HS-11: Mechanistic aspects and biological application. *Process Biochemistry, 51*(3), 374–383. https://doi.org/10.1016/j.procbio.2015.12.013
35. Gahlawat, G., Shikha, S., Chaddha, B. S., Chaudhuri, S. R., Mayilraj, S., & Choudhury, A. R. (2016). Microbial glycolipoprotein-capped silver nanoparticles as emerging antibacterial agents against cholera. *Microbial Cell Factories, 15*(1), 25. https://doi.org/10.1186/s12934-016-0422-x
36. Abd-Elnaby, H. M., Abo-Elala, G. M., Abdel-Raouf, U. M., & Hamed, M. M. (2016). Antibacterial and anticancer activity of extracellular synthesized silver nanoparticles from marine Streptomyces Rochei MHM13. *Egyptian Journal of Aquatic Research, 42*(3), 301–312. https://doi.org/10.1016/j.ejar.2016.05.004
37. Manikprabhu, D., Cheng, J., Chen, W., Sunkara, A. K., Mane, S. B., Kumar, R., das, M., N Hozzein, W., Duan, Y.-Q., & Li, W.-J. (2016). Sunlight mediated synthesis of silver nanoparticles by a novel actinobacterium (Sinomonas Mesophila MPKL 26) and its antimicrobial activity against multi drug resistant staphylococcus aureus. *Journal of Photochemistry and Photobiology. B, 158*, 202–205. https://doi.org/10.1016/j.jphotobiol.2016.01.018
38. Shunmugam, R., Balusamy, S. R., Kumar, V., Menon, S., Lakshmi, T., & Perumalsamy, H. (2021). Biosynthesis of gold nanoparticles using marine microbe (vibrio alginolyticus) and its anticancer and antioxidant analysis. *Journal of King Saud University-Science, 33*(1), 101260. https://doi.org/10.1016/j.jksus.2020.101260
39. Patil, M. P., Kang, M., Niyonizigiye, I., Singh, A., Kim, J.-O., Seo, Y. B., & Kim, G.-D. (2019). Extracellular synthesis of gold nanoparticles using the marine bacterium Paracoccus haeundaensis BC74171T and evaluation of their antioxidant activity and antiproliferative effect on normal and cancer cell lines. *Colloids and Surfaces. B, Biointerfaces, 183*, 110455. https://doi.org/10.1016/j.colsurfb.2019.110455
40. Jafari, M., Rokhbakhsh-Zamin, F., Shakibaie, M., Moshafi, M. H., Ameri, A., Rahimi, H. R., & Forootanfar, H. (2018). Cytotoxic and antibacterial activities of biologically synthesized gold nanoparticles assisted by micrococcus yunnanensis strain J2. *Biocatalysis and Agricultural Biotechnology, 15*, 245–253. https://doi.org/10.1016/j.bcab.2018.06.014
41. Camas, M., Sazak Camas, A., & Kyeremeh, K. (2018). Extracellular synthesis and characterization of gold nanoparticles using mycobacterium Sp. BRS2A-AR2 isolated from

the aerial roots of the Ghanaian mangrove plant, Rhizophora Racemosa. *Indian Journal of Microbiology, 58*(2), 214–221. https://doi.org/10.1007/s12088-018-0710-8
42. Bing, W., Sun, H., Wang, F., Song, Y., & Ren, J. (2018). Hydrogen-producing hyperthermophilic bacteria synthesized size-controllable fine gold nanoparticles with excellence for eradicating biofilm and antibacterial applications. *Journal of Materials Chemistry B, 6*(28), 4602–4609. https://doi.org/10.1039/C8TB00549D
43. Ranjitha, V. R., & Rai, V. R. (2017). Actinomycetes mediated synthesis of gold nanoparticles from the culture supernatant of streptomyces griseoruber with special reference to catalytic activity. *3 Biotech, 7*(5), 299. https://doi.org/10.1007/s13205-017-0930-3
44. Markus, J., Mathiyalagan, R., Kim, Y.-J., Abbai, R., Singh, P., Ahn, S., Perez, Z. E. J., Hurh, J., & Yang, D. C. (2016). Intracellular synthesis of gold nanoparticles with antioxidant activity by probiotic lactobacillus Kimchicus DCY51T isolated from Korean kimchi. *Enzyme and Microbial Technology, 95*, 85–93. https://doi.org/10.1016/j.enzmictec.2016.08.018
45. Bennur, T., Khan, Z., Kshirsagar, R., Javdekar, V., & Zinjarde, S. (2016). Biogenic gold nanoparticles from the Actinomycete Gordonia amarae: Application in rapid sensing of copper ions. *Sensors and Actuators, B: Chemical, 233*, 684–690. https://doi.org/10.1016/j.snb.2016.04.022
46. Hassan, S. E.-D., Salem, S. S., Fouda, A., Awad, M. A., El-Gamal, M. S., & Abdo, A. M. (2018). New approach for antimicrobial activity and bio-control of various pathogens by biosynthesized copper nanoparticles using endophytic actinomycetes. *Journal of Radiation Research and Applied Science, 11*(3), 262–270. https://doi.org/10.1016/j.jrras.2018.05.003
47. Rasool, U., & Hemalatha, S. (2017). Marine endophytic actinomycetes assisted synthesis of copper nanoparticles (CuNPs): Characterization and antibacterial efficacy against human pathogens. *Materials Letters, 194*, 176–180. https://doi.org/10.1016/j.matlet.2017.02.055
48. Eramabadi, P., Masoudi, M., Makhdoumi, A., & Mashreghi, M. (2020). Microbial cell lysate supernatant (CLS) alteration impact on platinum nanoparticles fabrication, characterization, antioxidant and antibacterial activity. *Materials Science and Engineering: C, 117*, 111292. https://doi.org/10.1016/j.msec.2020.111292
49. Xu, H., Xiao, Y., Xu, M., Cui, H., Tan, L., Feng, N., Liu, X., Qiu, G., Dong, H., & Xie, J. (2018). Microbial synthesis of Pd–Pt alloy nanoparticles using Shewanella oneidensis MR-1 with enhanced catalytic activity for nitrophenol and azo dyes reduction. *Nanotechnology, 30*(6), 065607. https://doi.org/10.1088/1361-6528/aaf2a6
50. Baskaran, B., Muthukumarasamy, A., Chidambaram, S., Sugumaran, A., Ramachandran, K., & Rasu Manimuthu, T. (2017). Cytotoxic potentials of biologically fabricated platinum nanoparticles from Streptomyces Sp. on MCF-7 breast cancer cells. *IET Nanobiotechnology, 11*(3), 241–246. https://doi.org/10.1049/iet-nbt.2016.0040
51. Tuo, Y., Liu, G., Dong, B., Yu, H., Zhou, J., Wang, J., & Jin, R. (2017). Microbial synthesis of bimetallic PdPt nanoparticles for catalytic reduction of 4-nitrophenol. *Environmental Science and Pollution Research, 24*(6), 5249–5258. https://doi.org/10.1007/s11356-016-8276-7
52. Taran, M., Rad, M., & Alavi, M. (2017). Biosynthesis of TiO2 and ZnO nanoparticles by Halomonas elongata IBRC-M 10214 in different conditions of medium. *Bio Impacts, 8*(2), 81–89. https://doi.org/10.15171/bi.2018.10
53. Saravanan, M., Gopinath, V., Chaurasia, M. K., Syed, A., Ameen, F., & Purushothaman, N. (2018). Green synthesis of anisotropic zinc oxide nanoparticles with antibacterial and cytofriendly properties. *Microbial Pathogenesis, 115*, 57–63. https://doi.org/10.1016/j.micpath.2017.12.039
54. Król, A., Railean-Plugaru, V., Pomastowski, P., Złoch, M., & Buszewski, B. (2018). Mechanism study of intracellular zinc oxide nanocomposites formation. *Colloids and Surfaces A: Physicochemical and Engineering Aspects, 553*, 349–358. https://doi.org/10.1016/j.colsurfa.2018.05.069
55. Rauf, M. A., Owais, M., Rajpoot, R., Ahmad, F., Khan, N., & Zubair, S. (2017). Biomimetically synthesized ZnO nanoparticles attain potent antibacterial activity against less susceptible S. Aureus skin infection in experimental animals. *RSC Advances, 7*(58), 36361–36373. https://doi.org/10.1039/C7RA05040B

56. Balraj, B., Senthilkumar, N., Siva, C., Krithikadevi, R., Julie, A., Potheher, I. V., & Arulmozhi, M. (2017). Synthesis and characterization of zinc oxide nanoparticles using marine Streptomyces Sp. with its investigations on anticancer and antibacterial activity. *Research on Chemical Intermediates, 43*(4), 2367–2376. https://doi.org/10.1007/s11164-016-2766-6
57. Venil, C. K., Devi, P. R., & Dufossé, L. (2020). Synthesis of pigment-mediated nanoparticles and its pharmacological applications. In J. K. Patra, L. F. Fraceto, G. Das, & E. V. R. Campos (Eds.), *Green nanoparticles. Nanotechnology in the life sciences* (pp. 331–346). Springer. https://doi.org/10.1007/978-3-030-39246-8_17
58. Singh, N., Naik, B., Kumar, V., Kumar, V., & Gupta, S. (2021). Actinobacterial pigment assisted synthesis of nanoparticles and its biological activity: Nanoparticles and its biological activity. *Journal of Microbiology, Biotechnology and Food Sciences, 10*(4), 604–608. https://doi.org/10.15414/jmbfs.2021.10.4.604-608
59. Khaksar, F., Rigi, G., & Mirdamadian, S. H. (2021). Creation of a violacein pigment hybrid with silver and titanium dioxide nanoparticles to produce multifunctional textiles with antimicrobial properties. *Nanomedicine Research Journal, 6*(1), 60–72. https://doi.org/10.22034/nmrj.2021.01.007
60. Muthukrishnan, L., Chellappa, M., Nanda, A., Thukkaram, S., Selvaraj, G., Muthiah, B., Sagadevan, S., & Anita Lett, J. (2019). Bio-fabrication of pigment-capped silver nanoparticles encountering antibiotic-resistant strains and their cytotoxic effect towards human epidermoid larynx carcinoma (HEp-2) cells. *RSC Advances, 9*(28), 15874–15886. https://doi.org/10.1039/C9RA01072F
61. El-Naggar, N. E.-A., Hussein, M. H., & El-Sawah, A. A. (2017). Bio-fabrication of silver nanoparticles by phycocyanin, characterization, in vitro anticancer activity against breast cancer cell line and in vivo cytotxicity. *Scientific Reports, 7*(1), 1–20. https://doi.org/10.1038/s41598-017-11121-3
62. Venil, C. K., Sathishkumar, P., Malathi, M., Usha, R., Jayakumar, R., Yusoff, A. R. M., & Ahmad, W. A. (2016). Synthesis of flexirubin-mediated silver nanoparticles using chryseobacterium artocarpi CECT 8497 and investigation of its anticancer activity. *Materials Science and Engineering: C, 59*, 228–234. https://doi.org/10.1016/j.msec.2015.10.019
63. Rahul, S., Chandrashekhar, P., Hemant, B., Bipinchandra, S., Mouray, E., Grellier, P., & Satish, P. (2015). In vitro antiparasitic activity of microbial pigments and their combination with phytosynthesized metal nanoparticles. *Parasitology International, 64*(5), 353–356. https://doi.org/10.1016/j.parint.2015.05.004
64. Patel, V., Berthold, D., Puranik, P., & Gantar, M. (2014). Screening of cyanobacteria and microalgae for their ability to synthesize silver nanoparticles with antibacterial activity. *Biotechnology Reports, 5*, 112–119. https://doi.org/10.1016/j.btre.2014.12.001
65. Lahiri, D., Nag, M., Sheikh, H. I., Sarkar, T., Edinur, H. A., Pati, S., & Ray, R. R. (2021). Microbiologically-synthesized nanoparticles and their role in silencing the biofilm signaling cascade. *Frontiers in Microbiology, 12*, 636588. https://doi.org/10.3389/fmicb.2021.636588
66. Emam, H. E., & Ahmed, H. B. (2016). Polysaccharides templates for assembly of nanosilver. *Carbohydrate Polymers, 135*, 300–307. https://doi.org/10.1016/j.carbpol.2015.08.095
67. Kanmani, P., & Lim, S. T. (2013). Synthesis and structural characterization of silver nanoparticles using bacterial exopolysaccharide and its antimicrobial activity against food and multidrug resistant pathogens. *Process Biochemistry, 48*(7), 1099–1106. https://doi.org/10.1016/j.procbio.2013.05.011
68. Hu, Y., He, L., Ding, J., Sun, D., Chen, L., & Chen, X. (2016). One-pot synthesis of dextran decorated reduced graphene oxide nanoparticles for targeted photo-chemotherapy. *Carbohydrate Polymers, 144*, 223–229. https://doi.org/10.1016/j.carbpol.2016.02.062
69. Zhang, R., & Edgar, K. J. (2014). Properties, chemistry, and applications of the bioactive polysaccharide curdlan. *Biomacromolecules, 15*(4), 1079–1096. https://doi.org/10.1021/bm500038g
70. Ahmed, K. B. A., Kalla, D., Uppuluri, K. B., & Anbazhagan, V. (2014). Green synthesis of silver and gold nanoparticles employing levan, a biopolymer from Acetobacter Xylinum

NCIM 2526, as a reducing agent and capping agent. *Carbohydrate Polymers, 112*, 539–545. https://doi.org/10.1016/j.carbpol.2014.06.033
71. Escárcega-González, C. E., Garza-Cervantes, J. A., Vázquez-Rodríguez, A., & Morones-Ramírez, J. R. (2018). Bacterial exopolysaccharides as reducing and/or stabilizing agents during synthesis of metal nanoparticles with biomedical applications. *International Journal of Polymeric Science, 2018*, e7045852. https://doi.org/10.1155/2018/7045852
72. Raveendran, S., Poulose, A. C., Yoshida, Y., Maekawa, T., & Kumar, D. S. (2013). Bacterial exopolysaccharide based nanoparticles for sustained drug delivery, cancer chemotherapy and bioimaging. *Carbohydrate Polymers, 91*(1), 22–32. https://doi.org/10.1016/j.carbpol.2012.07.079
73. Bankura, K. P., Maity, D., Mollick, M. M. R., Mondal, D., Bhowmick, B., Bain, M. K., Chakraborty, A., Sarkar, J., Acharya, K., & Chattopadhyay, D. (2012). Synthesis, characterization and antimicrobial activity of dextran stabilized silver nanoparticles in aqueous medium. *Carbohydrate Polymers, 89*(4), 1159–1165. https://doi.org/10.1016/j.carbpol.2012.03.089
74. Leung, T. C.-Y., Wong, C. K., & Xie, Y. (2010). Green synthesis of silver nanoparticles using biopolymers, carboxymethylated-curdlan and fucoidan. *Materials Chemistry and Physics, 121*(3), 402–405. https://doi.org/10.1016/j.matchemphys.2010.02.026
75. Zhang, T., Wang, W., Zhang, D., Zhang, X., Ma, Y., Zhou, Y., & Qi, L. (2010). Biotemplated synthesis of gold nanoparticle–bacteria cellulose nanofiber nanocomposites and their application in biosensing. *Advanced Functional Materials, 20*(7), 1152–1160. https://doi.org/10.1002/adfm.200902104
76. Kwon, C., Park, B., Kim, H., & Jung, S. (2009). Green synthesis of silver nanoparticles by sinorhizobial octasaccharide isolated from Sinorhizobium meliloti. *Bulletin of the Korean Chemical Society, 30*(7), 1651–1654. https://doi.org/10.5012/bkcs.2009.30.7.1651
77. Dhar, S., Reddy, E. M., Shiras, A., Pokharkar, V., & Prasad, B. L. V. (2008). Natural gum reduced/stabilized gold nanoparticles for drug delivery formulations. *Chemistry – A European Journal, 14*(33), 10244–10250. https://doi.org/10.1002/chem.200801093
78. Nadeem, M., Khan, R., Shah, N., Bangash, I. R., Abbasi, B. H., Hano, C., et al. (2022). A review of microbial mediated iron nanoparticles (IONPs) and its biomedical applications. *Nanomaterials, 12*(1), 130. https://doi.org/10.3390/nano12010130
79. Thomas, R., Soumya, K. R., Mathew, J., & Radhakrishnan, E. K. (2015). Inhibitory effect of silver nanoparticle fabricated urinary catheter on colonization efficiency of coagulase negative staphylococci. *Journal of Photochemistry and Photobiology B: Biology, 149*, 68–77. https://doi.org/10.1016/j.jphotobiol.2015.04.034
80. Faisal, S., Rizwan, M., Ullah, R., Alotaibi, A., Khattak, A., Bibi, N., & Idrees, M. (2022). Paraclostridiumbenzoelyticum bacterium-mediated zinc oxide nanoparticles and their in vivo multiple biological applications. *Oxidative Medicine and Cellular Longevity, 2022*. https://doi.org/10.1155/2022/5994033
81. Kalishwaralal, K., Deepak, V., Pandian, S. R. K., Kottaisamy, M., Barath ManiKanth, S., Kartikeyan, B., & Gurunathan, S. (2010). Biosynthesis of silver and gold nanoparticles using Brevibacteriumcasei. *Colloids and Surfaces B: Biointerfaces, 77*(2), 257–262. https://doi.org/10.1016/j.colsurfb.2010.02.007
82. Kousheh, S. A., Moradi, M., Tajik, H., & Molaei, R. (2020). Preparation of antimicrobial/ultraviolet protective bacterial nanocellulose film with carbon dots synthesized from lactic acid bacteria. *International Journal of Biological Macromolecules, 155*, 216–225. https://doi.org/10.1016/j.ijbiomac.2020.03.230
83. Mohd Yusof, H., Abdul Rahman, N. A., Mohamad, R., & Zaidan, U. H. (2020). Microbial mediated synthesis of silver nanoparticles by Lactobacillus Plantarum TA4 and its antibacterial and antioxidant activity. *Applied Sciences, 10*(19), 6973. https://doi.org/10.3390/app10196973
84. Singh, P., Kim, Y. J., Singh, H., Wang, C., Hwang, K. H., Farh, M. E. A., & Yang, D. C. (2015). Biosynthesis, characterization, and antimicrobial applications of silver nanoparticles. *International Journal of Nanomedicine, 10*, 2567. https://doi.org/10.2147/IJN.S72313

85. Li, Q., Feng, T., Li, H., Wang, Z., Wei, X., & Liu, J. (2022). Green synthesis of silver nanoparticles using endophytic bacterium Bacillus zanthoxyli GBE11 and their antimicrobial activity. *Biomass Conversion and Biorefinery, 1-13*. https://doi.org/10.1007/s13399-022-03266-7
86. Elbahnasawy, M. A., Shehabeldine, A. M., Khattab, A. M., Amin, B. H., & Hashem, A. H. (2021). Green biosynthesis of silver nanoparticles using novel endophytic Rothiaendophytica: Characterization and anticandidal activity. *Journal of Drug Delivery Science and Technology, 62*, 102401. https://doi.org/10.1016/j.jddst.2021.102401
87. Khalil, I., Yehye, W. A., Etxeberria, A. E., Alhadi, A. A., Dezfooli, S. M., Julkapli, N. B. M., et al. (2019). Nanoantioxidants: Recent trends in antioxidant delivery applications. *Antioxidants, 9*(1), 24. https://doi.org/10.3390/antiox9010024
88. Harish, V., Tewari, D., Gaur, M., Yadav, A. B., Swaroop, S., Bechelany, M., & Barhoum, A. (2022). Review on nanoparticles and nanostructured materials: Bioimaging, biosensing, drug delivery, tissue engineering, antimicrobial, and agro-food applications. *Nanomaterials, 12*(3), 457. https://doi.org/10.3390/nano12030457
89. Nirmala, C., & Sridevi, M. (2021). Characterization, antimicrobial and antioxidant evaluation of biofabricated silver nanoparticles from Endophytic Pantoeaanthophila. *Journal of Inorganic and Organometallic Polymers and Materials, 31*(9), 3711–3725. https://doi.org/10.1007/s10904-021-01974-7
90. Mujaddidi, N., Nisa, S., Al Ayoubi, S., Bibi, Y., Khan, S., Sabir, M., et al. (2021). Pharmacological properties of biogenically synthesized silver nanoparticles using endophyte Bacillus cereus extract of Berberis lyceum against oxidative stress and pathogenic multidrug-resistant bacteria. *Saudi Journal of Biological Sciences, 28*(11), 6432–6440. https://doi.org/10.1016/j.sjbs.2021.07.009
91. Salem, S. S., Ali, O. M., Reyad, A. M., Abd-Elsalam, K. A., & Hashem, A. H. (2022). Pseudomonas indica-mediated silver nanoparticles: Antifungal and antioxidant biogenic tool for suppressing mucormycosis fungi. *Journal of Fungi, 8*(2), 126. https://doi.org/10.3390/jof8020126
92. Torres, S. K., Campos, V. L., León, C. G., Rodríguez-Llamazares, S. M., Rojas, S. M., Gonzalez, M., et al. (2012). Biosynthesis of selenium nanoparticles by Pantoeaagglomerans and their antioxidant activity. *Journal of Nanoparticle Research, 14*, 1–9. https://doi.org/10.1007/s11051-012-1236-3
93. Ullah, A., Yin, X., Wang, F., Xu, B., Mirani, Z. A., Xu, B., et al. (2021). Biosynthesis of selenium nanoparticles (via Bacillus subtilis BSN313), and their isolation, characterization, and bioactivities. *Molecules, 26*(18), 5559. https://doi.org/10.3390/molecules26185559
94. Cherian, T., Maity, D., Rajendra Kumar, R. T., Balasubramani, G., Ragavendran, C., Yalla, S., et al. (2022). Green chemistry based gold nanoparticles synthesis using the marine bacterium Lysinibacillusodysseyi PBCW2 and their multitudinous activities. *Nanomaterials, 12*(17), 2940. https://doi.org/10.3390/nano12172940
95. Buttacavoli, M., Albanese, N. N., Di Cara, G., Alduina, R., Faleri, C., Gallo, M., et al. (2018). Anticancer activity of biogenerated silver nanoparticles: An integrated proteomic investigation. *Oncotarget, 9*(11), 9685. https://doi.org/10.18632/oncotarget.23859
96. Nandhini, J. T., Ezhilarasan, D., & Rajeshkumar, S. (2021). An ecofriendly synthesized gold nanoparticles induces cytotoxicity via apoptosis in HepG2 cells. *Environmental Toxicology, 36*(1), 24–32. https://doi.org/10.1002/tox.23007
97. Majeed, S., Danish, M., Mohamad Ibrahim, M. N., Sekeri, S. H., Ansari, M. T., Nanda, A., & Ahmad, G. (2021). Bacteria mediated synthesis of iron oxide nanoparticles and their antibacterial, antioxidant, cytocompatibility properties. *Journal of Cluster Science, 32*, 1083–1094. https://doi.org/10.1007/s10876-020-01876-7
98. Almalki, M. A., & Khalifa, A. Y. (2020). Silver nanoparticles synthesis from Bacillus sp KFU36 and its anticancer effect in breast cancer MCF-7 cells via induction of apoptotic mechanism. *Journal of Photochemistry and Photobiology B: Biology, 204*, 111786. https://doi.org/10.1016/j.jphotobiol.2020.111786
99. Maharani, V., Sundaramanickam, A., & Balasubramanian, T. (2016). In vitro anticancer activity of silver nanoparticle synthesized by Escherichia coli VM1 isolated from marine

sediments of Ennore southeast coast of India. *Enzyme and Microbial Technology, 95*, 146–154. https://doi.org/10.1016/j.enzmictec.2016.09.008

100. Rana, K. L., Kour, D., Yadav, N., & Yadav, A. N. (2020). Endophytic microbes in nanotechnology: Current development, and potential biotechnology applications. In *Microbial endophytes* (pp. 231–262). Woodhead Publishing. https://doi.org/10.1016/B978-0-12-818734-0.00010-3
101. Ramasamy, M., Lee, J. H., & Lee, J. (2016). Potent antimicrobial and antibiofilm activities of bacteriogenically synthesized gold–silver nanoparticles against pathogenic bacteria and their physiochemical characterizations. *Journal of Biomaterials Applications, 31*(3), 366–378. https://doi.org/10.1177/0885328216646910
102. Khaleghi, M., Khorrami, S., & Ravan, H. (2019). Identification of Bacillus thuringiensis bacterial strain isolated from the mine soil as a robust agent in the biosynthesis of silver nanoparticles with strong antibacterial and anti-biofilm activities. *Biocatalysis and Agricultural Biotechnology, 18*, 101047. https://doi.org/10.1016/j.bcab.2019.101047
103. Jayabalan, J., Mani, G., Krishnan, N., Pernabas, J., Devadoss, J. M., & Jang, H. T. (2019). Green biogenic synthesis of zinc oxide nanoparticles using Pseudomonas putida culture and its in vitro antibacterial and anti-biofilm activity. *Biocatalysis and Agricultural Biotechnology, 21*, 101327. https://doi.org/10.1016/j.bcab.2019.101327
104. Alam, H., Khatoon, N., Khan, M. A., Husain, S. A., Saravanan, M., & Sardar, M. (2020). Synthesis of selenium nanoparticles using probiotic bacteria Lactobacillus acidophilus and their enhanced antimicrobial activity against resistant bacteria. *Journal of Cluster Science, 31*, 1003–1011. https://doi.org/10.1007/s10876-019-01705-6
105. Abishad, P., Vergis, J., Unni, V., Ram, V. P., Niveditha, P., Yasur, J., et al. (2022). Green synthesized silver nanoparticles using lactobacillus acidophilus as an antioxidant, antimicrobial, and antibiofilm agent against multi-drug resistant enteroaggregative Escherichia coli. *Probiotics and Antimicrobial Proteins, 14*(5), 904–914. https://doi.org/10.1007/s12602-022-09961-1
106. Zahin, N., Anwar, R., Tewari, D., Kabir, M. T., Sajid, A., Mathew, B., et al. (2020). Nanoparticles and its biomedical applications in health and diseases: Special focus on drug delivery. *Environmental Science and Pollution Research, 27*, 19151–19168. https://doi.org/10.1007/s11356-019-05211-0
107. Zepeda-Cervantes, J., Cruz-Reséndiz, A., Sampieri, A., Carreón-Nápoles, R., Sánchez-Betancourt, J. I., & Vaca, L. (2019). Incorporation of ORF2 from porcine circovirus type 2 (PCV2) into genetically encoded nanoparticles as a novel vaccine using a self-aggregating peptide. *Vaccine, 37*(14), 1928–1937. https://doi.org/10.1016/j.vaccine.2019.02.044
108. Gomaa, E. Z. (2022). Microbial mediated synthesis of zinc oxide nanoparticles, characterization and multifaceted applications. *Journal of Inorganic and Organometallic Polymers and Materials, 32*(11), 4114–4132. https://doi.org/10.1007/s10904-022-02406-w
109. El-Naggar, N. E. A., Hussein, M. H., & El-Sawah, A. A. (2018). Phycobiliprotein-mediated synthesis of biogenic silver nanoparticles, characterization, in vitro and in vivo assessment of anticancer activities. *Scientific Reports, 8*(1), 1–20. https://doi.org/10.1038/s41598-018-27276-6
110. Shah, A., Aftab, S., Nisar, J., Ashiq, M. N., & Iftikhar, F. J. (2021). Nanocarriers for targeted drug delivery. *Journal of Drug Delivery Science and Technology, 62*, 102426. https://doi.org/10.1016/j.jddst.2021.102426
111. Paramasivam, G., Kayambu, N., Rabel, A. M., Sundramoorthy, A. K., & Sundaramurthy, A. (2017). Anisotropic noble metal nanoparticles: Synthesis, surface functionalization and applications in biosensing, bioimaging, drug delivery and theranostics. *Acta Biomaterialia, 49*, 45–65. https://doi.org/10.1016/j.actbio.2016.11.066
112. Tan, P., Li, H., Wang, J., & Gopinath, S. C. (2021). Silver nanoparticle in biosensor and bioimaging: Clinical perspectives. *Biotechnology and Applied Biochemistry, 68*(6), 1236–1242. https://doi.org/10.1002/bab.2045
113. Srivastava, P., & Kowshik, M. (2017). Fluorescent lead (IV) sulfide nanoparticles synthesized by Idiomarina sp. strain PR58-8 for bioimaging applications. *Applied and Environmental Microbiology, 83*(7), e03091–e03016. https://doi.org/10.1128/AEM.03091-16

114. Goel, M., Sharma, A., & Sharma, B. (2023). Recent advances in biogenic silver nanoparticles for their biomedical applications. *Sustainable Chemistry, 4*(1), 61–94. https://doi.org/10.3390/suschem4010007
115. Noah, N. M., & Ndangili, P. M. (2022). Green synthesis of nanomaterials from sustainable materials for biosensors and drug delivery. *Sensors International, 3*, 100166. https://doi.org/10.1016/j.sintl.2022.100166
116. Luo, P., Liu, Y., Xia, Y., Xu, H., & Xie, G. (2014). Aptamer biosensor for sensitive detection of toxin A of Clostridium difficile using gold nanoparticles synthesized by Bacillus stearothermophilus. *Biosensors and Bioelectronics, 54*, 217–221. https://doi.org/10.1016/j.bios.2013.11.013
117. Eid, A. M., Fouda, A., Niedbała, G., Hassan, S. E. D., Salem, S. S., Abdo, A. M., et al. (2020). Endophytic Streptomyces laurentii mediated green synthesis of Ag-NPs with antibacterial and anticancer properties for developing functional textile fabric properties. *Antibiotics, 9*(10), 641. https://doi.org/10.3390/antibiotics9100641
118. Shah, M. A., Pirzada, B. M., Price, G., Shibiru, A. L., & Qurashi, A. (2022). Applications of nanotechnology in smart textile industry: A critical review. *Journal of Advanced Research*. https://doi.org/10.1016/j.jare.2022.01.008
119. El-Bendary, M. A., Afifi, S. S., Moharam, M. E., Abo El-Ola, S. M., Salama, A., Omara, E. A., et al. (2021). Biosynthesis of silver nanoparticles using isolated Bacillus subtilis: Characterization, antimicrobial activity, cytotoxicity, and their performance as antimicrobial agent for textile materials. *Preparative Biochemistry & Biotechnology, 51*(1), 54–68. https://doi.org/10.1080/10826068.2020.1789992
120. Davidović, S., Miljković, M., Lazić, V., Jović, D., Jokić, B., Dimitrijević, S., & Radetić, M. (2015). Impregnation of cotton fabric with silver nanoparticles synthesized by dextran isolated from bacterial species Leuconostocmesenteroides T3. *Carbohydrate Polymers, 131*, 331–336. https://doi.org/10.1016/j.carbpol.2015.06.024
121. Tharani, S., Bharathi, D., & Ranjithkumar, R. (2020). Extracellular green synthesis of chitosan-silver nanoparticles using Lactobacillus reuteri for antibacterial applications. *Biocatalysis and Agricultural Biotechnology, 30*, 101838. https://doi.org/10.1016/j.bcab.2020.101838
122. Kundu, D., Hazra, C., Chatterjee, A., Chaudhari, A., & Mishra, S. (2014). Extracellular biosynthesis of zinc oxide nanoparticles using Rhodococcuspyridinivorans NT2: Multifunctional textile finishing, biosafety evaluation and in vitro drug delivery in colon carcinoma. *Journal of Photochemistry and Photobiology B: Biology, 140*, 194–204. https://doi.org/10.1016/j.jphotobiol.2014.08.001
123. Andra, S., Balu, S. K., Jeevanandam, J., & Muthalagu, M. (2021). Emerging nanomaterials for antibacterial textile fabrication. *Naunyn-Schmiedeberg's Archives of Pharmacology, 394*, 1355–1382. https://doi.org/10.1007/s00210-021-02064-8
124. Ahmed, T., Noman, M., Shahid, M., Niazi, M. B. K., Hussain, S., Manzoor, N., et al. (2020). Green synthesis of silver nanoparticles transformed synthetic textile dye into less toxic intermediate molecules through LC-MS analysis and treated the actual wastewater. *Environmental Research, 191*, 110142. https://doi.org/10.1016/j.envres.2020.110142
125. Kumar, A., Choudhary, A., Kaur, H., Mehta, S., & Husen, A. (2021a). Metal-based nanoparticles, sensors, and their multifaceted application in food packaging. *Journal of Nanobiotechnology, 19*(1), 256. https://doi.org/10.1186/s12951-021-00996-0
126. Nile, S. H., Baskar, V., Selvaraj, D., Nile, A., Xiao, J., & Kai, G. (2020). Nanotechnologies in food science: Applications, recent trends, and future perspectives. *Nano-Micro Letters, 12*, 1–34. https://doi.org/10.1007/s40820-020-0383-9
127. Dos Santos, C. A., Ingle, A. P., & Rai, M. (2020). The emerging role of metallic nanoparticles in food. *Applied Microbiology and Biotechnology, 104*, 2373–2383. https://doi.org/10.1007/s00253-020-10372-x
128. Shafiq, M., Anjum, S., Hano, C., Anjum, I., & Abbasi, B. H. (2020). An overview of the applications of nanomaterials and nanodevices in the food industry. *Food, 9*(2), 148. https://doi.org/10.3390/foods9020148

129. Bhople, S., Gaikwad, S., Deshmukh, S., Bonde, S., Gade, A., Sen, S., et al. (2016). Myxobacteria-mediated synthesis of silver nanoparticles and their impregnation in wrapping paper used for enhancing shelf life of apples. *IET Nanobiotechnology, 10*(6), 389–394. https://doi.org/10.1049/iet-nbt.2015.0111
130. Rajoka, M. S. R., Mehwish, H. M., Zhang, H., Ashraf, M., Fang, H., Zeng, X., et al. (2020). Antibacterial and antioxidant activity of exopolysaccharide mediated silver nanoparticle synthesized by lactobacillus brevis isolated from Chinese koumiss. *Colloids and Surfaces B: Biointerfaces, 186*, 110734. https://doi.org/10.1016/j.colsurfb.2019.110734
131. Basavegowda, N., Mandal, T. K., & Baek, K. H. (2020). Bimetallic and trimetallic nanoparticles for active food packaging applications: A review. *Food and Bioprocess Technology, 13*, 30–44. https://doi.org/10.1007/s11947-019-02370-3
132. Hazarika, A., Yadav, M., Yadav, D. K., & Yadav, H. S. (2022). An overview of the role of nanoparticles in sustainable agriculture. *Biocatalysis and Agricultural Biotechnology, 102399*. https://doi.org/10.1016/j.bcab.2022.102399
133. Singh, R. P., Handa, R., & Manchanda, G. (2021). Nanoparticles in sustainable agriculture: An emerging opportunity. *Journal of Controlled Release, 329*, 1234–1248. https://doi.org/10.1016/j.jconrel.2020.10.051
134. Sangeetha, J., Hospet, R., Thangadurai, D., Adetunji, C. O., Islam, S., Pujari, N., & Al-Tawaha, A. R. M. S. (2021). *Nanopesticides, nanoherbicides, and nanofertilizers: The greener aspects of agrochemical synthesis using nanotools and nanoprocesses toward sustainable agriculture* (pp. 1663–1677). Handbook of Nanomaterials and Nanocomposites for Energy and Environmental Applications. https://doi.org/10.1007/978-3-030-36268-3_44
135. Tripathi, A., & Prakash, S. (2022). *Nanobiotechnology: Emerging trends, prospects, and challenges* (pp. 1–21). Agricultural Nanobiotechnology. https://doi.org/10.1016/B978-0-323-91908-1.00006-7
136. Ndaba, B., Roopnarain, A., Haripriya, R. A. M. A., & Maaza, M. (2022). Biosynthesized metallic nanoparticles as fertilizers: An emerging precision agriculture strategy. *Journal of Integrative Agriculture, 21*(5), 1225–1242. https://doi.org/10.1016/S2095-3119(21)63751-6
137. Usman, M., Farooq, M., Wakeel, A., Nawaz, A., Cheema, S. A., ur Rehman, H., et al. (2020). Nanotechnology in agriculture: Current status, challenges and future opportunities. *Science of the Total Environment, 721*, 137778. https://doi.org/10.1016/j.scitotenv.2020.137778
138. Mahapatra, D. M., Satapathy, K. C., & Panda, B. (2022). Biofertilizers and nanofertilizers for sustainable agriculture: Phycoprospects and challenges. *Science of the Total Environment, 803*, 149990. https://doi.org/10.1016/j.scitotenv.2021.149990
139. Saqib, S., Nazeer, A., Ali, M., Zaman, W., Younas, M., Shahzad, A., & Nisar, M. (2022). Catalytic potential of endophytes facilitates synthesis of biometallic zinc oxide nanoparticles for agricultural application. *Bio Metals, 35*(5), 967–985. https://doi.org/10.1007/s10534-022-00417-1
140. de França Bettencourt, G. M., Degenhardt, J., Torres, L. A. Z., de Andrade Tanobe, V. O., & Soccol, C. R. (2020). Green biosynthesis of single and bimetallic nanoparticles of iron and manganese using bacterial auxin complex to act as plant bio-fertilizer. *Biocatalysis and Agricultural Biotechnology, 30*, 101822. https://doi.org/10.1016/j.bcab.2020.101822
141. Ahmed, T., Noman, M., Manzoor, N., Shahid, M., Hussaini, K. M., Rizwan, M., et al. (2021a). Green magnesium oxide nanoparticles-based modulation of cellular oxidative repair mechanisms to reduce arsenic uptake and translocation in rice (Oryza sativa L.) plants. *Environmental Pollution, 288*, 117785. https://doi.org/10.1016/j.envpol.2021.117785
142. Noman, M., Ahmed, T., Hussain, S., Niazi, M. B. K., Shahid, M., & Song, F. (2020a). Biogenic copper nanoparticles synthesized by using a copper-resistant strain Shigella flexneri SNT22 reduced the translocation of cadmium from soil to wheat plants. *Journal of Hazardous Materials, 398*, 123175. https://doi.org/10.1016/j.jhazmat.2020.123175
143. Noman, M., Ahmed, T., Shahid, M., Niazi, M. B. K., Qasim, M., Kouadri, F., et al. (2021). Biogenic copper nanoparticles produced by using the Klebsiella pneumoniae strain NST2 curtailed salt stress effects in maize by modulating the cellular oxidative repair mechanisms. *Ecotoxicology and Environmental Safety, 217*, 112264. https://doi.org/10.1016/j.ecoenv.2021.112264

144. Kumar, P., Pandhi, S., Mahato, D. K., Kamle, M., & Mishra, A. (2021b). Bacillus-based nano-bioformulations for phytopathogens and insect–pest management. *Egyptian Journal Of Biological Pest Control, 31*, 1–11. https://doi.org/10.1186/s41938-021-00475-6
145. Chhipa, H. (2017). Nanofertilizers and nanopesticides for agriculture. *Environmental Chemistry Letters, 15*, 15–22. https://doi.org/10.1007/s10311-016-0600-4
146. Tarafdar, J. C., & Rathore, I. (2016). *Microbial synthesis of nanoparticles for use in agriculture ecosystem. Microbes for plant stress management* (pp. 105–118). New India Publishing Agency.
147. Ahmed, T., Wu, Z., Jiang, H., Luo, J., Noman, M., Shahid, M., et al. (2021c). Bioinspired green synthesis of zinc oxide nanoparticles from a native Bacillus cereus strain RNT6: Characterization and antibacterial activity against rice panicle blight pathogens Burkholderiaglumae and B. gladioli. *Nanomaterials, 11*(4), 884. https://doi.org/10.3390/nano11040884
148. Ahmed, T., Ren, H., Noman, M., Shahid, M., Liu, M., Ali, M. A., et al. (2021b). Green synthesis and characterization of zirconium oxide nanoparticles by using a native Enterobacter sp. and its antifungal activity against bayberry twig blight disease pathogen Pestalotiopsis versicolor. *Nano Impact, 21*, 100281. https://doi.org/10.1016/j.impact.2020.100281
149. Sayed, A. M., Kim, S., & Behle, R. W. (2017). Characterisation of silver nanoparticles synthesised by Bacillus thuringiensis as a nanobiopesticide for insect pest control. *Biocontrol Science and Technology, 27*(11), 1308–1326. https://doi.org/10.1080/09583157.2017.1397597
150. Annamalai, J., Ummalyma, S. B., Pandey, A., & Bhaskar, T. (2021). Recent trends in microbial nanoparticle synthesis and potential application in environmental technology: A comprehensive review. *Environmental Science and Pollution Research, 28*, 49362–49382. https://doi.org/10.1007/s11356-021-15680-x
151. Gebre, S. H., & Sendeku, M. G. (2019). New frontiers in the biosynthesis of metal oxide nanoparticles and their environmental applications: An overview. *SN Applied Sciences, 1*, 1–28. https://doi.org/10.1007/s42452-019-0931-4
152. Kumari, V., & Tripathi, A. K. (2020). Remediation of heavy metals in pharmaceutical effluent with the help of Bacillus cereus-based green-synthesized silver nanoparticles supported on alumina. *Applied Nanoscience, 10*, 1709–1719. https://doi.org/10.1007/s13204-020-01351-9
153. Jeyaraj, A., & Subramanian, S. (2022). Synthesis, optimization, and characterization of biogenic manganese oxide (BioMnOx) by bacterial isolates from mangrove soils with sorbents property towards different toxic metals. *Biometals, 35*(3), 429–449. https://doi.org/10.1007/s10534-022-00378-5
154. Noman, M., Shahid, M., Ahmed, T., Niazi, M. B. K., Hussain, S., Song, F., & Manzoor, I. (2020b). Use of biogenic copper nanoparticles synthesized from a native Escherichia sp. as photocatalysts for azo dye degradation and treatment of textile effluents. *Environmental Pollution, 257*, 113514. https://doi.org/10.1016/j.envpol.2019.113514
155. Kulkarni, R., Harip, S., Kumar, A. R., Deobagkar, D., & Zinjarde, S. (2018). Peptide stabilized gold and silver nanoparticles derived from the mangrove isolate Pseudoalteromonaslipolytica mediate dye decolorization. *Colloids and Surfaces A: Physicochemical and Engineering Aspects, 555*, 180–190. https://doi.org/10.1016/j.colsurfa.2018.06.083
156. Saravanan, A., Kumar, P. S., Karishma, S., Vo, D. V. N., Jeevanantham, S., Yaashikaa, P. R., & George, C. S. (2021). A review on biosynthesis of metal nanoparticles and its environmental applications. *Chemosphere, 264*, 128580. https://doi.org/10.1016/j.chemosphere.2020.128580
157. Samuel, M. S., Jose, S., Selvarajan, E., Mathimani, T., & Pugazhendhi, A. (2020). Biosynthesized silver nanoparticles using Bacillus amyloliquefaciens; application for cytotoxicity effect on A549 cell line and photocatalytic degradation of p-nitrophenol. *Journal of Photochemistry and Photobiology B: Biology, 202*, 111642. https://doi.org/10.1016/j.jphotobiol.2019.111642
158. Abbas, A., Ahmad, T., Hussain, S., Noman, M., Shahzad, T., Iftikhar, A., et al. (2022). Immobilized biogenic zinc oxide nanoparticles as photocatalysts for degradation of methylene

blue dye and treatment of textile effluents. *International journal of Environmental Science and Technology, 19*(11), 11333–11346. https://doi.org/10.1007/s13762-021-03872-4

159. Raj, R., Dalei, K., Chakraborty, J., & Das, S. (2016). Extracellular polymeric substances of a marine bacterium mediated synthesis of CdS nanoparticles for removal of cadmium from aqueous solution. *Journal of Colloid and Interface Science, 462*, 166–175. https://doi.org/10.1016/j.jcis.2015.10.004
160. Balaji, S., & Shanmugam, V. K. (2021). Enhanced antibiofilm activity of endophytic bacteria mediated zirconium nanoparticles. *Biointerface Research in Applied Chemistry, 12*(1), 74–82. https://doi.org/10.33263/BRIAC121.074082
161. Sadeghi-Aghbash, M., Rahimnejad, M., & Pourali, S. M. (2020). Bio-mediated synthesis and characterization of zinc phosphate nanoparticles using enterobacter aerogenes cells for antibacterial and anti-corrosion applications. *Current Pharmaceutical Biotechnology, 21*(12), 1232–1241. https://doi.org/10.2174/1389201021666200506073534
162. Ameen, F., AlYahya, S., Govarthanan, M., ALjahdali, N., Al-Enazi, N., Alsamhary, K., et al. (2020). Soil bacteria Cupriavidus sp. mediates the extracellular synthesis of antibacterial silver nanoparticles. *Journal of Molecular Structure, 1202*, 127233. https://doi.org/10.1016/j.molstruc.2019.127233
163. Xu, C., Qiao, L., Guo, Y., Ma, L., & Cheng, Y. (2018). Preparation, characteristics and antioxidant activity of polysaccharides and proteins-capped selenium nanoparticles synthesized by Lactobacillus casei ATCC 393. *Carbohydrate Polymers, 195*, 576–585. https://doi.org/10.1016/j.carbpol.2018.04.110
164. El-Saadony, M. T., Saad, A. M., Taha, T. F., Najjar, A. A., Zabermawi, N. M., Nader, M. M., et al. (2021). Selenium nanoparticles from Lactobacillus paracasei HM1 capable of antagonizing animal pathogenic fungi as a new source from human breast milk. *Saudi Journal of Biological Sciences, 28*(12), 6782–6794. https://doi.org/10.1016/j.sjbs.2021.07.059
165. Galal, G. F., Abd-Elhalim, B. T., Abou-Taleb, K. A., Haroun, A. A., & Gamal, R. F. (2021). Toxicity assessment of green synthesized Cu nanoparticles by cell-free extract of Pseudomonas silesiensis as antitumor cancer and antimicrobial. *Annals of Agricultural Sciences, 66*(1), 8–15. https://doi.org/10.1016/j.aoas.2021.01.006
166. Fouad, H., Hongjie, L., Yanmei, D., Baoting, Y., El-Shakh, A., Abbas, G., & Jianchu, M. (2017). Synthesis and characterization of silver nanoparticles using Bacillus amyloliquefaciens and Bacillus subtilis to control filarial vector Culex pipienspallens and its antimicrobial activity. *Artificial Cells, Nanomedicine, and Biotechnology, 45*(7), 1369–1378. https://doi.org/10.1080/21691401.2016.1241793
167. Elbeshehy, E. K., Elazzazy, A. M., & Aggelis, G. (2015). Silver nanoparticles synthesis mediated by new isolates of Bacillus spp., nanoparticle characterization and their activity against bean yellow mosaic virus and human pathogens. *Frontiers in Microbiology, 6*, 453. https://doi.org/10.3389/fmicb.2015.00453
168. Rajkumar, K., Mvs, S., Koganti, S., & Burgula, S. (2020). Selenium nanoparticles synthesized using pseudomonas stutzeri (MH191156) show antiproliferative and anti-angiogenic activity against cervical cancer cells. *International Journal of Nanomedicine*, 4523–4540.
169. Ibrahim, E., Fouad, H., Zhang, M., Zhang, Y., Qiu, W., Yan, C., et al. (2019). Biosynthesis of silver nanoparticles using endophytic bacteria and their role in inhibition of rice pathogenic bacteria and plant growth promotion. *RSC Advances, 9*(50), 29293–29299. https://doi.org/10.1039/C9RA04246F
170. El-Saadony, M. T., Abd El-Hack, M. E., Taha, A. E., Fouda, M. M., Ajarem, J. S., Maodaa S, N., et al. (2020). Ecofriendly synthesis and insecticidal application of copper nanoparticles against the storage pest Triboliumcastaneum. *Nanomaterials, 10*(3), 587. https://doi.org/10.3390/nano10030587
171. Hosseini, F., Lashani, E., & Moghimi, H. (2023). Simultaneous bioremediation of phenol and tellurite by Lysinibacillus sp. EBL303 and characterization of biosynthesized Te nanoparticles. *Scientific Reports, 13*(1), 1243. https://doi.org/10.1038/s41598-023-28468-5

172. Momin, B., Rahman, S., Jha, N., & Annapure, U. S. (2019). Valorization of mutant Bacillus licheniformis M09 supernatant for green synthesis of silver nanoparticles: Photocatalytic dye degradation, antibacterial activity, and cytotoxicity. *Bioprocess and Biosystems Engineering, 42*, 541–553. https://doi.org/10.1007/s00449-018-2057-2
173. Mittal, D., Kumar, A., Balasubramaniam, B., Thakur, R., Siwal, S. S., Saini, R. V., et al. (2022). Synthesis of biogenic silver nanoparticles using plant growth-promoting bacteria: Potential use as biocontrol agent against phytopathogens. *Biomaterials and Polymers Horizon, 1*(1), 22–31. https://doi.org/10.37819/bph.001.01.0130
174. Sharma, S., Sharma, N., & Kaushal, N. (2023). Utilization of novel bacteriocin synthesized silver nanoparticles (AgNPs) for their application in antimicrobial packaging for preservation of tomato fruit. *Frontiers in Sustainable Food Systems, 7*, 1–11. https://doi.org/10.3389/fsufs.2023.1072738
175. Nazari, N., & Kashi, F. J. (2021). A novel microbial synthesis of silver nanoparticles: Its bioactivity, Ag/Ca-Alg beads as an effective catalyst for decolorization Disperse Blue 183 from textile industry effluent. *Separation and Purification Technology, 259*, 118117. https://doi.org/10.1016/j.seppur.2020.118117
176. Saravanan, C., Rajesh, R., Kaviarasan, T., Muthukumar, K., Kavitake, D., & Shetty, P. H. (2017). Synthesis of silver nanoparticles using bacterial exopolysaccharide and its application for degradation of azo-dyes. *Biotechnology Reports, 15*, 33–40. https://doi.org/10.1016/j.btre.2017.02.006
177. Zeinivand, M., Aghaei, S. S., Zargar, M., & Ghasemzadeh, M. A. (2023). Exopolysaccharide-mediated silver nanoparticles synthesized from Lactobacillus paracasei with antimicrobial, antibiofilm and antioxidant activities. *Archives of Microbiology, 205*(5), 210. https://doi.org/10.1007/s00203-023-03497-w

Chapter 2
Insights on Microbes-Mediated Greener Synthesis of Nanoparticles: Advantages and Challenges

Gita Singh and Sudeshna Chandra

2.1 Introduction

The unique size- and shape-dependent physicochemical properties of nanomaterials are indispensable in several areas of human activity ranging from medicine, biology, food technology to electronics, energy and aerospace engineering. However, their chemical route of synthesis has always been a concern due to use and handling of hazardous substances and generation of toxic by-products that are detrimental to human health and environment. Further, the synthesis process includes use of expensive instruments and techniques like laser ablation, chemical deposition, mechanical milling, spinning, etc. that makes the production of nanoparticles expensive, time-consuming and less sustainable [1]. Therefore, there is a necessity to develop a sustainable, economical, and environmentally benign approach for synthesizing nanoparticles.

Synthesis of nanoparticles may be triggered by naturally occurring compounds like terpenoids, phenols, flavones, amines, amides, carbonyls, proteins, pigments, and alkaloids that are abundantly found in microbial cells and plant extracts. Therefore, microbial synthesis of nanoparticles are envisaged as a good alternative to chemical methods of synthesis as they are cost-effective, environmentally safe, and do not require toxic chemicals for reduction of metal ions and external stabilizing agents for stabilizing the nanoparticles. Further, cultivation of microorganisms like bacteria, algae, fungi and yeast are easy because of their ability to grow rapidly at ambient environmental and laboratory conditions like temperature, pH, and pressure [2]. However, the mechanism of formation of nanoparticles varies from

G. Singh
School of Chemistry, University College Dublin, Dublin, Ireland

S. Chandra (✉)
Institute of Analytical Chemistry, University of Regensburg, Regensburg, Germany
e-mail: Chandra.Sudeshna@chemie.uni-regensburg.de

R. S. Chaughule, A. S. Lokur (eds.), *Applications of Nanotechnology in Microbiology*, https://doi.org/10.1007/978-3-031-49933-3_2

microbes to microbes because biological agents interact differently (both intra- and/or extracellularly) with different metal solutions. Commonly, metal ions are trapped on the cell surface of microorganisms followed by reduction of metal ions to nanoparticles in presence of biological reducing agents [3]. Figure 2.1 shows a schematic representation of microbes-mediated synthesis of nanoparticles.

Microbial synthesis of nanoparticles takes place through two processes viz., *Biosorption* and *Bioreduction*. In biosorption, absorption of metal cations onto the cell wall of microbial species takes place through physisorption, precipitation, ion-exchange, and complexation [4] (Fig. 2.2). Microbes usually secrete extracellular polysaccharides containing anionic glycoproteins and lipopolysaccharides. These anionic moieties have the potential to attract metal cations from the aqueous media or solutions resulting in metal binding to the cell wall of microbes. For example, the cell wall of bacteria contains peptidoglycan, liposaccharides, teichoic acids, and phospholipids; and these negatively charged components trap the positively charged metal ions and bind with them. Similarly, the fungal cell wall is essentially made up of chitin, which has the potential to undergo complexation reactions with the metal cation [5].

In bioreduction, metal salts are chemically reduced to a biologically stable form of nanoparticles. Chemical reduction of metal ions to zerovalent nanoparticles takes place through various functional groups like amines, amides, carbonyls, proteins, and alkaloids that are secreted by microbes such as bacteria, fungi, and algae [6]. Production of nanoparticles in this process can be either extracellular or intracellular. The extracellular process involves reduction of metal ions by microbial enzymes and proteins, bacterial or fungal cell wall components, or organic molecules present in the culture medium, whereas the intracellular process involves initial electrostatic attraction of metal ions by carboxyl groups of the microbial cell wall, resulting in passage of metal ions through the cells and reduction by intracellular proteins and

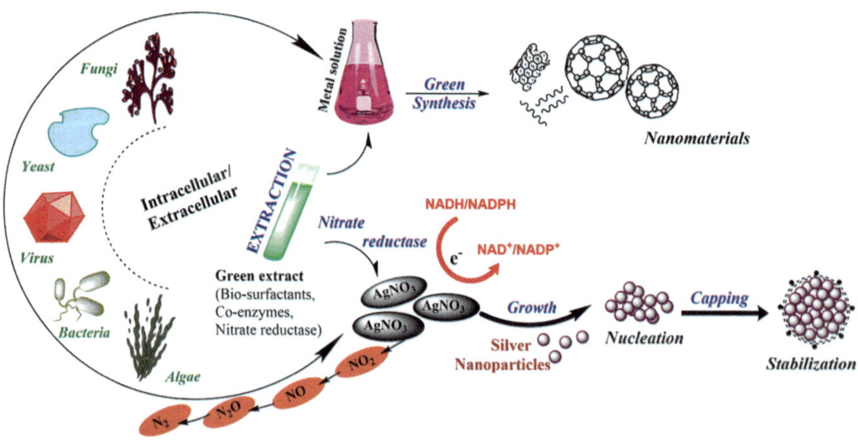

Fig. 2.1 Mechanistic representation of the green synthesis of metal nanoparticles and its bioreduction followed by stabilization from microbes

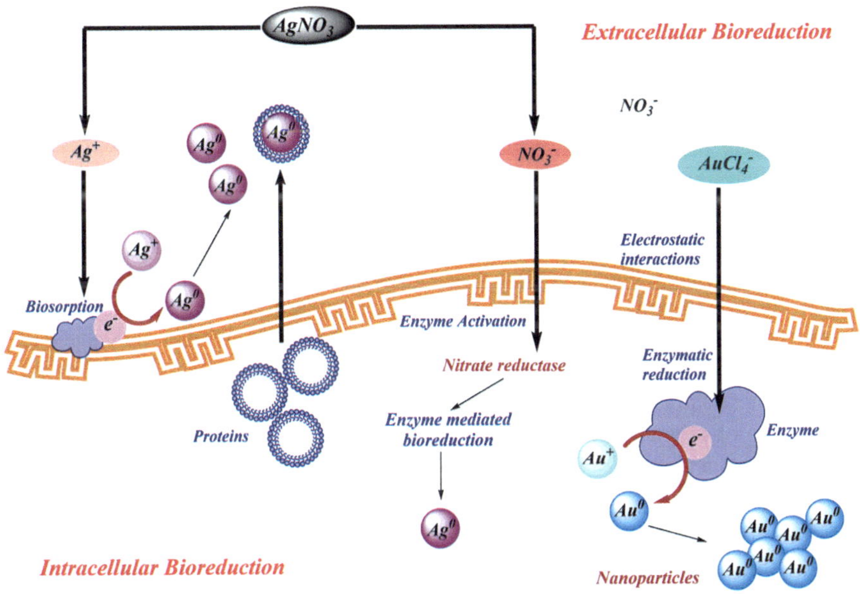

Fig. 2.2 Reduction of gold and silver nanoparticles in microorganism, viz. extracellular and intracellular mechanisms

cofactors to produce nanoparticles [7]. In other words, the extracellular process involves trapping (binding) of metal ions on the microbial cell surface and reducing the metal ions in presence of enzymes, whereas the intracellular process involves transport of metal ions into microbial cells and then reduction by intracellular components [8]. Figure 2.2 presents the graphical representation of the extracellular and intracellular microbial synthesis of nanoparticles through the bioreduction process.

Biological synthesis of nanoparticles has certain disadvantages like lack of control over shape and size of the nanoparticles, process scale up, polydispersity, biochemical activity, low production rate, and elucidation of mechanism of synthesis [9]. In this chapter, we will focus on microbial synthesis of metal nanoparticles, mechanism of formation of the nanoparticles, and the challenges in the synthesis of the nanoparticles. The following section deals with fungal, algal, and bacterial-mediated synthesis and their mechanism.

2.2 Fungus- and Yeast-Mediated Synthesis of Nanoparticles

Synthesis of nanoparticles from fungus are preferred over other microbial approaches due to high withstanding ability of fungal mycelial mesh to higher flow and agitation in bioreactors, high metal tolerance and bioaccumulation capacity, ease in handling of biomass, downstream processing, and possibility of upscaling the synthesis.

Kobashiwaga et al. [10] synthesized silver nanoparticles (AgNPs) from fungi *Trametestrogii* wherein silver nitrate solution was added to the fungal filtrate in an alkaline pH. Electrostatic adsorption of hydroxide ions on AgNPs was observed, which stabilized the nanoparticles. AgNPs were also prepared from the cell-free fungal filtrate of *Penicillium chrysogenum* NG85 and *Fusarium chlamydosporum* NG30, wherein reducing enzymes in the filtrate reduced Ag^+ ions to Ag^0 and the proteins in the filtrate stabilized the AgNPs [11]. Wanarska and Maliszewska in 2019 [12] reported the ability of live and dead fungal cells of *Penicillium cyclopium* to synthesize AgNPs due to the presence of proteins and saccharides on the surface of the fungal cell wall. On similar lines, Seetharaman et al. also reported that the proteins in the extracellular filtrate of *Phomopsis liquidambaris* can reduce and cap AgNPs efficiently [13]. The pH of the medium carrying the fungal biomass and the substrate concentration strongly influences the secretion of the proteins [14]. The mechanism of biosynthesis of AgNPs can be explained as follows: first, bioreduction of components or biomolecules present in cell-free supernatant, followed by its absorption onto mycelial pellet or through mycelium and then getting released in the silver nitrate solution. Enzymes such as nitrate reductase, hydrogenase present in the extract mediate the synthesis. Biomolecules like reducing sugars, proteins (e.g. glyceraldehyde-3-phosphate dehydrogenase, ATPase), etc. are involved in energy metabolism of fungal cells for synthesis of nanoparticles. Endophytic fungus isolated from the roots of *Chonemorpha fragrans* [15] and Phytopathogen fungus *Fusarium oxysporum* [16] were reported for synthesis of gold nanoparticles (AuNPs) of varied shapes and sizes. Mesophilous filamentous fungi, *Cladosporium cladosporiodes* [17], isolated from seaweeds, and cell-free extract of *Aspergillus* sp. [18] were used to synthesize stable AuNPs. Other nanoparticles like ZnO and CdS were also prepared from various fungal strains. Ganesan and group reported ZnO nanoparticles prepared from an endophytic fungus *Periconium* sp. isolated from the leaves of *Balanites aegyptiaca*. The fungus was grown in potato dextrose broth and the mycelium was recovered after 21 days of incubation. The recovered mycelium was dried at 60 °C for 12 h and the aqueous extract was prepared from the dried fungal biomass. Metal precursor zinc nitrate was added to the fungal extract and evaporated, which resulted in formation of a sol. Upon further evaporation of water molecules from the sol, a gel was formed, which was dried under calcination at 700 °C for 4 h to form the zinc oxide (ZnO) nanoparticles [19]. Similarly, copper oxide nanoparticles were synthesized from an endogenous fungi, *Trichoderma asperellum*. The aqueous mycelial-free water extract of the fungi was treated with the metal precursor (copper nitrate solution) during which copper hydroxide was formed and the hydroxyl ions reacted with the enzymes and proteins in the fungal extract to convert copper hydroxide into copper oxide nanoparticles [20]. Gudikandula et al. reported that presence of large amounts of enzymes and proteins in free and pure fungal extract helps in the downstream process in the extracellular synthesis of nanoparticles [21]. Extracellular synthesis of magnesium oxide [22], platinum [23], cobalt [24], and cadmium sulphide [25] nanoparticles were reported from *Pencillium chrysogenum*, *Fusarium oxysporum*, *Aspergillus nidulans*, and *Coriolus versicolor*, respectively. In all the synthesis, presence of natural proteins and enzymes in the fungal strain acted as a capping or stabilizing agent eliminating the need of external stabilizing agents. Table 2.1 lists fungus-mediated synthesis of nanoparticles.

Simple nutrient requirement, easy culture and control of yeast growth in laboratory conditions, increased function, and stability makes it an exceptional source for the synthesis of nanoparticles. Jha et al. [37] reported a greener, reproducible, and low-cost synthesis of Sb_2O_3 nanoparticles using Baker's yeast (*Saccharomyces cerevisiae*). Yeast cells were grown as culture media to which 0.025 M $SbCl_3$ solution was added and heated on a steam bath up to 60 °C for 10–20 min. The culture media was cooled after the appearance of the metal nanoparticles as white deposition. The culture solution was further incubated for 3–4 days at room temperature to obtain white clusters of Sb_2O_3 nanoparticles. Majority of the particles were well dispersed and spherical shaped having a size ranging from 2 to 10 nm. *Saccharomyces cerevisiae* efficiently produce zinc sulphide (ZnS) quantum dot [38] and silica nanoparticles [39] under controlled parameters like temperature, pH and incubation time. Cell-free water extract of *Rhodotorula species* and *Yarrowialipolytica* were used to synthesize AgNPs [40] and AuNPs [41], respectively. Reduction of M^{n+} to M^0 nanoparticles took place in the cell wall of the yeast and was influenced by the concentration of metal ions and the biomass. Proteins present in the biomass efficiently capped and stabilized the nanoparticles.

Table 2.1 Fungus-mediated synthesis of nanoparticles of various shapes and sizes

Fungi	Nanoparticles (NPs)	Shapes and sizes	References
Macrophomina phaseolina	Ag	Spherical; 5–40 nm	[26]
Trichoderma viride	Ag	Spherical; 2–5 nm Rectangular; 40–65 nm Penta/hexagonal; 50–100 nm	[27]
Pichia pastoris	Ag and Se	Spherical; 70–180 nm	[28]
Trichoderma harzianum	CdS	Spherical; 3–8 nm	[29]
Pichia kudriavzevii	ZnO	Hexagonal wurtzite structure; 10–60 nm	[30]
Metarhizium anisopliae	Ag	Rod-shaped; 28–38 nm	[31]
Saccharomyces cerevisiae	Ag	Spherical; 2–20 nm	[32]
Saccharomyces cerevisiae	Pd	Hexagonal; 32 nm	[33]
Magnusiomyces ingens LH-F1	Au	Spherical and pseudospherical; 20.3–28.3 nm	[34]
Colletotrichum sp.	Al_2O_3	Spherical; 30–50 nm	[35]
Pleurotus ostreatus	Au	Spherical and prism shaped; 10–30 nm	[36]

2.3 Algae-Mediated Synthesis of Nanoparticles

Algae constitute a major class of marine ecosystem and contribute to a large number of secondary metabolites. Carbohydrates, minerals, proteins, phenols, and pigments (chlorophyll, carotenoids, phycobilins, etc.) that are present in algal species can act as reducing and stabilizing agents in the synthesis of nanoparticles [42]. Algae can produce nanoparticles both extracellularly [43] and intracellularly [44]. Typically, synthesis of nanoparticles from algae involves simple incubation of metal precursor solution with algal extract wherein the controlling factors are concentration of the extract and the precursor, pH, temperature, and the incubation time.

AgNPs can be synthesized using soluble polysaccharides that are extracted from various marine algal strains like green algae (e.g. *Ulva faciata*), red algae (e.g. *Pterocladiacapillaciae*, *Janiarubins*), and brown algae (e.g. *Colpmeniasinusa*). The polysaccharide extracts are incubated with metal precursors during which reducing sugars that are present in the extract reduces the metal ions to form nanoparticles [45]. Gonzalez-Ballesterosetal et al. reported the use of *Cystoceriabaccatta* (Brown algae) for preparing polycrystalline gold nanoparticles [46]. Functional groups like amino acids and carbonyls in red algae, *Portieriahornemannii* [47], and *Gelidium corneum* [48] acted as reducing agents for synthesis of AgNPs. Algal extract from *Egregia* sp. [49] served as both reducing and stabilizing agents in the preparation of AuNPs. The sugar in the algal extract facilitated the formation of AuNPs by reducing the Au^{3+} ions and stabilizing the nanoparticles.

Koopi and Buazar et al. reported a novel method of synthesis of pure aluminium nanoparticle from macroalgae *Sargassum ilicifolium*. The synthesis was carried out in three primary steps: activation, growth, and termination. In the activation phase, the metal precursor, aluminium sulphate, dissociated into aluminium ions and sulphate ions in water. Thereafter, a redox (reduction–oxidation) process took place between the Al^{3+} ions and the electron rich biomolecules of the algal species. The electron donating functional groups (R-COO^- and R-NH_2) reduced the Al^{3+} ions to zerovalent Al metal nanoparticles [50]. Table 2.2 lists algae-mediated synthesis of nanoparticles.

2.4 Bacteria-Mediated Synthesis of Nanoparticles

Several bacterial strains have the potential to adsorb/bind metal ions and reduce them into NPs in presence of enzymes that are produced during extra- and intracellular metabolism. Faster reproduction rate and relatively easier cultivation of bacteria makes them an ideal choice for microbial synthesis via bioreduction mechanism [59]. Jorge de Souza et al. [60] reported the synthesis of silver nanoparticles (AgNPs) from the *Pseudomonas stutzeri* AG259 strain isolated from silver mine. The nanoparticles were synthesized using the bacterial biomass and silver nitrate ($AgNO_3$) solution under ambient pressure and temperature. Bioreduction by

Table 2.2 Algae-mediated synthesis of nanoparticles of various shapes and sizes

Algae	Nanoparticles (NPs)	Shapes and sizes	References
Sargassum bovinum	Pd	Octahedral; 5–10 nm	[51]
Cystoseirabaccata	Au	Spherical; 8.4 nm	[46]
Sargassum muticum	ZnO	Hexagonal; 30–57 nm	[52]
Padina tetrastromatica and *Turbinaria conoides*	ZnO	Spherical, pentagonal, hexagonal and triangles; 90–120 nm	[53]
Laminaria japonica	Ag	Spherical to oval; 30 nm	[54]
Ecklonia cava	Au	Spherical and triangular; 30 nm	[55]
Gelidium amansii	Ag	Spherical; 27–54 nm	[56]
Caulerpa racemosa	Ag	Spherical and triangular; 5–25 nm	[57]
Spirogyra varians	Ag	Quasi-sphere; 35 nm	[58]

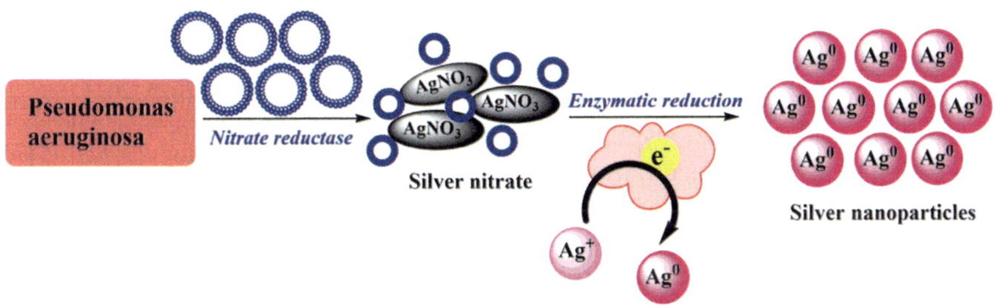

Fig. 2.3 Mechanism showing reduction of silver nitrate to produce silver nanoparticles

bacteria took place in the presence of reductase enzymes, which reduced the silver ions (Ag^+) to AgNPs. The mechanism of simultaneous gaining of electrons from NADH by NADH-dependent reductase enzymes and reduction of Ag^+ to AgNPs was later elucidated by Javaid et al. [61]. A schematic representation of the mechanism is shown in Fig. 2.3.

Microbial synthesis of AgNPs were also reported from various strains of bacteria viz., *Bacillus brevis* (NCIM 2533) [62], *Pseudomonas aeroginosa* [63], *Streptomyces* sp. [64], and *Cupriavidus* strain [65]. The average size of the nanoparticles were 10–12 nm. Gram-negative bacteria like *Escherichia fergusonii*, *Enterobacter cloacae*, *Klebsiella sp*, *Shigella* sp. and gram-positive bacteria such as *Bacillus* and *Paenibacillus* sp. were also used to produce AgNPs of various shapes (spherical, hexagonal, etc.) and sizes [66]. Similarly, gold nanoparticles (AuNPs) were also synthesized using *Bacillus marisflavi* [67] wherein the AuNPs were formed extracellularly from the cell-free extract of the bacteria. Formation of the nanoparticles were confirmed by the change in colour from pale yellow to bluish purple indicating surface plasmon resonance (SPR) of the AuNPs. Extracellular biosynthesis of AuNPs from *Rhodopseudomonas capsulata* was reported by He et al. [68] wherein cofactors NADH and NADH-dependent enzymes induce reduction of Au^{3+} to Au^0

Table 2.3 Bacteria-mediated synthesis of nanoparticles of various shapes and sizes

Bacteria	Nanoparticles (NPs)	Shapes and sizes	References
Gordonia amicalis HS-11	Ag	Spherical; 5–25 nm	[77]
Lactobacillus rhamnosus GG ATCC 53103	Ag	Spherical, triangular, rod-shaped and hexagonal; 2–15 nm	[78]
Kocuria ava	Cu	Spherical; 5–30 nm	[79]
Shewanella loihica PV-4	Pd and Pt	Spherical; 2–7 nm	[80]
Ochrobactrum sp. *MPV1*	Se and Te	Roughly spherical and rods	[81]
Streptomyces laurentii	Ag	Spherical; 7–15 nm	[82]
Streptomyces griseoruber	Au	Spherical, hexagonal and triangular; 5–50 nm	[83]
Klebsiella pneumoniae, Escherichia coli, Pseudomonas jessinii	Ag	Cubic and star like shapes; 50–100 nm	[84]

forming the AuNPs. Various sizes (2–500 nm) and shapes (spherical, trigonal, pentagonal, hexagonal, and nanosheets) of AuNPs can be biosynthesized from *Trichoderma viride* filtrate by tuning the physical parameters like pH, temperature, incubation time, and concentration of culture filtrate [69]. Further, extracellular biosynthesis of spherical and crystalline AuNPs using *Bacillus megaterium* MSBN04 were reported wherein the exopolysaccharide (EPS) produced by the bacteria acted as both reducing and stabilizing agents of the nanoparticles [70]. Kunoh et al. [71] reported synthesis of spherical AuNPs using *Leptothrix* (iron-oxidizing bacteria) that released extracellular RNA, which would reduce Au^{3+} ions to form the nanoparticles.

Besides AgNPs and AuNPs, bacterial strains have been used to produce magnetic (Fe_3O_4) [72], cadmium sulphide (CdS) [73], zinc oxide (ZnO) [74], selenium (Se) [75], and molybdenum (Mo) [76] nanoparticles. Table 2.3 lists bacteria-mediated synthesis of nanoparticles.

2.5 Mechanism of Microbial Synthesis

The microbial biosynthesis of nanoparticles involves multiple, coordinated reactions within a microbial cell and the efficiency of the pathway is greatly influenced by the interactions of the nanoparticles and the bioactives produced by microbes. Different microbes have different mechanisms of forming the nanoparticles, which may be routed either by the intracellular or extracellular pathway.

The synthetic process basically involves capture of metal precursors, reduction process, and capping/stabilization of the formed nanoparticles. In the intracellular synthesis pathway, distinctive ion transportation in the microbial cell in the presence of enzymes, coenzymes, etc. takes place by electrostatic attraction. The cell wall of microbes consists of a variety of polysaccharides and protein, which provides active sites for binding of the metal ions. Metal ions (M^{n+}) are first adsorbed on the surface or inside the microbial cells and reduced to the metallic M^0 state to form clusters of nanoparticles. The bioactive molecules provide electrons to the metal ions for the reduction process in which the metal ion (M^+) is converted into their metallic form (M^0). The bioactive molecules also serve as a nucleation site for subsequent growth and accumulation of the nanoparticles. Kalishwaralal and co-workers reported the involvement of nitrate reductase enzyme in the synthesis of silver nanoparticles from *Bacillus licheniformis* [85]. The nitrate reductase enzymes, which are NADH and NADH-dependent, aid the reduction of M^{n+} to M^0 through an electron shuttle mechanism [86]. The protein, peptides, amino acids, etc. that are present inside the cell caps and stabilizes the nanoparticles. Figure 2.4 gives an overview of intracellular synthesis of nanoparticles.

The extracellular pathway involves enzyme-mediated synthesis that are located on cell membranes and are released to the growth medium as an extracellular enzyme. For instance, the bioreduction of Zn^{2+} by the electron transfer from NADH by NADH-dependent reductase that acts as an electron carrier [3]. Zn^{2+} gains an electron and gets reduced to Zn^0, subsequently forming ZnO NPs. The protein and other bioactives secreted by microbes act as a reducing as well as a capping agent. The extracellular synthesis mechanism is illustrated in Fig. 2.4.

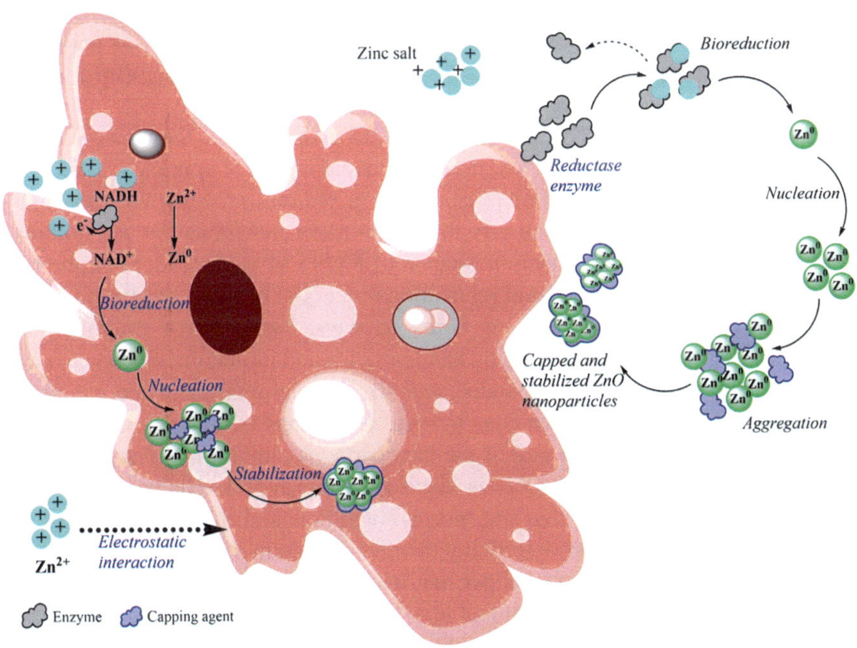

Fig. 2.4 Illustration of the intracellular and extracellular pathways for the bioreduction of zinc ions to produce zinc oxide nanoparticles and the dual role of microbes acting as reducing and capping agents

Extracellular synthesis has an edge over the intracellular route as it could be used to synthesize large quantities and it involves simple downstream processing that eliminates various steps of synthesis, easy separation, and industrialization. In contrast, recovery of nanoparticles in the intracellular synthesis requires additional downstream processes like harvesting the cell biomass by centrifugation and ultrasonication to obtain the purified nanoparticles.

2.6 Challenges and Future Perspectives

Microbial biosynthesis of nanoparticles are envisaged as a greener, economical, and sustainable approach. Due to their distinctive properties, they are increasingly being used in environmental remediation. Further, microorganisms are also used for removal and degradation of pollutants. Therefore, simultaneous production of nanoparticles along with reduction in contaminants has attracted researchers in the recent past. Further, the biogenic route of synthesis eliminates the additional step of capping and stabilizing nanoparticles because the bioactive compounds in the microbes themselves act as capping/stabilizing agents [87]. Also, in such green synthesis, since stabilization of the nanoparticles is achieved by biocompatible compounds, toxicity is reduced significantly. Moreover, requirement of lesser time in biosynthesis than chemical or physical route of synthesis is also an advantage. For instance, Arsiya et al. [88] reported one-step and fast biosynthesis of palladium nanoparticles using *Chlorella vulgaris*. The nanoparticles were formed within 10 min at room temperature.

However, there are certain challenges and disadvantages of the microbial synthetic process that limit the use and commercialization of the approach. For example, bacterial culture and its maintenance is a tedious process wherein safety measures need to be followed strictly to prevent mass contamination and outbreak of diseases. Secondly, the reduction process of M^{n+} to M^0 is dependent on the biochemical activity, growth rate, and replication process of bacterial strain and, therefore, is slow and may take from hours to days. Selection of microbes and biomolecules for synthesis of specific nanoparticles and its stabilization is also crucial. Control on monodispersity, shape, and size of microbial-mediated nanoparticles is also an important parameter. In spite of various advantages of microbial synthesis, polydispersity and size of the nanoparticles remains a big and challenging issue and, therefore, much work is required to improve the efficiency of synthesis, and control the particle size and morphology. Size and shape of nanoparticles could be controlled by either optimizing the process parameters or modifying these parameters. It is also vital to consider that use of pathogenic microorganisms can negatively impact the environment and may lead to implications in large-scale synthetic processes. Maintenance of crucial parameters for microbial-mediated synthesis may also result in increase in the production cost [3]. Nevertheless, there are several reports that show that the nanoparticles synthesized through microbial

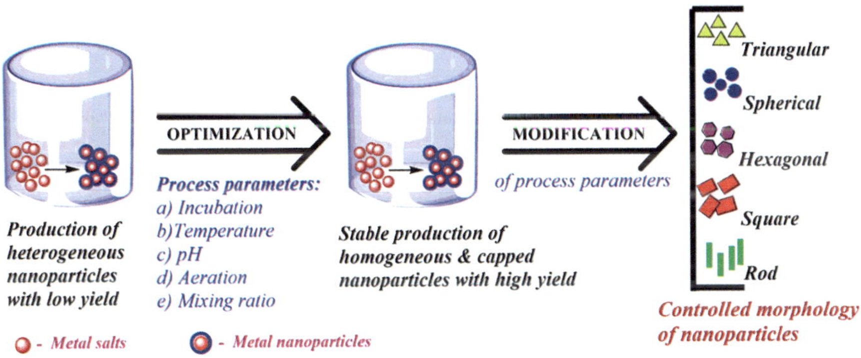

Fig. 2.5 Illustration of parameters for the bulk production of stable, monodispersed biological nanoparticles

processes have better properties than those synthesized through conventional chemical processes [89, 90].

Current research should focus on controlling morphology of biosynthesized nanoparticles, and their stabilization by optimizing the reaction parameters and judicious selection of microbial strain. Figure 2.5 gives an overview of various parameters that can be considered for modulating the physicochemical properties of the nanoparticles.

Large-scale production of nanoparticles is a bottleneck in their development and commercialization. It is now important to focus on large-scale synthesis of nanoparticles, which are not only scalable and reproducible with narrow size distribution. Therefore, the bulk cultivation approach of the synthetic process and downstream processing techniques need to be improved. Scaling up of nanoparticles is hampered by high cost, high energy requirement, polydispersity, and low yield. Production of nanoparticles using microbes at room temperatures without any reducing/capping/stabilizing agents would definitely make large-scale production cost-effective and energy sustainable. This would help to sustain the biosynthesized nanoparticles and lead to their commercialization.

References

1. Saravanan, A., Senthil Kumar, P., Karishma, S., Vo, D. N., Jeevanantham, S., Yaashika, P. R., & George, C. S. (2021). A review on biosynthesis of metal nanoparticles and its environmental applications. *Chemosphere, 264*, 128580(16pp). https://doi.org/10.1016/j.chemosphere.2020.128580
2. Fariq, A., Khan, T., & Yasmin, A. (2017). Microbial synthesis of nanoparticles and their potential applications in biomedicine. *Journal of Applied Biomedicine, 15*, 241–248. https://doi.org/10.1016/j.jab.2017.03.004
3. Hulkoti, N. I., & Taranath, T. (2014). Biosynthesis of nanoparticles using microbes – A review. *Colloids and Surfaces B: Biointerfaces, 121*, 474–483. https://doi.org/10.1016/j.colsurfb.2014.05.027

4. Pantidos, N., & Horsfall, L. E. (2014). Biological synthesis of metallic nanoparticles by bacteria, fungi and plants. *Journal of Nanomedicine and Nanotechnology, 5*, 233(10pp). https://doi.org/10.4172/2157-7439.1000233
5. Wang, L., Liu, X., Lee, D. J., Tay, J. H., Zhang, Y., Wan, C. L., & Chen, X. F. (2018). Recent advances on biosorption by aerobic granular sludge. *Journal of Hazardous Materials, 357*, 253–270. https://doi.org/10.1016/j.jhazmat.2018.06.010
6. Asmathunisha, N., & Kathiresan, K. (2013). A review on biosynthesis of nanoparticles by marine organisms. *Colloids and Surfaces B: Biointerfaces, 103*, 283–287. https://doi.org/10.1016/j.colsurfb.2012.10.030
7. Siddiqi, K. S., Husen, A., & Rao, R. A. K. (2018). A review on biosynthesis of silver nanoparticles and their biocidal properties. *Journal of Nanbiotechnology, 16*, 14(28pp). https://doi.org/10.1186/s12951-018-0334-5
8. Li, X., Xu, H., Chen, Z. S., & Chen, G. (2011). Biosynthesis of nanoparticles by microorganisms and their applications. *Journal of Nanomaterials, 2011*, 270974(17pp). https://doi.org/10.1155/2011/270974
9. Raouf, H. M., & Nasiri, S. M. (2015). Recent achievements in the microbial synthesis of semiconductor metal sulfide nanoparticles. *Materials Science in Semiconductor Processing, 40*, 293–301. https://doi.org/10.1016/j.mssp.2015.06.003
10. Kobashigawa, J. M., Robles, C. A., Martínez Ricci, M. L., & Carmaran, C. C. (2018). Influence of strong bases on the synthesis of silver nanoparticles (AgNPs) using the ligninolytic fungi *Trametes trogii*. *Saudi Journal of Biological Sciences, 26*, 1331–1337. https://doi.org/10.1016/j.sjbs.2018.09.006
11. Khalil, N. M., Abd El-Ghany, M. N., & Rodríguez-Couto, S. (2018). Antifungal and antimycotoxin efficacy of biogenic silver nanoparticles produced by *Fusarium chlamydosporum* and *Penicillium chrysogenum* at non-cytotoxic doses. *Chemosphere, 218*, 477–486. https://doi.org/10.1016/j.chemosphere.2018.11.129
12. Wanarska, E., & Maliszewska, I. (2019). The possible mechanism of the formation of silver nanoparticles by *Penicillium cyclopium*. *Bioorganic Chemistry, 93*, 102803(9pp). https://doi.org/10.1016/j.bioorg.2019.02.028
13. Seetharaman, P. K., Chandrasekaran, R., Gnanasekar, S., Chandrakasan, G., Gupta, M., Manikandan, D. B., & Sivaperumal, S. (2018). Antimicrobial and larvicidal activity of eco-friendly silver nanoparticles synthesized from endophytic fungi *Phomopsis liquidambaris*. *Biocatalysis and Agricultural Biotechnology, 16*, 22–30. https://doi.org/10.1016/j.bcab.2018.07.006
14. Shareef, J. U., Navya Rani, M., Anand, S., & Rangappa, D. (2017). Synthesis and characterization of silver nanoparticles from *Penicillium sps*. *Materials Today, 4*, 11923–11932. https://doi.org/10.1016/j.matpr.2017.09.113
15. Clarance, P., Luvankar, B., Sales, J., Khusro, A., Agastian, P., Tack, J. C., Al- Shwaiman, H. A., Al Khulaifi, M. M., Elgorban, A. M., Syed, A., & Kim, H. J. (2019). Green synthesis and characterization of gold nanoparticles using endophytic fungi *Fusarium solani* and its in-vitro anticancer and biomedical applications. *Saudi Journal of Biological Sciences, 27*, 706–712. https://doi.org/10.1016/j.sjbs.2019.12.026
16. Naimi-Shamel, N., Pourali, P., & Dolatabadi, S. (2019). Green synthesis of gold nanoparticles using *Fusarium oxysporum* and antibacterial activity of its tetracycline conjugant. *Journal de Mycologie Médicale, 29*, 7–13. https://doi.org/10.1016/j.mycmed.2019.01.005
17. Vago, A., Szakacs, G., Safran, G., Horvath, R., Pecz, B., & Lagzi, I. (2016). One-step green synthesis of gold nanoparticles by mesophilic filamentous fungi. *Chemical Physics Letters, 645*, 1–4. https://doi.org/10.1016/j.cplett.2015.12.019
18. Shen, W., Qu, Y., Pei, X., Li, S., You, S., Wang, J., Zhang, Z., & Zhou, J. (2017). Catalytic reduction of 4-nitrophenol using gold nanoparticles biosynthesized by cell-free extracts of *Aspergillus* sp. WL-Au. *Journal of Hazardous Materials, 321*, 299–306. https://doi.org/10.1016/j.jhazmat.2016.07.051

19. Ganesan, V., Hariram, M., Vivekanandhan, S., & Muthuramkumar, S. (2020). *Periconium* sp. (endophytic fungi) extract mediated sol-gel synthesis of ZnO nanoparticles for antimicrobial and antioxidant applications. *Materials Science in Semiconductor Processing, 105*, 104739(12pp. https://doi.org/10.1016/j.mssp.2019.104739
20. Saravanakumar, K., Shanmugam, S., Varukattu, N. B., MubarakAli, A., Kathiresan, K., & Wang, M. H. (2018). Biosynthesis and characterization of copper oxide nanoparticles from indigenous fungi and its effect of photothermolysis on human lung carcinoma. *Journal of Photochemistry and Photobiology A, 190*, 103–109. https://doi.org/10.1016/j.jphotobiol.2018.11.017
21. Gudikandula, K., Vadapally, P., & Singara Charya, M. A. (2017). Biogenic synthesis of silver nanoparticles from white rot fungi: Their characterization and antibacterial studies. *Open, 2*, 64–78. https://doi.org/10.1016/j.onano.2017.07.002
22. El-Sayyad, G. S., Mosallam, F. M., & El-Batal, A. I. (2018). One-pot green synthesis of magnesium oxide nanoparticles using *Penicillium chrysogenum* melanin pigment and gamma rays with antimicrobial activity against multidrug-resistant microbes. *Advanced Powder Technology, 29*, 2616–2625. https://doi.org/10.1016/j.apt.2018.07.009
23. Syed, A., & Ahmad, A. (2012). Extracellular biosynthesis of platinum nanoparticles using the fungus *Fusarium oxysporum. Colloids and Surfaces B: Biointerfaces, 97*, 27–31. https://doi.org/10.1016/j.colsurfb.2012.03.026
24. Vijayanandan, A. S., & Balakrishnan, R. M. (2018). Biosynthesis of cobalt oxide nanoparticles using endophytic fungus *Aspergillus nidulans. Journal of Environmental Management, 218*, 442–450. https://doi.org/10.1016/j.jenvman.2018.04.032
25. Sanghi, R., & Verma, P. (2019). A facile green extracellular biosynthesis of CdS nanoparticles by immobilized fungus. *The Chemical Engineering Journal, 155*, 886–891. https://doi.org/10.1016/j.cej.2009.08.006
26. Chowdhury, S., Basu, A., & Kundu, S. (2014). Green synthesis of protein capped silver nanoparticles from phytopathogenic fungus *Macrophomina phaseolina* (Tassi) Goid with antimicrobial properties against multidrug-resistant bacteria. *Nanoscale Research Letters, 9*, 365(11pp). https://doi.org/10.1186/1556-276X-9-365
27. Kumari, M., Pandey, S., Giri, V. P., Bhattacharya, A., Shukla, R., Mishra, A., & Nautiyal, C. S. (2017). Tailoring shape and size of biogenic silver nanoparticles to enhance antimicrobial efficacy against MDR bacteria. *Microbial Pathogenesis, 105*, 346–355. https://doi.org/10.1016/j.micpath.2016.11.012
28. Elahian, F., Reiisi, S., Shahidi, A., & Mirzaei, S. A. (2017). High-throughput bioaccumulation, biotransformation, and production of silver and selenium nanoparticles using genetically engineered *Pichia pastoris. Nanomedicine, 13*, 853–861. https://doi.org/10.1016/j.nano.2016.10.009
29. Bhadwal, A. S., Tripathi, R., Gupta, R. K., Kumar, N., Singh, R., & Shrivastav, A. (2014). Biogenic synthesis and photocatalytic activity of CdS nanoparticles. *RSC Advances, 4*, 9484–9490. https://doi.org/10.1039/C3RA46221H
30. Moghaddam, A. B., Moniri, M., Azizi, S., Rahim, R. A., Ariff, A. B., Saad, W. Z., Namvar, F., Navaderi, M., & Mohamad, R. (2017). Biosynthesis of ZnO nanoparticles by a new Pichia kudriavzevii yeast strain and evaluation of their antimicrobial and antioxidant activities. *Molecules, 22*, 872(18pp). https://doi.org/10.3390/molecules22060872
31. Amerasan, D., Nataraj, T., Murugan, K., Panneerselvam, C., Madhiyazhagan, P., Nicoletti, M., & Benelli, G. (2016). Myco-synthesis of silver nanoparticles using *Metarhizium anisopliae* against the rural malaria vector *Anopheles culicifacies* Giles (Diptera: Culicidae). *Journal of Pest Science, 89*, 249–256. https://doi.org/10.1007/s10340-015-0675-x
32. Korbekandi, H., Mohseni, S., MardaniJouneghani, R., Pourhossein, M., & Iravani, S. (2016). Biosynthesis of silver nanoparticles using *Saccharomyces cerevisiae. Artificial Cells Blood Substitutes and Biotechnology, 44*, 235–239. https://doi.org/10.3109/21691401.2014.937870
33. Sriramulu, M., & Sumathi, S. (2018). Biosynthesis of palladium nanoparticles using *Saccharomyces cerevisiae* extract and its photocatalytic degradation behaviour. *Advances*

in Natural Sciences: Nanoscience and Nanotechnology, 9, 025018(7pp). https://doi.org/10.1088/2043-6254/aac506
34. Qu, Y., You, S., Zhang, X., Pei, X., Shen, W., Li, Z., Li, S., & Zhang, Z. (2018). Biosynthesis of gold nanoparticles using cell-free extracts of *Magnusiomyces ingens* LH-F1 for nitrophenols reduction. *Bioprocess and Biosystems Engineering, 41*, 359–367. https://doi.org/10.1007/s00449-017-1869-9
35. Suryavanshi, P., Pandit, R., Gade, A., Derita, M., Zachino, S., & Rai, M. (2017). Colletotrichum sp.- mediated synthesis of sulphur and aluminium oxide nanoparticles and its *in vitro* activity against selected food-borne pathogens. *LWT Food Science and Technology, 81*, 188–194. https://doi.org/10.1016/j.lwt.2017.03.038
36. El Domany, E. B., Essam, T. M., Ahmed, A. E., & Farghali, A. A. (2018). Biosynthesis physico-chemical optimization of gold nanoparticles as anti-cancer and synergetic antimicrobial activity using *Pleurotus ostreatus* fungus. *Journal of Applied Pharmaceutical Science, 8*, 119–128. https://doi.org/10.7324/JAPS.2018.8516
37. Jha, A. K., Prasad, K., & Prasad, K. (2009). A green low-cost biosynthesis of Sb_2O_3 nanoparticles. *Biochemical Engineering Journal, 43*, 303–306. https://doi.org/10.1016/j.bej.2008.10.016
38. Sandanamala, J. G., & Rose, C. (2014). Facile production of ZnS quantum dot nanoparticles by *Saccharomyces cerevisiae* MTCC 2918. *Journal of Biotechnology, 170*, 73–78. https://doi.org/10.1016/j.jbiotec.2013.11.017
39. Zamani, H., Jafari, A., Mousavii, S. M., & Darezereshki, E. (2020). Biosynthesis of silica nanoparticle using *Saccharomyces cerevisiae* and its application on enhanced oil recovery. *Journal of Petroleum Science and Engineering, 190*, 107002(10pp). https://doi.org/10.1016/j.petrol.2020.107002
40. Soliman, H., Elsayed, A., & Dyaa, A. (2018). Antimicrobial activity of silver nanoparticles biosynthesised by *Rhodotorula* sp. strain ATL72. *Egyptian Journal of Basic and Applied Sciences, 5*, 228–233. https://doi.org/10.1016/j.ejbas.2018.05.005
41. Pimprikar, P. S., Joshi, S. S., Kumar, A. R., Zinjarde, S. S., & Kulkarni, S. K. (2009). Influence of biomass and gold salt concentration on nanoparticle synthesis by the tropical marine yeast *Yarrowia lipolytica* NCIM 3589. *Colloids and Surfaces B: Biointerfaces, 74*, 309–316. https://doi.org/10.1016/j.colsurfb.2009.07.040
42. Jacob, J. M., Ravindran, R., Narayanan, M., Samuel, S. M., Pugazhendhi, A., & Kumar, G. (2020). Microalgae: A prospective low cost green alternative for nanoparticle synthesis. *Current Opinion in Environmental Science & Health, 20*, 100163. https://doi.org/10.1016/j.coesh.2019.12.005
43. Aboelfetoh, E. E., El-Shenody, R. A., & Ghobara, M. M. (2017). Eco-friendly synthesis of silver nanoparticles using green algae (*Caulerpa serrulata*): Reaction optimization, catalytic and antibacterial activities. *Environmental Monitoring and Assessment, 189*, 349(15pp). https://doi.org/10.1007/s10661-017-6033-0
44. Lengke, M. F., Fleet, M. E., & Southam, G. (2007). Synthesis of palladium nanoparticles by reaction of filamentous cyanobacterial biomass with a palladium(II) chloride complex. *Langmuir, 23*, 8982–8987. https://doi.org/10.1021/la7012446
45. El-Rafie, H. M., El-Rafie, M. H., & Zahran, M. K. (2013). Green synthesis of silver nanoparticles using polysaccharides extracted from marine macro algae. *Carbohydrate Polymers, 96*, 403–410. https://doi.org/10.1016/j.carbpol.2013.03.071
46. Gonzalez-Ballesteros, N., Prado-Lopez, S., Rodríguez-Gonzalez, J. B., Lastra, M., & Rodríguez-Argüelles, M. C. (2017). Green synthesis of gold nanoparticles using brown algae *Cystoseira baccata*: Its activity in colon cancer cells. *Colloids and Surfaces B: Biointerfaces, 153*, 190–198. https://doi.org/10.1016/j.colsurfb.2017.02.020
47. Fatima, R., Priya, M., Indurthi, L., Radhakrishnan, V., & Sudhakaran, R. (2020). Biosynthesis of silver nanoparticles using red algae *Portieria hornemannii* and its antibacterial activity against fish pathogens. *Microbial Pathogenesis, 138*, 103780(25pp). https://doi.org/10.1016/j.micpath.2019.103780

48. Oztürk, B. Y., Gürsu, B. Y., & Dag, I. (2019). Antibiofilm and antimicrobial activities of green synthesized silver nanoparticles using marine red algae *Gelidium corneum*. *Process Biochemistry, 89*, 208–219. https://doi.org/10.1016/j.procbio.2019.10.027
49. Colin, J. A., Pech-Pech, I. E., Oviedo, M., Aguila, S. A., Romo-Herrera, J. M., & Contreras, O. E. (2018). Gold nanoparticles synthesis assisted by marine algae extract: Biomolecules shells from a green chemistry approach. *Chemical Physics Letters, 708*, 210–215. https://doi.org/10.1016/j.cplett.2018.08.022
50. Koopi, H., & Buazar, F. (2018). A novel one-pot biosynthesis of pure alpha aluminum oxide nanoparticles using the macroalgae *Sargassum ilicifolium*: A green marine approach. *Ceramics International, 44*, 8940–8945. https://doi.org/10.1016/j.ceramint.2018.02.091
51. Momeni, S., & Nabipour, I. (2015). A simple green synthesis of palladium nanoparticles with *Sargassum* alga and their electrocatalytic activities towards hydrogen peroxide. *Applied Biochemistry and Biotechnology, 176*, 1937–1949. https://doi.org/10.1007/s12010-015-1690-3
52. Sanaeimehr, Z., Javadi, I., & Namvar, F. (2018). Antiangiogenic and antiapoptotic effects of green-synthesized zinc oxide nanoparticles using *Sargassum muticum* algae extraction. *Cancer Nanotechnology, 9*, 3(16pp). https://doi.org/10.1186/s12645-018-0037-5
53. Rajeshkumar, S. (2018). Synthesis of zinc oxide nanoparticles using algal formulation (*Padina tetrastromatica* and *Turbinaria conoides*) and their antibacterial activity against fish pathogens. *Research Journal of Biotechnology, 13*, 15–19.
54. Kim, D. Y., Saratale, R. G., Shinde, S., Syed, A., Ameen, F., & Ghodake, G. (2018). Green synthesis of silver nanoparticles using *Laminaria japonica* extract: Characterization and seedling growth assessment. *Journal of Cleaner Production, 172*, 2910–2918. https://doi.org/10.1016/j.jclepro.2017.11.123
55. Venkatesan, J., Manivasagan, P., Kim, S. K., Kirthi, A. V., Marimuthu, S., & Rahuman, A. A. (2014). Marine algae-mediated synthesis of gold nanoparticles using a novel *Ecklonia cava*. *Bioprocess and Biosystems Engineering, 37*, 1591–1597. https://doi.org/10.1007/s00449-014-1131-7
56. Pugazhendhi, A., Prabakar, D., Jacob, J. M., Karuppusamy, I., & Saratale, R. G. (2018). Synthesis and characterization of silver nanoparticles using *Gelidium amansii* and its antimicrobial property against various pathogenic bacteria. *Microbial Pathogenesis, 114*, 41–45. https://doi.org/10.1016/j.micpath.2017.11.013
57. Kathiraven, T., Sundaramanickam, A., Shanmugam, N., & Balasubramanian, T. (2015). Green synthesis of silver nanoparticles using marine algae *Caulerpa racemosa* and their antibacterial activity against some human pathogens. *Applied Nanoscience, 5*, 499–504. https://doi.org/10.1007/s13204-014-0341-2
58. Salari, Z., Danafar, F., Dabaghi, S., & Ataei, S. A. (2016). Sustainable synthesis of silver nanoparticles using macroalgae *Spirogyra varians* and analysis of their antibacterial activity. *Journal of Saudi Chemical Society, 20*, 459–464. https://doi.org/10.1016/j.jscs.2014.10.004
59. Saratale, R. G., Karuppusamy, I., Saratale, G. D., Pugazhendhi, A., Kumar, G., Park, Y., Ghodake, G. S., Bhargava, R. N., Banu, J. R., & Shin, H. S. (2018). A comprehensive review on green nanomaterials using biological systems: Recent perception and their future applications. *Colloids and Surfaces B: Biointerfaces, 170*, 20–35. https://doi.org/10.1016/j.colsurfb.2018.05.045
60. Jorge de Souza, T. A., Rosa Souza, L. R., & Franchi, L. P. (2019). Silver nanoparticles: An integrated view of green synthesis methods, transformation in the environment, and toxicity. *Ecotoxicology and Environmental Safety, 171*, 691–700. https://doi.org/10.1016/j.ecoenv.2018.12.095
61. Javaid, A., Oloketuyi, S. F., Khan, M. M., & Khan, F. (2018). Diversity of bacterial synthesis of silver nanoparticles. *Bionanoscience, 8*, 43–59. https://doi.org/10.1007/s12668-017-0496-x
62. Saravanan, M., Barik, S. K., MubarakAli, D., Prakash, P., & Pugazhendhi, A. (2018). Synthesis of silver nanoparticles from *Bacillus brevis* (NCIM 2533) and their antibacterial activity against pathogenic bacteria. *Microbial Pathogenesis, 116*, 221–226. https://doi.org/10.1016/j.micpath.2018.01.038

63. Quinteros, M. A., Bonilla, J. O., Albores, S. V., Villegas, L. B., & Paez, P. L. (2019). Biogenic nanoparticles: Synthesis, stability and biocompatibility mediated by proteins of *Pseudomonas aeruginosa*. *Colloids and Surfaces B: Biointerfaces, 184*, 110517. https://doi.org/10.1016/j.colsurfb.2019.110517
64. Al-Dhabi, N. A., Ghilan, A. K. M., Arasu, M. V., & Duraipandiyan, V. (2018). Green biosynthesis of silver nanoparticles produced from marine *Streptomyces* sp. Al-Dhabi-89 and their potential applications against wound infection and drug resistant clinical pathogens. *The Journal of Photochemistry and Photobiology A, 189*, 176–184. https://doi.org/10.1016/j.jphotobiol.2018.09.012
65. Ameen, F., AlYahia, S., Govarthanan, M., Aljahdali, N., Al-Enazi, N., Alsamhari, K., Alshehri, W. A., Alwakeel, S. S., & Alharbi, S. A. (2019). Soil bacteria *Cupriavidus* sp. mediates the extracellular synthesis of antibacterial silver nanoparticles. *Journal of Molecular Structure, 1202*, 127233(25pp). https://doi.org/10.1016/j.molstruc.2019.127233
66. Pourali, P., & Yahyaei, B. (2016). Biological production of silver nanoparticles by soil isolated bacteria and preliminary study of their cytotoxicity and cutaneous wound healing efficiency in rat. *Journal of Trace Elements in Medicine and Biology, 34*, 22–31. https://doi.org/10.1016/j.jtemb.2015.11.004
67. Nadaf, N. Y., & Kanase, S. S. (2019). Biosynthesis of gold nanoparticles by *Bacillus marisflavi* and its potential in catalytic dye degradation. *Arabian Journal of Chemistry, 12*, 4806–4814. https://doi.org/10.1016/j.arabjc.2016.09.020
68. He, S., Guo, Z., Zhang, Y., Zhang, S., Wang, J., & Gu, N. (2007). Biosynthesis of gold nanoparticles using the bacteria *Rhodopseudomonas capsulata*. *Materials Letters, 61*, 3984–3987. https://doi.org/10.1016/j.matlet.2007.01.018
69. Kumari, M., Mishra, A., Pandey, S., Singh, S. P., Chaudhry, V., Mudiam, M. K. R., Shukla, S., Kakkar, P., & Nautiyal, C. S. (2016). Physico-chemical condition optimization during biosynthesis lead to development of improved and catalytically efficient gold nanoparticles. *Scientific Reports, 6*, 27575(14pp). https://doi.org/10.1038/srep27575
70. Sathiyanarayanan, G., Kiran, G. S., Selvin, J., & Saibaba, G. (2013). Optimization of polyhydroxybutyrate production by marine *Bacillus megaterium* MSBN04 under solid state culture. *International Journal of Biological Macromolecules, 60*, 253–261. https://doi.org/10.1016/j.ijbiomac.2013.05.031
71. Kunoh, T., Takeda, M., Matsumoto, S., Suzuki, I., Takano, M., Kunoh, H., & Takada, J. (2018). Green synthesis of gold nanoparticles coupled with nucleic acid oxidation. *ACS Sustainable Chemistry & Engineering, 6*, 364–373. https://doi.org/10.1021/acssuschemeng.7b02610
72. Fatemi, M., Mollania, N., Momeni-Moghaddam, M., & Sadeghifar, F. (2018). Extracellular biosynthesis of magnetic iron oxide nanoparticles by Bacillus cereus strain HMH1: Characterization and *in vitro* cytotoxicity analysis on MCF-7 and 3T3 cell lines. *Journal of Biotechnology, 270*, 1–11. https://doi.org/10.1016/j.jbiotec.2018.01.021
73. Bai, H. J., Zhang, Z. M., Guo, Y., & Yang, G. E. (2009). Biosynthesis of cadmium sulfide nanoparticles by photosynthetic bacteria *Rhodopseudomonas palustris*. *Colloids and Surfaces B: Biointerfaces, 70*, 142–146. https://doi.org/10.1016/j.colsurfb.2008.12.025
74. Jayaseelan, C., Rahuman, A. A., Kirthi, A. V., Marimuthu, S., Santhoshkumar, T., Bagavan, A., Gaurav, K., Karthik, L., & Bhaskara Rao, K. V. (2012). Novel microbial route to synthesize ZnO nanoparticles using *Aeromonas hydrophila* and their activity against pathogenic bacteria and fungi. *Spectrochimica Acta A, 90*, 78–84. https://doi.org/10.1016/j.saa.2012.01.006
75. Shoeibi, S., & Mashreghi, M. (2017). Biosynthesis of selenium nanoparticles using *Enterococcus faecalis* and evaluation of their antibacterial activities. *Journal of Trace Elements in Medicine and Biology, 39*, 135–139. https://doi.org/10.1016/j.jtemb.2016.09.003
76. Nordmeier, A., Merwin, A., Roeper, D. E., & Chidambaram, D. (2018). Microbial synthesis of metallic molybdenum nanoparticles. *Chemosphere, 203*, 521–525. https://doi.org/10.1016/j.chemosphere.2018.02.079
77. Sowani, H., Mohite, P., Munot, H., Shouche, Y., Bapat, T., Kumar, A. R., Kulkarni, M., & Zinjarde, S. (2016). Green synthesis of gold and silver nanoparticles by an actinomy-

cete *Gordonia amicalis* HS-11: Mechanistic aspects and biological application. *Process Biochemistry, 51*, 374–383. https://doi.org/10.1016/j.procbio.2015.12.013

78. Kanmani, P., & Lim, S. T. (2013). Synthesis and structural characterization of silver nanoparticles using bacterial exopolysaccharide and its antimicrobial activity against food and multidrug resistant pathogens. *Process Biochemistry, 48*, 1099–1106. https://doi.org/10.1016/j.procbio.2013.05.011
79. Kaur, H., Dolma, K., Kaur, N., Malhotra, A., Kumar, N., Dixit, P., Sharma, D., Mayilraj, S., & Choudhury, A. R. (2015). Marine microbe as nano-factories for copper biomineralization. *Biotechnology and Bioprocess Engineering, 20*, 51–57. https://doi.org/10.1007/s12257-014-0432-7
80. Ahmed, E., Kalathil, S., Shi, L., Alharbi, O., & Wang, P. (2018). Synthesis of ultra-small platinum, palladium and gold nanoparticles by *Shewanella loihica PV-4* electrochemically active biofilms and their enhanced catalytic activities. *Journal of Saudi Chemical Society, 22*, 919–929. https://doi.org/10.1016/j.jscs.2018.02.002
81. Zonaro, E., Piacenza, E., Presentato, A., Monti, F., Dell'Anna, R., Lampis, S., & Vallini, G. (2017). *Ochrobactrum* sp. MPV1 from a dump of roasted pyrites can be exploited as bacterial catalyst for the biogenesis of selenium and tellurium nanoparticles. *Microbial Cell Factories, 16*, 215(17pp). https://doi.org/10.1186/s12934-017-0826-2
82. Eid, A. M., Fouda, A., Niedbała, G., El-Din Hassan, S., Salem, S. S., Abdo, A. M., Hetta, H. F., & Shaheen, T. I. (2020). Endophytic *Streptomyces laurentii* mediated green synthesis of Ag-NPs with antibacterial and anticancer properties for developing functional textile fabric properties. *Antibiotics (Basel), 9*, 641(18pp). https://doi.org/10.3390/antibiotics9100641
83. Abd-Elnaby, H. M., Abo-Elala, G. M., Abdel-Raouf, U. M., & Hamed, M. M. (2016). Antibacterial and anticancer activity of extracellular synthesized silver nanoparticles from marine *Streptomyces rochei* MHM13. *Egyptian Journal of Basic and Applied Sciences, 42*, 301–312. https://doi.org/10.1016/j.ejar.2016.05.004
84. Muëller, A., Behsnilian, D., Walz, E., Graf, V., Hogekamp, L., & Greiner, R. (2016). Effect of culture medium on the extracellular synthesis of silver nanoparticles using *Klebsiella pneumoniae, Escherichia coli* and *Pseudomonas jessinii*. *Biocatalysis and Agricultural Biotechnology, 6*, 107–115. https://doi.org/10.1016/j.bcab.2016.02.012
85. Kalishwaralal, K., Deepak, V., Ramkumarpandian, S., Nellaiah, H., & Sangiliyandi, G. (2008). Extracellular biosynthesis of silver nanoparticles by the culture supernatant of *Bacillus licheniformis*. *Materials Letters, 62*, 4411–4413. https://doi.org/10.1016/j.matlet.2008.06.051
86. Husseiny, M. I., El-Aziz, M. A., Badr, Y., & Mahmoud, M. A. (2007). Biosynthesis of gold nanoparticles using *Pseudomonas aeruginosa*. *Spectrochimica Acta Part A, 67*, 1003–1006. https://doi.org/10.1016/j.saa.2006.09.028
87. Yumei, L., Yamei, L., Qiang, L., & Jie, B. (2017). Rapid biosynthesis of silver nanoparticles based on flocculation and reduction of an exopolysaccharide from *Arthrobacter* sp. B4: Its antimicrobial activity and phytotoxicity. *Journal of Nanomaterials, 1*, 8pp. https://doi.org/10.1155/2017/9703614
88. Arsiya, F., Sayadi, M. H., & Sobhani, S. (2017). Green synthesis of palladium nanoparticles using *Chlorella vulgaris*. *Materials Letters, 186*, 113–115. https://doi.org/10.1016/j.matlet.2016.09.101
89. Li, J., Li, Q., Ma, X., Tian, B., Li, T., Yu, J., Dai, S., Weng, Y., & Hua, Y. (2016). Biosynthesis of gold nanoparticles by the extreme bacterium *Deinococcus radiodurans* and an evaluation of their antibacterial properties. *International Journal of Nanomedicine, 11*, 5931–5944. https://doi.org/10.2147/IJN.S119618
90. Gholami-Shabani, M., Shams-Ghahfarokhi, M., Gholami-Shabani, Z., Akbarzadeh, A., Riazi, G., Ajdari, S., Amani, A., & Razzaghi-Abyaneh, M. (2015). Enzymatic synthesis of gold nanoparticles using sulfite reductase purified from *Escherichia coli*: A green eco-friendly approach. *Process Biochemistry, 50*, 1076–1085. https://doi.org/10.1016/j.procbio.2015.04.004

Chapter 3
Use of Nanomaterials as an Antimicrobial and Antiviral Regimen

Ashok Chakraborty, Anil Diwan, and Jayant Tatake

3.1 Introduction

Infectious diseases, in recent times, are a great concern on public health. Various health issues those are caused by microbes can turn to mortality of millions of people every year around the world [1–4]. Though the antibiotics are the first choice to treat the infections, microbes can turn to a resistant type due to their mutation and morphological changes [5–7]. Nanoparticle (NP)-based treatment can overcome the microbial resistance. Besides, NPs have large surface areas that allows for efficient microbial attachment and rapid penetration into the cell and render innate antimicrobial activities [8, 9]. NPs can also generate reactive oxygen species (ROS), which damage the DNA and proteins conformation, and cause the loss of cellular integrity, thereby the growth inhibition of bacteria, fungus, and also the viral growth. Conjugated use of antibiotics with nanoparticles was, therefore, thought to enhance the inhibition capacity as well as the developments of resistance by organisms [10]. Gold nanoparticles (52–22 nm) attached with the antibiotic, and *Cefaclor*, showed significant microbial biofilm formation and achieve bactericidal and fungicidal activity [11–13]. Selenium nanoparticles possess antibiofilm activity and retard the growth of gram-negative *Pseudomonas aeruginosa* [14]. Similar result was also observed with TiO_2 nanoparticles, which inhibit the formation of fungal biofilms [15].

Here we discussed the several aspects of using nanoparticles in the microbial field, and their prospects in future.

A. Chakraborty (✉) · J. Tatake
AllExcel, Inc, Shelton, CT, USA
e-mail: ashok.chakraborty@allexcel.com

A. Diwan
Nanoviricide, Inc, Shelton, CT, USA

R. S. Chaughule, A. S. Lokur (eds.), *Applications of Nanotechnology in Microbiology*, https://doi.org/10.1007/978-3-031-49933-3_3

3.2 Antimicrobial Properties of Nanoparticles/Nanocomposites

Nanoparticles (NPs) belong to a group of substances with a diameter ranging from 1 to 100 nm, and may originate from plant or animal, or may be synthetic [16–19]. The main characteristics of a antimicrobials is the ability of penetration of the bacterial cell wall made up of peptidoglycan, a rigid layer. Both Gram (+) and Gram (−) bacteria possess peptidoglycan; however, Gram (+) has a very thick peptidoglycan coat than that of Gram (−) organisms. NPs can dismantle the peptidoglycan layer, and also overcome antimicrobial resistance [5, 20, 21].

3.2.1 *Antibacterial Activities of NPs*

Several nanoparticles with their antibacterial properties are listed in Table 3.1.

Table 3.1 Antibacterial nanoparticles and their functions

Antibacterial nanoparticles	Functions
Phosphatidylcholine–chitosan nanoparticles coated with the gentamycin antibiotic [22]	Work against Gram (+) and Gram (−) bacteria [23]
Supramolecular polyelectrolyte complexes between the (+)vely charged-NH_3^+ groups of the β-cyclodextrin-grafted chitosan and the (−)vely charged-SO_3^- groups [24]	β-Cyclodextrin complexes with the silver sulfadiazine molecules release silver ions to penetrate bacterial cells [25]
Polymersomes encapsulated Vancomycin antibiotic [26]	Antibacterial activity against methicillin-resistant *S. aureus* [27]
Mannose-functionalized chitosan nanoparticles [28]	Intrinsic bactericidal properties [28]
PLGA nanoparticles attached with teicoplanin [29]	Potential antibacterial effect against *S. aureus* [29]
PLA nanoparticles encapsulating the *Pistacialentiscus L.* var. chia essential oil [30]	Work against Gram (+) and Gram (−) [30]
Silver nanoparticles/PLA nanocoatings [31] Silver nanoparticles/polyethylene terephthalate nanofibers [32]	Work against Gram (+) and Gram (−) [31, 32]

3.2.2 *Application of NPs in Viral Infection*

Viruses can infect eukaryotes, and prokaryotes, also. Smallpox and paralytic poliomyelitis like viral diseases can be managed by using vaccination programs at the right time. However, nanotechnology can offer further a successful strategy to overcome the antiviral drug resistance [33]. Some nanoparticles with their antiviral properties are shown in Table 3.2.

Table 3.2 Functions of antiviral nanoparticles

Antiviral nanoparticles	Functions
HIV-1 P24 protein-derived peptides adsorbed onto the surface of chitosan nanoparticles [34]	Results a controlled peptide drug release [35]
Dolutegravir sodium-loaded nanoparticles results from cross-linking hydroxypropyl-β-cyclodextrin with diphenyl carbonate [36]	This nanoparticles-attached medicine can permeate through the nasal mucosa without causing any damage there [37]

3.2.3 Application of NPs in Fungal and Parasite Infection (Tables 3.3 and 3.4)

Table 3.3 Nanoparticles and their antifungal functions

Antifungal nanoparticles	Functions
Chitosan nanoparticles can deliver miconazole and farnesol [38]	Inhibits the growth of *C. albicans* [39]
Chitosan nanoparticles–mediated delivery of itraconazole [40]	Potentially inhibits *A*. fumigatus, *C. neoformans*, and *C. albicans* [41]
Delivery of amphotericin nanocapsules with modified polysaccharide [42]	This joint administration of drugs with this nanosystem can destroy *C. albicans* strains, with lower MIC values than the free drug [43]

Table 3.4 Antiparasitic nanoparticles and their functions

Antiparasitic nanoparticles	Functions
Chitosan nanocapsules containing triclabendazole [44]	Therapeutic use for *fascioliasis* has been proved [45]
Previously mentioned and study performed by [40]	Antiparasitic effects against the *Leishmania promastigotes* protozoan were identified [46]

3.3 Nanoparticles (NPs) Are Biologically Compatible (Table 3.5)

NPs when they come into contact of blood they can cause various biological effects, which could be either beneficial or destructive. Hence, it is important to determine the blood-NP compatibility before they can be used in human [47]. Blood cell aggregation and hemolysis are some of the examples that can be used to detect the blood-NPs compatibility [48]. Few observations are as follows:

- The biological compatibility of NPs depends on their structure, size, and formulation [49, 50]
- Polymeric NPs are biocompatible and are used in the treatment of asthma, tuberculosis, and lung cancer [51, 52].

Table 3.5 The comparative biocompatibility of several NPs

NPs	In vitro and In vivo toxicity studies
Dendrimers	No toxicity [53]
AuNPs	Nontoxic [53]
Carbon nanotubes	No toxicity [53]
Super-paramagnetic Fe_3O_4 (SPIONs)	No toxicity [53]
Silica-based NPs	No effects on cell proliferation [54]
AgNPs	Induce cell death; and affect membrane damage [55–57] Release free radicals, which induce damages to cell and DNA [58] Immunotoxicity in rats [59] AgNP-biopolymer showed antibacterial activity but no effects on eukaryotic cell lines [42, 60]
Fe_3O_4-AuNPs	Negligible cytotoxicity was observed in the cell lines [61]
TiO_2 NPs	Nontoxic at low doses (5 mg/kg body weight) [62, 63]
Manganese ferrite ($MnFe_2O_4$) NPs	They are nontoxic and biocompatible too at 125 μg/mL or below in 4 T1 cells (a murine breast cancer cell line) [64]
Fe_3O_4, $ZnFe_3O_4$, and $NiFe_3O_4$ - nanoparticles	Those NPs did not show any toxicity on HeLa cells when used at less than 100 μg/mL [65]
TiO_2 NPs	At lower concentrations (100 μg/mL) it is harmless to humans [66]
$CaFe_2O_4$ NPs	This is safe in humans when used at concentrations below 200 μg/mL [67]

3.4 Biodegradability and Encapsulation of Nanoparticles

Biodegradable NPs (BNPs) are always with advantages than non-degradable NPs because of the pollution issues, which is less or none with the former one. Accumulation of the later in the liver and spleen may have toxic side effects in the individual [68]. Other significances are as follows:

- Polymer-based BNPs are colloidal particles and can encapsulate other therapeutic agent to make it available to the action site [69].
- Proteins, polysaccharides, and synthetic biodegradable polymers are the sources of BNPs.

3.4.1 The Selection of Basic Polymers and the Synthesis of BNPs

The selection of polymers and the synthesis of BNPs depends on the end-application of the NPs, in addition to their degree of biocompatibility, biodegradability, the size and the surface characteristics of NPs and the properties of the encapsulated drugs [40]. Some of the biodegradable polymers used in the preparation of BNPs, and their merits are listed in Table 3.6.

Table 3.6 Basic polymers for the synthesis of BNPs

PLGA	PLGA is hydrolyzed in the body to lactic and glycolic acids which are biodegradable [70] PLGA can be used for manufacturing protein- and peptide-based nanomedicines, as well as in the production of nanovaccines and gene delivery [70, 71]
PLA	PLA like PLGA is a biocompatible and biodegradable polymer, and it is catabolized in the body to lactic acid [40]
Gelatin	Gelatin is a polyampholyte and is used in medicine and food industry Gelatin NPs are nontoxic, biodegradable, and bioactive and can be used for drug delivery
PACs	PACs is also a basic polymer but cannot be used in either medicine or food industries as they produce several toxic compounds that may damage the central nervous system

3.5 Use of Nanoparticles for Microbial Targeting Strategies

Human body recognizes NPs as foreign particles; hence body's immune cells can remove those from the circulation. Therefore, in order to bypass the immune system, NPs should be modified before administration in the body [72]. Several polymers like PEG, dextran, polysorbate 20, polysorbate 80, and tocopheryl polyethylene glycol 1000 succinate are available to protect the NPs. Leroux et al. [73] showed that PEGylation of NPs make them to bypass the phagocytes and can sustain in the circulation system. Similarly, tocopheryl polyethylene glycol 1000 succinate can modify NPs and that allow them for increased adhesion on tumor cell surfaces [40].

Further, NPs-mediated drug delivery are more specific to the target, have increased biodistribution, and less side effects compared to conventional drug delivery system [74–76].

3.6 Limitations

- The major challenge is to control the size and shape of NPs during their synthesis.
- NPs may gather in different biological tissues and ultimately can impair the biological functions [77, 78].
- NPs since may bypass the immune challenges, they may cause swelling and toxicity in the body [78]
- NPs can generate the ROS species, which in turn may cause oxidative stress, inflammation, and apoptosis.
- Air pollution by NPs is another concern that may cause harm to the other biological species and thereby the ecosystem [21].

3.7 "Nanoviricide", A New Antiviral Regimen Could Be an Antimicrobial Agent, Too

"TheraCour" platform polymer NV-387 consists of polyethylene glycol (PEG) within its monomer unit, which is functionalized with a designed linker unit that has covalently connected aliphatic chains and a site-targeting ligand [79–81] (Fig. 3.1).

These materials (a) are capable of protecting active pharmaceutical ingredient (API) by encapsulation, (b) can do the direct delivery of encapsulated APIs into the target-specific cell or virus, (c) are biodegradable, biocompatible, nontoxic, and non-immunogenic [79–81].

This model is vastly applicable in the antiviral field. In particular, a drug, NV-CoV-2 targeted for SARS-CoV-2 that causes COVID-19, has completed pre-clinical studies including GLP Safety/Toxicology, and at present entered in human clinical trials. Another drug candidate, NV-CoV-2-R made by us that encapsulates remdesivir found to be more effective than the standard remdesivir formulation (Table 3.7).

3.7.1 Experimental Proof of TheraCour Biopolymer for Their Antiviral Activity

TheraCour model consists of a special design where chemicals bind to the glycoprotein of the virus at the same site where the virus uses for binding to its cognate cell surface receptor [70, 82]. The resulting self-assembled nanoviricide micelle

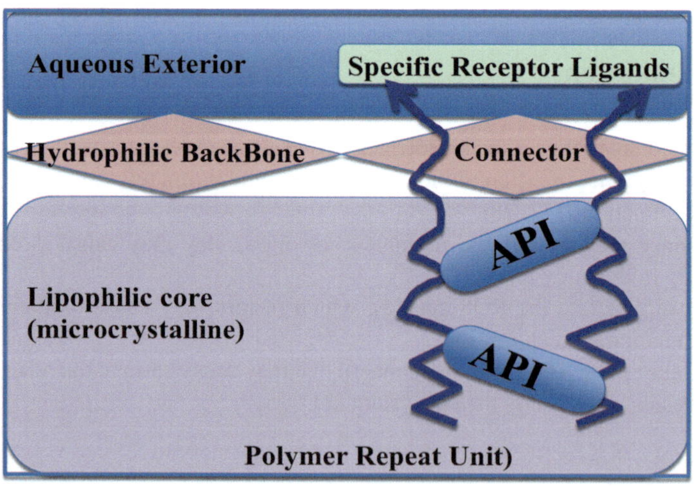

Fig. 3.1 TheraCour biopolymer

Table 3.7 TheraCour approaches are unique compared to other nanomedicine approaches

Vehicle	TheraCour	Dendrimer	PLA/PLGA	Virus-Based	Nano-shells, Mettalics
Nano-scale Velcro effect with wrap-on	Yes	No	No	No	No
Technology complexity	Simple	Complex	Medium	Complex	Complex
Safety	Safe	No	Medium	No	Medium
Specific targeting	Yes: Flexible Wrap-On	Yes: Limited by Hard Bal	No	No	May be
Detection	Yes	Yes	No	No	May be
Extended release	Yes	May be	Yes	Yes	Accumulate
Controlled release	Yes	May be	Yes	No	No
Efficacy improvements	Yes, Very Large	Yes	No (slow release only)	Yes but infectious	May be

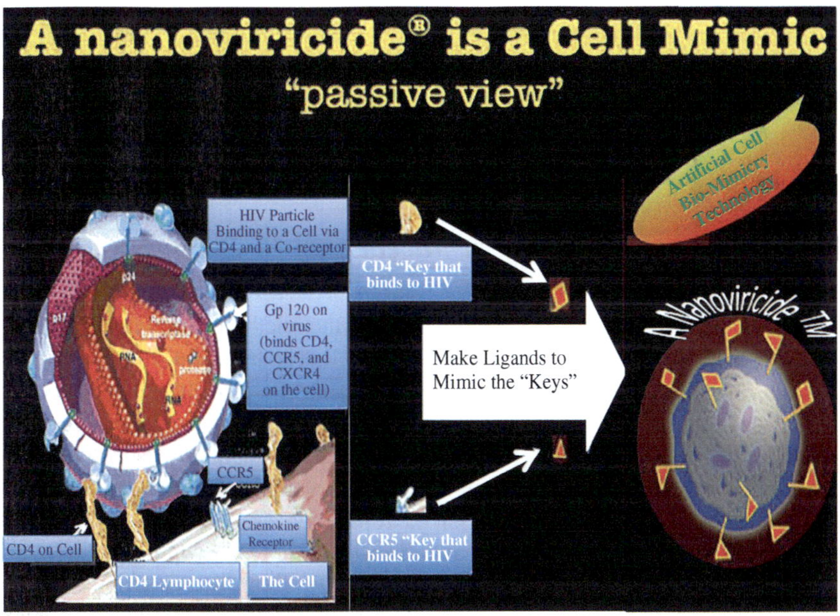

Fig. 3.2 Nanoviricide is a Cell Mimic. A nanoviricide "looks like" a human cell to the virus. A nanoviricide micelle wraps the virus particle and encapsulates it

thereupon fuse with the viral membrane by lipid–lipid mixing, and simultaneously dismantle virus capsids required for entering into cells (Fig. 3.2) [80, 81].

A *nanoviricide* unlike antibodies and vaccines does not allow any virus to escape. Therefore, this approach promises do not allow to cause drug-resistance evolution in viruses. A graphical representation of *nanoviricide* action-mechanism is shown in Fig. 3.3.

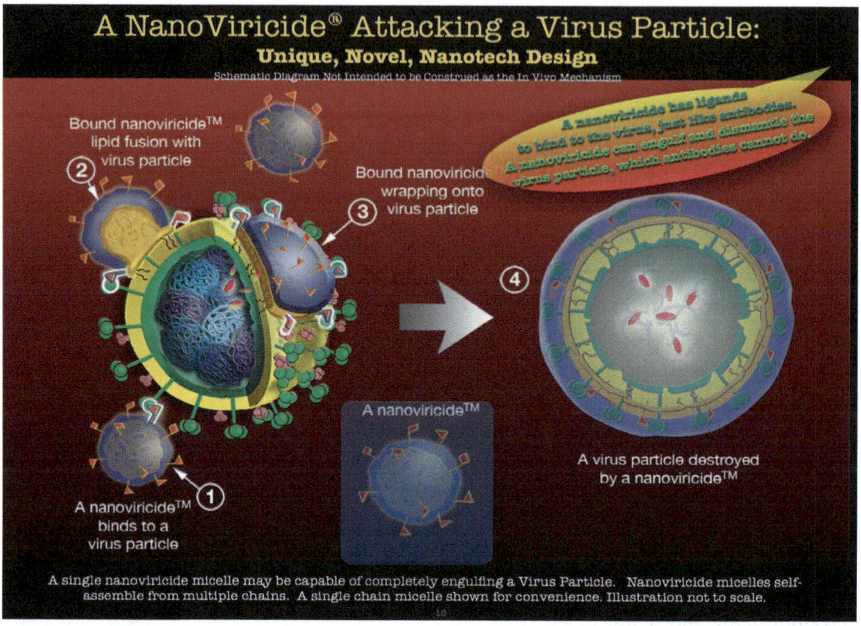

Fig. 3.3 Graphic representation of the action-mechanism of *Nanoviricide*. **1:** A *nanoviricide*™ binds to virus particle; **2:** Bound *nanoviricide*™ fused with the virus particles; **3:** Bound *nanoviricide*™ wrapping onto virus particles; **4:** Virus particle gets destroyed by *nanoviricide*™

3.7.2 Nanoviricide Polymeric Micelle Can Be Developed as a Drug Against SARS-CoV-2

SARS-CoV-2 belongs to beta family of human respiratory coronavirus and causes the severe lower-tract communicable lung disease termed as COVID-19 [83–85].

The accessible drugs, available at present time are remdesivir (Gilead), molnupiravir (Merck), and Paxlovid™ (Pfizer), which have significant limitations and limited efficiency. Molnupiravir has very low efficiency and also mutagenic. Paxlovid is virostatic, but the virus rebounds upon the removal of drugs. Remdesivir is highly effective in cell cultures, but its efficacy in vivo is very poor primarily due to the instability in the system. Thus, a novel drug development targeting SARS-CoV-2 is desperately needed. NV-387 is very effective in cell cultures against coronavirus hCoV-NL63 that binds to ACE2 cell receptor, as well as against coronavirus hCoV-229E that binds to a different cellular receptor, Aminopeptidase N (APN) [85–89].

3.7.3 *Encapsulation of the Virus by NV-CoV-2 Leads to Its Disintegration*

NV-CoV-2 is designed in such a manner that it can effectively bind to free virion particles at various sites through interactions between viral glycoprotein receptors and ligands, ultimately leading to the encapsulation of the virus particle. The key chemical process underlying the function of the nanoviricide involves the fusion of lipids in the alkyl chains of the nanoviricide micelle with the lipid envelope of the virus [70].

Support for this process is demonstrated in the electron photomicrographs presented in Fig. 3.4. In this study, murine cytomegalovirus (MCMV) was incubated with a nanoviricide displaying sialic acid as a ligand. MCMV contains multiple virus capsids enclosed within a single lipid membrane that houses the viral glycoproteins. In Fig. 3.4b, the bright area at the top right corner indicates the binding of the nanoviricide micelle to the lipid coat of the virus, leading to the evident loss of the viral envelope. Figure 3.4c further illustrates that only virion capsids remain as a result of this attack.

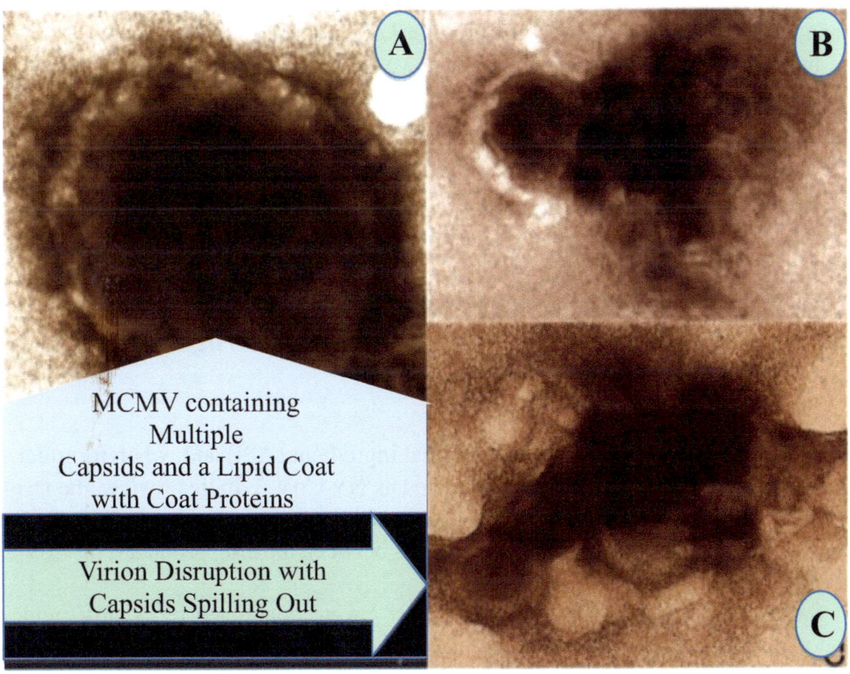

Fig. 3.4 Effects of two different nanoviricides on Murine Cytomegalovirus (MCMV). (**a**) Control Treated virion: MCMV containing multiple capsids and a lipid coat with coat proteins. (**b** & **c**) MCMV virions treated with two different nanoviricides. Virion disruption with capsids spilling out are visible

We have successfully demonstrated the efficacy of our drug candidate, NV-CoV-2-encapsulated remdesivir, against NL-63 infection in an animal model [71, 72].

3.7.4 Protection of Small Antiviral Drugs by Encapsulation with NV-CoV-2

Remdesivir (RDV), the sole drug approved by the FDA for the treatment of hospitalized adult patients, has shown promising results. In cell culture studies, the effective concentration for 50% inhibition (EC50) of Remdesivir on SARS-CoV-2-infected primary human airway epithelial (HAE) cells and vero cells was found to be **9.9 nM** and **750 nM**, respectively, 48 hours after treatment [82]. Furthermore, RDV exhibits a remarkable safety profile, boasting a therapeutic index greater than 10 in cell culture studies.

Its approval was based on clinical trials lasting nine days, resulting in an average reduction of SARS-CoV-2 patient hospitalization stays by approximately 7–8 days (Clinical Trials.gov number NCT04280705) [84]. However, in another clinical trial (Clinical Trials.gov Identifier: NCT04365725), no statistically significant difference was observed compared to the standard of care arm [82, 86, 87].

In pursuit of enhanced antiviral effects, we hypothesized that encapsulating RDV within NV-CoV-2 might offer greater protection against bodily metabolism. We explored this possibility in a rat model system and discovered that the antiviral efficacy of NV-CoV-2-encapsulated-RDV far exceeded that of either NV-CoV-2 or RDV alone, particularly in the context of lethal lung infection with NL-63 [84, 88, 89] (see Fig. 3.5).

3.7.5 Safety Studies of NV-387 Polymer, and Drug Product NV-CoV-2

NV-387 serves as the active pharmaceutical ingredient (API) and, when formulated into an antiviral drug product, is designated as NV-CoV-2. In this review, the terms NV-387 and NV-CoV-2 are used interchangeably. It is important to note that there are regulatory and compositional distinctions between the drug product (NV-CoV-2) and the API (NV-387).

Safety pharmacological studies of NV-387 were conducted by NanoViricides, following the core battery tests outlined in the ICH S7A guidelines. Additionally, assessments were made regarding its impact on the central nervous system and cardiovascular system using conscious rat and non-human primate models (specifically cynomolgus monkeys - Macaca fascicularis). Notably, no significant adverse effects

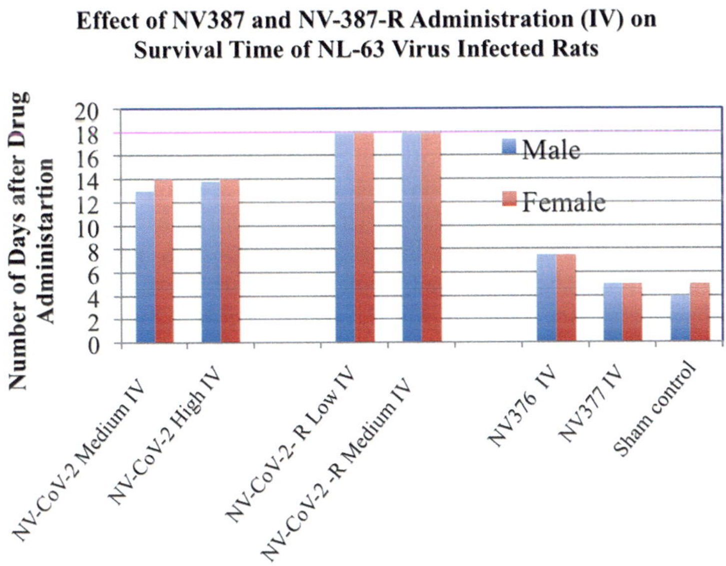

Fig. 3.5 NV-CoV-2-R rescues the rats from NL-63 Infection

were observed in respiratory function or neurobehavioral effects after 1-hour of drug treatment across all dosed groups of rats. The intravenous administration of NV-CoV-2 had no effect on the body temperature of rats.

When administered intravenously to conscious telemetered cynomolgus monkeys, NV-CoV-2 did not elicit any biologically significant effects on heart rate, arterial blood pressure, cardiac rhythm, or ECG parameters. All monkeys maintained sinus rhythm throughout the research study.

Moreover, it was determined that NV-387/NV-CoV-2 is non-immunogenic in a rat model and non-allergenic in several animal models when given via injections or orally. The NV-387 polymer was found to be non-mutagenic (as demonstrated by the Ames test) and non-genotoxic as well.

3.8 Future Prospects

NPs/nanocomposites may be the rescue of difficulties to manage any complicated disease. However, safety and efficacy issues of NPs need to be addressed before their use in humans. Such issues may include the understanding of their action mechanism, development of eco-friendly methods for their synthesis, and also the environmental and social aspects of their use.

3.9 Discussions and Conclusion

The mutation of microorganisms to a drug-resistant type offers a great challenge for the treatment of the infectious diseases. Therefore, discovery of some new concept of treatments deserves a great priority. NP-based bioimaging and early detection of disease become a great advantage in the treatment procedure. Nanoparticles show advantageous activities because of their size, shape, charge, and surface area. From the recent studies, it is obvious that the nanoparticles possess significant antimicrobial activities. Beside, NPs are equally applicable in various other fields of biomedicine such as bioimaging, targeted-drug delivery, and gene delivery. However, there are some limitations and health risks in using NPs in medicine as well as in foods.

Acknowledgements We thank our colleagues and secretary for their help during the preparation of this manuscript. Thanks to Ms. Bethany Pond and Dr. Smita Guha (St. John's University, NY) for their editorial assistance.

Conflict of Interest (COI) Authors, Anil Diwan and Jayant Tatake, are employed by Nanoviricides, Inc. The authors declare no conflicts of interest among the companies, AllExcel, Inc., TheraCour Pharma, Inc., and NanoViricides, Inc. Further, the authors declare that the research was conducted in the absence of any commercial or financial relationships that could be construed as a potential conflict of interest.

Source of Funding NanoViricides, Inc., stock symbol NNVC, has financed the development of nanoviricides platform and antiviral drugs based on it.

Certification We certify that all the figures and schematic designs are our own.

References

1. Sethi, S., & Murphy, T. F. (2001). Bacterial infection in chronic obstructive pulmonary disease in 2000: A state-of-the-art review. *Clinical Microbiology Reviews, 14*(2), 336–363.
2. Durand, M. L., Calderwood, S. B., Weber, D. J., Miller, S. I., Southwick, F. S., Caviness, V. S., Jr., & Swartz, M. N. (1993). Acute bacterial meningitis in adults–A review of 493 episodes. *The New England Journal of Medicine, 328*(1), 21–28.
3. Lara, H. H., Ayala-Nuñez, N. V., Ixtepan-Turrent, L., & Rodriguez-Padilla, C. (2010). Mode of antiviral action of silver nanoparticles against HIV-1. *Journal of Nanbiotechnology, 8*(1), 1–10.
4. Singh, S. R., Krishnamurthy, N. B., & Mathew, B. B. (2014). A review on recent diseases caused by microbes. *Journal of Applied & Environmental Microbiology, 2*(4), 106–115.
5. Laxminarayan, R., Duse, A., Wattal, C., Zaidi, A. K., Wertheim, H. F., Sumpradit, N., & Greko, C. (2013). Antibiotic resistance—The need for global solutions. *The Lancet Infectious Diseases, 13*(12), 1057–1098.
6. Friedman, H., Newton, C., & Klein, T. W. (2003). Microbial infections, immunomodulation, and drugs of abuse. *Clinical Microbiology Reviews, 16*(2), 209–219.
7. Atolani, O., Baker, M. T., Adeyemi, O. S., Olanrewaju, I. R., Hamid, A. A., Ameen, O. M., & Usman, L. A. (2020). Covid-19: Critical discussion on the applications and implications of chemicals in sanitizers and disinfectants. *EXCLI Journal, 19*, 785–799.

8. Gurunathan, S., Han, J. W., Kwon, D. N., & Kim, J. H. (2014). Enhanced antibacterial and antibiofilm activities of silver nanoparticles against Gram-negative and Gram-positive bacteria. *Nanoscale Research Letters, 9*(1), 1–17.
9. Yien, L., Zin, N. M., Sarwar, A., & Katas, H. (2012). Antifungal activity of chitosan nanoparticles and correlation with their physical properties. *International Journal of Biomaterials, 2012*, 632698. https://doi.org/10.1155/2012/632698
10. Sobhani, Z., Samani, S. M., Montaseri, H., & Khezri, E. (2017). Nanoparticles of chitosan loaded ciprofloxacin: Fabrication and antimicrobial activity. *Advanced Pharmaceutical Bulletin, 7*(3), 427.
11. Rai, A., Prabhune, A., & Perry, C. C. (2010). Antibiotic mediated synthesis of gold nanoparticles with potent antimicrobial activity and their application in antimicrobial coatings. *Journal of Materials Chemistry, 20*(32), 6789–6798.
12. Alshammari, F., Alshammari, B., Moin, A., Alamri, A., Al Hagbani, T., Alobaida, A., Baker, A., Khan, S., & Rizvi, S. M. D. (2021). Ceftriaxone mediated synthesized gold nanoparticles: A nano-therapeutic tool to target bacterial resistance. *Pharmaceutics, 13*(11), 1896. https://doi.org/10.3390/pharmaceutics13111896
13. Cepas, V., López, Y., Gabasa, Y., Martins, C. B., Ferreira, J. D., Correia, M. J., Santos, L. M. A., Oliveira, F., Ramos, V., Reis, M., Castelo-Branco, R., Morais, J., Vasconcelos, V., Probert, I., Guilloud, E., Mehiri, M., & Soto, S. M. (2019). Inhibition of bacterial and fungal biofilm formation by 675 extracts from microalgae and cyanobacteria. *Antibiotics (Basel), 8*(2), 77. https://doi.org/10.3390/antibiotics8020077
14. Cremonini, E., Zonaro, E., Donini, M., Lampis, S., Boaretti, M., Dusi, S., & Vallini, G. (2016). Biogenic selenium nanoparticles: Characterization, antimicrobial activity and effects on human dendritic cells and fibroblasts. *Microbial Biotechnology, 9*(6), 758–771.
15. Haghighi, F., Roudba, M. S., Mohammadi, P., Hossinkhani, S., & Shidpour, R. (2013). Antifungal acitivity of TiO_2 nanoparticles and EDTA on Candida abicans biofilms. *Infection, Epidemiology and Microbiology, 1*, 133–138.
16. Khan, I., Saeed, K., & Khan, I. (2017). Nanoparticles: Properties, applications and toxicities. *Arabian Journal of Chemistry, 12*(7), 908–931.
17. Song, X., Bayati, P., Gupta, M., Elahinia, M., & Haghshenas, M. (2021). Fracture of magnesium matrix nanocomposites - A review. *International Journal of Lightweight Materials and Manufacture, 4*(1), 67–98.
18. Vert, M., Doi, Y., Hellwich, K. H., Hess, M., Hodge, P., Kubisa, P., & Schue, F. (2012). Terminology for biorelated polymers and applications (IUPAC recommendations 2012). *Pure and Applied Chemistry, 84*(2), 377–410.
19. Von Nussbaum, F., Brands, M., Hinzen, B., Weigand, S., & Habich, D. (2006). Antibacterial natural products in medicinal chemistry—Exodus or revival? *Angewandte Chemie, International Edition, 45*, 5072–5129.
20. Akhtar, M., Swamy, M. K., Umar, A., Sahli, A., & Abdullah, A. (2015). Biosynthesis and characterization of silver nanoparticles from methanol leaf extract of Cassia didymobotyra and assessment of their antioxidant and antibacterial activities. *Journal of Nanoscience and Nanotechnology, 15*, 9818–9823.
21. Fröhlich, E., & Salar-Behzadi, S. (2014). Toxicological assessment of inhaled nanoparticles: Role of in vivo, ex vivo, in vitro, and in silico studies. *International Journal of Molecular Sciences, 15*, 4795–4822.
22. Qiu, Y., Xu, D., Sui, G., Wang, D., Wu, M., Han, L., Mu, H., & Duan, J. (2020). Gentamicin decorated phosphatidylcholine-chitosan nanoparticles against biofilms and intracellular bacteria. *International Journal of Biological Macromolecules, 156*, 640–647.
23. Wu, P., Xie, R., Imlay, K., & Shang, J. K. (2010). Visible-light-induced bactericidal activity of titanium dioxide codoped with nitrogen and silver. *Environmental Science & Technology, 44*, 6992–6997.
24. Evangelista, T. F., Andrade, G. R., Nascimento, K. N., Dos Santos, S. B., Santos, M. D. F. C., D'Oca, C. D. R. M., Estevam, C. D. S., Gimenez, I. F., & Almeida, L. E. (2020). Supramolecular

polyelectrolyte complexes based on cyclodextrin-grafted chitosan and carrageenan for controlled drug release. *Carbohydrate Polymers, 245*, 116592.
25. Walvekar, P., Gannimani, R., Salih, M., Makhathini, S., Mocktar, C., & Govender, T. (2019). Self-assembled oleylamine grafted hyaluronic acid polymersomes for delivery of vancomycin against methicillin resistant Staphylococcus aureus (MRSA). *Colloids and Surfaces. B, Biointerfaces, 182*, 110388.
26. Ejaz, S., Ihsan, A., Noor, T., Shabbir, S., & Imran, M. (2020). Mannose functionalized chitosan nanosystems for enhanced antimicrobial activity against multidrug resistant pathogens. *Polymer Testing, 91*, 106814.
27. Ucak, S., Sudagidan, M., Borsa, B. A., Mansuroglu, B., & Ozalp, V. C. (2020). Inhibitory effects of aptamer targeted teicoplanin encapsulated PLGA nanoparticles for Staphylococcus aureus strains. *World Journal of Microbiology and Biotechnology, 36*, 69.
28. Vrouvaki, I., Koutra, E., Kornaros, M., Avgoustakis, K., Lamari, F. N., & Hatziantoniou, S. (2020). Polymeric nanoparticles of Pistacialentiscus var. chia essential oil for cutaneous applications. *Pharmaceutics, 12*, 353.
29. Gherasim, O., Grumezescu, A. M., Grumezescu, V., Iordache, F., Vasile, B. S., & Holban, A. M. (2020). Bioactive surfaces of polylactide and silver nanoparticles for the prevention of microbial contamination. *Materials, 13*, 768.
30. Grumezescu, A. M., Stoica, A. E., Dima-Balcescu, M.-S., Chircov, C., Gharbia, S., Balta, C., Ros, M., Herman, H., Holban, A. M., Ficai, A., et al. (2019). Electrospun polyethylene terephthalate nanofibers loaded with silver nanoparticles: Novel approach in anti-infective therapy. *Journal of Clinical Medicine, 8*, 1039.
31. Khandelwal, N., Kaur, G., Kumara, N., & Tiwari, A. (2014). Application of silver nanoparticles in viral inhibition: A new hope for antivirals. *Digest Journal of Nanomaterials and Biostructures, 9*, 175–186.
32. Alamdaran, M., Movahedi, B., Mohabatkar, H., & Behbahani, M. (2018). In-vitro study of the novel nanocarrier of chitosan-based nanoparticles conjugated HIV-1 P24 protein-derived peptides. *Journal of Molecular Liquids, 265*, 243–250.
33. Belgamwar, A. V., Khan, S. A., & Yeole, P. G. (2019). Intranasal dolutegravir sodium loaded nanoparticles of hydroxypropyl-beta-cyclodextrin for brain delivery in neuro-AIDS. *Journal of Drug Delivery Science and Technology, 52*, 1008–1020.
34. Costa, A. F., Araujo, D. E., Cabral, M. S., Brito, I. T., Leite, L. B. D. M., Pereira, M., & Amaral, A. C. (2018). Development, characterization, and in vitro–in vivoevaluation of polymeric nanoparticles containing miconazole and farnesol for treatment of vulvovaginal candidiasis. *Medical Mycology, 57*, 52–62.
35. Reddy, Y. C. (2018). Formulation and evaluation of chitosan nanoparticles for improved efficacy of itraconazole antifungal drug. *Asian Journal of Pharmaceutical and Clinical Research, 11*, 147–152.
36. Sombra, F. M., Richter, A. R., De Araújo, A. R., Ribeiro, F. D. O. S., Mendes, J. D. F. S., Fontenelle, R. O. D. S., Da Silva, D. A., Paula, H. C., Feitosa, J. P., Goycoolea, F. M., et al. (2020). Development of amphotericin B-loaded propionate Sterculiastriata polysaccharide nanocarrier. *International Journal of Biological Macromolecules, 146*, 1133–1141.
37. Real, D., Hoffmann, S., Leonardi, D., Salomon, C., & Goycoolea, F. M. (2018). Chitosan-based nanodelivery systems applied to the development of novel triclabendazole formulations. *PLoS One, 13*, e0207625.
38. Durak, S., Arasoglu, T., Ates, S. C., & Derman, S. (2020). Enhanced antibacterial and antiparasitic activity of multifunctional polymeric nanoparticles. *Nanotechnology, 31*, 175705.
39. Williams, D. F. (2008). On the mechanisms of biocompatibility. *Biomaterials, 2*, 2941–2953.
40. Li, X., Wang, L., Fan, Y., Feng, Q., & Cui, F. Z. (2012). Biocompatibility and toxicity of nanoparticles and nanotubes. *Journal of Nanomaterials, 2012*, 1–22. https://doi.org/10.1155/2012/548389
41. Flower, N. A., Brabu, B., Revathy, M., Gopalakrishnan, C., Raja, S. V., Murugan, S. S., & Kumaravel, T. S. (2012). Characterization of synthesized silver nanoparticles and assess-

ment of its genotoxicity potentials using the alkaline comet assay. *Mutation Research/Genetic Toxicology and Environmental Mutagenesis, 742*, 61–65.
42. Naahidi, S., Jafari, M., Edalat, F., Raymond, K., Khademhosseini, A., & Chen, P. (2013). Biocompatibility of engineered nanoparticles for drug delivery. *Journal of Controlled Release, 166*, 182 191.
43. Sosnik, A., Carcaboso, A. M., Glisoni, R. J., Moretton, M. A., & Chiappetta, D. A. (2010). New old challenges in tuberculosis: Potentially effective nanotechnologies in drug delivery. *Advanced Drug Delivery Reviews, 62*, 547–559.
44. Sarfati, G., Dvir, T., Elkabets, M., Apte, R. N., & Cohen, S. (2011). Targeting of polymeric nanoparticles to lung metastases by surface-attachment of YIGSR peptide from laminin. *Biomaterials, 32*, 152–161.
45. Reddy, N. N., Varaprasad, K., & Ravindra, S. (2011). Evaluation of blood compatibility and drug release studies of gelatin based magnetic hydrogel nanocomposites. *Colloids and Surfaces A: Physicochemical and Engineering Aspects., 385*, 20–27.
46. Li, P. W., Kuo, T. H., Chang, J. H., Yeh, J. M., & Chan, W. H. (2010). Induction of cytotoxicity and apoptosis in mouse blastocysts by silver nanoparticles. *Toxicology Letters, 197*, 82–87.
47. Hadrup, N., & Lam, H. R. (2014). Oral toxicity of silver ions, silver nanoparticles and colloidal silver—A review. *Regulatory Toxicology and Pharmacology, 68*, 1–7.
48. Hadrup, N., Loeschner, K., & Mortensen, A. (2012). The similar neurotoxic effects of nanoparticulate and ionic silver in vivo and in vitro. *Neurotoxicology, 33*, 416–423.
49. Rim, K. T., Song, S. W., & Kim, H. Y. (2013). Oxidative DNA damage from nanoparticle exposure and its application to workers' health: A literature review. *Safety and Health at Work, 4*, 177–186.
50. De Jong, W. H., Van Der Ven, L. T., Sleijffers, A., Park, M. V., Jansen, E. H., Van Loveren, H., & Vandebriel, R. J. (2013). Systemic and immunotoxicity of silver nanoparticles in an intravenous 28 days repeated dose toxicity study in rats. *Biomaterials, 34*, 8333–8343.
51. Espinosa-Cristobal, L. F., Martinez-Castanon, G. A., Loyola-Rodriguez, J. P., Patino-Marin, N., Reyes-Macias, J. F., Vargas-Morales, J. M., & Ruiz, F. (2013). Toxicity, distribution, and accumulation of silver nanoparticles in Wistar rats. *Journal of Nanoparticle Research, 15*, 1702–1704.
52. Travan, A., Pelillo, C., Donati, I., Marsich, E., Benincasa, M., Scarpa, T., Semeraro, S., Turco, G., Gennaro, R., & Paoletti, S. (2009). Non-cytotoxic silver nanoparticle-polysaccharide nanocomposites with antimicrobial activity. *Biomacromolecules, 10*, 1429–1435.
53. Barnett, C. M., Gueorguieva, M., Lees, M. R., McGarvey, D. J., & Hoskins, C. (2013). Physical stability, biocompatibility and potential use of hybrid iron oxide-gold nanoparticles as drug carriers. *Journal of Nanoparticle Research, 15*, 1706–1720.
54. Gamucci, O., Bertero, A., Gagliardi, M., & Bardi, G. (2014). Biomedical nanoparticles: Overview of their surface immune-compatibility. *Coatings, 4*, 139–159.
55. Fabian, E., Landsiedel, R., Ma-Hock, L., Wiench, K., Wohlleben, W., & van Ravenzwaay, B. (2008). Tissue distribution and toxicity of intravenously administered titanium dioxide nanoparticles in rats. *Archives of Toxicology, 82*, 151–157.
56. Kanagesan, S., Hashim, M., Aziz, S. A. B., Ismail, I., Tamilselvan, S., Alitheen, N. B., Swamy, M. K., & Rao, B. P. C. (2016). Synthesis, characterization and in vitro evaluation of manganese ferrite (MnFe2O4) nanoparticles for their biocompatibility with murine breast cancer cells (4T1). *Molecules, 21*, 312–328. https://doi.org/10.3390/molecules21030312
57. Tomitaka, A., Jeun, M., Bae, S., & Takemura, Y. (2011). Evaluation of magnetic and thermal properties of ferrite nanoparticles for biomedical applications. *Journal of Magnetics, 16*, 164–168.
58. Iavicoli, I., Leso, V., Fontana, L., & Bergamaschi, A. (2011). Toxicological effects of titanium dioxide nanoparticles: A review of in vitro mammalian studies. *European Review for Medical and Pharmacological Sciences, 15*, 481–508.

59. Khanna, L., & Verma, N. K. (2013). Synthesis, characterization and in vitro cytotoxicity study of calcium ferrite nanoparticles. *Materials Science in Semiconductor Processing, 16*, 1842–1848.
60. Labhasetwar, V., Song, C., & Levy, R. J. (1997). Nanoparticle drug delivery system for restenosis. *Advanced Drug Delivery Reviews, 24*, 63–85.
61. Mahapatro, A., & Singh, D. K. (2011). Biodegradable nanoparticles are excellent vehicle for site directed in vivo delivery of drugs and vaccines. *Journal of Nanbiotechnology, 9*, 55–65. https://doi.org/10.1186/1477-3155-9-55
62. Kumari, A., Yadav, S. K., & Yadav, S. C. (2010). Biodegradable polymeric nanoparticles based drug delivery systems. *Colloids and Surfaces. B, Biointerfaces, 75*, 1–8. https://doi.org/10.1016/j.colsurfb.2009.09.001
63. Carrasquillo, K. G., Stanley, A. M., Aponte-Carro, J. C., de Jésus, P., Costantino, H. R., Bosques, C. J., & Griebenow, K. (2001). Non-aqueous encapsulation of excipient-stabilized spray-freeze dried BSA into poly(lactide-co-glycolide) microspheres results in release of native protein. *Journal of Controlled Release, 76*, 199–208.
64. Kim, D., El-Shall, H., Dennis, D., & Morey, T. (2005). Interaction of PLGA nanoparticles with human blood constituents. *Colloids and Surfaces. B, Biointerfaces, 40*, 83–91.
65. Leroux, J. C., Allémann, E., de Jaeghere, F., Doelker, E., & Gurny, R. (1996). Biodegradable nanoparticles from sustained release formulations to improved site specific drug delivery. *Journal of Controlled Release, 39*, 339–350.
66. Nevozhay, D., Kañska, U., Budzyñska, R., & Boratyñski, J. (2007). Current status of research on conjugates and related drug delivery systems in the treatment of cancer and other diseases (polish). *Advances in Hygiene and Experimental Medicine, 61*, 350–360.
67. Wilczewska, A. Z., Katarzyna, N. K., Markiewicz, K. H., & Car, H. (2012). Nanoparticles as drug delivery systems. *Pharmacological Reports, 64*, 1020–1037.
68. Suri, S. S., Fenniri, H., & Singh, B. (2007). Nanotechnology-based drug delivery systems. *Journal of Occupational Medicine and Toxicology*, 2.
69. Buzea, C., Pacheco, I. I., & Robbie, K. (2007). Nanomaterials and nanoparticles: Sources and toxicity. *Biointerphases, 2*, 17–71.
70. NanoViricides, Inc. *Pan-coronavirus COVID-19 drug candidates are highly effective in pre-clinical animal studies in support of FDA pre IND application*. March 9, 2021, 6:45 AM EST. https://www.bloomberg.com/press-releases/2021-03-09/nanoviricides-inc-pan-coronavirus-covid-19-drug-candidates-are-highly-effective-in-pre-clinical-animal-studies-i
71. Barton, R. W., Tatake, J. G., & Diwan, A. R. (2011). Nanoviricides- a novel approach to antiviral therapeutics. In D. E. Reisner (Ed.), *Bionanotechnology II* (pp. 141–154). CRC Press. Taylor and Francis Group, (www.crcpress.com).
72. Barton, R. W., Tatake, J. G., & Diwan, A. R. *Nanoviricides: Targeted anti-viral nanomaterials handbook of clinical nanomedicine, nanoparticles, imaging, therapy, and clinical applications*. R. Bawa, G. F. Audette (Eds.), Israel Rubinstein. ISBN 9789814669207. Published February 24, 2016 by Jenny Stanford Publishing, pp. 1039–1046.
73. NanoViricides is developing drugs against SARS-CoV-2 with an integrated approach to combat COVID-19, as reported at The LD 500 Virtual Conference. Friday, September 4, 2020 6:50 AM. https://www.accesswire.com/604794/NanoViricides-is-Developing-Drugs-Against-SARS-CoV-2-with-an-Integrated-Approach-to-Combat-COVID-19-as-Reported-at-The-LD-500-Virtual-Conference
74. DGAP-News:NanoViricides, Inc.: *Broad-spectrum direct antiviral nanomedicine should remain effective against COVID-19 virus*. January 27, 2021, 6:15 AM EST. https://www.bloomberg.com/press-releases/2021-01-27/dgap-news-nanoviricides-inc-broad-spectrum-direct-antiviral-nanomedicine-should-remain-effective-against-covid-19-virus
75. Feng, Y., Ling, Y., Bai, T., Xie, Y., Huang, J., Li, J., et al. (2020). COVID-19 with different severities: A multicenter study of clinical features. *American Journal of Respiratory and Critical Care Medicine, 201*(11), 1380–1388. https://doi.org/10.1164/rccm.202002-0445O

76. Spencer Kimbal. *Health and Science. More contagious omicron BA.2 subvariant now dominant in the U.S., CDC says*. Published Tue, Mar 29 20221:38 PM Edtupdated Tue, Mar 29 20226:24 PM EDT. https://www.cnbc.com/2022/03/29/more-contagious-omicron-bapoint2-covid-subvariant-dominant-in-the-us-cdc-says.html
77. Belouzard, S., Millet, J. K., Licitra, B. N., & Whittaker, G. R. (2012). Mechanisms of corona virus cell entry mediated by the viral spike protein. *Viruses, 4*(6), 1011–1033. https://doi.org/10.3390/v4061011
78. Li, W., Sui, J., Huang, I. C., Kuhn, J. H., Radoshitzky, M., et al. (2017). The S proteins of human coronavirus NL63 and severe acute respiratory syndrome coronavirus bind overlapping regions of ACE2. *Virology, 367*(2), 367–374. https://doi.org/10.1016/j.virol.2007.04.035
79. Chakraborty, A., & Diwan, A. (2020). NL-63: A better surrogate virus for studying SARS-CoV-2. *Integrative Molecular Medicine, 2020*(7), 1–9. https://doi.org/10.15761/IMM.1000408
80. Hofmann, H., Pyrc, K., van der Hoek, L., Geier, M., Berkhout, B., & Pöhlmann, S. (2005). Human corona virus NL63 employs the severe acute respiratory syndrome coronavirus receptor for cellular entry. *Proceedings of the National Academy of Sciences, 102*, 7988–7993.
81. Liu, D. X., Liang, J. Q., & Fung, T. S. (2021). Human Coronavirus-229E, -OC43, NL63, and HKU1 (Coronaviridae). *Encyclopedia of Virology*, 428–440. https://doi.org/10.1016/B978-0-12-809633-8.21501-X
82. https://hospitals.vchca.org/images/medical_staff/Remdesivir_Fact_Sheet_Health_ Care_Providers_6-15-20.pdf
83. Beigel, J. M., Tomashek, K. M., Dodd, L. E., Mehta, A. E., Zingman, B. S., Kalil, A. C., et al. (2020). Remdesivir for the treatment of Covid-19 — Final report. *The New England Journal of Medicine, 383*, 1813–1826. https://doi.org/10.1056/NEJMoa2007764
84. Davis, M. E., Chen, Z. G., & Shin, D. M. (2008). Nanoparticle therapeutics: An emerging treatment modality for cancer. *Nature Reviews. Drug Discovery, 7*(9), 771–782.
85. ClinicalTrials.gov. *Multicenter, retrospective study of the effects of remdesivir in the treatment of severe Covid-19 infections*. Last Update Posted: April 19, 2021. https://clinicaltrials.gov/ct2/show/NCT04365725
86. Zhang, L., Gu, F. X., Chan, J. M., Wang, A. Z., Langer, R. S., & Farokhzad, O. C. (2008). Nanoparticles in medicine: Therapeutic applications and developments. *Clinical Pharmacology and Therapeutics, 83*(5), 761–769.
87. Ferrari, M. (2005). Cancer nanotechnology: Opportunities and challenges. *Nature Reviews. Cancer, 5*(3), 161–171.
88. Lammers, T., Kiessling, F., Hennink, W. E., & Storm, G. (2012). Drug targeting to tumors: Principles, pitfalls and (pre-) clinical progress. *Journal of Controlled Release, 161*(2), 175–187.
89. Chakraborty, A., Diwan, A., Arora, V., Thakur, Y., Chiniga, V., Tatake, J., et al. (2022). Encapsulation of remdesivir in nanoviricide's platform technology based NV-CoV-2 polymer protects the drug and improves its pharmacokinetics. *EC Pharmacology and Toxicology, 10*(2), 108–118.

Chapter 4
Vaccine Nanotechnology for the Prevention of Infectious Diseases

Sumit Sontakke, Poonam Sharma, Rebecca Worsham, Kishan Kadur, Chandana Bhargavi, and Olugbenga Morebise

4.1 Introduction

Infectious diseases continue to be a major global health concern, causing significant morbidity and mortality worldwide [1]. The development of effective preventive and therapeutic measures, such as vaccines and antibiotics, has played a pivotal role in reducing the burden of infectious diseases. A vaccine is a biological preparation that induces body's immune response against a particular infectious agent or malignant disease [2, 3]. Typically, a vaccine contains a component that resembles a pathogenic microorganism. This component is often derived from killed or weakened forms of the microorganism, its toxins, or one of its surface proteins [4]. Upon administration, the vaccine stimulates the body's immune system to recognize the component as a threat, initiate an immune response, eliminate the component, and destroy any microbes associated with that component that may be encountered in the future [5]. Vaccines can serve as prophylactic measures, preventing or mitigating the impact of future infections caused by natural or wild-type pathogens [4, 6]. Additionally, they can also function therapeutically, combating diseases that have already manifested, such as cancer [7]. The significance of vaccines in combating infectious diseases cannot be overstated. Vaccines play a crucial role in preventing, controlling, and even eradicating infectious diseases that pose a significant threat to

S. Sontakke (✉) · K. Kadur · C. Bhargavi
American University of Antigua College of Medicine, Osbourn, Antigua and Barbuda

P. Sharma
University of Mississippi Medical Center, Jackson, MS, USA

R. Worsham
Trinity School of Medicine, Kingstown, Saint Vincent

O. Morebise
All Saints University School of Medicine, Roseau, Dominica

R. S. Chaughule, A. S. Lokur (eds.), *Applications of Nanotechnology in Microbiology*, https://doi.org/10.1007/978-3-031-49933-3_4

global health. Some key reasons why vaccines are of utmost importance in this ongoing battle include protection of vulnerable populations, control and eradication of the diseases, establishing herd immunity, preventing antibiotic resistance, as well as offering a cost-effective approach to disease prevention.

Traditional vaccine strategies encompass various approaches, including live attenuated vaccines, inactivated whole organism vaccines, subunit vaccines, and toxoid vaccines [8] (Fig. 4.1). Live-attenuated vaccines elicit descent immune memory and the production of neutralizing antibodies. Their effectiveness stems from the presence of pathogen-associated molecular patterns (PAMPs) on the surface and interior of the live pathogens, which engage pattern-recognition receptors (PRRs) of the innate immune system like nucleotide-binding oligomerization domain (NOD)-like receptors and toll-like receptors [9]. Generally, live vaccines do not need additional adjuvants or periodic booster shots for reactivating the immune memory response due to their live nature and diverse PAMPs [10]. However, caution is needed with live vaccines as they can pose risks to individuals with compromised immune systems, necessitating alternative vaccine options [11]. Inactivated or killed vaccines, achieved by subjecting pathogens to heat or chemical treatments, provoke some memory response. Adjuvants, such as aluminum hydroxide, are typically employed to mimic artificial PAMPs and engage the adaptive immune system. Although inactivated vaccines often generate a strong immune memory response,

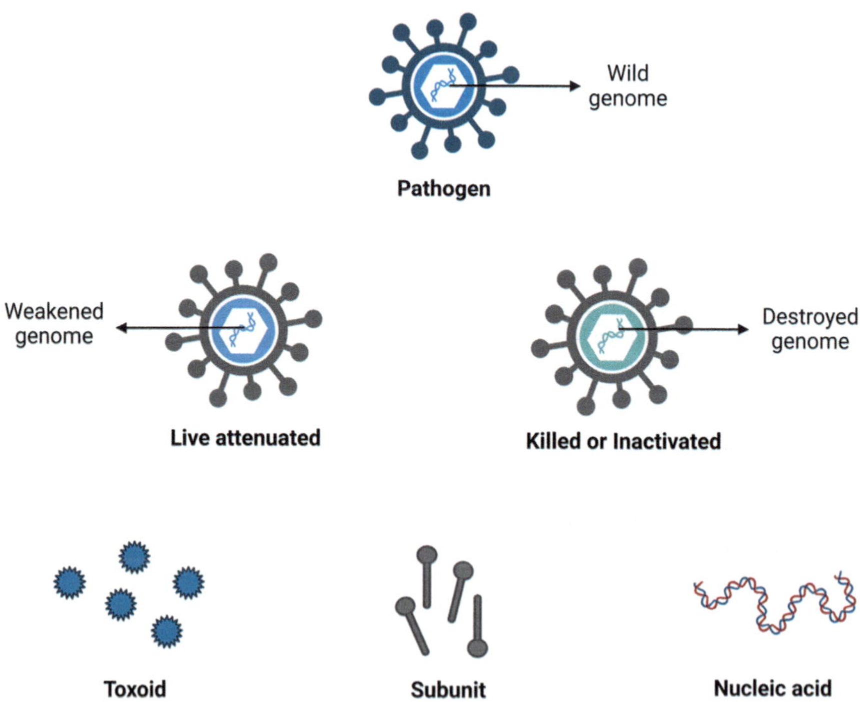

Fig. 4.1 Types of vaccine approaches

they may not always confer lifelong protection. Consequently, many inactivated vaccines necessitate booster doses to reactivate the immune system and restore protective responses [10]. Subunit and toxoid vaccines function similar to inactivated vaccines but are composed of specific components of the microorganisms, like proteins or toxins [12]. These vaccines offer improved safety profiles compared to whole-pathogen vaccines because of their non-replicative and non-infectious nature. However, since these pathogen-derived antigens represent only a component of the microorganism in many formulations, they often exhibit poor immunogenicity. As a result, adjuvants and booster shots are commonly required to enhance their effectiveness.

In recent years, there has been a focus on developing innovative vaccines that offer enhanced safety and effectiveness. As a result, various new vaccine platforms have emerged, with RNA vaccines now joining the list of approved human vaccines following remarkable progress in the development of coronavirus disease 2019 (COVID-19) vaccines. The COVID-19 vaccines developed by Pfizer-BioNTech and Moderna utilize synthetic lipid components to protect and encapsulate messenger RNA (mRNA). Once administered, the mRNA enters the nearby cells, prompting them to produce a viral protein that activates the immune system. The immune response generated by mRNA vaccines persists for more than 6 months [13, 14], albeit with a gradual decrease from the peak response after the initial 6 months [14, 15]. However, alongside this diminishing immune response, viral variants have posed challenges to the efficacy of these vaccines, necessitating the administration of booster doses [16].

Different approaches for vaccine development may be based on the whole-pathogen, killed-inactivated or live-attenuated vaccines, or on utilizing subsets of the pathogen, like protein subunits or toxins produced by the pathogen. Additionally, nucleic acid vaccines that utilize DNA or RNA are another strategy for vaccine development. The figure was created with the assistance of Biorender.com.

4.2 Nanotechnology in Vaccine Design

Developing vaccines that provide long-lasting and effective immune responses without the use of live pathogens has presented a significant challenge. However, nanotechnology offers numerous advantages that can be utilized to enhance the effectiveness of vaccine formulations. An important advantage of nanoparticle-based formulations is their capacity to deliver both antigenic materials and immunostimulatory adjuvants concurrently. This co-delivery is crucial for optimal immune stimulation, as the spatial co-localization of these two components ensures a prompt immune response specifically targeting the desired antigen [17]. Achieving co-delivery can be accomplished by encapsulating the adjuvant within the nanoparticle core, attaching it to the nanoparticle surface, or employing the nanoparticle material itself as the immunostimulant. Additionally, nanoparticles facilitate the controlled release of antigens and adjuvants [18].

Nanoparticles have been used in different areas of antimicrobial therapies. The recent COVID-19 pandemic has brought attention to the role of nanoparticles against severe acute respiratory syndrome coronavirus 2 (SARS-CoV-2) [19, 20]. Further, heparin and various marine sulfated glycans have demonstrated their ability to inhibit various viral infections [21–28] and to enhance the antiviral efficacy of these inhibitory substances, they can be conjugated with different nanoparticles. Nanoparticles provide multiple benefits not only in therapeutics, but also can assist in creating better vaccine formulations. These advantages include preserving the bioactivity of encapsulated loads, co-localizing the antigens and adjuvants for coordinated delivery to immune cells, as well as targeting specific cell groups by incorporating functional surface ligands. Also, their tiny dimensions allow efficient transport through the lymphatic system, which supports important processes like antigen presentation and enhances immune response [29].

Vaccine antigens exhibit a wide range of sizes, spanning from subunit antigens measuring less than 10 nm, to biological or synthetic nanoparticle systems with sizes ranging from 20 to 200 nm, and even whole-cell vaccines that can reach sizes of up to 20 μm (Fig. 4.2). Nanoparticles within the size range of 20–200 nm possess the ability to freely migrate to lymph nodes and are presented to dendritic cells (DCs) that reside there. This characteristic makes them well-suited for delivering antigens and immunostimulants [29, 30].

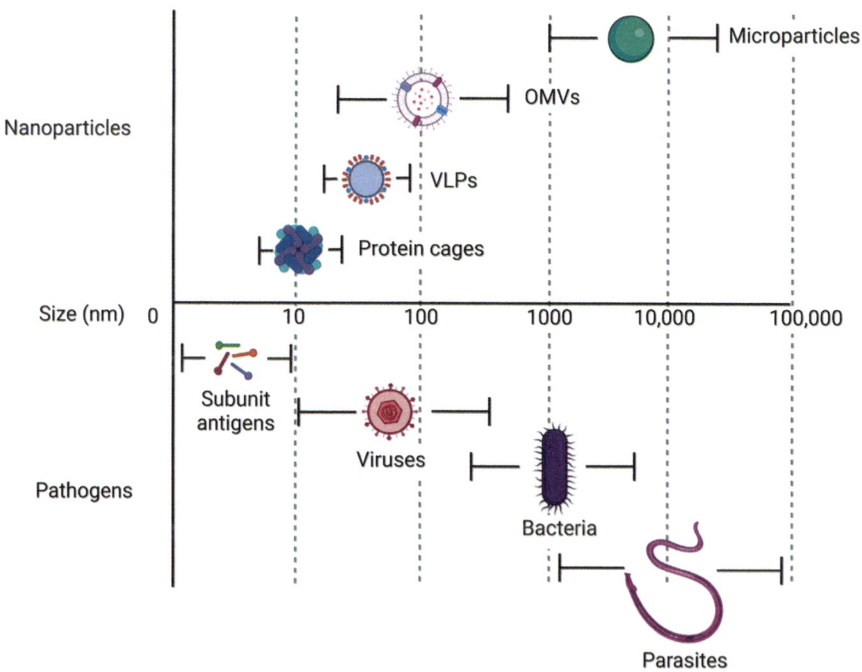

Fig. 4.2 Size ranges of vaccine nanoparticles and pathogens

The size ranges of vaccine delivery systems and the microorganisms are compared in this figure, with dimensions given in nanometers. The size ranges reflect efficient entry into lymphatic vessels and uptake by antigen presenting cells, highlighting the importance of size in effective antigen delivery, VLPs – virus-like particles, and OMVs – outer membrane vesicles. The figure was created with the assistance of Biorender.com.

The antigen-presenting cells (APCs), including DCs, macrophages, and B cells, capture the extracellular proteins and break them down into peptides for presentation to CD4+ T helper cells. This interaction stimulates the generation of humoral immune responses. The interaction between CD4+ T cells and B cells is critical for the production of long-lasting plasma cells that produce antigen-specific antibodies. These antibodies can neutralize the infectious agent thereby preventing infection or target it for elimination by APCs through phagocytosis. Additionally, DCs can trigger cytotoxic CD8+ T cell responses, leading to the direct elimination of infected cells [31] (Fig. 4.3).

The interaction between the immune system and nanoparticle vaccines results in the establishment of immunological memory and provides protection against subsequent infections. Nanovaccine platforms have the capacity to stimulate the host's immune response, leading to the generation of both cellular responses involving CD8+ T cells and humoral responses involving antibodies specific to the antigens included in the vaccine formulation. This immune activation induced by nanoparticle vaccines plays a crucial role in conferring immunity and preventing future infections. The figure was created with the assistance of Biorender.com.

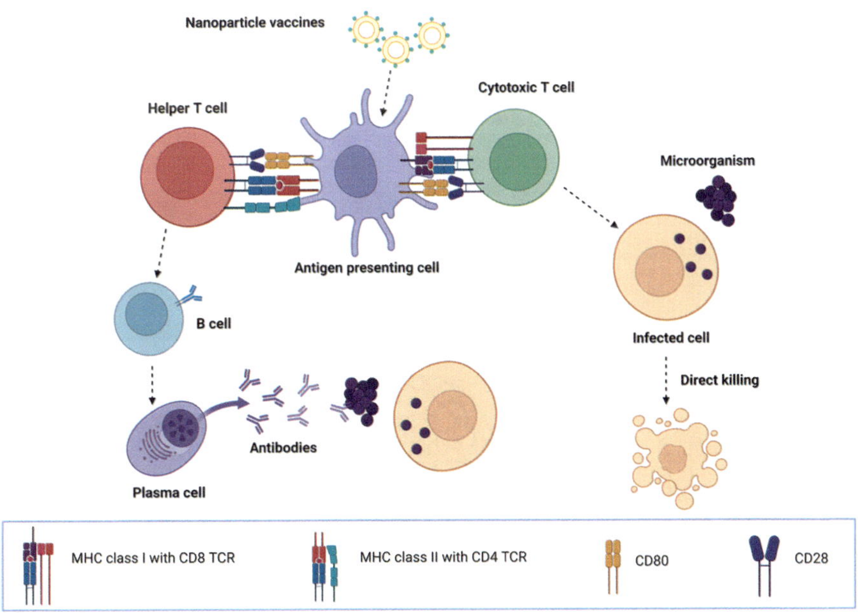

Fig. 4.3 Cellular and humoral immune responses by nanoparticle vaccines

4.3 Nanomaterial-Based Vaccine Platforms

In recent years, there has been significant exploration into diverse nanomaterials for vaccine development. This includes protein nanoparticles, lipid-based nanoparticles, inorganic nanocarriers, polymeric nanoparticles, and biomimetic nanoparticles. Each type of nanocarrier possesses unique physicochemical characteristics and exhibits specific behaviors in vivo, which in turn impacts the vaccination process. Here, we provide a concise overview of the various types of nanomaterials utilized for vaccines, highlighting their distinctive features and properties (Fig. 4.4).

The figure gives an illustration of different types of nanoparticle scaffolds including self-assembling (ferritin and virus-like particles), lipid, inorganic (gold), polymeric, and biomimetic (outer membrane vesicles) nanoparticles. The figure was created with the assistance of Biorender.com.

4.3.1 Self-Assembling Protein Nanoparticles

Natural nanomaterials possess an exceptional biocompatibility and biodegradability, making them highly desirable for various applications. Among these, protein nanoparticles derived from natural sources have been extensively studied for antigen delivery [32]. Self-assembling protein nanoparticles, including heat shock

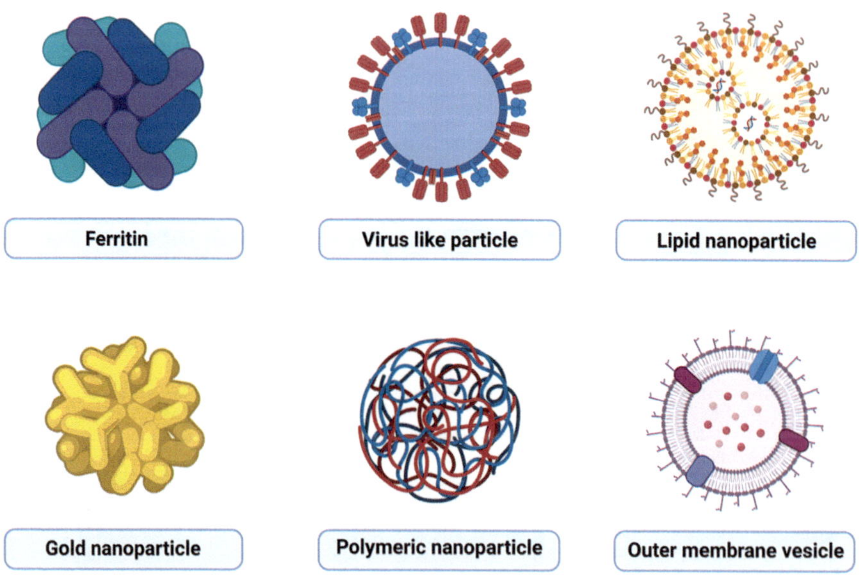

Fig. 4.4 Nanoparticle scaffold platforms for vaccines

proteins (Hsp), pyruvate dehydrogenase (E2), ferritin family proteins, and virus-like particles (VLPs), are being used in the field of nanovaccines.

Hsp are ubiquitous chaperones that bind to non-native or misfolded proteins thereby preventing their aggregation. The small Hsp16.5, which is derived from the primitive thermophilic archaeon, *Methanocaldoccus jannaschii* is one example of Hsp. This protein assembles into a 24-subunit caged nanoparticles [33]. The *Geobacillus stearothermophilus* pyruvate dehydrogenase complex includes a protein nanoparticle. It exhibits icosahedral symmetry and is formed from the E2 subunit of the complex. E2 is a 60-mer that assembles to create a hollow particle with a diameter of approximately 24 nm [34, 35].

The VLPs are a type of subunit vaccines consisting of self-assembling viral capsid proteins without genetic components or replication ability, making them safe and efficient platforms for antigen delivery [36]. Antigens can be coupled chemically or engineered genetically to VLPs with high density. VLPs exhibit favorable immunological properties, acting as self-adjuvants and being recognized by the immune system due to their virus-like size and repetitive surface geometry. VLPs can be efficiently taken up by APCs and induce immune responses [37].

In addition to exogenous viral proteins, endogenous self-assembled proteins can serve as promising nanovaccine platforms. These protein nanoparticles, also known as caged protein nanoparticles, offer unique advantages due to their highly organized structures [38]. One such example is ferritin, a spherical caged protein nanoparticle extensively utilized for antigen delivery, drug delivery, imaging, and diagnostics [39]. Ferritin consists of 24 subunits that form a central cavity structure of 12 nm × 8 nm dimensions capable of storing iron. Antigen proteins can be modified genetically into subunits for forming ferritins or incorporated into the ferritins to facilitate efficient phagocytosis by APCs. Ferritin has demonstrated the ability to target the lymph nodes passively and induce robust immune responses with a prolonged retention time [40].

These natural protein-based nanovaccine platforms offer immense potential for the development of effective and targeted immunization strategies. Their utilization holds promise in combating various diseases and advancing the field of nanovaccines.

4.3.2 Lipid Nanoparticles

Lipid nanoparticles (LNPs) are nanoscale vesicles composed of amphipathic phosphor lipid molecules. They offer a considerable potential as nanocarriers for nucleic acid delivery, due to their controlled release properties, high biocompatibility, and less toxicity [41]. Furthermore, LNPs play a crucial role in the field of mRNA drugs and vaccines. The significance of LNPs lies in its important function of protecting mRNA vaccines against degradation by nucleases, thereby ensuring the effective delivery. Their size, shape, and charge can be precisely controlled, which are critical factors influencing immune activation efficacy. Thus LNPs can be modified to achieve optimal immune responses [29]. Also, LNPs enable the co-delivery of different antigens and adjuvants.

LNPs have demonstrated significant potential in the development of nanovaccines, with successful applications in both preclinical and clinical settings [42–44]. Various components, including cationic lipids, ionizable lipids, and various other types of lipids, can be the appropriate components of the LNPs. Moreover, lipid functionalization, such as polyethylene glycol (PEG) conjugation- PEGylation, further enhances the versatility and efficacy of LNPs in vaccine development [45].

4.3.3 Inorganic Nanoparticles

Metallic oxides, non-metal oxides, and inorganic salts are commonly employed in nanomedicine. Inorganic materials exhibit strong structural stability and low biodegradability. Many inorganic nanomaterials possess inherent adjuvant activity [46]. However, when considering their application in nanovaccines, it is necessary to modify the physicochemical properties of inorganic nanomaterials to enhance their biocompatibility. The primary inorganic materials utilized for antigen delivery are gold, iron, and silica nanoparticles [47–49].

Gold nanoparticles (GNPs) have a spherical shape and carry a positive charge. They demonstrate favorable biocompatibility, low immunogenicity, and a high capacity for loading antigens. GNPs exhibit toxicity that depends on their size [50]. Nevertheless, GNPs also possess a strong affinity for sulfhydryl groups [51], which can be exploited for surface engineering by coupling them with cysteine residues to produce polypeptide antigens with enhanced safety and pharmacokinetic parameters. Furthermore, GNPs possess intrinsic immunostimulatory effects that promote the production of inflammatory cytokines [52]. Consequently, GNPs can serve not only as antigen transport carriers but also as stimulators of immune responses [53].

Silica nanoparticles are also promising candidates as carrier materials for nanovaccines [54]. Controlling the morphology [55] and pore size [49, 56] of silicon particles enables them to possess variable porosity, thereby increasing their effective load capacity for diverse antigens and adjuvants. The porous structure of silicon particles allows for the encapsulation of various active biomolecules or direct wrapping on their surface, enhancing the targeting as well as the uptake of the nanovaccine. Silica nanoparticles have been employed to target lymph nodes and thereby accumulating in APCs for the delivery of antigens and adjuvants [57].

4.3.4 Polymeric Nanoparticles

Polymeric nanoparticles exhibit a wide size range, spanning from 10 to 1000 nm [58]. These nanoparticles possess marked immunogenicity and stability, enabling efficient encapsulation as well as display of antigens that can be either loaded within the core of nanoparticle or conjugated to its surface. The polymeric nanoparticles are the colloidal and despite being primarily solid, these nanoparticles can be precisely controlled in terms of size and possess inherent adjuvant properties [59]. These nanoparticles can enhance the uptake of antigens by the APCs by endocytosis or phagocytosis [60].

Both natural and synthetic polymeric nanomaterials offer valuable options for the development of nanovaccines. Natural polymeric nanoparticles, such as chitosan and dextran, as well as synthetic polymeric nanoparticles like polylactic acid (PLA) and poly(lactic-*co*-glycolic) acid (PLGA), serve as useful tools. Polymeric nanoparticles derived from natural sources exhibit high biocompatibility, water solubility, and cost-effectiveness. Chitosan, for instance, is a linear cationic polysaccharide derived from chitin that can be employed for vaccine delivery. Its bioadhesive properties and cationic charge make chitosan a compelling candidate for gene delivery and as a coating for other polymeric nanoparticles to enhance adhesion and immunogenicity [61]. Furthermore, chitosan can be tailored for specific purposes through the introduction of functional groups [62].

In comparison, synthetic polymeric nanoparticles generally offer greater reproducibility and control over molecular weight compositions and degradation rates [63]. For instance, PLGA nanoparticles exhibit high biodegradability and can be finely adjusted to achieve desired properties. By coupling PLGA with PEG, it is possible to self-assemble them into polymeric micelles for the delivery of hydrophobic peptide antigens, resulting in enhanced T-cell responses [64].

4.3.5 Biomimetic Nanoparticles

Biomimetic nanomaterials are gaining prominence in the field of nanovaccine development due to their effective and intricate biofunctions [65, 66]. The multifunctional nature of these biomimetic nanomaterials allows for efficient delivery to the intended site or effective interaction with biological systems. They demonstrate high biocompatibility, extended circulation, and possess unique antigenic properties, making them valuable for the development of effective vaccine formulations.

A simple biomimetic approach involves the use of natural ligands or peptides to modify nanoparticles and increase their binding capabilities for better targeting and efficient delivery. Another biomimetic strategy involves the utilization of biomembranes to coat nanoparticles, enhancing their interaction with biological systems. Exploiting this principle, cell membrane-coated nanoparticles have been utilized as decoys capable of neutralizing toxins and protecting healthy cells from their harmful effects [67, 68]. This concept was initially demonstrated using nanoparticles coated with red blood cell membranes, effectively countering the activity of hemolytic toxins [69]. Recently, nanotoxoids have been developed by the application of cell membrane coating nanotechnology in toxoid vaccines resulting in improved efficacy and safety profiles [70].

Several other biomimetic strategies are emerging in the design of nanovaccines to combat infections. Virosomes, lipidic uni-lamellar nanocarriers (60–200 nm), are based on the concept of liposomes but possess a structure similar to enveloped viruses with the nucleocapsid removed [71]. Virosomes serve as promising biomimetic nanoparticles for the development of nanovaccines against viral infections. They can be engineered with different antigen epitopes to target specific host cells and modified with polymers to enhance their pharmacokinetic profiles [72].

Outer membrane vesicles (OMVs) are nanostructures with sizes ranging from approximately 20 to 250 nm, formed through the blebbing of the bacterial cell envelope [73]. OMVs comprise a diverse array of antigenic materials, including various outer membrane proteins, as well as inner membrane and cytoplasmic proteins. Also, the distinctive profile of membrane proteins within OMVs stimulates both the innate and adaptive immune systems, initiating immune responses through the presentation of PAMPs that bind to the PRRs on APCs [74]. The presence of multiple antigens in OMVs as well as the ability to enter lymph nodes efficiently due to their small size, make OMVs highly appealing for antibacterial vaccination.

4.4 Applications of Vaccine Nanotechnology

Vaccine nanotechnology offers exciting prospects for combating viral, bacterial, parasitic, and mycotic infections.

4.4.1 Viral Infections

In the context of viral infections, nanotechnology-based approaches have been employed to develop vaccines against different viruses (Table 4.1). Nanoparticles can be engineered to present viral antigens in a controlled and efficient manner, stimulating robust immune responses. Additionally, the versatility of nanoparticles allows for the incorporation of multiple viral antigens, enabling the development of broad-spectrum vaccines against viral strains with high genetic variability.

4.4.2 Bacterial Infections

For bacterial infections, vaccine nanotechnology has shown potential in the development of vaccines against pathogens such as *Streptococcus pneumoniae*, *Staphylococcus aureus*, and *Neisseria meningitidis*. Nanoparticles can encapsulate bacterial antigens and adjuvants, facilitating their targeted delivery to APCs and enhancing immune recognition (Table 4.2).

Table 4.1 Viral nanovaccine platforms

Pathogens	Nanovaccine	References
Influenza virus	sHSP	[75, 76]
	Ferritin	[77]
	P22 bacteriophage derived VLP	[78, 79]
	Qβ bacteriophage derived VLP	[80, 81]
	Potato X virus derived VLP	[82]
	Tobacco mosaic virus derived VLP	[83]
	Papaya mosaic virus derived VLP	[84–86]
Severe acute respiratory syndrome coronavirus 2	Lipid nanoparticles	[87]
	AP205 bacteriophage derived VLP	[88]
	Qβ bacteriophage derived VLP	[89]
	Tobacco mosaic virus derived VLP	[90]
	Cucumber mosaic virus derived VLP	[91, 92]
Human immunodeficiency virus 1	Ferritin	[77, 93, 94]
	E2	[34, 93, 95]
	Qβ bacteriophage derived VLP	[96]
	Cucumber mosaic virus derived VLP	[97]
	Potato X virus derived VLP	[98]
	Tomato bushy stunt virus derived VLP	[99]
Hepatitis B virus	Ferritin	[100]
	λ bacteriophage derived VLP	[101]
Hepatitis C virus	Ferritin	[102]
	Potato X virus derived VLP	[103]
	Cucumber mosaic virus derived VLP	[104, 105]
	Papaya mosaic virus derived VLP	[106]
Rotavirus	Ferritin	[107]
Epstein Barr virus	Ferritin	[108]
Respiratory syncytial virus	Ferritin	[109]
	Alfalfa mosaic virus derived VLP	[110]
Dengue virus	E2	[111]
Middle East respiratory syndrome coronavirus	Ferritin	[112]
Human papilloma virus	MS2 bacteriophage derived VLP	[113, 114]
Zika virus	Cucumber mosaic virus derived VLP	[115]

Abbreviations: *sHSP* small heat shock proteins, *VLP* virus-like particle

Table 4.2 Bacterial nanovaccine platforms

Pathogens	Nanovaccine	References
Bacillus anthracis	Qβ bacteriophage derived VLP	[116]
	Tobacco mosaic virus derived VLP	[117]
Yersinia pestis	Qβ bacteriophage derived VLP	[116]
	Tobacco mosaic virus derived VLP	[118]
Staphylococcus aureus	Cowpea mosaic virus derived VLP	[119, 120]
	Potato X virus derived VLP	[120]
Pseudomonas aeruginosa	Cowpea mosaic virus derived VLP	[121]
	Tobacco mosaic virus derived VLP	[122]
Group B *Streptococcus*	Cowpea mosaic virus derived VLP	[123]
Francisella tularensis	Tobacco mosaic virus derived VLP	[124–126]
Neisseria meningitides	OMV	[127]
Haemophilus influenzae	OMV	[128]

Abbreviations: *OMV* outer membrane vesicle, *VLP* virus-like particle

Table 4.3 Parasitic nanovaccine platforms

Pathogens	Nanovaccine	References
Plasmodium species	E2	[129]
	Qβ bacteriophage derived VLP	[130]
	Cucumber mosaic virus derived VLP	[115]
	Alfalfa mosaic virus derived VLP	[131, 132]
Leishmania species	Qβ bacteriophage derived VLP	[133]
Trypanosoma cruzi	Filamentous bacteriophage	[134]

Abbreviation: *VLP* virus-like particle

4.4.3 Parasitic Infections

For parasitic infections, nanotechnology-based vaccines have been investigated for parasites like *Plasmodium* and *Leishmania* species. Nanoparticles can improve the stability and immunogenicity of parasitic antigens, while also enabling controlled release and antigen presentation to immune cells. This approach holds promise for developing effective vaccines against complex parasitic diseases that have evasive immune mechanisms and antigenic variation (Table 4.3).

4.4.4 Fungal Infections

In the case of mycotic infections, nanotechnology offers innovative strategies for vaccines targeting fungal pathogens such as *Candida* and *Sporothrix* species. The use of nanoparticle-based vaccines may overcome the challenges associated with fungal infections, such as weak immune responses and antigen variability (Table 4.4).

Table 4.4 Fungal nanovaccine platforms

Pathogens	Nanovaccine	References
Candida albicans	Filamentous bacteriophage	[135]
Sporothrix globosa	Filamentous bacteriophage	[136]

4.5 Challenges and Outlook

The ongoing research efforts should focus on optimizing the design and formulation of nanovaccines. The development of versatile nanovaccine platforms that can be easily customized for different pathogens or diseases is a promising direction. Understanding the potential adverse effects, biodistribution, and long-term safety profiles of nanovaccines is essential for their clinical translation and acceptance by regulatory authorities.

Further, collaboration between researchers, clinicians, industry partners, and regulatory agencies will be crucial for the successful translation of nanovaccines from the laboratory to clinical practice. This collaboration can help address technical challenges, facilitate regulatory processes, and ensure the efficient translation of nanovaccine technologies into real-world applications, benefiting global public health.

Overall, vaccine nanotechnology presents exciting opportunities to develop effective vaccines against a wide range of infectious agents, including viruses, bacteria, parasites, and fungi. By leveraging the unique properties of nanoparticles, such as their size, surface functionality, and controlled release capabilities, researchers can design vaccines that improve antigen stability, enhance immune recognition, and elicit potent immune responses, ultimately contributing to the prevention and control of these infectious diseases.

References

1. Bloom, D. E., & Cadarette, D. (2019). Infectious disease threats in the twenty-first century: Strengthening the global response. *Frontiers in Immunology, 10*, 549.
2. Dictionary, M.-W. (2023, May 5). *Vaccine definition*. [Cited 2023 17 June 2023]. Available from https://www.merriam-webster.com/dictionary/vaccine
3. Liu, J., et al. (2022). Cancer vaccines as promising immuno-therapeutics: Platforms and current progress. *Journal of Hematology & Oncology, 15*(1), 28.
4. Pan, C., et al. (2021). Prophylactic vaccine delivery systems against epidemic infectious diseases. *Advanced Drug Delivery Reviews, 176*, 113867.
5. Amanna, I. J., & Slifka, M. K. (2020). Successful vaccines. In L. Hangartner & D. R. Burton (Eds.), *Vaccination strategies against highly variable pathogens* (pp. 1–30). Springer.
6. Brotherton, J. M. (2015). HPV prophylactic vaccines: Lessons learned from 10 years experience. *Future Virology, 10*(8), 999–1009.
7. Melief, C. J., et al. (2015). Therapeutic cancer vaccines. *The Journal of Clinical Investigation, 125*(9), 3401–3412.
8. Foged, C., et al. (2015). *Subunit vaccine delivery*. Springer.

9. Curley, S. M., & Putnam, D. (20221). Biological nanoparticles in vaccine development. *Frontiers in Bioengineering and Biotechnology, 10*, 867119.
10. Abdelaal, A., et al. (2022). Preventing the next pandemic: Is live vaccine efficacious against monkeypox, or is there a need for killed virus and mRNA vaccines? *Vaccines (Basel), 10*(9), 1419.
11. Clem, A. S. (2011). Fundamentals of vaccine immunology. *Journal of Global Infectious Diseases, 3*(1), 73–78.
12. Heidary, M., et al. (2022). A comprehensive review of the protein subunit vaccines against COVID-19. *Frontiers in Microbiology, 13*, 927306.
13. Doria-Rose, N., et al. (2021). Antibody persistence through 6 months after the second dose of mRNA-1273 vaccine for Covid-19. *New England Journal of Medicine, 384*(23), 2259–2261.
14. Pegu, A., et al. (2021). Durability of mRNA-1273 vaccine-induced antibodies against SARS-CoV-2 variants. *Science, 373*(6561), 1372–1377.
15. Naaber, P., et al. (2021). Dynamics of antibody response to BNT162b2 vaccine after six months: A longitudinal prospective study. *The Lancet Regional Health-Europe, 10*, 100208.
16. Choi, A., et al. (2021). Safety and immunogenicity of SARS-CoV-2 variant mRNA vaccine boosters in healthy adults: An interim analysis. *Nature Medicine, 27*(11), 2025–2031.
17. Fischer, N. O., et al. (2013). Colocalized delivery of adjuvant and antigen using nanolipoprotein particles enhances the immune response to recombinant antigens. *Journal of the American Chemical Society, 135*(6), 2044–2047.
18. Zhu, M., Wang, R., & Nie, G. (2014). Applications of nanomaterials as vaccine adjuvants. *Human Vaccines & Immunotherapeutics, 10*(9), 2761–2774.
19. Pramanik, A., et al. (2022). Human ACE2 peptide-attached plasmonic-magnetic heterostructure for magnetic separation, surface enhanced Raman spectroscopy identification, and inhibition of different variants of SARS-CoV-2 infections. *ACS Applied Bio Materials, 9*, 4454–4464.
20. Pramanik, A., et al. (2022). Blocking SARS-CoV-2 delta variant (B.1.617.2) spike protein receptor-binding domain binding with the ACE2 receptor of the host cell and inhibiting virus infections using human host defense peptide-conjugated graphene quantum dots. *ACS Omega, 7*(9), 8150–8157.
21. Dwivedi, R., et al. (2023). The sea cucumber *Thyonella gemmata* contains a low anticoagulant sulfated fucan with high anti-SARS-CoV-2 actions against wild-type and delta variants. *Journal of Natural Products, 86*(6), 1463–1475.
22. Farrag, M., et al. (2023). Structural requirements of *Holothuria floridana* fucosylated chondroitin sulfate oligosaccharides in anti-SARS-CoV-2 and anticoagulant activities. *PLoS One, 18*(5), e0285539.
23. Sharma, P., et al. (2023). Inhibition of cytomegalovirus by *Pentacta pygmaea* fucosylated chondroitin sulfate depends on its molecular weight. *Viruses, 15*(4), 859.
24. Maurya, A. K., et al. (2023). Structure, anti-SARS-CoV-2, and anticoagulant effects of two sulfated galactans from the red alga Botryocladia occidentalis. *International Journal of Biological Macromolecules, 238*, 124168.
25. Dwivedi, R., et al. (2023). Anti-SARS-CoV-2 and anticoagulant properties of *Pentacta pygmaea* fucosylated chondroitin sulfate depend on high molecular weight structures. *Glycobiology, 33*(1), 75–85.
26. Dwivedi, R., et al. (2022). Inhibition of SARS-CoV-2 wild-type (Wuhan-Hu-1) and Delta (B.1.617.2) strains by marine sulfated glycans. *Glycobiology, 32*(10), 849–854.
27. Dwivedi, R., et al. (2021). Structural and kinetic analyses of holothurian sulfated glycans suggest potential treatment for SARS-CoV-2 infection. *The Journal of Biological Chemistry, 297*(4), 101207.
28. Tandon, R., et al. (2021). Effective inhibition of SARS-CoV-2 entry by heparin and enoxaparin derivatives. *Journal of Virology, 95*(3), e01987-20.
29. Bachmann, M. F., & Jennings, G. T. (2010). Vaccine delivery: A matter of size, geometry, kinetics and molecular patterns. *Nature Reviews. Immunology, 10*(11), 787–796.

30. Manolova, V., et al. (2008). Nanoparticles target distinct dendritic cell populations according to their size. *European Journal of Immunology, 38*(5), 1404–1413.
31. Look, M., et al. (2010). Application of nanotechnologies for improved immune response against infectious diseases in the developing world. *Advanced Drug Delivery Reviews, 62*(4–5), 378–393.
32. O'Neill, C. L., et al. (2021). Peptide-based supramolecular vaccine systems. *Acta Biomaterialia, 133*, 153–167.
33. Shi, J., et al. (2013). Cryoelectron microscopy analysis of small heat shock protein 16.5 (Hsp16.5) complexes with T4 lysozyme reveals the structural basis of multimode binding. *The Journal of Biological Chemistry, 288*(7), 4819–4830.
34. Krebs, S. J., et al. (2014). Multimeric scaffolds displaying the HIV-1 envelope MPER induce MPER-specific antibodies and cross-neutralizing antibodies when co-immunized with gp160 DNA. *PLoS One, 9*(12), e113463.
35. Milne, J. L., et al. (2006). Molecular structure of a 9-MDa icosahedral pyruvate dehydrogenase subcomplex containing the E2 and E3 enzymes using cryoelectron microscopy. *The Journal of Biological Chemistry, 281*(7), 4364–4370.
36. Frietze, K. M., Peabody, D. S., & Chackerian, B. (2016). Engineering virus-like particles as vaccine platforms. *Current Opinion in Virology, 18*, 44–49.
37. Mohsen, M. O., et al. (2017). Major findings and recent advances in virus-like particle (VLP)-based vaccines. *Seminars in Immunology, 34*, 123–132.
38. Neek, M., Kim, T. I., & Wang, S. W. (2019). Protein-based nanoparticles in cancer vaccine development. *Nanomedicine, 15*(1), 164–174.
39. Tesarova, B., et al. (2020). Taking advantage of cellular uptake of ferritin nanocages for targeted drug delivery. *Journal of Controlled Release, 325*, 176–190.
40. Boone, C. E., et al. (2020). Active microneedle administration of plant virus nanoparticles for cancer in situ vaccination improves immunotherapeutic efficacy. *ACS Applied Nano Materials, 3*(8), 8037–8051.
41. Samaridou, E., Heyes, J., & Lutwyche, P. (2020). Lipid nanoparticles for nucleic acid delivery: Current perspectives. *Advanced Drug Delivery Reviews, 154–155*, 37–63.
42. Hou, X., et al. (2021). Lipid nanoparticles for mRNA delivery. *Nature Reviews Materials, 6*(12), 1078–1094.
43. Polack, F. P., et al. (2020). Safety and efficacy of the BNT162b2 mRNA Covid-19 vaccine. *The New England Journal of Medicine, 383*(27), 2603–2615.
44. Anderson, E. J., et al. (2020). Safety and immunogenicity of SARS-CoV-2 mRNA-1273 vaccine in older adults. *The New England Journal of Medicine, 383*(25), 2427–2438.
45. Hajj, K. A., & Whitehead, K. A. (2017). Tools for translation: Non-viral materials for therapeutic mRNA delivery. *Nature Reviews Materials, 2*(10), 1–17.
46. Hess, K. L., Medintz, I. L., & Jewell, C. M. (2019). Designing inorganic nanomaterials for vaccines and immunotherapies. *Nano Today, 27*, 73–98.
47. Dykman, L. A. (2020). Gold nanoparticles for preparation of antibodies and vaccines against infectious diseases. *Expert Review of Vaccines, 19*(5), 465–477.
48. Zhao, Y., et al. (2018). Iron oxide nanoparticles-based vaccine delivery for cancer treatment. *Molecular Pharmaceutics, 15*(5), 1791–1799.
49. Hong, X., et al. (2020). The pore size of mesoporous silica nanoparticles regulates their antigen delivery efficiency. *Science Advances, 6*(25), eaaz4462.
50. Deng, G., et al. (2018). Cell-membrane immunotherapy based on natural killer cell membrane coated nanoparticles for the effective inhibition of primary and abscopal tumor growth. *ACS Nano, 12*(12), 12096–12108.
51. Sharma, M., et al. (2013). Gold nanoparticles induce transcriptional activity of NF-κB in a B-lymphocyte cell line. *Nanoscale, 5*(9), 3747–3756.
52. González-Ballesteros, N., et al. (2019). Immunostimulant and biocompatible gold and silver nanoparticles synthesized using the *Ulva intestinalis* L. aqueous extract. *Journal of Materials Chemistry B, 7*(30), 4677–4691.

53. Mateu Ferrando, R., Lay, L., & Polito, L. (2020). Gold nanoparticle-based platforms for vaccine development. *Drug Discovery Today: Technologies, 38*, 57–67.
54. Mody, K. T., et al. (2013). Mesoporous silica nanoparticles as antigen carriers and adjuvants for vaccine delivery. *Nanoscale, 5*(12), 5167–5179.
55. Lee, J. Y., et al. (2020). Hollow mesoporous silica nanoparticles with extra-large mesopores for enhanced cancer vaccine. *ACS Applied Materials & Interfaces, 12*(31), 34658–34666.
56. Maria de Souza Morais, S., et al. (2018). Serum albumin nanoparticles vaccine provides protection against a lethal *Pseudomonas aeruginosa* challenge. *Vaccine, 36*(43), 6408–6415.
57. An, M., et al. (2017). Silica nanoparticle as a lymph node targeting platform for vaccine delivery. *ACS Applied Materials & Interfaces, 9*(28), 23466–23475.
58. Kang, H., et al. (2020). Size-dependent EPR effect of polymeric nanoparticles on tumor targeting. *Advanced Healthcare Materials, 9*(1), e1901223.
59. Simón-Vázquez, R., Peleteiro, M., & González-Fernández, Á. (2020). Polymeric nanostructure vaccines: Applications and challenges. *Expert Opinion on Drug Delivery, 17*(7), 1007–1023.
60. Das, A., & Ali, N. (2021). Nanovaccine: An emerging strategy. *Expert Review of Vaccines, 20*(10), 1273–1290.
61. Wu, D., et al. (2020). Chitosan-based colloidal polyelectrolyte complexes for drug delivery: A review. *Carbohydrate Polymers, 238*, 116126.
62. Butkovich, N., et al. (2021). Advancements in protein nanoparticle vaccine platforms to combat infectious disease. *Wiley Interdisciplinary Reviews. Nanomedicine and Nanobiotechnology, 13*(3), e1681.
63. Karlsson, J., Vaughan, H. J., & Green, J. J. (2018). Biodegradable polymeric nanoparticles for therapeutic cancer treatments. *Annual Review of Chemical and Biomolecular Engineering, 9*, 105–127.
64. Rietscher, R., et al. (2016). Antigen delivery via hydrophilic PEG-b-PAGE-b-PLGA nanoparticles boosts vaccination induced T cell immunity. *European Journal of Pharmaceutics and Biopharmaceutics, 102*, 20–31.
65. Hu, H., et al. (2021). A versatile and robust platform for the scalable manufacture of biomimetic nanovaccines. *Advanced Science (Weinh), 8*(15), 2002020.
66. Yang, J., et al. (2021). Biologically modified nanoparticles as theranostic bionanomaterials. *Progress in Materials Science, 118*, 100768.
67. Kroll, A. V., Fang, R. H., & Zhang, L. (2017). Biointerfacing and applications of cell membrane-coated nanoparticles. *Bioconjugate Chemistry, 28*(1), 23–32.
68. Fang, R. H., et al. (2017). Cell membrane-derived nanomaterials for biomedical applications. *Biomaterials, 128*, 69–83.
69. Hu, C. M., et al. (2013). A biomimetic nanosponge that absorbs pore-forming toxins. *Nature Nanotechnology, 8*(5), 336–340.
70. Angsantikul, P., Fang, R. H., & Zhang, L. (2018). Toxoid vaccination against bacterial infection using cell membrane-coated nanoparticles. *Bioconjugate Chemistry, 29*(3), 604–612.
71. Asadi, K., & Gholami, A. (2021). Virosome-based nanovaccines; a promising bioinspiration and biomimetic approach for preventing viral diseases: A review. *International Journal of Biological Macromolecules, 182*, 648–658.
72. Felnerova, D., et al. (2004). Liposomes and virosomes as delivery systems for antigens, nucleic acids and drugs. *Current Opinion in Biotechnology, 15*(6), 518–529.
73. Schwechheimer, C., & Kuehn, M. J. (2015). Outer-membrane vesicles from Gram-negative bacteria: Biogenesis and functions. *Nature Reviews. Microbiology, 13*(10), 605–619.
74. Tan, K., et al. (2018). Outer membrane vesicles: Current status and future direction of these novel vaccine adjuvants. *Frontiers in Microbiology, 9*, 783.
75. Richert, L. E., et al. (2013). Inducible bronchus-associated lymphoid tissue (iBALT) synergizes with local lymph nodes during antiviral CD4+ T cell responses. *Lymphatic Research and Biology, 11*(4), 196–202.

76. Richert, L. E., et al. (2012). A virus-like particle vaccine platform elicits heightened and hastened local lung mucosal antibody production after a single dose. *Vaccine, 30*(24), 3653–3665.
77. Georgiev, I. S., et al. (2018). Two-component ferritin nanoparticles for multimerization of diverse trimeric antigens. *ACS Infectious Diseases, 4*(5), 788–796.
78. Patterson, D. P., et al. (2013). Biomimetic antigenic nanoparticles elicit controlled protective immune response to influenza. *ACS Nano, 7*(4), 3036–3044.
79. Sharma, J., et al. (2020). A self-adjuvanted, modular, antigenic VLP for rapid response to influenza virus variability. *ACS Applied Materials & Interfaces, 12*(16), 18211–18224.
80. Jegerlehner, A., et al. (2013). Bacterially produced recombinant influenza vaccines based on virus-like particles. *PLoS One, 8*(11), e78947.
81. Skibinski, D. A., et al. (2013). Enhanced neutralizing antibody titers and Th1 polarization from a novel *Escherichia coli* derived pandemic influenza vaccine. *PLoS One, 8*(10), e76571.
82. Lico, C., et al. (2009). Plant-produced potato virus X chimeric particles displaying an influenza virus-derived peptide activate specific CD8+ T cells in mice. *Vaccine, 27*(37), 5069–5076.
83. Mallajosyula, J. K., et al. (2014). Single-dose monomeric HA subunit vaccine generates full protection from influenza challenge. *Human Vaccines & Immunotherapeutics, 10*(3), 586–595.
84. Denis, J., et al. (2008). Development of a universal influenza A vaccine based on the M2e peptide fused to the papaya mosaic virus (PapMV) vaccine platform. *Vaccine, 26*(27–28), 3395–3403.
85. Bolduc, M., et al. (2018). The quest for a nanoparticle-based vaccine inducing broad protection to influenza viruses. *Nanomedicine, 14*(8), 2563–2574.
86. Thérien, A., et al. (2017). A versatile papaya mosaic virus (PapMV) vaccine platform based on sortase-mediated antigen coupling. *Journal of Nanobiotechnology, 15*(1), 54.
87. Guerrini, G., et al. (2022). Characterization of nanoparticles-based vaccines for COVID-19. *Nature Nanotechnology, 17*(6), 570–576.
88. Fougeroux, C., et al. (2021). Capsid-like particles decorated with the SARS-CoV-2 receptor-binding domain elicit strong virus neutralization activity. *Nature Communications, 12*(1), 324.
89. Ortega-Rivera, O. A., et al. (2021). Trivalent subunit vaccine candidates for COVID-19 and their delivery devices. *Journal of the American Chemical Society, 143*(36), 14748–14765.
90. Royal, J. M., et al. (2021). Development of a SARS-CoV-2 vaccine candidate using plant-based manufacturing and a tobacco mosaic virus-like nano-particle. *Vaccines (Basel), 9*(11), 1347.
91. Zha, L., et al. (2021). Development of a vaccine against SARS-CoV-2 based on the receptor-binding domain displayed on virus-like particles. *Vaccines (Basel), 9*(4), 395.
92. Mohsen, M. O., et al. (2022). A scalable and highly immunogenic virus-like particle-based vaccine against SARS-CoV-2. *Allergy, 77*(1), 243–257.
93. He, L., et al. (2016). Presenting native-like trimeric HIV-1 antigens with self-assembling nanoparticles. *Nature Communications, 7*(1), 12041.
94. Sliepen, K., et al. (2015). Presenting native-like HIV-1 envelope trimers on ferritin nanoparticles improves their immunogenicity. *Retrovirology, 12*, 82.
95. Caivano, A., et al. (2010). HIV-1 Gag p17 presented as virus-like particles on the E2 scaffold from *Geobacillus stearothermophilus* induces sustained humoral and cellular immune responses in the absence of IFNγ production by CD4+ T cells. *Virology, 407*(2), 296–305.
96. Purwar, M., et al. (2018). Design, display and immunogenicity of HIV1 gp120 fragment immunogens on virus-like particles. *Vaccine, 36*(42), 6345–6353.
97. McInerney, T. L., et al. (1999). Analysis of the ability of five adjuvants to enhance immune responses to a chimeric plant virus displaying an HIV-1 peptide. *Vaccine, 17*(11–12), 1359–1368.

98. Marusic, C., et al. (2001). Chimeric plant virus particles as immunogens for inducing murine and human immune responses against human immunodeficiency virus type 1. *Journal of Virology, 75*(18), 8434–8439.
99. Joelson, T., et al. (1997). Presentation of a foreign peptide on the surface of tomato bushy stunt virus. *The Journal of General Virology, 78*(Pt 6), 1213–1217.
100. Wang, W., et al. (2020). Dual-targeting nanoparticle vaccine elicits a therapeutic antibody response against chronic hepatitis B. *Nature Nanotechnology, 15*(5), 406–416.
101. March, J. B., Clark, J. R., & Jepson, C. D. (2004). Genetic immunisation against hepatitis B using whole bacteriophage lambda particles. *Vaccine, 22*(13–14), 1666–1671.
102. Yan, Y., et al. (2020). A nanoparticle-based hepatitis C virus vaccine with enhanced potency. *The Journal of Infectious Diseases, 221*(8), 1304–1314.
103. Uhde-Holzem, K., et al. (2010). Immunogenic properties of chimeric potato virus X particles displaying the hepatitis C virus hypervariable region I peptide R9. *Journal of Virological Methods, 166*(1–2), 12–20.
104. Piazzolla, G., et al. (2005). Immunogenic properties of a chimeric plant virus expressing a hepatitis C virus (HCV)-derived epitope: New prospects for an HCV vaccine. *Journal of Clinical Immunology, 25*(2), 142–152.
105. Nuzzaci, M., et al. (2007). Cucumber mosaic virus as a presentation system for a double hepatitis C virus-derived epitope. *Archives of Virology, 152*(5), 915–928.
106. Denis, J., et al. (2007). Immunogenicity of papaya mosaic virus-like particles fused to a hepatitis C virus epitope: Evidence for the critical function of multimerization. *Virology, 363*(1), 59–68.
107. Li, Z., et al. (2019). A milk-based self-assemble rotavirus VP6-ferritin nanoparticle vaccine elicited protection against the viral infection. *Journal of Nanobiotechnology, 17*(1), 13.
108. Kanekiyo, M., et al. (2015). Rational design of an Epstein-Barr virus vaccine targeting the receptor-binding site. *Cell, 162*(5), 1090–1100.
109. Swanson, K. A., et al. (2020). A respiratory syncytial virus (RSV) F protein nanoparticle vaccine focuses antibody responses to a conserved neutralization domain. *Science Immunology, 5*(47), eaba6466.
110. Yusibov, V., et al. (2005). Peptide-based candidate vaccine against respiratory syncytial virus. *Vaccine, 23*(17–18), 2261–2265.
111. McBurney, S. P., et al. (2016). Evaluation of protection induced by a dengue virus serotype 2 envelope domain III protein scaffold/DNA vaccine in non-human primates. *Vaccine, 34*(30), 3500–3507.
112. Kim, Y. S., et al. (2018). Chaperna-mediated assembly of ferritin-based middle east respiratory syndrome-coronavirus nanoparticles. *Frontiers in Immunology, 9*, 1093.
113. Tumban, E., et al. (2012). VLPs displaying a single L2 epitope induce broadly cross-neutralizing antibodies against human papillomavirus. *PLoS One, 7*(11), e49751.
114. Tyler, M., et al. (2014). Immunization with a consensus epitope from human papillomavirus L2 induces antibodies that are broadly neutralizing. *Vaccine, 32*(34), 4267–4274.
115. Cabral-Miranda, G., et al. (2017). Virus-like particle (VLP) plus microcrystalline tyrosine (MCT) adjuvants enhance vaccine efficacy improving T and B cell immunogenicity and protection against Plasmodium berghei/vivax. *Vaccine, 5*(2), 10.
116. Tao, P., et al. (2018). A bacteriophage T4 nanoparticle-based dual vaccine against anthrax and plague. *MBio, 9*, e01926-18.
117. McComb, R. C., et al. (2015). Presentation of peptides from *Bacillus anthracis* protective antigen on Tobacco Mosaic Virus as an epitope targeted anthrax vaccine. *Vaccine, 33*(48), 6745–6751.
118. Arnaboldi, P. M., et al. (2016). Intranasal delivery of a protein subunit vaccine using a Tobacco Mosaic Virus platform protects against pneumonic plague. *Vaccine, 34*(47), 5768–5776.
119. Rennermalm, A., et al. (2001). Antibodies against a truncated *Staphylococcus aureus* fibronectin-binding protein protect against dissemination of infection in the rat. *Vaccine, 19*(25–26), 3376–3383.

120. Brennan, F. R., et al. (1999). Immunogenicity of peptides derived from a fibronectin-binding protein of *S. aureus* expressed on two different plant viruses. *Vaccine, 17*(15–16), 1846–1857.
121. Brennan, F. R., et al. (1999). A chimaeric plant virus vaccine protects mice against a bacterial infection. *Microbiology (Reading), 145*(Pt 8), 2061–2067.
122. Staczek, J., et al. (2000). Immunization with a chimeric tobacco mosaic virus containing an epitope of outer membrane protein F of *Pseudomonas aeruginosa* provides protection against challenge with *P. aeruginosa*. *Vaccine, 18*(21), 2266–2274.
123. Pomwised, R., et al. (2016). Coupling peptide antigens to virus-like particles or to protein carriers influences the Th1/Th2 polarity of the resulting immune response. *Vaccine, 4*(2), 15.
124. McCormick, A. A., et al. (2018). Intranasal administration of a two-dose adjuvanted multi-antigen TMV-subunit conjugate vaccine fully protects mice against *Francisella tularensis* LVS challenge. *PLoS One, 13*(4), e0194614.
125. Mansour, A. A., et al. (2018). An improved Tobacco Mosaic Virus (TMV)-conjugated multiantigen subunit vaccine against respiratory tularemia. *Frontiers in Microbiology, 9*, 1195.
126. Banik, S., et al. (2015). Development of a multivalent subunit vaccine against tularemia using Tobacco Mosaic Virus (TMV) based delivery system. *PLoS One, 10*(6), e0130858.
127. Petousis-Harris, H. (2018). Impact of meningococcal group B OMV vaccines, beyond their brief. *Human Vaccines & Immunotherapeutics, 14*(5), 1058–1063.
128. Vella, P. P., et al. (1990). Immunogenicity of a new Haemophilus influenzae type b conjugate vaccine (meningococcal protein conjugate) (PedvaxHIB). *Pediatrics, 85*(4 Pt 2), 668–675.
129. Domingo, G. J., Orru, S., & Perham, R. N. (2001). Multiple display of peptides and proteins on a macromolecular scaffold derived from a multienzyme complex. *Journal of Molecular Biology, 305*(2), 259–267.
130. Khan, F., et al. (2015). Head-to-head comparison of soluble vs. Qβ VLP circumsporozoite protein vaccines reveals selective enhancement of NANP repeat responses. *PLoS One, 10*(11), e0142035.
131. Chichester, J. A., et al. (2018). Safety and immunogenicity of a plant-produced Pfs25 virus-like particle as a transmission blocking vaccine against malaria: A Phase 1 dose-escalation study in healthy adults. *Vaccine, 36*(39), 5865–5871.
132. Jones, R. M., et al. (2013). A plant-produced Pfs25 VLP malaria vaccine candidate induces persistent transmission blocking antibodies against *Plasmodium falciparum* in immunized mice. *PLoS One, 8*(11), e79538.
133. Moura, A. P. V., et al. (2017). Virus-like particle display of the α-gal carbohydrate for vaccination against Leishmania infection. *ACS Central Science, 3*(9), 1026–1031.
134. Gomes-Neto, J. F., et al. (2018). Vaccination with recombinant filamentous fd phages against parasite infection requires TLR9 expression. *Frontiers in Immunology, 9*, 1173.
135. Wang, Y., et al. (2014). Hybrid phage displaying SLAQVKYTSASSI induces protection against *Candida albicans* challenge in BALB/c mice. *Human Vaccines & Immunotherapeutics, 10*(4), 1057–1063.
136. Chen, F., et al. (2017). Recombinant phage elicits protective immune response against systemic *S. globosa* infection in mouse model. *Scientific Reports, 7*, 42024.

Chapter 5
Exploring the Application, Safety, and Challenges of Free Versus Immobilized Antimicrobial Nanomaterials

Sharda Bharti, Pooja Nag, Kapil Sadani, Soumyo Mukherji, and Suparna Mukherji

5.1 Introduction

Antimicrobial nanoparticles have been the subject of intense study in recent years due to their potential to reduce the prevalence of infectious diseases and increase overall system security. Both free and immobilized nanomaterials have shown promise in antibacterial applications. In contrast to immobilized antimicrobial nanomaterials, which are glued to a surface or substrate, free antimicrobial nanomaterials are freely dispersed in solution. There is a need for understanding the possible applications, hazards, and limitations of these two kinds of antimicrobial nanomaterials. Nanoscale antimicrobial materials can be used in a variety of applications, healthcare, and wound healing, coating of medical devices, and antimicrobial fabrics [1]. Antimicrobial nanoparticles can safeguard public health and improve water quality when they are incorporated in water filtration and disinfection units. The shelf life of perishable goods can be prolonged by incorporating antimicrobial nanoparticles in food packaging. Their inclusion in commonplace

S. Bharti
Department of Biotechnology, NIT Raipur, Raipur, India

P. Nag
Department of Mechatronics, Manipal Institute of Technology, Manipal Academy of Higher Education (MAHE), Manipal, India

K. Sadani
Department of Instrumentation and Control, Manipal Institute of Technology, Manipal Academy of Higher Education (MAHE), Manipal, India

S. Mukherji
Department of Bioscience and Bioengineering (BSBE), IIT Bombay, Mumbai, India

S. Mukherji (✉)
Environmental Science and Engineering Department (ESED), IIT Bombay, Mumbai, India
e-mail: mitras@iitb.ac.in

R. S. Chaughule, A. S. Lokur (eds.), *Applications of Nanotechnology in Microbiology*, https://doi.org/10.1007/978-3-031-49933-3_5

consumer goods such as cosmetics and cleaning supplies can protect against microbial contamination and make homes and communities safer.

It is important to distinguish between free and immobilized forms of antimicrobial nanoparticles when considering their antibacterial efficacy, durability, and user-friendliness. Better contact with pathogens is possible when freely dispersed nanoparticles are employed. Thus, they may have higher capacity to inactivate microorganisms. However, there are concerns regarding the toxicity, bioaccumulation, reusability, and environmental impact of free nanoparticles. Immobilized nanomaterials, on the other hand, may remain effective against microorganisms over a longer period as they remain attached on to a surface. Attachment ensures better retention of nanoparticles within a system and prevents their uptake by microorganisms. It also prevents increase in size of the nanoparticles through agglomeration. Antimicrobial coatings on medical equipment or food packaging materials are examples where immobilized nanoparticles are commonly applied. Tailoring the performance and stability of nanomaterials relies heavily on the immobilization mechanisms such as physical adsorption, chemisorption, entrapment, electrostatic immobilization, and covalent immobilization.

The use of antimicrobial nanoparticles has the potential to improve public health, but they also have certain risks. To promote their practical application, issues of scalability and cost-effectiveness must be resolved. As a corollary, ensuring the security and performance of antimicrobial nanoparticles across sectors requires strict adherence to regulatory rules and licensing processes. When designing nanomaterials with antibacterial capabilities, it is also crucial to consider their long-term stability and endurance. Furthermore, the risks posed by antimicrobial nanomaterials to human beings and other organisms in the ecosystem are important considerations. To ensure the safety of users and workers, it is crucial to assess the potential toxicity and human health concerns associated with the application of the nanomaterials. To avoid unfavourable effects on ecosystems, it is essential to understand the environmental implications of antimicrobial nanoparticles and take appropriate measures to mitigate any potential ecotoxicological effects.

This study examines free versus immobilized antimicrobial nanoparticles in terms of applications, risks, and other limitations. This chapter begins with antibacterial nanoparticle characteristics and processes. Silver, copper oxide, and zinc oxide nanoparticles are widely reported for their antimicrobial effects due to their distinct physicochemical characteristics. Transition metal oxides, nanostructured bimetallic systems, polymeric nanosystems, and drug-encapsulated liposomal solid lipid-based and carbon-based nanoparticles are reported for antimicrobial characteristics. Antimicrobial nanotechnology has been used in traditional Indian alternative medicine, for generations. The second section covers free antibacterial nanomaterial design, production, and characterization. The subsequent section discusses free and immobilized antimicrobial nanoparticles, their long-term usage, and immobilization techniques and substrates. Application of immobilized antimicrobial nanoparticles in healthcare, water treatment, food packaging, and consumer items are discussed together with the risks posed by them in terms of bioaccumulation, absorption, and ecotoxicity. Free nanoparticle bioaccumulation, absorption, and ecotoxicity are discussed. Toxicity, health, and environmental effects affect the safe usage of antimicrobial nanomaterials. Regulatory requirements, risk assessment, and mitigation are specifically being studied to make antimicrobial nanoparticles safe. This chapter discusses the challenges faced in the use of antimicrobial nanomaterials such as scalability, cost-effectiveness, regulatory compliance and approval, and long-term stability and durability. Antimicrobial nanoparticles must overcome these obstacles before widespread adoption. The antimicrobial nanomaterial research trends, current advances, possible breakthroughs, and research prospects are also discussed.

This study explored the characteristics, mechanisms, and uses of antimicrobial nanomaterials in healthcare, water purification, food packaging, and consumer products. Widespread application of antimicrobial nanoparticles will only be realized when issues related to toxicity, bioaccumulation, and environmental impact are addressed. With a thorough understanding of scalability, cost-effectiveness, regulatory compliance, and long-term stability, one can readily create and execute these technologies. An understanding of the benefits, risks, and limitations of antimicrobial nanoparticles will enable knowledge-based decisions.

5.2 Fundamentals of Antimicrobial Nanomaterials

The term "antimicrobial nanomaterials" refer to nanoscale materials that may suppress or destroy microorganisms including bacteria, viruses, and fungi. The distinctive physicochemical features of these materials and the mechanism by which they interact with microbial cells make them effective growth inhibitors. Several distinct types of antimicrobial nanoparticles have been developed and studied. The widespread antibacterial action of silver nanoparticles (Ag NPs) is well-documented; Ag NPs kill bacteria by rupturing their cell membrane and by interacting with other components in the cell. Zinc oxide nanoparticles (ZnO NPs) produce reactive oxygen species (ROS), which are harmful to microbial cells [2]. Copper nanoparticles (Cu NPs) and copper oxide nanoparticles (CuO NPs) damage cell membranes, interfere with enzymes, and cause oxidative stress [3]. Titanium dioxide nanoparticles (TiO_2 NPs) exhibit photocatalytic properties and generate reactive oxygen species (ROS) when exposed to ultraviolet light, damaging microbial cells in the process. Cell membranes can be damaged by carbon-based nanoparticles, including carbon nanotubes (CNTs) and graphene-based nanomaterials [4, 5]. Thus, antimicrobial nanoparticles work through physical interactions, ROS generation, ion release, surface charge-based binding, and by mimicking antibacterial peptides. These mechanisms lead to cell death by creating oxidative stress, interfering with cellular processes, and by altering the microbial membranes [6, 7]. Since effective antimicrobial agents are needed in a variety of areas, antimicrobial nanoparticles have attracted attention for use in medicine, food safety, water treatment, and other diverse areas [8–10]. Different types of nanomaterials are described in the subsequent sections.

5.2.1 Metal Nanoparticles

Metal nanoparticles have been extensively investigated as antimicrobials over the years. The use of bulk copper metal as antimicrobial is not new and the earliest use of copper for sterilizing wounds and drinking water dates back to 2500 BC [11]. The antimicrobial action of metallic copper is principally brought on by copper ions. The copper ions can enter microbial cells and interfere with biomolecules thereby adversely affecting cellular functions and causing death of the microorganisms. DNA damage affects DNA replication and transcription, and protein denaturation inhibits enzyme activity. To exhibit its antibacterial properties, metallic copper needs to come into direct contact with microbes. Compared to bulk metal, nanoparticles have a much greater surface area due to their high surface-to-volume ratio. Improved contact and interaction with microorganisms are made possible by the larger surface area, which boosts the effectiveness of antimicrobial agents. The antibacterial activity of NPs is affected by various physical and chemical factors. Larger NPs have the potential to be harmful even though smaller NPs are typically more efficient. Metal/metalloid nanoparticles, such as Ag [12], Cu [13], Au [14], and Bi [15] nanoparticles, work via multiple antimicrobial mechanisms. To begin with, they bind to bacterial cells, altering the structure of the cell membrane and obstructing the transport channels, and this is followed by ionization within the cell and damage to intracellular structures [16]. Furthermore, metal NPs can release reactive oxidative species (ROS) that can harm proteins, DNA, mRNA, ribosomes, peptidoglycan, and cell membranes [17]. Additionally, metal ions can damage DNA molecules by interfering with hydrogen bonds and inactivate enzymes by attaching to thiol groups [18]. Figure 5.1 depicts the mechanism of antimicrobial action of silver nanoparticles. Apart from size, the shape of the nanoparticles also determines their antimicrobial efficacy [19]. While spherical nanoparticles are the most common ones, other shapes, such as nanotriangles [20], nanocubes [21], nanowires [21], nanoclusters [16], rods [22] and decahedral [23] NPs have also been reported. The bactericidal effects of nanoparticles have been reported to be enhanced by the exposure of certain crystal facets. For example, based on experimental investigations Pal et al. [20] reported that triangular silver nanoparticles were more bactericidal than silver nanospheres. This was because of the presence of {111} facets in nanotriangles, compared to other shapes such as spheres, cubes and rods, which mostly had the {100} facet exposed. Bharti et al. [23] also demonstrated a tenfold increase in bactericidal activity of decahedral silver nanoparticles as compared to spherical silver nanoparticles.

5.2.2 Transition Metal Oxides

Transition metal oxide nanoparticles are metal oxide nanoparticles made up of transition metal elements bound with oxygen atoms. TiO_2, ZnO, and CuO nanoparticles are the transition metal oxide nanoparticles that have drawn the most interest because of their distinct chemical and physical characteristics [25]. Because of their

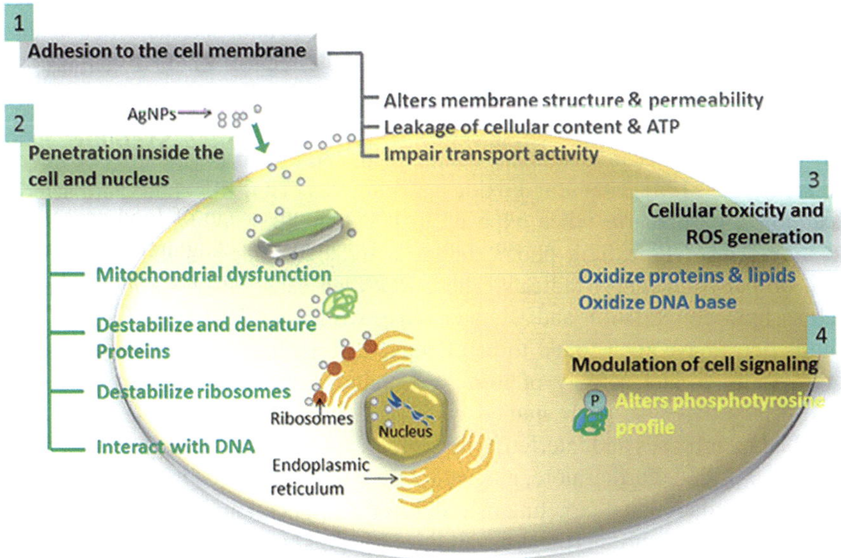

Fig. 5.1 Mechanism of antimicrobial action of silver nanoparticles. (Adapted with permission from Dakal et al. [24] under the terms and conditions of the Creative Commons Attribution license)

tiny size and high surface-to-volume ratio, transition metal oxide nanoparticles have distinct characteristics when compared to their bulk counterparts. These qualities make them very appealing for a wide range of applications, including catalysis, energy storage, electronics, and biomedicine [25–28]. Transition metal oxides, including copper oxide, zinc oxide, and titanium dioxide, employ diverse antimicrobial mechanisms [29–32]. These processes involve the generation of reactive oxygen species (ROS) in response to exposure to UV or visible light, including superoxide and hydroxyl radicals. Damage to cell components, membrane disruption, enzyme inhibition, DNA damage, and adherence to microbes are all possible outcomes of ROS production and accumulation [30]. Furthermore, copper oxide releases copper ions that inhibit the membrane and enzyme function of microorganisms [29]. The properties and application of transition metal oxide nanoparticles depend on the metal oxide under consideration, particle size, shape, surface chemistry, and the methods used for synthesis. Several transition metal oxides display antibacterial activity and have been investigated for potential applications in bacterial infection control. Copper oxide nanoparticles (CuO NPs) have shown significant antimicrobial action against a variety of pathogens, including bacteria, viruses, and fungus [31]. The antibacterial activity of CuO NPs is linked to the release of copper ions (Cu^{2+}) and the development of oxidative stress. The infiltration of copper ions within the cell can lead to several other effects in addition to membrane damage. CuO NPs can also stimulate the generation of reactive oxygen species (ROS), causing oxidative damage to microbial cells [33, 34]. As a result of their multiple mechanisms, transition metal oxides are attractive as potential agents for fighting microbial infections and overcoming antimicrobial resistance.

Antimicrobial properties of zinc oxide nanoparticles (ZnO NPs) have also been extensively explored. ZnO nanoparticles have extensive antibacterial action against bacteria, viruses, and fungus [35, 36]. Antibacterial effects of ZnO NPs are accompanied by the production of reactive oxygen species (ROS), such as hydrogen peroxide and superoxide radicals, which can lead to oxidative stress and damage to cells [37]. ZnO NPs can potentially interact with and impair the integrity of microbial membranes [32]. Titanium dioxide nanoparticles (TiO_2 NPs) have antibacterial capabilities, particularly when exposed to UV light [38]. When exposed to UV radiation, TiO_2 NPs exhibit photocatalytic activity, resulting in the formation of ROS, primarily hydroxyl radicals. These ROS can cause microbial cell damage, such as lipid peroxidation and DNA damage, which can lead to cell death. TiO_2 nanoparticles have been found to be effective against bacteria, viruses, and some fungi. Antimicrobial activity of iron oxide nanoparticles, particularly magnetite (Fe_3O_4) NPs, has also been studied widely [39]. Fe_3O_4 nanoparticles can break microbial membranes and interfere with biological functions, causing bacteria to be inhibited or killed. Furthermore, magnetic characteristics of Fe_3O_4 NPs enable controlled manipulation and targeting in antibacterial applications.

To improve the antimicrobial capabilities of transition metal oxide NPs, silver (Ag)-modified transition metal oxides such as Ag-doped TiO_2 or Ag-doped ZnO have been created [30, 40]. Because silver has high antibacterial action, incorporating silver nanoparticles or ions in transition metal oxides can boost their antimicrobial efficiency synergistically [41]. The antibacterial actions of transition metal oxides involve several components, including membrane rupture, oxidative stress creation, ROS-induced damage, and interference with cellular functions (as illustrated in Fig. 5.2). Collectively, these strategies contribute to the suppression or elimination of microbes. With their various compositions and characteristics, transition metal oxides offer interesting pathways for the creation of antimicrobial nanomaterials. However, aspects such as nanoparticle size, concentration, and stability must be considered to maximize antibacterial activity while minimizing potential harmful consequences.

5.2.3 Nanostructured Bimetallic Systems

Antimicrobial monometallic nanoformulations such as silver and copper have been extensively studied [43, 44]. Two of the most common mechanisms for antimicrobial resistance (AMR) to monometallic nanoparticles typically involve the uptake of the formulation followed by its efflux or its gradual intracellular bioaccumulation [45, 46]. To circumvent the limitations of monometallic nano/ionic systems, researchers have tried to explore various combinations, such as bimetallic, trimetallic, and core shell nanoparticles. Such systems can achieve enhanced synergistic antimicrobial properties thereby reducing the chances of AMR development [47]. The mechanisms of synergistic antibacterial effects are similar to that of

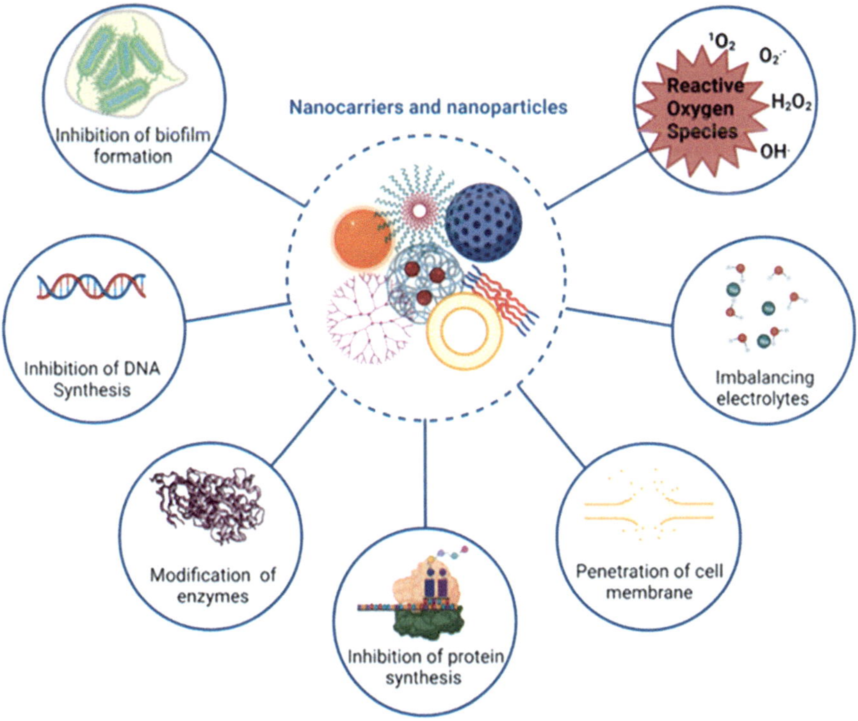

Fig. 5.2 The multiple mechanism of action of antimicrobial nanoparticles and nanocomposites. (Adapted with permission from Skwarczynski et al. [42] under the terms and conditions of the Creative Commons Attribution license)

monometallic systems, including physical interactions with bacterial cells, the release of metal ions, and production of reactive oxygen species (ROS). However, bimetallic systems may be tailored to achieve higher catalytic activities through surface redox reactions such as Fenton reaction on the surface of the nanoparticle to enhance the contact inactivation kinetics. Another significant benefit is in the reusability of systems containing magnetic materials even in the free from using magnetic collection methods [48]. Such systems can potentially reduce adverse effects [49]. A researcher with thorough knowledge regarding the mechanisms and kinetics of action of the constituent metallic systems, and the target matrix of end use can thus design various multi-metallic nanosystems. These typically involve devising varied sizes and morphologies of nanoparticles by various bottom-up synthesis approaches, by varying the ratios of nanoparticles and ensuring stable surface groups by use of suitable capping and stabilizing agents. The kinetics of antimicrobial action on susceptible and resistant microbes is subsequently analysed. A genomic study of the bacteria exposed to such systems could reveal further molecular insights into their mode of action. Thus, multi-metallic systems have good potential for sustainable use as broad spectrum antimicrobials.

5.2.4 Polymeric Nanosystems

Polymeric nanosystems are any nanoscale structures or particles that are entirely or primarily made of polymers. These systems are designed to have unique features and functions at the nanoscale, which may be used in a variety of applications such as medication administration, imaging, and sensing [50, 51]. Polymeric nanosystems have emerged as potential antimicrobial nanomaterials because of properties such as tunability and ability to encapsulate and disseminate antimicrobial compounds. These systems have several advantages, including controlled release, better stability, and lower toxicity [52]. Antimicrobial compounds such as antibiotics, peptides, or essential oils can be delivered using manufactured polymeric nanoparticles with diameters between 10 and 200 nm. The antibacterial compounds contained within the polymeric matrix are protected from degradation and are released slowly and steadily. Common polymers for creating antimicrobial nanoparticles include poly (lactic-*co*-glycolic acid) (PLGA), chitosan, and polyethylene glycol (PEG) [50]. Antimicrobial medications can be delivered more effectively when they are encapsulated in polymeric nanoparticles. Polymeric micelles are self-assembled nanostructures with amphiphilic properties. These micelles have a hydrophobic core that can encapsulate hydrophobic antimicrobial chemicals and a hydrophilic shell that provides stability and stops aggregation. Antimicrobials are more readily soluble in polymeric micelles, and their bioavailability and sustained release can also be controlled. They may be designed to target specific locations or illnesses, so as to enhance the therapeutic efficacy [52]. Of all the natural polymers, chitosan, a cationic polysaccharide, has demonstrated an exceptional ability to encapsulate antibiotics and enhance their effectiveness. The characteristics of nanocarriers based on chitosan can be regulated by altering the hydrophobicity of the side chains to reveal the positively charged components on the surface. The chitosan carriers possessing positive charges have an affinity for bacteria, which possess negatively charged membrane surfaces. This interaction facilitates the rupture of the bacterial membrane and enables the effective transport of medications into the bacterial cells [42].

Electrospun polymeric nanofibers are gaining popularity as antibacterial materials [53]. These nanofibers can be loaded with antimicrobial medicines or directly functionalized with antimicrobial components such as peptides or metal nanoparticles [54]. The high surface area-to-volume ratio of nanofibers allows for effective antimicrobial interactions. Furthermore, the nanofiber structure can mimic the extracellular matrix, increasing cell adhesion and delivering antimicrobial compounds directly at the site of infection [55]. Drug–polymer conjugates are also being widely explored. In polymer–drug conjugates, antimicrobial drugs are covalently bonded to polymer backbones [50]. The stability, solubility, and targeted administration of the antimicrobial agents are all improved by polymer conjugation. Furthermore, the polymer can behave as a prodrug, releasing the active antimicrobial agent in response to enzymatic or chemical signals in the target environment.

Polymer–drug conjugates have antibacterial activity that lasts for a long time and have low systemic toxicity. Thus, polymeric nanosystems have the potential to be used to create antimicrobial coatings for a wide range of surfaces, including medical equipment, implants, and textiles [51, 52, 55]. These coatings can inhibit microbial adherence, biofilm production, and microbial growth. Antimicrobial chemicals can be functionalized using polymers, or polymers having inherent antibacterial characteristics, such as cationic polymers that destroy microbial membranes, can be developed. Polymeric nanosystems offer a versatile framework for the development of antimicrobial agents with increased efficacy, controlled release, and targeted dispersion. More research is still being done to enhance the design, stability, and therapeutic potential of polymeric nanosystems as effective antibacterial agents in a wide range of medicinal and industrial applications.

5.2.5 Drug Encapsulated Nanoparticles

As antimicrobial nanomaterials, drug-encapsulated nanoparticles such as liposomal nanoparticles, solid lipid nanoparticles (SLNs), and carbon-based nanoparticles have shown potential [56–58]. These systems enable the tailored distribution of antimicrobial drugs, which improves efficacy, decreases adverse effects, and combats drug resistance. These drug-encapsulated nanoparticles as antimicrobial nanomaterials include the following type of materials:

Liposomes Liposomes are spherical vesicles made up of lipid bilayers. Within their watery cores or lipid bilayers, they can encapsulate hydrophilic and hydrophobic antimicrobial drugs, respectively. The advantages of liposomal nanoparticles include high drug-loading capacity, biocompatibility, and controlled release. They can target microbial cells, merge with their membranes, and deliver the antimicrobial chemicals within the cells [56]. Antibiotics, antifungal drugs, and antimicrobial peptides have been studied using liposomal formulations to improve their pharmacokinetics and therapeutic efficacy [59, 60].

Solid Lipid Nanoparticles (SLNs) At room temperature, solid lipid nanoparticles are colloidal nanoscale particles consisting of lipids. Hydrophobic antimicrobial compounds can be effectively encapsulated by SLNs. They provide regulated release, increased stability, and resistance to deterioration. Antimicrobial activity of SLNs against numerous microorganisms, including bacteria and fungus, has been demonstrated [61, 62]. The lipid matrix of SLNs can interact with bacteria membranes, enabling the delivery of antimicrobial chemicals contained in the SLNs.

Carbon-Based Nanoparticles Carbon nanotubes (CNTs) and graphene-based materials have showed promise as antimicrobial agent carriers [63, 64]. These nanomaterials have a large surface area. They are programmable and can interact with microbial cells. Antimicrobial drugs can be functionalized into carbon-based nanoparticles, thereby enabling targeted distribution and increased antimicrobial efficacy. They can physically damage microbial membranes, infiltrate cells, and cause oxidative stress, which leads to microbial cell death. Furthermore, antibiotics or antimicrobial peptides can be incorporated into carbon-based nanoparticles, enhancing their stability, solubility, and bioavailability.

The encapsulation of antimicrobial within nanoparticles has various advantages, including increased drug stability, controlled release, and targeted delivery. These drug-encapsulated nanoparticles can improve the therapeutic efficacy of the antimicrobial agent, reduce systemic toxicity, and slow down the development of drug resistance [56]. To obtain desirable antibacterial characteristics while minimizing potential harmful effects, parameters such as particle size, drug loading, and surface modifications must be optimized. Continued research and development in this sector can have a high potential for advancing drug-encapsulated nanoparticles as effective antimicrobial nanomaterials for treating microbial illnesses.

5.2.6 Traditional Indian Nanoparticle Systems (Mineral Organic Bhasma)

Indian traditional medicine has been using various nanoformulations, such as tamrabhasma (copper nanoformulations), rajatbhasma (silver nanoformulations), and swarnabhasma (gold nanoparticles), for treatment of various bacterial infectious ailments both as topical and internal medicine as detailed in the Charakh Samhita. The synthesis of these systems is generally in a top-down approach where a metallic sheet or foil is heated and systematically incubated in various carbonaceous plant or animal extract where the surfaces are oxidized and converted to nanoforms. The use of lipids of animal origin such as ghee and navneet provides for various surface groups that stabilize the nanosystems as well as make them biocompatible. Although known by the name of a single metal, their synthesis involves the use of secondary heavy metals such as arsenic, thereby enhancing their antimicrobial action. Their very use for centuries is testimony to the fact that these do not have cytotoxic effects as anticipated by modern science. Cytotoxicity and stability of these nanoformulations were potentially ensured by the incorporation of surface ligands. In hospital settings, some of these nanosystems are very effective in treating diseases of gastrointestinal system caused by multidrug-resistant microbes. However, the use of mineral organic bhasmas needs to be systematically evaluated for global acceptance and use.

5.3 Free Antimicrobial Nanomaterials

5.3.1 Considerations for Design of Free Nanoparticle-Based Systems

To ensure the efficacy and safety of free nanoparticle-based systems as antimicrobials, there are several important considerations. First and foremost is the choice of the nanomaterial and its composition. Metals such as silver, copper, and zinc oxide are being studied for their antibacterial properties due to their propensity to generate metal ions that inhibit the growth of bacteria. Metals (such as silver and copper), metal oxides (such as zinc oxide and titanium dioxide), and carbon-based materials (such as graphene and carbon nanotubes) all have diverse antibacterial properties [12, 41, 65, 66]. It is crucial to select a nanomaterial having antimicrobial activity against the target microorganisms. While the antimicrobial activity should be high, toxicity to non-target organisms such as human beings should be low. Stability, biocompatibility, and cost-effectiveness are additional aspects to be kept in mind when selecting materials. The antibacterial activity of nanoparticles is also affected by their size, shape, and surface area [19, 23, 67]. Smaller nanoparticles have a larger surface area-to-volume ratio, which enhances interactions with bacteria and hence increases antimicrobial activity. Nanoparticles that are too small, on the other hand, may agglomerate or become unstable. As a result, the optimal size that ensures high antibacterial activity and ensures stability is preferred for practical applications. Furthermore, nanoparticles of specific shapes, such as nanospheres, nanorods, nano-decahedron, or nanotubes, exhibit unique antibacterial properties according to their morphological properties [23, 68]. The capacity to modify the size and shape of nanoparticles allows for improved antimicrobial activity against certain illnesses.

A surface modification of nanoparticles is another important consideration for the creation of effective antibacterial nanomaterials. Antimicrobial substances with inherent antimicrobial activity, such as peptides or polymers, can be functionalized onto the surface of nanoparticles to boost their killing power and allow them to target certain types of microorganisms [69]. Furthermore, nanoparticle surface modification can increase nanoparticle stability, dispersibility, and bioavailability, prevent agglomeration, and lessen potential toxicity to the host cells or animals [70]. The controlled release of antibacterial compounds from nanoparticles while limiting side effects is another key concern. Embedding antimicrobials in nanoparticle matrixes or on their surfaces sustains their antibacterial action [12]. Nanoparticle stability is also essential for long-term antibacterial usage. Nanoparticle aggregation can reduce their antibacterial activity. Surface modifications, stabilizing agents, and encapsulation inside matrices can all help to prevent aggregation and maintain optimal antibacterial action.

It is also important to consider environmental factors and regulatory constraints. The fate, ecological effects, and potential for antibiotic resistance of unbound antimicrobial nanoparticles must also be investigated. Antimicrobial nanoparticles must also meet local regulations. Nanoparticle scalability and manufacturability should be considered. Fabricating nanoparticles with constant properties and antibacterial activity requires reproducible procedures. Finally, creating free antimicrobial nanoparticle systems must include nanoparticle composition, size, shape, surface changes, controlled release mechanisms, cytotoxicity and biocompatibility, scalability, and cost-effectiveness. Researchers can design highly efficient and safe nanomaterials that minimize risk to non-target organisms and prevent environmental antimicrobial resistance by addressing these properties.

5.3.2 Techniques for Synthesis of Nanoparticles

Nanoparticle synthesis methods include solvochemical method, hydrothermal method, microwave-assisted method, ultrasonication-based method, photo-assisted method, and traditional methods. Some of the industrial mass production methods and standard top-down synthesis approaches are discussed here.

5.3.2.1 Hydrothermal/Solvothermal Methods

Hydrothermal/solvothermal process is one of the methods used to synthesize nanoparticles using solvents in a sealed vessel. In this technique, the temperature of the solvent can be brought to their critical point through heating under pressure. When water is used as a solvent the method is referred as a "hydrothermal" process and when organic solvents are used it is referred as "solvothermal" process. The synthesis process generally consists of the precursor or the reactants in the form of solution, gel or suspension, the mineralizer or inorganic/organic additives present at a high concentration (used for controlling the pH) and the organic/inorganics additives present at a low concentration (used for controlling the morphology of the crystals or for enhancing the particle dispersion). Advantages of this method are that it produces large amount of nanomaterials at a lower cost, and it also yields highly crystalline nanoparticles with controlled dimensions. Factors that affect nanoparticles synthesis include the type of solvent, duration of the reaction and reaction temperature. Giannousi et al. [71] reported a hydrothermal based synthesis of Cu, Cu_2O, and Cu/Cu_2O in the presence of surfactants such as Tween 20 and PEG8000. The experimental conditions affect the size of nanoparticles and varying size from 10 to 44 nm has been reported. Out of the synthesized nanoparticles, Cu_2O nanoparticles of size 12 and 16 nm were found to exhibit the lowest IC_{50} values of 3.7 μg/mL and 2.13 μg/ml, respectively.

5.3.2.2 Microwave-Assisted Methods

Microwave (MW) is an electromagnetic wave with a frequency of 0.3–300 GHz. It is an environment-friendly method for synthesizing metal nanoparticles. A soluble salt of metal can be used as a source of metal and a reducing reagent is used to reduce the metallic ions to form a metal nanostructure in the presence of microwave (in the presence/absence of additives or surfactants). Water, being a polar solvent is commonly used in MW synthesis. Water is non-flammable, non-toxic, and has a relatively lower vapour pressure in comparison with organic solvents. Loss tangent of a solvent is dependent on the relaxation time of the molecule. Ethylene glycol/glycerol/1,3-propanediol are polyalcohol with high boiling points that are commonly used as solvent for MW-assisted synthesis since they have high loss tangents. Due to the tendency of these highly viscous solvents to form H-bonds, these solvents have a long relaxation time. Other solvents such as alcohols, dimethyl formamide, low molecular weight PEG, toluene, and dimethyl sulfoxide can also be used as solvents in MW-assisted synthesis.

Gold, silver, platinum, palladium, and copper are the metal nanoparticles that are frequently used. Platinum, palladium, gold, silver, copper, iron, nickel, and bismuth are synthesized by using microwave-assisted polyol method. In the presence of ionic liquids, microwave-assisted rapid synthesis of gold, cobalt, manganese, chromium, and molybdenum is possible. Ionic character and the high polarizability of the ionic liquids make them susceptible to microwave irradiation. Hong et al. [21] prepared Ag nanocubes, nanospheres, and nanowires using microwave-assisted method and compared the shape-dependent antimicrobial activities of the nanoparticles. The synthesis method involved reduction of silver nitrate in the presence of polyvinyl pyrrolidone and ethylene glycol in a microwave oven heated at 320 W. Silver nanocubes were found to have the least minimum inhibitory concentration (MIC) of 37.5 μg/mL and silver nanowires were found to have the worst MIC of 100 μg/ml.

5.3.2.3 Ultrasonication-Based Methods

Metal nanoparticles can be obtained without using a reducing agent by passing ultrasound [72]. Sonolysis of the metal salt solution (aqueous/ethanolic) generate H• and other secondary radicals that act as reducing agents to form metal nanoparticles. Surfactants are used to control the size of the nanoparticles. Core-shell metal nanoparticles can be synthesized using sonochemical reduction of 2 metallic salts that have a difference in their reduction potential. Sonochemical co-reduction of Au(III) and Ag(I) ions in aqueous solution was used to produce bimetallic Au–Ag nanoparticles [73]. Polymers and surfactants also act as capping agent and structure directing agent, this leads to non-spherical structures like nanorod and nanobelt. An advantage of this method is increased surface area, better distribution of size and shorter reaction time.

5.3.2.4 Photo-Assisted Methods

Photo-assisted nanomaterial synthesis employs light as a driving factor to guide and regulate the production of nanoparticles or nanostructures [74]. When compared to standard chemical synthesis methods, these techniques have various advantages, including the capacity to tailor the size, shape, and composition of nanomaterials, as well as lower energy usage and softer reaction conditions. Photoreduction, in which light reduces metal ions to generate nanoparticles, and photochemical vapour deposition, in which light-induced processes deposit nanostructures on surfaces, are two extensively used photo-assisted technologies. Photothermal synthesis uses light-induced heating to promote nanomaterial nucleation and growth, whereas photocatalytic synthesis uses light-sensitive catalysts to begin reactions that result in the formation of nanoscale products. In addition, photoinduced self-assembly, photochemical etching, and photolithography are used to build ordered nanostructures and templates for nanomaterial fabrication [75, 76]. These adaptable photo-assisted techniques allow researchers to modify nanomaterial characteristics for a wide range of applications, including catalysis, electronics, photonics, biomedicine, and environmental remediation, and are predicted to further advance nanotechnology and photonics.

5.3.2.5 Traditional Methods

Grinding is one of the tradition physical process and mortar and pestle are the traditional tools used in this method. In grinding/high-energy ball milling, mechanical force is applied. This force causes weakening of the chemical bonds, creates defects, destabilizes, and make the nanoparticles reactive [77].

5.3.3 Applications in Healthcare Settings, Water Treatment, Food Packaging, and Consumer Products

Antimicrobial nanoparticles are used extensively in healthcare settings (a) for wound healing applications particularly in certain critical cases such as trauma wounds, diabetic foot ulcers, and burns; (b) as prosthetic implants to prevent biofilm formation, (c) in sustained, advanced drug delivery applications (d) for centralized water and air handling units of critical care facilities to prevent the spread of nosocomial infections and (e) as general disinfectants in pathological and biochemistry laboratories. In water treatment facilities metal and metal oxide-coated pipes are used for systemic inactivation of bacteria, and in packed bed filters as nanocomposites, nanofibers and as photocatalytic nanocomposites. Advancements in membrane technology have seen an increase in use of copper and silver nanomaterials for enhanced pathogen inactivation.

Bacteria such as *E.coli*, *Salmonella*, and *Listeria monocytogenes* that are known to cause severe illness are often found to contaminate food and water. A surge in nanotechnology research has made a significant impact in the food packaging industry. The beneficial effect is seen to extend the shelf life of the packaged food, preserve the nutritional value by preventing microorganism aided oxidation, and as smart indicators for detection of spoilage. The use of these nanostructures is also explored as alternative chemical-free food preservatives. Personal care products such as soaps, toothpastes, and deodorants have seen an increase in use of nanoparticles to prevent bacterial and fungal infections. Incorporation of antimicrobial nanoparticles in personal care products have enhanced the overall hygiene.

5.4 Immobilized Antimicrobial Nanomaterials

5.4.1 Considerations in the Design of Nanoparticle Immobilized Systems for Sustained Use

The primary consideration in design of immobilized nanoparticle systems is the selection of the nanoparticles with defined elemental composition, surface ligands, surface charge, and morphology. The next and most important step is the selection and modification of the target where the nanoparticles/nanocomposites need to be immobilized. The target must be suitably modified with various surface chemistries to immobilize the nanoparticles on it for providing stable antimicrobial action. Laboratory studies typically focus on analysis of antimicrobial kinetics, targeted delivery or release kinetics, biocompatibility, and cytotoxicity studies. However, for most real-world antimicrobial applications in complex aqueous matrices, system specific variables such as ionic strength, pH, temperature, and dissolved gases and others may affect the kinetics of antimicrobial action. The immobilization of the nanoparticles should ideally be unaffected by changes in solution chemistry. Scalability and manufacturing at an industrial scale, quality control, and compliance to environmental and medical regulations are some of the additional critical considerations for deploying these nanoparticle-based systems. However, these aspects are usually neglected in routine laboratory research.

5.4.2 Nanoparticle Immobilization Strategies

(i) *Physical adsorption*: Physical adsorption or physisorption involves simple adsorption or attachment of the nanoparticle on the substrate. The substrate is usually solid, and the adsorbate is dissolved in a solvent. Physical adsorption makes use of weak forces such as van der Waals interactions, hydrophobic interactions, or electrostatic interactions. The process of physical adsorption is quick and reversible. It is not very selective, and adsorbate multilayer formation occurs. Nanoparticles are loosely attached since they involve weaker forces. Thus, even temperature or pH variations may enhance their detachment or desorption.

(ii) *Entrapment*: In this mechanism, the target is entrapped within a network rather than being directly bonded on to the surface of the substrate [78, 79]. The pore size of the matrix significantly affects the capacity of the substrate for loading and retaining the target. Active species diffusion through a thick polymer matrix may control the inactivation kinetics. Thus, reducing the size of the entrapping polymer matrix may significantly reduce mass transfer limitations. Sol-gel techniques, polymerization for generation of insoluble polymers, cross-linking of polymers, and supramolecular assembly are the four basic approaches for entrapment [80].

(iii) *Electrostatic immobilization*: In this approach, the charged target molecules interact electrostatically with a substrate carrying opposite charge [81, 82]. For instance, negatively charged molecules may be immobilized on a positively charged substrate. Compared to other approaches, electrostatic immobilization is easy to implement but may provide lower stability.

(iv) *Covalent immobilization*: In this approach, covalent bond is formed between the target nanoparticle and the substrate [83–85]. Covalent bond formation occurs between the surface functional groups of the target and the substrate. In general, the process of attaching a target to a substrate involves two steps: first, the surface is activated using linker molecules such as glutaraldehyde or thiol, and subsequently the target is covalently coupled to the substrate. Since strong bonds are formed, release of the target from the substrate is avoided. However, occasionally, covalent immobilization may cause deactivation of the substrate. Park et al. (2012) reported covalent immobilization of colloidal silver and palladium nanoparticles on thiol-modified cellulose fabric. The nanoparticle-modified fabric was found to have high antimicrobial efficacy with minimum release of nanoparticles [85].

5.4.3 Applications in Healthcare Settings, Water Treatment, Food Packaging, and Consumer Products

Because of their unique features and capacity to suppress the development of bacteria, immobilized antimicrobial nanoparticles have found a variety of uses in healthcare settings, water treatment, food packaging, and consumer items. Some examples of applications in each of these fields are discussed in the following subsections.

5.4.3.1 Application in Healthcare Facilities

Hospital-acquired illnesses have increased, threatening world health. Biomedical implants are the main cause of contamination. Hence, researchers worldwide are developing implants coated with antimicrobial materials. Biofilms on implants make antibiotic treatment difficult [86]. The rapid return of antibiotic resistance is another concern. Several metal and metal oxide nanoparticles have antimicrobial activity. To prevent bacterial colonization and minimize the risk of healthcare-associated infections, immobilized antimicrobial nanoparticles can be incorporated into a variety of medical equipment such as catheters, implants, wound dressings, and surgical tools [87]. Nanomaterials can be immobilized onto surfaces such as hospital walls, floors, and furniture to generate antimicrobial coatings that limit pathogen development and spread, thereby lowering the risk of cross-contamination [88]. A variety of metal and metal oxide nanoparticles, and two-dimensional nanomaterials, and their composites have been utilized as antimicrobial coating agents for biomedical implants. Such implants can effectively battle biofilm-associated illness and disorders [89].

5.4.3.2 Application in Water Purification

Immobilized antimicrobial nanoparticles have the potential to inactivate harmful microbes in water and can thus be utilized for water purification. Silver, copper, zinc, and titanium dioxide nanoparticles have been used for water purification. These nanoparticles have been reported to cause efficient elimination of bacteria, viruses, protozoa, and fungi commonly found in water [66, 90–92]. Silver nanoparticles are widely reported to possess potent antibacterial capabilities, and they have been incorporated in various devices for disinfection of water [92]. Immobilized antimicrobial nanoparticles can be employed in water disinfection systems to inactivate and remove microbes and generate safe drinking water. To improve antimicrobial capabilities, nanomaterials can be immobilized on filtration membranes. This allows bacteria, viruses, and other microbes to be removed from water sources, thereby enhancing water quality and safety [66, 92]. Moreover, immobilized antimicrobial nanoparticles can prevent the formation of harmful biofilms in water distribution systems. These nanoparticles maintain the performance of the equipment by either limiting the adhesion of microorganisms or by destroying the structure of biofilms [93]. To suppress the formation of biofilm and regulate the proliferation of microorganisms in water systems, various nanoparticles including silver and copper-based nanoparticles have been utilized [88, 94]. Antimicrobial nanoparticles may be integrated into water filtration systems to improve the removal of microorganisms from water. Nanoparticles of titanium dioxide have been included in water filters to improve microbial inactivation during water treatment [95]. Immobilized antibacterial nanoparticles may also be used for purifying drinking water. They eliminate or considerably reduce waterborne microbes, making drinking water safe. Antimicrobial nanoparticle-based point-of-use water treatment devices are widely used to control microbes and improve water quality. However, a thorough examination of their efficacy, toxicity, and long-term viability is needed to ensure if their usage in water treatment systems is safe and effective.

5.4.3.3 Application in Food Packaging

Nanomaterials, as films and coatings, can be immobilized onto food packaging materials as antimicrobial films and coatings to limit the development of spoilage-causing bacteria and increase the shelf life of packaged food items [96, 97]. This contributes to the preservation of product quality and the reduction of food waste. Silver nanoparticles, zinc oxide nanoparticles, and chitosan nanoparticles are some of the antimicrobial agents that may be incorporated into food packaging [96, 98]. Nanoparticle-incorporation techniques can be used to integrate cutting-edge technology into food manufacture, development, fabrication, packaging, storage, and distribution. Food science using nanomaterials can use many nanostructured materials. Antimicrobial nanoparticles (Cu/CuO, Ag, MgO, TiO_2, ZnO, carbon dots, mesoporous particles, and graphene) are used in food preservation, additives, and packaging. Nanoparticles outperform standard packaging with respect to the creation of mechanical barrier, heat resistance, and biodegradability. Nanomaterials enable nanosensors to detect food deterioration. The active substance can permeate food or may be in the form of nanostructured polymeric films or nanoparticle-encapsulating polymers to limit microbial activity. Active packaging with antimicrobial nanoparticles kills harmful germs to maintain food quality and shelf life. Active packing with nanoparticles is stronger, lighter, and less O_2 accessible [97]. These antimicrobials can prevent food spoilage by inhibiting bacteria, fungus, and other pathogens. Thus, food contamination is reduced and packaged goods last longer.

Antimicrobial nanoparticles can be adhered to a substrate via electrostatic, hydrogen bonding, and covalent interactions. In food packaging these interactions are utilized to form antimicrobial nanoparticle films or coatings such that extremely thin, translucent, and flexible packaging may be synthesized using materials such as plastic, paper, and metal [97, 99]. These films and coatings prepared using nanomaterials can inhibit gas penetration thereby acting as a barrier preventing the permeation of oxygen and moisture. Exposure to other environmental variables that might hasten the decomposition of food is also prevented [99]. In addition, the nanoscale size of the nanoparticles found in these materials offers a higher surface area, which boosts the antibacterial activity of the material. Nanoparticles have the potential to interact with the microbes, and lead to growth inhibition and eventually inactivate them due to the disruption of the cell membranes and other causes [96, 99]. More extensive research and risk evaluations should be carried out to ensure that nanomaterials in food packaging would not increase health hazard. Regulatory agencies all over the globe, such as the Food and Drug Administration (FDA) in the United States and the European Food Safety Authority (EFSA), are responsible for analysing the safety and possible dangers associated with the use of nanomaterials in food packaging [100, 101]. The use of nanoparticles in food packaging offers promise for enhancing food safety and increase the shelf life of packaged food products by preventing the proliferation of bacteria and other microbes that cause food to decay.

5.4.3.4 Application in Consumer Goods

Antibacterial properties can be added to fabrics and textiles by immobilizing antibacterial nanoparticles [102]. Several approaches can be used to accomplish this. Coating is one strategy, in which nanoparticles are applied as a thin layer on the surface of the fabric using processes such as dip coating or electrostatic deposition. For example, silver nanoparticles can be deposited into sporting materials, facemasks, and other fabrics to suppress the development of odour due to bacterial activity [1, 103, 104]. Copper nanoparticles may be immobilized within hospital linens, thereby preventing the proliferation of pathogens [105, 106]. Another approach involves the incorporation of nanoparticles into inks or dyes used to print graphics onto cloth. Titanium dioxide nanoparticles can be printed on curtains to provide antibacterial capabilities against airborne infections [106]. Fabrics with antibacterial nanoparticles can resist bacteria growth, making them beneficial in healthcare applications and in sports goods. Nanoparticle integration is being optimized to improve efficacy, durability, and environmental and health safety. Healthcare, athletics, and other clothes that inhibit microbes and odour would benefit from nanoparticle insertion.

Nanomaterials can significantly improve the antibacterial efficiency and long-term protection of personal care products such as soaps, hand sanitizers, and cosmetics. Table 5.1 summarizes various antimicrobial applications of various nanoparticles. These materials have unique properties that fight bacteria, fungi, and other microorganisms. Personal care products use nanosilver because of its potent antibacterial properties against many diseases. Titanium dioxide nanoparticles, which have photocatalytic capabilities and kill bacteria when exposed to light, are utilized in sunscreens and cosmetics. Deodorants, antiperspirants, and lotions contain zinc oxide nanoparticles to inhibit bacterial growth and manage body odour. Lotions, creams, and shampoos use nanoemulsion with tiny droplets of antimicrobial compounds to boost antibacterial capabilities [107, 108]. By adding these nanoparticles, manufacturers may boost antimicrobial efficacy and protect skin and sanitary goods from harmful microorganisms. To ensure the safe and responsible use of nanomaterials, their environmental and health implications must be assessed throughout the product lifecycle. Immobilized antimicrobial nanoparticles extend antibacterial action, minimize antimicrobial resistance, and improve product performance. However, nanomaterial safety and regulatory compliance are essential to protect human health and the environment. R&D involves optimizing the design, performance, and scalability of immobilized antimicrobial nanoparticles to meet specific application needs while minimizing dangers.

Table 5.1 Application of immobilized nano-antimicrobials for healthcare settings, water treatment, food packaging, and consumer products

S. No.	Types of nanomaterials	Characteristics	Effect on microbes[a]	Application	References
1.	Silver Oxide (Ag_2O) Coating	Mean particle size = 1.5 µm Shape = lotus-leaf-like-feature	99.3% of SARS-CoV-2 and > 99.5% of *P. aeruginosa*, *S. aureus*, and *methicillin-resistant S. Aureus*	Antimicrobial coating of silver oxide particles in a silicate matrix on glass	[109]
2.	Cu Doped Carbon Dots(Cu-CDs)	Average particle size = 4.5 nm	*E. coli* *S. aureus* *S. mutans*	Daily strategy for oral healthcare	[110]
3.	Nano-ZnO	Oval shaped	*S. mutans*: inhibited;*P. gingivalis*: lower than 20%	Polycarboxylate cement (dental cement)	[111]
4.	Nano-TiO_2	Spherical shape	*S. mutans* biofilm: affected the viability	To overcome the problem of secondary caries as dental adhesives in adhesive dentistry	[112]
5.	Ag/ZnO-Decorated Micro-/Nanoporous Sulfonated Polyetheretherketone	Ag = 50–70 nm ZnO = 120–150 nm	*S. aureus*: ~99.3% decrease *E. coli:* more than a 99.2% decrease	A dual therapy implant coating on the 3D micro/nanoporous sulfonated PEEK	[113]
6.	Ag/ZnO	Rod-like morphologylength: 300–500 nm width: 10–20 nm	*S. mutans*: inhibited 91% of bacterial growth	Potentially effective dental antibacterial agents	[114]
7.	CeO_2–Ti	Octa-CeO_2 rod-CeO_2,cube-CeO_2	*S. sanguinis*, *P. gingivalis*, *F. nucleatum*: decreased up to 2 orders of magnitude; octa>cube>rod	Implant surface-modification strategy	[115]
8.	Nano-TiO_2-litchi peel extract (LPE)- chitosan (CS) matrix	Opacity = 5.71 ± 0.15 mm^{-1} CTL film = rough surface Thickness = 63.50 ± 3.81 µm		Novel active packaging film and apple coating	[116]

9.	Mahua oil-based polyurethane/chitosan/nano-ZnO composite film	Average film thickness = 0.1 mm	Zone of inhibition: *E. coli* = 25 mm *S. aureus* = 20 mm	Biodegradable food packaging for carrot	[117]
10.	Polylactic acid (PLA)/TiO_2/GO nanofibrous films	Diameter Pure PLA = 331 ± 155 nm After adding TiO_2 NPs = 297 ± 128 nm	Inhibition rates *E. coli* = 94.4 ± 1.8% *S. aureus* = 92.6 ± 1.7%	Preservation effect on green peppers	[118]
11.	PLA / Bergamot essential oils (BEO)/nano-TiO_2 + nano-Ag film	Thickness = 50 μm	Microbial growth inhibition;increased shelf-life up to 15 days	Mango preservation	Chi et al. (2019)
12.	Bacterial cellulose (BC)/polypyrrole (PPy)/ZnO	BC = compact surface Size = 40–60 nm. BV/PPy/ZnO diameter = ≈60–150 nm	BC-PPy-ZnO film controlled microbial growth in the chicken thigh meat	Intelligent and active packaging of chicken thigh meat	[119]
13.	CuO NPs	Size = 21–47 nm Shape = spherical	*L. monocytogenes*, *Bacillus cereus*, *E. coli*, *S. typhi*, *P. aeruginosa* and *E. faecalis*	Nano bioremediation of textile industry wastewater	[120]
14.	B-F-TiO_2 NP: Flower-like TiO_2(F-TiO_2) immobilized on basil seed	Size Commercial TiO_2 = 0.11 μm F-TiO_2 = 18.80 nm	Antibacterial ring area 1.83 mm^2 in *E. coli* Methylene blue degradation = 98.95%	Super-adsorbent for dye wastewater	[121]

(continued)

Table 5.1 (continued)

S. No.	Types of nanomaterials	Characteristics	Effect on microbes[a]	Application	References
15.	Ag doped Sn_3O_4 (Immobilized on hyperbranched polypyrrole)	Sn_3O_4crystallite size 36.3 nm	*S. pneumonia* *P. aeruginosa*	Photocatalytic, antibacterial agent and microbial detection	[110]
16.	TiO_2/Ag_2O immobilized on cellulose paper	Size = 100–150 nm	*E. coli* = 3 mm 97% of aniline photodegradation	Enhanced photocatalytic and antibacterial activity	[122]
17.	Graphitic carbon nitride ($g\text{-}C_3N_4$) immobilized Bi_2S_3/Ag_2WO_4 (BS/AW/CN)	Bi_2S_3 = 0.37 nm; Ag_2WO_4 = 0.31 nm $g\text{-}C_3N_4$ = 0.35 nm BS/AW/CN BET surface area = 20.201 m^2/g Total pore vol = 0.1624 cm^3/g Average pore size = 30.012 nm	Photocatalytic antimicrobial activity against *S. aureus* (~1×10^7 CFU/mL) under 140 W	Disinfection of *S. aureus* cells in wastewater	[123]
18.	Zinc oxide NPs ($ZnO_{nano\text{-}synth}$) immobilized on white chitosan-coated silica gel beads ($ZnO_{nano\text{-}com}$)	Average size of ZnO:5 ± 1.7 nm (spherical) $ZnO_{nano\text{-}com}$ = 30.8 ± 0.75 nm (sheet-like shape)	*S. aureus* = 33.3 ± 0.07 mm *E. coli* = 29.54 ± 0.33 mm	Antimicrobial air filter	[124]
19.	Polyaniline/AgNPs	Ag NPs diameter = 5–18 nm	Reduction (%) against *S. aureus* and *E. coli* = 31–49 and 35–58, respectively	Antimicrobial, nonwoven fabrics comprising of polyaniline/Ag NPs	[125]
20.	Ag NPs	Size = 20 nm Shape = spherical	Highest antibacterial efficiency against both *E. Coli* and *S. aureus* 99.99% after 24 h with 160 μg/ml	Antimicrobial agent in pig leather	[126]
21.	Ag doped ZnO	Size: Ag NPs = 40–50 nm ZnO = 300–400 nm	Antibacterial activity: *E. coli* and *S. aureus* up to 64.72% and 58.90% decrease, respectively	Boosting the antibacterial performance of natural rubber latex foam	[127]
22.	Chitosan-based Schiff-base TiO_2-ZnO NCs	TiO_2 = 48 nm ZnO = 56 nm	*S. aureus*, *E. coli*, and *C. albicans*	Nano-bio finishing of cotton fabric	[128]
23.	Ag NPs	Diameter = 32–144 nm	*E. coli* = 28 mm *S. aureus* = 15 mm *P. aeruginosa* = 22 mm *B. subtilis* = 20 mm	Ag NPs deposited on cotton fabrics with antimicrobial properties	[129]

[a]Effect on microbes is expressed as %reduction of microbes in batch studies and clearing zone in disc diffusion studies (mm)

5.5 Bioaccumulation, Environmental Uptake, and Ecotoxicology

When analyzing the possible environmental implications of either free or immobilized antimicrobial nanoparticles, their bioaccumulation, uptake, and ecotoxicology are critical concerns for understanding their environmental impact. Nanomaterials are reported to bioaccumulate in living tissues [130]. Bioaccumulation is the gradual build-up of a substance in living tissues. Bioaccumulation is likely to be affected by the physicochemical features of nanomaterials, exposure pathways, and nature of the organisms. Small, and stable nanomaterials can bioaccumulate in aquatic and terrestrial creatures and get biomagnified along the food chain. Additionally, immobilized antibacterial nanoparticles must be tested to see if they can accumulate in organisms and cause harm. Bioaccumulation can occur by eating, inhalation, and absorption through the skin [131]. Size of the nanomaterial also affects bioavailability and bioaccumulation. Smaller particles are more likely to be absorbed and accumulate in tissues. Charge and functionalization on nanomaterial surfaces may change their interactions with biological systems and may alter their bioaccumulation [132]. Bioaccumulation potential is also affected by the environmental stability and degradation of antimicrobial nanoparticles. Absorption and bioaccumulation may depend on their release kinetics, aggregation, and transformation. The metabolic and elimination pathways of organisms can also affect nanomaterial bioaccumulation. However, persistent nanoparticles can bioaccumulate in tissues. Laboratory studies can examine immobilized antibacterial nanomaterial bioaccumulation using animals from different trophic levels. Bioconcentration factors (BCFs) and biomagnification potential can be determined by monitoring the nanomaterial concentration in tissues. Bioaccumulation potential can also be assessed using mathematical modelling based on physicochemical and environmental variables [130].

Another concern is environmental absorption, or antimicrobial nanoparticle penetration and dispersion. Understanding how nanomaterials are released, transported, and distributed in water, soil, and air is crucial. Environmental absorption of nanoparticles is affected by the application processes, release mechanisms, ambient conditions, and physicochemical features of the nanomaterial [133]. Nanoparticles enter the ecosystem with water and effluents, air deposition, and from plant or animal biomass. Physicochemical properties, environmental matrix interactions, and ecological variables impact nanomaterial fate and mobility. Knowledge of environmental absorption pathways and processes is required to comprehend the potential risks posed by free antibacterial nanoparticles. Understanding nanomaterial absorption is crucial to analyzing their impact on ecosystems [134]. Nanomaterial-containing goods, leaching from treated surfaces, erosion or deterioration of nanomaterial-coated surfaces, and production or disposal can release nanoparticles. Immobilized antibacterial nanoparticles can be released through several processes. Discharged immobilized antimicrobial nanoparticles may be affected by various fate and transport processes. They may get accumulated in plants and soil. Water, wind, and soil erosion can transport them to different environments. Agglomeration, sedimentation, and water dispersion affect nanomaterial behaviour and dispersion [135].

Furthermore, soil contamination from nanoparticles can occur through a variety of entry sites, including agricultural use of nanomaterial-containing commodities, sewage sludge or biosolids discharged from wastewater treatment plants, and air deposition [133]. Thus, nanoparticles may be transferred across living organisms in various trophic levels [136]. Antimicrobial nanoparticles, both free and immobilized, can enter food webs via trophic transfer. If species at lower trophic levels acquire nanoparticles, they can be transmitted to higher trophic levels by predation or ingestion. This can result in biomagnifications [130]. Thus, predators who consume prey would be exposed to more nanoparticles and if these are stored in their tissues the nanoparticle content in the tissues may increase and they may manifest greater adverse effects. Assessing the environmental risks of free antimicrobial nanoparticles necessitates understanding their absorption. Investigations need to be conducted on the fate, behaviour, and interactions of nanoparticles with organisms under a variety of circumstances.

Ecotoxicology studies on how pollutants affect ecosystems, and their inhabitants should be taken up. When assessing ecotoxicological effects of antimicrobial nanoparticles, aquatic, plant, and soil species must be considered [137]. Nanomaterials can disrupt membranes, cause oxidative stress, or release toxic ions and reactive species [138]. Study on the impacts of nanoparticles on growth, reproduction, behaviour, and survival of exposed species, as well as ecosystem processes and biodiversity are crucial. Free antimicrobial nanoparticles pose environmental risks, hence comprehensive risk assessment methods are needed. This includes understanding nanomaterial exposure mechanisms, environmental absorption, and ecotoxicology. The concentration, time of exposure, and type of nanomaterial must be determined to assess potential risks [139]. Antimicrobial nanoparticles, both free and immobilized, can enter organisms by ingestion, inhalation, or skin contact. By analyzing these channels of contact, we may get insight into the many ways in which nanomaterials could exert their effects on living things. Reduced mortality, stunted development, impaired reproduction, altered conduct, and changes in physiological markers have been used as measures of toxicity [137, 140]. These benchmarks and dose–response plots provide data on the potential adverse effects of nanomaterials. Well-established experimental designs and methods have been suggested for use in ecotoxicology studies such that the results may be reliably compared. Scientists and regulatory agencies developed these standards. For more accurate ecotoxicological assessments of immobilized antimicrobial nanoparticles, guidelines can specify test settings, concentrations, duration, and outcomes. Researchers and manufacturers of nanoparticle-containing products should be environmentally conscious throughout the product life. Toxicity can be reduced through careful design and synthesis, production and use confinement, and waste management. Continuous monitoring and investigations are needed to understand the long-term environmental impact of free antimicrobial nanoparticles and ensure their safe and sustainable use.

5.5.1 Safety Considerations of Antimicrobial Nanomaterials

Antimicrobial nanomaterials have recently been recognized as a potentially useful weapon in the ongoing war against illnesses caused by microbes. However, safety issues cannot be ignored. They should be carefully considered because of the potential toxicity of these nanomaterials and their potential impact on human health. Due to their unusual properties, such as their small size and huge surface area, nanomaterials can exhibit abnormal behaviour and pose a threat to the cells and tissues in the human body. Silver nanoparticles (Ag NPs) are often used as antibacterial agents due to their unique properties. Ag NPs may also cause cytotoxicity and genotoxicity in human cells [141]. Certain nanomaterials can also cause oxidative stress and inflammation, which can lead to a range of illnesses. A few research have examined human exposure mechanisms. Food containing TiO_2NPs are the main source of oral exposure. Most in vivo and in vitro skin exposure studies reported that TiO_2 NPs could not penetrate the stratum corneum (SC). Nanomedicine using intravenous injection can deliver TiO_2 nanoparticle carriers into the body. When given intravenously, TiO_2 NPs can damage the liver, spleen, kidneys, and brain [142]. They further demonstrated that most of these effects may be attributed to the use of extremely high dosages of TiO_2 NPs. Despite rising production and usage, there is a significant dearth of epidemiological data on TiO_2 NPs. Long-term inhalation experiments in rats, on the other hand, have revealed lung tumours. The accumulation of these elements in ecosystems might occur because of their introduction into the environment. Hence, it is essential to conduct exhaustive research on the possible toxicity of antimicrobial nanoparticles to guarantee that their usage in medical applications is risk-free.

5.5.2 Regulatory Guidelines and Standards

Several countries have produced regulatory guidelines and regulations to address safety concerns with antimicrobial nanoparticles. Nanomaterials should be created, manufactured, and used in an environmentally friendly way to reduce health and environmental risks. The US Food and Drug Administration (FDA) regulate nanotechnology in medical product development. These standards emphasize physicochemical characterization of nanomaterials and toxicity evaluations for product quality and safety. Global regulatory bodies have created guidelines for medical use of nanomaterials. The purpose of the Scientific Committee on Consumer Safety (SCCS) Guidance is to aid applicants in putting together safety dossiers, and to aid risk assessors and risk managers in carrying out the requirements of article 16 of Cosmetics Regulation (EC) No 1223/2009 [143]. To aid in the implementation of EU chemicals legislation for nanomaterials that may have an effect on the environment, European Chemical Agency (ECHA) works closely with competent authorities from Member States, the European Commission, non-governmental organizations (NGOs), and industry associations, as well as international organizations such as the Organisation for Economic Cooperation and Development (OECD) (https://echa.europa.eu/regulations/nanomaterials-under-bpr). Compliance with these recommendations and standards guarantees that the possible risks associated with antimicrobial nanoparticles are effectively assessed and managed.

5.5.3 Risk Assessment and Mitigation Strategies

To ensure that antimicrobial nanoparticles may be used in a safe manner, conducting risk assessments is an essential step [133]. It involves finding hazardous circumstances, determining how individuals may be exposed to them, and assessing their risks. If complete risk assessment studies are done, risk mitigation techniques can lower injury risk. Engineering controls are a proven risk-reduction strategy. Enclosed production systems can prevent worker exposure to nanoparticles and their environmental release during manufacture. Segregating nanomaterial waste and disposing it properly can also lessen environmental impacts. PPE may also be needed to protect workers from nanoparticles. Gloves, respirators, and protective clothing prevent contact and inhalation injuries. Risk mitigation can be enhanced by adopting occupational health and safety protocols, training workers, and installing monitoring systems. Current research focuses on developing safer nanomaterials with lower toxicity and environmental safety. Researchers are exploring innovative nanomaterial surface modifications to promote biocompatibility and prevent damage. These materials can be produced and used safely, and the antibacterial nanoparticles reduce human and other organism dangers. In conclusion, a thorough risk assessment and appropriate procedures can assure the safe use of these compounds and optimize their benefits in fighting microbial illnesses.

5.5.4 Challenges in the Application of Antimicrobial Nanomaterials

Scalability, cost-effectiveness, regulatory compliance, and long-term stability are significant challenges to the widespread usage of antibacterial nanoparticles. These factors pose substantial challenges to the development of effective antibacterial nanoparticles. Continuous research and development initiatives are needed to properly address these challenges. Scientists and engineers are now working to develop scalable and cost-effective production procedures, improve nanoparticle stability, and learn more about the risks and benefits associated with the use of nanoparticles. Some of these challenges are addressed below.

5.5.4.1 Scalability and Cost-Effectiveness

Scalability and cost-effectiveness are two major obstacles to using antimicrobial nanoparticles. Despite promising antibacterial properties in the lab, scaling up manufacturing of these materials may be challenging to fulfil the industrial or commercial demands. Particle size, shape, and composition must be carefully controlled in nanomaterial synthesis. Repeatable findings may be hard to obtain at a large scale. The necessity for using specific machinery and techniques would further increase production costs. For instance, silver nanoparticles (Ag NPs) are effective against bacteria. Even though laboratory scale synthesis of Ag NPs is simple and may be done using several chemical and physical methods, large-scale manufacturing is challenging since the size and dispersion of the nanoparticles have to be reproducible. Therefore, cost-effective solutions through manufacturing process optimization are needed to overcome this hurdle. The cost-effectiveness of antimicrobial nanoparticles is also important. Using lots of resources, machinery, and quality inspections can raise the manufacturing cost. Antimicrobial nanoparticles may be too expensive for low-resource or large-scale procedures. If the product is widely accepted, cost-effective manufacturing procedures, alternative raw materials, and production efficiency must be developed.

5.5.4.2 Regulatory Compliance and Approval Processes

Complying with regulations and gaining the requisite permissions presents significant challenges for the usage of antimicrobial nanoparticles. For ensuring the safety of products based on nanomaterials, regulatory agencies worldwide have their own distinct set of regulations covering testing, labelling, and registration. Due to their unique properties, nanoparticles may not fit into typical material regulation frameworks. Nanomaterial efficacy and safety evaluation may require more testing and data, which can be time-consuming and expensive. Drug regulators including the US Food and Drug Administration (FDA) and the European Medicines Agency (EMA) have developed nanomaterial-based medicinal product evaluation and approval guidelines. Physicochemical characterization, toxicological research, and environmental impact assessments are typical criteria. Meeting these criteria may require expert knowledge and testing. The launch of products incorporating nanoparticles onto the market can be delayed because it can be difficult to successfully traverse the complicated regulatory environment and meet the regulatory requirements. Hence, the scientific and production teams must work closely with government bodies, communicate clearly, and meet all standards.

5.5.4.3 Long-Term Stability and Durability

Long-term stability and efficacy of antibacterial nanoparticles is an additional problem. Physical and chemical changes over time may diminish the effectiveness of nanoparticles as antibacterial agents. For instance, the breakdown of antimicrobial surface coatings or ligands on nanoparticles can reduce their efficiency as antibacterial agents. Temperature, humidity, light, and chemical interactions are just a few causes of such degradation. Nanoparticle aggregation changes their characteristics and reduces their efficacy. For antimicrobial nanoparticles to last longer, they need durable covering materials or encapsulating techniques that prevent core material degradation. Encapsulation can do this. The coatings should prevent antimicrobial release and leakage of hazardous species and resist such circumstances during the product's lifespan. Long-term performance and reliability of antibacterial nanoparticles require comprehensive stability testing under relevant settings. Temperature, humidity, pH, and light exposure are some aspects to consider during testing. By understanding how antimicrobial nanomaterials degrade and inventing more durable materials, researchers can overcome this obstacle and maintain their potency. Finally, scalability, cost-effectiveness, regulatory compliance and licensing processes, long-term stability, and durability are all issues that must be addressed to safely employ antimicrobial nanoparticles.

5.6 Future Directions and Emerging Trends

Extensive research is being undertaken by researchers worldwide to tackle AMR both for healthcare and environmental applications. Recent years have seen significant impetus in development of theragnostic nanoparticle-based antimicrobials where conventional metallic, metal oxide, and polymer systems are tagged with antimicrobial peptides, antibiotics, and photoactive labels to achieve target specific antimicrobial action thereby minimalizing systemic toxicity [144, 145]. In some cases, the nanoparticles are directed to their desired site of action externally with magnetic fields [146]. Customized plasmonic core shell nanoparticle systems excited by a suitable wavelength of light can cause hyperthermia and controlled release of drug thus allowing for site-specific antimicrobial action. These photothermal therapies are also being explored with newer carbonaceous polymer substrates loaded with suitable antimicrobials for sustained release of drugs [147]. In recent times, nanoparticles have also been used as carriers of gene-editing tools such as CRISPER-Cas9 for modifying resistance-specific genes in bacterial genome, thus restoring bacterial susceptibility [148].

Environmental applications of antimicrobial systems typically involve engineering nanoparticle coatings on piping, textiles, and surfaces for prevention and/or disruption of biofilm formations. Some of these materials tagged with biomarkers are also used for devising sensors that can potentially evaluate the burden of resistant bacteria and antibiotics in environmental samples [149–151]. There remains significant scope for research on antimicrobial material for tackling antimicrobial resistance as (a) Diagnostics or point of use systems to evaluate AMR (b) as therapeutics to attain enhanced and effective antimicrobial action and (c) for limiting the persistence of antimicrobials and bacteria in ecological pools. This requires interdisciplinary research focussed on understanding the complexities of real-world systems such as the physiochemical effects, matrix and environmental effects of real-world environments.

5.7 Conclusion and Future Perspectives

Considerable research efforts are currently being dedicated to the exploration and development of antimicrobial nanomaterials. These materials hold great potential in addressing the pressing issue of antimicrobial resistance, enhancing infection control measures, advancing medicinal approaches, ensuring the safe preservation of food and water, and several other applications. The bioaccumulation and environmental toxicity of nanomaterials necessitate careful examination of their widespread utilization across various sectors. Although there is an expectation that free antimicrobials may possess more antimicrobial potency, current research indicates that immobilized nanoparticles also have the potential to achieve comparable or improved and prolonged antibacterial activity. There is still significant potential for research to advance antimicrobial nanomaterials and their effective integration onto surfaces that are susceptible to microbial contamination. Examples of such surfaces include air ducts and filters in operating theatres, railings and essential supply tubing in hospitals, and assembly lines in the food and beverage industry, among others. Therefore, it is imperative for scientists, industry stakeholders, and policy makers to collaborate diligently and with a clear objective to effectively use the inherent capabilities of nanomaterials and mitigate the issue of antimicrobial resistance.

References

1. Abazari, M., Badeleh, S. M., Khaleghi, F., Saeedi, M., & Haghi, F. (2023). Fabrication of silver nanoparticles-deposited fabrics as a potential candidate for the development of reusable facemasks and evaluation of their performance. *Scientific Reports, 13*, 1–16.
2. Gudkov, S. V., Burmistrov, D. E., Serov, D. A., Rebezov, M. B., Semenova, A. A., & Lisitsyn, A. B. (2021). A mini review of antibacterial properties of ZnO nanoparticles. *Frontiers of Physics, 9*, 1–12.
3. Valodkar, M., Modi, S., Pal, A., & Thakore, S. (2011). Synthesis and anti-bacterial activity of Cu, Ag and Cu-Ag alloy nanoparticles: A green approach. *Materials Research Bulletin, 46*, 384–389.
4. Singhal, S. K., Lal, M., Kabi, S. R., & Mathur, R. B. (2012). Synthesis of Cu/CNTs nanocomposites for antimicrobial activity. *Advances in Natural Sciences: Nanoscience and Nanotechnology, 3*, 045011.
5. Li, C., Wang, X., Chen, F., Zhang, C., Zhi, X., Wang, K., & Cui, D. (2013). The antifungal activity of graphene oxide-silver nanocomposites. *Biomaterials, 34*, 3882–3890.
6. Baptista, P. V., McCusker, M. P., Carvalho, A., Ferreira, D. A., Mohan, N. M., Martins, M., & Fernandes, A. R. (2018). Nano-strategies to fight multidrug resistant bacteria-"A battle of the titans". *Frontiers in Microbiology, 9*, 1–26.

7. Firouzjaei, M. D., Shamsabadi, A. A., Sharifian, G. M., Rahimpour, A., & Soroush, M. (2018). A novel nanocomposite with superior antibacterial activity: A silver-based metal organic framework embellished with graphene oxide. *Advanced Materials Interfaces, 5*, 1701365.
8. Padmanaban, V. C., Giri Nandagopal, M. S., Madhangi Priyadharshini, G., Maheswari, N., Janani Sree, G., & Selvaraju, N. (2016). Advanced approach for degradation of recalcitrant by nanophotocatalysis using nanocomposites and their future perspectives. *International journal of Environmental Science and Technology, 13*, 1591–1606.
9. Shalaby, T., Mahmoud, O., & Al-Oufy, A. (2016). Antibacterial silver embedded nanofibers for water disinfection. *International Journal of Materials Science and Applications, 4*, 293.
10. Al-Sherbini, A. S. A., Ghannam, H. E. A., El-Ghanam, G. M. A., El-Ella, A. A., & Youssef, A. M. (2019). Utilization of chitosan/ag bionanocomposites as eco-friendly photocatalytic reactor for bactericidal effect and heavy metals removal. *Heliyon, 5*, e01980.
11. Grass, G., Rensing, C., & Solioz, M. (2011). Metallic copper as an antimicrobial surface. *Applied and Environmental Microbiology, 77*, 1541–1547.
12. Agnihotri, S., Mukherji, S., & Mukherji, S. (2013). Immobilized silver nanoparticles enhance contact killing and show highest efficacy: Elucidation of the mechanism of bactericidal action of silver. *Nanoscale, 5*, 7328.
13. Ramyadevi, J., Jeyasubramanian, K., Marikani, A., Rajakumar, G., & Rahuman, A. A. (2012). Synthesis and antimicrobial activity of copper nanoparticles. *Materials Letters, 71*, 114–116.
14. Li, X., Robinson, S. M., Gupta, A., Saha, K., Jiang, Z., Moyano, D. F., Sahar, A., Riley, M. A., & Rotello, V. M. (2014). Functional gold nanoparticles as potent antimicrobial agents against multi-drug-resistant bacteria. *ACS Nano, 8*, 10682–10686.
15. Vazquez-Munoz, R., Arellano-Jimenez, M. J., & Lopez-Ribot, J. L. (2020). Bismuth nanoparticles obtained by a facile synthesis method exhibit antimicrobial activity against Staphylococcus aureus and Candida albicans. *BMC Biomedical Engineering, 2*, 11.
16. Prabhu, S., & Poulose, E. K. (2012). Silver nanoparticles: Mechanism of antimicrobial action, synthesis, medical applications, and toxicity effects. *International Nano Letters, 2*, 32.
17. Pelgrift, R. Y., & Friedman, A. J. (2013). Nanotechnology as a therapeutic tool to combat microbial resistance. *Advanced Drug Delivery Reviews, 65*, 1803–1815.
18. Jung, W. K., Koo, H. C., Kim, K. W., Shin, S., Kim, S. H., & Park, Y. H. (2008). Antibacterial activity and mechanism of action of the silver ion in Staphylococcus aureus and Escherichia coli. *Applied and Environmental Microbiology, 74*, 2171–2178.
19. Mukherji, S., Bharti, S., Shukla, G., & Mukherji, S. (2019). Synthesis and characterization of size- and shape-controlled silver nanoparticles. *Physical Sciences Reviews, 4*, 20170082.
20. Pal, S., Tak, Y. K., & Song, J. M. (2007). Does the antibacterial activity of silver nanoparticles depend on the shape of the nanoparticle? A study of the gram-negative bacterium escherichia coli. *Applied and Environmental Microbiology, 73*, 1712–1720.
21. Hong, X., Wen, J., Xiong, X., & Hu, Y. (2016). Shape effect on the antibacterial activity of silver nanoparticles synthesized via a microwave-assisted method. *Environmental Science and Pollution Research, 23*, 4489–4497.
22. Shaheen, T. I., & Fouda, A. (2018). Green approach for one-pot synthesis of silver nanorod using cellulose nanocrystal and their cytotoxicity and antibacterial assessment. *International Journal of Biological Macromolecules, 106*, 784–792.
23. Bharti, S., Mukherji, S., & Mukherji, S. (2021). Enhanced antibacterial activity of decahedral silver nanoparticles. *Journal of Nanoparticle Research, 23*, 36.
24. Dakal, T. C., Kumar, A., Majumdar, R. S., & Yadav, V. (2016). Mechanistic basis of antimicrobial actions of silver nanoparticles. *Frontiers in Microbiology, 7*, 1–17.
25. Huang, Y. W., Wu, C. H., & Aronstam, R. S. (2010). Toxicity of transition metal oxide nanoparticles: Recent insights from in vitro studies. *Materials (Basel), 3*, 4842–4859.
26. Liga, M. V., Bryant, E. L., Colvin, V. L., & Li, Q. (2011). Virus inactivation by silver doped titanium dioxide nanoparticles for drinking water treatment. *Water Research, 45*, 535–544.
27. Hasan, A., Morshed, M., Memic, A., Hassan, S., Webster, T. J., & Marei, H. E. S. (2018). Nanoparticles in tissue engineering: Applications, challenges and prospects. *International Journal of Nanomedicine, 13*, 5637–5655.

28. Mba, I. E., & Nweze, E. I. (2021). Nanoparticles as therapeutic options for treating multidrug-resistant bacteria: Research progress, challenges, and prospects. *World Journal of Microbiology and Biotechnology, 37*, 1–30.
29. Rubilar, O., Rai, M., Tortella, G., Diez, M. C., Seabra, A. B., & Durán, N. (2013). Biogenic nanoparticles: Copper, copper oxides, copper sulphides, complex copper nanostructures and their applications. *Biotechnology Letters, 35*, 1365–1375.
30. Younas, H., Qazi, I. A., Hashmi, I., Awan, M. A., Mahmood, A., & Qayyum, H. A. (2014). Visible light photocatalytic water disinfection and its kinetics using Ag-doped titania nanoparticles. *Environmental Science and Pollution Research International, 21*, 740–752.
31. Sánchez-López, E., Gomes, D., Esteruelas, G., Bonilla, L., Lopez-Machado, A. L., Galindo, R., Cano, A., Espina, M., Ettcheto, M., Camins, A., et al. (2020). Metal-based nanoparticles as antimicrobial agents: An overview. *Nanomaterials, 10*, 1–39.
32. Sirelkhatim, A., Mahmud, S., Seeni, A., Kaus, N. H. M., Ann, L. C., Bakhori, S. K. M., Hasan, H., & Mohamad, D. (2015). Review on zinc oxide nanoparticles: Antibacterial activity and toxicity mechanism. *Nano-Micro Letters, 7*, 219–242.
33. Moschini, E., Colombo, G., Chirico, G., Capitani, G., Dalle-Donne, I., & Mantecca, P. (2023). Biological mechanism of cell oxidative stress and death during short-term exposure to nano cuo. *Scientific Reports, 13*, 1–18.
34. Nguyen, K. V. T., Ameer, F. S., Anker, J. N., Brumaghim, J. L., & Minh, H. C. (2017). Reactive oxygen species generation by copper(ii) oxide nanoparticles determined by DNA damage assays and EPR spectroscopy. *Nanotoxicology, 11*, 278–288.
35. Siddiqi, K. S., & ur Rahman A, Tajuddin, Husen A. (2018). Properties of zinc oxide nanoparticles and their activity against microbes. *Nanoscale Research Letters, 13*, 141.
36. Zhang, L., Qi, H., Yan, Z., Gu, Y., Sun, W., & Zewde, A. A. (2017). Sonophotocatalytic inactivation of e. coli using ZnO nanofluids and its mechanism. *Ultrasonics Sonochemistry, 34*, 232–238.
37. Mendes, C. R., Dilarri, G., Forsan, C. F., Sapata, V. D. M. R., Lopes, P. R. M., de Moraes, P. B., Montagnolli, R. N., Ferreira, H., & Bidoia, E. D. (2022). Antibacterial action and target mechanisms of zinc oxide nanoparticles against bacterial pathogens. *Scientific Reports, 12*, 1–10.
38. Irshad, M. A., Nawaz, R., ur Rehman, M. Z., Adrees, M., Rizwan, M., Ali, S., Ahmad, S., & Tasleem, S. (2021). Synthesis, characterization and advanced sustainable applications of titanium dioxide nanoparticles: A review. *Ecotoxicology and Environmental Safety, 212*, 111978.
39. Kearns, H., Goodacre, R., Jamieson, L. E., Graham, D., & Faulds, K. (2017). SERS detection of multiple antimicrobial-resistant pathogens using nanosensors. *Analytical Chemistry, 89*, 12666–12673.
40. Gamage McEvoy, J., & Zhang, Z. (2014). Antimicrobial and photocatalytic disinfection mechanisms in silver-modified photocatalysts under dark and light conditions. *Journal of Photochemistry and Photobiology C: Photochemistry Reviews, 19*, 62–75.
41. Bharti, S., Mukherji, S., & Mukherji, S. (2018). Photocatalytic action of silver-titanium dioxide nanocomposite against pathogenic bacteria. *Journal of the Indian Chemical Society, 95*, 1–8.
42. Skwarczynski, M., Bashiri, S., Yuan, Y., Ziora, Z. M., Nabil, O., Masuda, K., Khongkow, M., Rimsueb, N., Cabral, H., Ruktanonchai, U., et al. (2022). Antimicrobial activity enhancers: Towards smart delivery of antimicrobial agents. *Antibiotics, 11*, 1–28.
43. Trevors, J. T. (1987). Silver resistance and accumulation in bacteria. *Enzyme and Microbial Technology, 9*, 331–333.
44. Cooksey, D. A. (1993). Copper uptake and resistance in bacteria. *Molecular Microbiology, 7*, 1–5.
45. Bondarczuk, K., & Piotrowska-Seget, Z. (2013). Molecular basis of active copper resistance mechanisms in Gram-negative bacteria. *Cell Biology and Toxicology, 29*, 397–405.
46. Niño-Martínez, N., Salas Orozco, M. F., Martínez-Castañón, G.-A., Torres Méndez, F., & Ruiz, F. (2019). Molecular mechanisms of bacterial resistance to metal and metal oxide nanoparticles. *International Journal of Molecular Sciences, 20*, 2808.

47. Hemmatian, T., Lee, H., & Kim, J. (2021). Bacteria adhesion of textiles influenced by wettability and pore characteristics of fibrous substrates. *Polymers (Basel), 13*, 1–14.
48. Leishangthem, D., Yumkhaibam, M. A. K., Henam, P. S., & Nagarajan, S. (2018). An insight into the effect of composition for enhance catalytic performance of biogenic Au/Ag bimetallic nanoparticles. *Journal of Physical Organic Chemistry, 31*, e3815.
49. Sharaf, E. M., Hassan, A., FA, A. L.-S., Albalwe, F. M., Albalawi, H. M. R., Darwish, D. B., & Fayad, E. (2022). Synergistic antibacterial activity of compact silver/magnetite core-shell nanoparticles core shell against Gram-negative foodborne pathogens. *Frontiers in Microbiology, 13*, 1–12.
50. Álvarez-Paino, M., Muñoz-Bonilla, A., & Fernández-García, M. (2017). Antimicrobial polymers in the nano-world. *Nanomaterials, 7*, 48.
51. Spirescu, V. A., Chircov, C., Grumezescu, A. M., & Andronescu, E. (2021). Polymeric nanoparticles for antimicrobial therapies: An up-to-date overview. *Polymers (Basel), 13*, 724.
52. Wang, Y., & Sun, H. (2021). Polymeric nanomaterials for efficient delivery of antimicrobial agents. *Pharmaceutics, 13*, 2108.
53. Maliszewska, I., & Czapka, T. (2022). Electrospun polymer nanofibers with antimicrobial activity. *Polymers (Basel), 14*, 1–32.
54. Yang, J., Wang, K., Yu, D. G., Yang, Y., Bligh, S. W. A., & Williams, G. R. (2020). Electrospun janus nanofibers loaded with a drug and inorganic nanoparticles as an effective antibacterial wound dressing. *Materials Science and Engineering: C, 111*, 110805.
55. Duan, X., Chen, H., & Guo, C. (2022). Polymeric nanofibers for drug delivery applications: A recent review. *Journal of Materials Science. Materials in Medicine, 33*, 78.
56. Ferreira, M., Ogren, M., Dias, J. N. R., Silva, M., Gil, S., Tavares, L., Aires-da-Silva, F., Gaspar, M. M., & Aguiar, S. I. (2021). Liposomes as antibiotic delivery systems: A promising nanotechnological strategy against antimicrobial resistance. *Molecules, 26*, 2047.
57. Mubeen, B., Ansar, A. N., Rasool, R., Ullah, I., Imam, S. S., Alshehri, S., Ghoneim, M. M., Alzarea, S. I., Nadeem, M. S., & Kazmi, I. (2021). Nanotechnology as a novel approach in combating microbes providing an alternative to antibiotics. *Antibiotics, 10*, 1473.
58. Bharti, S., Anant, P. S., & Kumar, A. (2023). Nanotechnology in stem cell research and therapy. *Journal of Nanoparticle Research, 25*, 6.
59. Pham, S. H., Choi, Y., & Choi, J. (2020). Stimuli-responsive nanomaterials for application in antitumor therapy and drug delivery. *Pharmaceutics, 12*, 1–19.
60. Werengowska-Ciećwierz, K., Wiśniewski, M., Terzyk, A. P., & Furmaniak, S. (2015). The chemistry of bioconjugation in nanoparticles-based drug delivery system. *Advances in Condensed Matter Physics, 2015*, 1–27.
61. Pandey, S., Shaikh, F., Gupta, A., Tripathi, P., & Yadav, J. S. (2022). A recent update: Solid lipid nanoparticles for effective drug delivery. *Advanced Pharmaceutical Bulletin, 12*, 17–33.
62. Mohanta, Y. K., Chakrabartty, I., Mishra, A. K., Chopra, H., Mahanta, S., Avula, S. K., Patowary, K., Ahmed, R., Mishra, B., Mohanta, T. K., et al. (2023). Nanotechnology in combating biofilm: A smart and promising therapeutic strategy. *Frontiers in Microbiology, 13*, 1–30.
63. Ganie, A. S., Bano, S., Khan, N., Sultana, S., Rehman, Z., Rahman, M. M., Sabir, S., Coulon, F., & Khan, M. Z. (2021). Nanoremediation technologies for sustainable remediation of contaminated environments: Recent advances and challenges. *Chemosphere, 275*, 130065.
64. Zare, H., Ahmadi, S., Ghasemi, A., Ghanbari, M., Rabiee, N., Bagherzadeh, M., Karimi, M., Webster, T. J., Hamblin, M. R., & Mostafavi, E. (2021). Carbon nanotubes: Smart drug/gene delivery carriers. *International Journal of Nanomedicine, 16*, 1681–1706.
65. Ma, J., Xiong, Z., David Waite, T., Ng, W. J., & Zhao, X. S. (2011). Enhanced inactivation of bacteria with silver-modified mesoporous TiO_2 under weak ultraviolet irradiation. *Microporous and Mesoporous Materials, 144*, 97–104.
66. Bharti, S., Mukherji, S., & Mukherji, S. (2021). Antiviral application of colloidal and immobilized silver nanoparticles. *Nanotechnology, 32*, 205102.
67. Agnihotri, S., Mukherji, S., & Mukherji, S. (2014). Size-controlled silver nanoparticles synthesized over the range 5–100 nm using the same protocol and their antibacterial efficacy. *RSC Advances, 4*, 3974–3983.

68. Kan, C. X., Zhu, J. J., & Zhu, X. G. (2008). Silver nanostructures with well-controlled shapes: Synthesis, characterization and growth mechanisms. *Journal of Physics D: Applied Physics, 41*, 155304.
69. Vassallo, A., Silletti, M. F., Faraone, I., & Milella, L. (2020). Nanoparticulate antibiotic systems as antibacterial agents and antibiotic delivery platforms to fight infections. *Journal of Nanomaterials, 2020*, 1–31.
70. Bharti, S., Mukherji, S., & Mukherji, S. (2020). Extracellular synthesis of silver nanoparticles by Thiosphaera pantotropha and evaluation of their antibacterial and cytotoxic effects. *3 Biotech, 10*, 1–12.
71. Giannousi, K., Lafazanis, K., Arvanitidis, J., Pantazaki, A., & Dendrinou-Samara, C. (2014). Hydrothermal synthesis of copper based nanoparticles: Antimicrobial screening and interaction with dna. *Journal of Inorganic Biochemistry, 133*, 24–32.
72. Zhu, Y. P., Wang, X. K., Guo, W. L., Wang, J. G., & Wang, C. (2010). Sonochemical synthesis of silver nanorods by reduction of sliver nitrate in aqueous solution. *Ultrasonics Sonochemistry, 17*, 675–679.
73. Anandan, S., Grieser, F., & Ashokkumar, M. (2008). Sonochemical synthesis of Au−Ag core−shell bimetallic nanoparticles. *Journal of Physical Chemistry C, 112*, 15102–15105.
74. Wuppaladhodi, V., Yang, S., Pouri, H., & Zhang, J. (2023). Laser-assisted process for the deposition of nanostructured anti-microbial coatings on hydrogels. *Optics and Laser Technology, 164*, 109485.
75. Agostino, A. D., Taglietti, A., Desando, R., Bini, M., Patrini, M., Dacarro, G., Cucca, L., Pallavicini, P., & Grisoli, P. (2017). Bulk surfaces coated with triangular silver nanoplates: Antibacterial action based on silver release and photo-thermal effect. *Nanomaterials, 7*, 1–10.
76. Park, M. H., Duan, X., Ofir, Y., Creran, B., Patra, D., Ling, X. Y., Huskens, J., & Rotello, V. M. (2010). Chemically directed immobilization of nanoparticles onto gold substrates for orthogonal assembly using dithiocarbamate bond formation. *ACS Applied Materials & Interfaces, 2*, 795–799.
77. Khayati, G. R., & Janghorban, K. (2012). The nanostructure evolution of Ag powder synthesized by high energy ball milling. *Advanced Powder Technology, 23*, 393–397.
78. Esfandyari-Manesh, M., Ghaedi, Z., Asemi, M., Khanavi, M., Manayi, A., Jamalifar, H., Atyabi, F., & Dinarvand, R. (2013). Study of antimicrobial activity of anethole and carvone loaded plga nanoparticles. *Journal of Pharmacy Research, 7*, 290–295.
79. Wei, C., Fan, C., Xie, D., Zhou, S., Zhang, H., Du, Q., & Jin, P. (2023). Fabrication of cinnamaldehyde-entrapped ethosome nanoparticles as antimicrobial agent. *LWT, 181*, 114760.
80. Imam, H. T., Marr, P. C., & Marr, A. C. (2021). Enzyme entrapment, biocatalyst immobilization without covalent attachment. *Green Chemistry, 23*, 4980–5005.
81. Li, W., Li, Y., Sun, P., Zhang, N., Zhao, Y., Qin, S., & Zhao, Y. (2020). Antimicrobial peptide-modified silver nanoparticles for enhancing the antibacterial efficacy. *RSC Advances, 10*, 38746–38754.
82. Smoukov, S. K., Bishop, K. J. M., Kowalczyk, B., Kalsin, A. M., & Grzybowski, B. A. (2007). Electrostatically "patchy" coatings via cooperative adsorption of charged nanoparticles. *Journal of the American Chemical Society, 129*, 15623–15630.
83. Qi, X., Poernomo, G., Wang, K., Chen, Y., Chan-Park, M. B., Xu, R., & Chang, M. W. (2011). Covalent immobilization of nisin on multi-walled carbon nanotubes: Superior antimicrobial and anti-biofilm properties. *Nanoscale, 3*, 1874.
84. Stillger, L., & Müller, D. (2022). Peptide-coating combating antimicrobial contaminations: A review of covalent immobilization strategies for industrial applications. *Journal of Materials Science, 57*, 10863–10885.
85. Park, S. Y., Chung, J. W., Priestley, R. D., & Kwak, S.-Y. (2012). Covalent assembly of metal nanoparticles on cellulose fabric and its antimicrobial activity. *Cellulose, 19*, 2141–2151.
86. Sathyanarayanan, M. B., Balachandranath, R., Genji Srinivasulu, Y., Kannaiyan, S. K., & Subbiahdoss, G. (2013). The effect of gold and iron-oxide nanoparticles on biofilm-forming pathogens. *ISRN Microbiology, 2013*, 1–5.

87. Agnihotri, S., & Dhiman, N. K. (2017). Development of nano-antimicrobial biomaterials for biomedical applications. In A. Tripathi & J. S. Melo (Eds.), *Advances in biomaterials for biomedical applications* (Advanced structured materials) (Vol. 66, pp. 479–545). Springer Singapore. ISBN 978-981-10-3327-8.
88. Greenhalgh, R., Dempsey-Hibbert, N. C., & Whitehead, K. A. (2019). Antimicrobial strategies to reduce polymer biomaterial infections and their economic implications and considerations. *International Biodeterioration & Biodegradation, 136*, 1–14.
89. Sahoo, J., Sarkhel, S., Mukherjee, N., & Jaiswal, A. (2022). Nanomaterial-based antimicrobial coating for biomedical implants: New age solution for biofilm-associated infections. *ACS Omega, 7*, 45962–45980.
90. Zhang, L., Yin, W., Shen, S., Feng, Y., Xu, W., Sun, Y., & Yang, Z. (2022). ZnO nanoparticles interfere with top-down effect of the protozoan paramecium on removing microcystis. *Environmental Pollution, 310*, 119900.
91. Alhazmi, N. M. (2022). Fungicidal activity of silver and silica nanoparticles against Aspergillus sydowii isolated from the soil in western Saudi Arabia. *Microorganisms, 11*, 86.
92. Bharti, S., Mukherji, S., & Mukherji, S. (2019). Water disinfection using fixed bed reactors packed with silver nanoparticle immobilized glass capillary tubes. *Science of The Total Environment, 689*, 991–1000.
93. Sawant, S. N., Selvaraj, V., Prabhawathi, V., & Doble, M. (2013). Antibiofilm properties of silver and gold incorporated PU, PCLM, PC and PMMA nanocomposites under two shear conditions. *PLoS One, 8*, 1–9.
94. Hulme, J. (2022). Application of nanomaterials in the prevention, detection, and treatment of methicillin-resistant Staphylococcus aureus (MRSA). *Pharmaceutics, 14*, 805.
95. Leong, S., Razmjou, A., Wang, K., Hapgood, K., Zhang, X., & Wang, H. (2014). TiO2 based photocatalytic membranes: A review. *Journal of Membrane Science, 472*, 167–184.
96. Dash, K. K., Deka, P., Bangar, S. P., Chaudhary, V., Trif, M., & Rusu, A. (2022). Applications of inorganic nanoparticles in food packaging: A comprehensive review. *Polymers (Basel), 14*, 521.
97. Biswas, R., Alam, M., Sarkar, A., Haque, M. I., Hasan, M. M., & Hoque, M. (2022). Application of nanotechnology in food: Processing, preservation, packaging and safety assessment. *Heliyon, 8*, e11795.
98. Kaur, J., Sood, K., Bhardwaj, N., Arya, S. K., & Khatri, M. (2020). Nanomaterial loaded chitosan nanocomposite films for antimicrobial food packaging. *Materials Today Proceedings, 28*, 1904–1909.
99. Nile, S. H., Baskar, V., Selvaraj, D., Nile, A., Xiao, J., & Kai, G. (2020). Nanotechnologies in food science: Applications, recent trends, and future perspectives. *Nano-Micro Letters, 12*, 45.
100. Hardy, A., Benford, D., Halldorsson, T., Jeger, M. J., Knutsen, H. K., More, S., Naegeli, H., Noteborn, H., Ockleford, C., Ricci, A., et al. (2018). Guidance on risk assessment of the application of nanoscience and nanotechnologies in the food and feed chain: Part 1, human and animal health. *EFSA Journal, 16*, 5327.
101. More, S., Bampidis, V., Benford, D., Bragard, C., Halldorsson, T., Hernández-Jerez, A., Bennekou, S. H., Koutsoumanis, K., Lambré, C., Machera, K., et al. (2021). Guidance on technical requirements for regulated food and feed product applications to establish the presence of small particles including nanoparticles. *EFSA Journal, 19*, 6769.
102. Salleh, A., Naomi, R., Utami, N. D., Mohammad, A. W., Mahmoudi, E., Mustafa, N., & Fauzi, M. B. (2020). The potential of silver nanoparticles for antiviral and antibacterial applications: A mechanism of action. *Nanomaterials, 10*, 1566.
103. Montes-Hernandez, G., Di Girolamo, M., Sarret, G., Bureau, S., Fernandez-Martinez, A., Lelong, C., & Eymard, V. E. (2021). In situ formation of silver nanoparticles (Ag-NPs) onto textile fibers. *ACS Omega, 6*, 1316–1327.
104. Ribeiro, A. I., Shvalya, V., Cvelbar, U., Silva, R., Marques-Oliveira, R., Remião, F., Felgueiras, H. P., Padrão, J., & Zille, A. (2022). Stabilization of silver nanoparticles on

polyester fabric using organo-matrices for controlled antimicrobial performance. *Polymers (Basel), 14*, 1138.
105. Abraham, J., Dowling, K., & Florentine, S. (2021). Can copper products and surfaces reduce the spread of infectious microorganisms and hospital-acquired infections? *Materials (Basel), 14*, 1–21.
106. Imani, S. M., Ladouceur, L., Marshall, T., Maclachlan, R., Soleymani, L., & Didar, T. F. (2020). Antimicrobial nanomaterials and coatings: Current mechanisms and future perspectives to control the spread of viruses including sars-cov-2. *ACS Nano, 14*, 12341–12369.
107. Kaci, M., Belhaffef, A., Meziane, S., Dostert, G., Menu, P., Velot, D. S., & Arab-Tehrany, E. (2018). Nanoemulsions and topical creams for the safe and effective delivery of lipophilic antioxidant coenzyme Q10. *Colloids Surfaces B Biointerfaces, 167*, 165–175.
108. Gupta, V., Mohapatra, S., Mishra, H., Farooq, U., Kumar, K., & Iqbal, Z. (2022). Nanotechnology in cosmetics and cosmeceuticals — A review. *Gels, 8*, 1–31.
109. Hosseini, M., Chin, A. W. H., Williams, M. D., Behzadinasab, S., Falkinham, J. O., Poon, L. L. M., & Ducker, W. A. (2022). Transparent anti-SARS-CoV-2 and antibacterial silver oxide coatings. *ACS Applied Materials & Interfaces, 14*, 8718–8727.
110. Liu, M., Huang, L., Xu, X., Wei, X., Yang, X., Li, X., Wang, B., Xu, Y., Li, L., & Yang, Z. (2022). Copper doped carbon dots for addressing bacterial biofilm formation, wound infection, and tooth staining. *ACS Nano, 16*, 9479–9497.
111. Nguyen, T. M. T., Wang, P. W., Hsu, H. M., Cheng, F. Y., Bin, S. D., Wong, T. Y., & Chang, H. J. (2019). Dental cement's biological and mechanical properties improved by ZnO nanospheres. *Materials Science and Engineering: C, 97*, 116–123.
112. Esteban Florez, F. L., Hiers, R. D., Larson, P., Johnson, M., O'Rear, E., Rondinone, A. J., & Khajotia, S. S. (2018). Antibacterial dental adhesive resins containing nitrogen-doped titanium dioxide nanoparticles. *Materials Science and Engineering: C, 93*, 931–943.
113. Deng, Y., Yang, L., Huang, X., Chen, J., Shi, X., Yang, W., Hong, M., Wang, Y., Dargusch, M. S., & Chen, Z. G. (2018). Dual Ag/ZnO-decorated micro−/nanoporous sulfonated polyetheretherketone with superior antibacterial capability and biocompatibility via layer-by-layer self-assembly strategy. *Macromolecular Bioscience, 18*, 1–12.
114. Wang, S., Wu, J., Yang, H., Liu, X., Huang, Q., & Lu, Z. (2017). Antibacterial activity and mechanism of Ag/ZnO nanocomposite against anaerobic oral pathogen streptococcus mutans. *Journal of Materials Science. Materials in Medicine, 28*, 1–8.
115. Li, X., Qi, M., Sun, X., Weir, M. D., Tay, F. R., Oates, T. W., Dong, B., Zhou, Y., Wang, L., & Xu, H. H. K. (2019). Surface treatments on titanium implants via nanostructured ceria for antibacterial and anti-inflammatory capabilities. *Acta Biomaterialia, 94*, 627–643.
116. Liu, Z., Du, M., Liu, H., Zhang, K., Xu, X., Liu, K., Tu, J., & Liu, Q. (2021). Chitosan films incorporating litchi peel extract and titanium dioxide nanoparticles and their application as coatings on watercored apples. *Progress in Organic Coatings, 151*, 106103.
117. Fadiji, T., Rashvand, M., Daramola, M. O., & Iwarere, S. A. (2023). A review on antimicrobial packaging for extending the shelf life of food. *PRO, 11*, 1–30.
118. Dong, X., Liang, X., Zhou, Y., Bao, K., Sameen, D. E., Ahmed, S., Dai, J., Qin, W., & Liu, Y. (2021). Preparation of polylactic acid/ TiO2/GO nano-fibrous films and their preservation effect on green peppers. *International Journal of Biological Macromolecules, 177*, 135–148.
119. Pirsa, S., & Shamusi, T. (2019). Intelligent and active packaging of chicken thigh meat by conducting nano structure cellulose-polypyrrole-zno film. *Materials Science and Engineering: C, 102*, 798–809.
120. Darwesh, O. M., Li, H., & Matter, I. A. (2023). Nano-bioremediation of textile industry wastewater using immobilized cuo-nps myco-synthesized by a novel cu-resistant Fusarium oxysporum osf18. *Environmental Science and Pollution Research, 30*, 16694–16706.
121. Alias, S. S., Harun, Z., Azhar, F. H., Ibrahim, S. A., & Johar, B. (2020). Comparison between commercial and synthesised nano flower-like rutile TiO2 immobilised on green super adsorbent towards dye wastewater treatment. *Journal of Cleaner Production, 251*, 119448.

122. Sboui, M., Lachheb, H., Bouattour, S., Gruttadauria, M., La Parola, V., Liotta, L. F., & Boufi, S. (2021). TiO2/Ag2O immobilized on cellulose paper: A new floating system for enhanced photocatalytic and antibacterial activities. *Environmental Research, 198*, 111257.
123. Jabbar, Z. H., Okab, A. A., Graimed, B. H., Issa, M. A., & Ammar, S. H. (2023). Fabrication of Ag-C3N4 nanosheets immobilized Bi2S3/Ag2WO4 nanorods for photocatalytic disinfection of Staphylococcus aureus cells in wastewater: Dual s-scheme charge separation pathway. *Journal of Photochemistry and Photobiology A: Chemistry, 438*, 114556.
124. Hidayat, D., Lestari, W. W., Dendy, D., Khoerunnisa, F., Handayani, M., Sanjaya, E. H., & Gunawan, T. (2023). Adsorption studies of anionic and cationic dyes on mil-100(cr) synthesized using facile and green mechanochemical method. *Journal of Inorganic and Organometallic Polymers and Materials, 33*, 1548–1561.
125. Nafady, A., Albaqami, M. D., & Alotaibi, A. M. (2023). Recycled polypropylene waste as abundant source for antimicrobial, superhydrophobic and electroconductive nonwoven fabrics comprising polyaniline/silver nanoparticles. *Journal of Inorganic and Organometallic Polymers and Materials, 33*, 1306–1316.
126. Nguyen, N.-T., Vu, T.-H., Bui, V.-H., Phan, D.-N., Nguyen, T.-H., & Nguyen, T.-M.-L. (2023). Investigation of the antimicrobial and physico-mechanical properties of nature-friendly nanosilver-loaded pig lining leather prepared using exhaustion method. *PRO, 1891*, 11.
127. Masa, A., Jehsoh, N., Dueramae, S., & Hayeemasae, N. (2023). Boosting the antibacterial performance of natural rubber latex foam by introducing silver-doped zinc oxide. *Polymers (Basel), 15*, 1040.
128. Refaee, A. A., El-Naggar, M. E., Mostafa, T. B., Elshaarawy, R. F. M., & Nasr, A. M. (2022). Nano-bio finishing of cotton fabric with quaternized chitosan schiff base-TiO2-ZnO nanocomposites for antimicrobial and uv protection applications. *European Polymer Journal, 166*, 111040.
129. Aravind, H. P., Jadhav, S. A., More, V. B., Sonawane, K. D., & Patil, P. S. (2019). Novel one step sonosynthesis and deposition technique to prepare silver nanoparticles coated cotton textile with antibacterial properties. *Colloid Journal, 81*, 720–727.
130. Kuehr, S., Kosfeld, V., & Schlechtriem, C. (2021). Bioaccumulation assessment of nanomaterials using freshwater invertebrate species. *Environmental Sciences Europe, 33*, 9.
131. Abbas, Q., Yousaf, B., Amina, A. M. U., Munir, M. A. M., El-Naggar, A., Rinklebe, J., & Naushad, M. (2020). Transformation pathways and fate of engineered nanoparticles (ENPs) in distinct interactive environmental compartments: A review. *Environment International, 138*, 105646.
132. Devasena, T., Iffath, B., Renjith Kumar, R., Muninathan, N., Baskaran, K., Srinivasan, T., & John, S. T. (2022). Insights on the dynamics and toxicity of nanoparticles in environmental matrices. *Bioinorganic Chemistry and Applications, 2022*, 1–21.
133. Singh, K., Thakur, S. S., Ahmed, N., Alharby, H. F., Al-Ghamdi, A. J., Al-Solami, H. M., Bahattab, O., & Yadav, S. (2022). Ecotoxicity assessment for environmental risk and consideration for assessing the impact of silver nanoparticles on soil earthworms. *Heliyon, 8*, e11167.
134. Pérez-de-Luque, A. (2017). Interaction of nanomaterials with plants: What do we need for real applications in agriculture? *Frontiers in Environmental Science, 5*, 1–7.
135. Gupta, R., & Xie, H. (2018). Nanoparticles in daily life: Applications, toxicity and regulations. *Journal of Environmental Pathology, Toxicology and Oncology, 37*, 139–148.
136. Wu, J., Bosker, T., Vijver, M. G., & Peijnenburg, W. J. G. M. (2021). Trophic transfer and toxicity of (mixtures of) ag and tio2 nanoparticles in the lettuce–terrestrial snail food chain. *Environmental Science & Technology, 55*, 16563–16572.
137. Boros, B.-V., & Ostafe, V. (2020). Evaluation of ecotoxicology assessment methods of nanomaterials and their effects. *Nanomaterials, 10*, 610.
138. Pareek, V., Gupta, R., & Panwar, J. (2018). Do physico-chemical properties of silver nanoparticles decide their interaction with biological media and bactericidal action? A review. *Materials Science and Engineering: C, 90*, 739–749.

139. Mishra, S., & Sundaram, B. (2023). Fate, transport, and toxicity of nanoparticles: An emerging pollutant on biotic factors. *Process Safety and Environment Protection, 174*, 595–607.
140. Angel, B. M., Batley, G. E., Jarolimek, C. V., & Rogers, N. J. (2013). The impact of size on the fate and toxicity of nanoparticulate silver in aquatic systems. *Chemosphere, 93*, 359–365.
141. Ferdous, Z., & Nemmar, A. (2020). Health impact of silver nanoparticles: A review of the biodistribution and toxicity following various routes of exposure. *International Journal of Molecular Sciences, 21*, 2375.
142. Shi, H., Magaye, R., Castranova, V., & Zhao, J. (2013). Titanium dioxide nanoparticles: A review of current toxicological data. *Particle and Fibre Toxicology, 10*, 15.
143. Henkler, F., Tralau, T., Tentschert, J., Kneuer, C., Haase, A., Platzek, T., Luch, A., & Götz, M. E. (2012). Risk assessment of nanomaterials in cosmetics: A european union perspective. *Archives of Toxicology, 86*, 1641–1646.
144. Xie, S., Manuguri, S., Proietti, G., Romson, J., Fu, Y., Inge, A. K., Wu, B., Zhang, Y., Häll, D., Ramström, O., et al. (2017). Design and synthesis of theranostic antibiotic nanodrugs that display enhanced antibacterial activity and luminescence. *Proceedings of the National Academy of Sciences of the United States of America, 114*, 8464–8469.
145. Ye, X., Feng, T., Li, L., Wang, T., Li, P., & Huang, W. (2021). Theranostic platforms for specific discrimination and selective killing of bacteria. *Acta Biomaterialia, 125*, 29–40.
146. El-Zowalaty, M. E., Al-Ali, S. H. H., Husseiny, M. I., Geilich, B. M., Webster, T. J., & Hussein, M. Z. (2015). The ability of streptomycin-loaded chitosan-coated magnetic nanocomposites to possess antimicrobial and antituberculosis activities. *International Journal of Nanomedicine, 10*, 3269–3274.
147. Song, J., Liu, H., Lei, M., Tan, H., Chen, Z., Antoshin, A., Payne, G. F., Qu, X., & Liu, C. (2020). Redox-channeling polydopamine-ferrocene (PDA-FC) coating to confer context-dependent and photothermal antimicrobial activities. *ACS Applied Materials & Interfaces, 12*, 8915–8928.
148. Zohra, T., Numan, M., Ikram, A., Salman, M., Khan, T., Din, M., Salman, M., Farooq, A., Amir, A., & Ali, M. (2021). Cracking the challenge of antimicrobial drug resistance with crispr/cas9, nanotechnology and other strategies in eskape pathogens. *Microorganisms, 9*, 954.
149. Sadani, K., Muthuraj, L., Nag, P., Fernandes, M., Kondabagil, K., Mukhopadhyay, C., & Mukherji, S. (2020). A point of use sensor assay for detecting purely viral versus viral-bacterial samples. *Sensors and Actuators B: Chemical, 322*, 128562.
150. Nag, P., Sadani, K., Mukherji, S., & Mukherji, S. (2020). Beta-lactam antibiotics induced bacteriolysis on LSPR sensors for assessment of antimicrobial resistance and quantification of antibiotics. *Sensors and Actuators B: Chemical, 311*, 127945.
151. Sadani, K., Nag, P., Thian, X. Y., & Mukherji, S. (2022). Enzymatic optical biosensors for healthcare applications. *Biosensors and Bioelectronics: X, 12*, 100278.

Web References

https://echa.europa.eu/regulations/nanomaterials-under-bpr, Last accessed on 9 Sept 2023.

Chapter 6
Application of Nanotechnology in Food Microbiology: Implication on Public Health

Smita Guha, Ashok Chakraborty, and Debjit Chakraborty

6.1 Introduction

The word "nano" actually came from the Greek word, "dwarf". One nanometer (10^{-9} m) is about 60,000 times smaller than a human hair in diameter. A red blood cell is about 2000–5000 nm in size, and the diameter of DNA is in the range of 2.5 nm; and nanotechnology deals with particle size below 100 nanometers [1]. These particles since having a greater ratio of their surface area to volume possess a unique phenomenon like solubility, diffusivity, color, and magnetic, optical, thermodynamic, etc. enable novel applications and benefits [2, 3]. Nanotechnology deals with the technique to manipulate the structure of matter on a near-atomic scale to produce new structures, materials and devices. This technology has already been used in modern science, including medical science, environmental science, energy and electronics, etc., recently reviewed by us [4]. Since last few decades, nanotechnology became indispensable to revolutionize the food industries.

The rising consumer concerns about food quality and health benefits are the driving force of studying the nanomaterials to increase not only the food quality but also the safety and health benefits issues that food delivers [5, 6]. In this review, we discussed the application of nanotechnology in the area of food microbiology, and discussed some of its positive as well as negative impacts on human health.

S. Guha (✉)
St. John's University, Queens, NY, USA
e-mail: guhas@stjohns.edu

A. Chakraborty
AllExcel, Inc., Shelton, CT, USA

D. Chakraborty
ICMR-NICED, Kolkata, India

R. S. Chaughule, A. S. Lokur (eds.), *Applications of Nanotechnology in Microbiology*, https://doi.org/10.1007/978-3-031-49933-3_6

6.2 Nanotechnology in Food Microbiology

Nanotechnology has potential applications in all aspects of food industry, including food processing, quality monitoring, food packaging, and storage. Food microbiologists are concerned about safety and quality assurance programs of food products with free of pathogens. Nanomaterials can contribute to food safety, inhibit the growth of food spoilage microorganisms by recruiting novel and unique agents that are involved in removal of microbes from foods or prevent adhesion of microbial cells to food surfaces. The main applications of nanotechnology in food industry are to detect the pathogens, if any, by nanosensors, and to exert antimicrobial effect on contaminating microorganisms.

6.2.1 *Antimicrobial Effect of Nanoparticles*

- Silver nanoparticles exhibited better antimicrobial activity against *Escherichia coli* and *Penicillium* spp. [7]. Silver-silica nanocomposite-containing polystyrene demonstrated that the inhibition of microbial growth [8].
- Cationic peptides nanoparticles have strong antimicrobial properties, including bacteria, yeasts, and fungi [9].
- Metal oxide, especially TiO_2 and Ag_2O, nanoparticles are effective against eukaryotic infectious agents [10].

6.2.2 *Nanoencapsulation Can Be Used for Antimicrobial Activity*

- Nanoencapsulation systems in food industries may improve the processing stage of the food via enabling the delivery and controlled release (within specific time and zone) of the materials/compounds that exist in different foods [11, 12].
- Encapsulation of the valuable microbial strains by nanoparticles improves the survival of probiotics under harsh conditions such as extreme levels of temperature, pH, and salinity during the processing of food products and within the GIT tract.
- Self-assembled nanotubes from hydrolyzed milk protein α-lactalbumin, a potential new carrier for nanoencapsulation of nutrients, supplements, and pharmaceuticals, have been reported [13].
- Casein micelles may be useful as nanovehicles for entrapment, protection, and delivery of sensitive hydrophobic nutraceuticals within other food products [14]. These have been eaten safely for generations. In fact, some of food's most important raw materials [proteins, starches, and fats] undergo structural changes at the nanometer and micrometer scales during normal food processing [15].

- Self-assembled nanotubes from hydrolyzed milk protein α-lactalbumin, a potential new carrier for nanoencapsulation of nutrients, supplements, and pharmaceuticals, have been reported [16].
- The dairy industry utilizes three basic microsized and nanosized structures (casein micelles, fat globules, and whey proteins) to build all sorts of emulsions (butter), foams (ice cream and whipped cream), complex liquids (milk), plastic solids (cheese), and gel networks (yogurt) [17]. In fact, dairy technology is not just a microtechnology but also a nanotechnology, and it has existed for a long time.
- Incorporation of encapsulated essential via nanovehicles into fruit juices exerts their antimicrobial activity without compromising the quality of the product [18]. Also, Ravichandran et al. (2011) showed that inhibition of *Listeria monocytogenes*, *Salmonella typhimurium*, and *Escherichia coli* in raw chicken is possible with encapsulated benzoic acid (1100 μg/mL) in polylactic-*co*-glycolic acid nanoparticles [19].
- Moreover, evaluation of the antibacterial activity of nisin-loaded chitosan/carrageenan nanocapsules showed better antibacterial effect on microbe's (*Micrococcus luteus*, *Pseudomonas aeruginosa*, *Salmonella enterica*, and *Enterobacter aerogenes*) in vitro compared to the components evaluated separately [20].

6.2.3 Microencapsulation Has Many Applications in Food Technology

- Numerous compounds are being used to coat probiotics, which the most common are alginate, waxes, milk protein, xanthan gelatin, chitosan, starch, cellulose, and carrageenan [21].
- These coatings are subject to particular conditions such as pH, the pressure to release their contents and melting temperature.
- Recently, both chitosan and EU S100 for double coating of probiotic strains when used together the viability of bacteria increases dramatically in yogurt and gastrointestinal tract [22].
- Microencapsulation occurs in a variety of ways, the most important of them about probiotics are emulsion, drying, and extrusion through sublimation, fluid bedding, and spraying.
- The risk of bacterial infection using this method deeply depends on changes in temperature, as well as on the phase. This method recruits edible films to coat the food, which increases the protection of probiotics.

6.2.4 Different Coating Methods Used to Encapsulate Probiotics

Generally, probiotics refer to live yeast and bacteria that bring numerous benefits for their hosts when they are on a sufficient scale in the body [23]. In the milk fermentation process, different probiotics strains play an important role, including Bifidobacterium, Lactococcus, and Lactobacillus owing to beneficial impacts on human health. Probiotics must survive sufficient against extreme conditions in the body and they can be easily replicated in the gut and reveal their health benefits to the host such as intestinal system maintenance, immune system stimulation, cholesterol, blood pressure regulation, and anticancer effects. Thus, the stability of probiotics strains against the environmental circumstances of the product and the digestive system is particularly significant.

Probiotics in the food factories are being developed quickly due to their health properties. Scientists have developed a new method for enhancing the stability of probiotics, which employs microencapsulation technology [24]. In other words, microencapsulation defined as a process in which microorganism mainly bacteria is captured and coated by hydro-colloidal material physicochemically. This technique protects them from extreme conditions like antimicrobial agents that inhibit cell damages, and other conditions such as cold shock, heat shock, high acidity, and salt concentrations (caused by spray drying), as well as oxygen for obligatory anaerobic bacteria [25].

- In the microencapsulation technique, bacteria are surrounded in a special coating for protection of bacterial cells.
- Encapsulated forms of ingredients achieve longer shelf life of the product.
- Nanoencapsulation of probiotic bacterial preparations may act as a de novo vaccines, with the capability of modulating immune responses [26].
- The health benefits of curcumin, the natural pigment that gives the spice turmeric its yellow color, could be enhanced by encapsulation in nanoemulsions [27].
- Nanoemulsions could improve stability and oral bioavailability of epigallocatechin gallate and curcumin [28].
- A stearin-rich milk fraction was used, alone or in combination with α-tocopherol, for the preparation of oil-in-water sodium caseinate-stabilized nanoemulsions.
- Immobilization of α-tocopherol in fat droplets, composed by high melting temperature milk fat triglycerides, provided protection against degradation [29].
- Nanocomposites are characterized by extremely high surface-to-volume ratio, making them highly reactive in comparison to their macroscale counterparts, and thus presenting fundamentally different properties [30].
- Nanocomposites can improve mechanical strength; reduce weight; increase heat resistance; and improve barrier against oxygen, carbon dioxide, ultraviolet radiation, moisture, and volatiles of food package materials.
- Polymer nanocomposites are thermoplastic polymers that have nanoscale inclusions, 2%–8% by weight.

6.2.5 Nanotechnology in Food Processing

- Protein–polysaccharide mixed solutions can spontaneously separate into a phase with nano- or microsized droplets dispersed in a continuous phase.
- Starch granules expand when heated and hydrated releasing biopolymers that can be recrystallized into nanosized structures (e.g., recrystallized amylose regions may be about 10–20 nm); dextrins and other degradation products of extrusion can be used to encapsulate bioactive substances in microregions.
- In the case of fats, monoglycerides, for example, can self-assemble into many morphologies at the nanoscale level, and hierarchically structured into triglycerides can be crystallites (10–100 nm), followed by arrangement into large clusters, then flocs, and finally, fat crystal networks [31].
- Screening of pea varieties using microscopic methods has identified commercial varieties, such as Green shaft, that contain blends of normal starches and these 'naturally resistant' novel starches.
- The nanostructured food ingredients are being developed with the claims that they offer improved taste, texture, and consistency [32].
- For example, low-fat nanostructured mayonnaise spreads and ice creams claim to be as "creamy" as their full fat alternatives and, hence, offer a healthier option to the consumer [32].
- Natural nanostructures found in foods are produced by some proteins and carbohydrates. Examples include digestion of food and food structure building processes, such as the arrangement of amylose and amylopectin in a starch granule.
- A cow's udder was given as an example of a nano device synthesizing, assembling, and dispensing proteins and fat into an aqueous phase, where they later become building blocks for myriad of protein structures.
- Such processes cause microstructural changes in food, such as homogenization and fine milling. Homogenized milk has a nanostructure of 100 nm sized droplets in it [33].
- The dairy industry utilizes three basic microsized and nanosized structures (casein micelles, fat globules, and whey proteins) to build all sorts of emulsions (butter), foams (ice cream and whipped cream), complex liquids (milk), plastic solids (cheese), and gel networks (yogurt) [17].
- In fact, dairy technology is not just a microtechnology but also a nanotechnology, and it has existed for a long time.

6.2.6 Preservation or Shelf Life of Foods

Nanotechnology can be applied for increasing the shelf life of foods by decreasing the spoilage due to microbial growth [34]. Nanoencapsulation of the bioactive components of foods can slow down the degradation processes or prevents degradation due to the hostile environment, and thus can extend the shelf life of the food products. Edible nano-coatings prevent gas exchange and also a barrier to moisture, and thus increase the shelf life of foods, in addition to contribution of colors and flavors [35, 36].

Encapsulation of functional components with nanoparticles often enables a slowdown of chemical degradation processes of the food particles. For example, curcumin the most active and least stable bioactive component of turmeric was found to be stable at different ionic strength and even after pasteurization upon encapsulation [37].

6.2.7 Increase in Nutritional Value

There are different techniques, like nanocomposite, nano-emulsification, and nano-structuration, which can be applied to encapsulate the nutrients like protein and antioxidants for health benefits. Polymeric nanoparticles are found suitable for the encapsulation of bioactive compounds (e.g., flavonoids and vitamins) in order to protect them during transport to the target [38].

- **Food quality:** Nanotechnology provides an improvement of the food quality, food taste, texture, and appearance of food. These nanostructures are used as food additives, carriers of nutrients, antimicrobial agents, and fillers for improving mechanical strength and durability of the packaging material [5, 39].
- **Thermal stability and photo stability:** Encapsulation of cyanidin-3-O-glucoside (C3G) molecules within the inner cavity of apo-recombinant soybean seed H-2 (rH-2) can improve the thermal and photo stability of C3G [40]. *Rutin*, a dietary flavonoid with important pharmacological activities is poorly soluble in water, makes its application limited in the food industry. However, encapsulation of *rutin* in ferritin nanocages enhanced the solubility, and also the stability from thermal and UV radiation [41].
- **Water-dispersion and bioavailability:** The nanoemulsions can be produced using natural food ingredients to deliver lipid-soluble bioactive compounds through enhancement of water-dispersion and bioavailability [42].
- **NPs adds color or flavor:** Many metallic oxides such as titanium dioxide and silicon dioxide (SiO_2) are being used to add color and flavor in food items [43, 44]. Undoubtedly, taste is a significant characteristic in any food and plays a crucial role in consumption, sensory properties, and the quality of food products, such as stability of the flavors and aromas during production and the storage [45]. Therefore, to limit flavor loss or degradation during storage and production, it would be better to provide controlled release of encapsulated flavor before application in food, resulting in improving its chemical stability.

Encapsulation with a protective carrier preserves the foods against interactions occurring between flavors and oxidation, as well as reactions triggered by light [46]. The most popular carriers are chitosan, proteins (e.g., gelatin and whey proteins), biopolymers or carbohydrates (e.g., starch, dextrose, and maltodextrins), gums (e.g., carrageenan, gum arabic, and alginates) [45]. Factors considered for designing an ideal encapsulation system are the carrier (viscosity), and the physicochemical properties of the flavor (solubility). Generally, nanoencapsulation inserts all types of materials into some nanocarriers; therefore, a suitable function of the ultimate product is guaranteed like the release of core substances [47]. The release of flavor materials may be influenced by communications between nanocarriers and core materials.

6.2.8 Improving the Bioavailability of Food Materials

In recent years, nanocarriers have been designed to increase the bioavailability of activated mixtures by target delivery systems. Due to poor absorption in the gastrointestinal fluids, lipophilic activated mixtures show low bioavailability within the GIT [48]. For resistance, owing to the high acidity in the duodenum and stomach, bioactive compounds are mainly encapsulated [49]. Biologically active materials and molecules cannot be employed in the design of food delivery systems in the body due to their potential effects on the human system.

6.2.9 Safety Issues

Nanotechnology promises big benefits for food safety, quality, and shelf life, provided the challenges it brings can be Nanotechnology, Science, and Applications overcome [50]. However, since the nanomaterials are fundamentally different substances that introduce new functionalities into foods, may create new and unique risks to the consumers.

According to the current information, nanomaterials used in food applications include both inorganic and organic substances [51]. Table 6.1 shows the use of inorganic and organic nanomaterials in food industries.

6.2.10 Biosensors to Detect Specific Bacteria Strain

Nanosensors or nanobiosensors are being used in food industries for the detection of microbes in food products [56, 57]. Nanosensor responds to any changes in environmental conditions such as humidity or temperature, microbial contamination, or products degradation [58, 59]. The recognized benefits of nanosensors are as follows:

- High sensitivity and excellent selectivity.
- Thin film-based optical immunosensors can detect microbial substances or cells in food items. In these immunosensors, protein molecules, antigen, and specific

Table 6.1 Some inorganic and organic nanomaterials used in food industries

Inorganic nanomaterials	Uses	Ref.
Nanosilver	As a safety materials for food, water, and packaging materials. Antimicrobial, antiodorant	[52]
Amorphous nanosilica	It is known to be used in food contact surfaces and food packaging applications.	
Nanoselenium	Marketed as an additive to a green tea product, with a number of (proclaimed) health benefits resulting from enhanced uptake of selenium. Health supplements	
Nanocalcium	Subject of patent applications for intended use in chewing gums. Health supplements	[53]
Nano-iron	A health supplement, and is used to decontaminate water by breaking down organic pollutants and killing microbial pathogens.	
Organic nanomaterials	Benzoic acid, citric acid, ascorbic acid, vitamins A and E, isoflavones, beta-carotene, lutein, omega-3 fatty acids, and coenzyme-Q10 are some of the examples of organic nanomaterials. They are used in food/feed products for their increased uptake and absorption, and improved bioavailability of vitamins, antioxidants in the body. Proteins, fat, and sugar molecules, as well as nutraceuticals consisting of food additives derived from plants are examples of organic nanomaterials. Certain enzymes with antimicrobial activity could be covalently immobilized on to amino- or carboxyl-plasma-activated bio-oriented polypropylene films via suitable coupling agents.	[54, 55]

antibodies are immobilized on thin nanofilms or sensor chips, which can detect the target molecules and emit signals [9].

- Nanotechnology can also assist in the detection of pesticides [60], pathogens [61], and toxins [62] in food products, and thereby monitoring the quality of foods.
- Carbon nanotubes is an example of a simple, cost-effective biosensor that can successfully detect microorganisms, toxins, and other degraded products in food and beverages [63]. These nanotubes if attached with toxin antibodies can also be used for detection of waterborne toxins [64].
- Using quantum spots in the identification of the pesticide 2,4-dichlorophenoxyacetic is an example of semiconductor applications [65]. Interestingly, Kraft Food Company provides some sorts of nanocrystal sensors that are directly used in the packaging sector of factories.
- When the foodborne pathogens produce gas due to their activities leading to food spoilage, these highly sensitive devices detect the produced gases, as a result, the packaged food would be placed away from the packaging line.
- The color changes in strips inserted on the sensors are a visual signal to detect whether the food is fresh or not, immediately alerting the operators [66].

- A group of nanosensors called nanoelectromechanical systems (NEMS) have previously been recruited in food companies, which can, specifically, detect hazardous materials in food.
- Nano-cantilever is the most recent type of nanosensors, which has emerged as the fundamental components of the micro (nano)-electromechanical systems such as nanomechanical-based mass sensors [67]. The main core of these nanosensors is composed of substances derived from small pieces of silicon that are able to detect microbial pathogens and toxic proteins in food.
- The selective difference of actual cell concentrations and bacterial strains is of great significance in all parts of microbial research like food safety and environmental science, toxicity monitoring, and clinical diagnostics [68].
- Applying immobilized antibodies as selective capture agents, cells, or bacterial substances bind selectively to the immuno-activated sensor surface. Moreover, whole the surface of the immuno-sensor for microbes, organic microcontaminants, and proteins is sensitized with an antigen or its relevant protein conjugate, the microbial analyte material in the specimen prevents binding of the antibodies to the immuno-activated surface.
- The signal measured by the immunosensor can be strengthened by nanoparticles coupled to the antibody [69].
- Subramanian (2006) developed a polyethylene glycol on a surface plasmon resonance (SPR) immunosensor using polyclonal antibodies or purified monoclonal on an activated sensor chip against *E. coli* [70].
- An important difference was observed between heat-treated cells and sensor signals of living, so that the rate of living and immobilized cells and the mass of the immobilized cells were determined [71].
- One kind of biosensors called SPR-based biosensor was applied for the rapid detection of *Campylobacter jejuni* in the specimen of broiler chicken. The sensitivity and specificity of antibodies against *C. jejuni* were examined with Campylobacter and non-Campylobacter bacterial strains [72].
- **Recognition of bacterial biofilm by different sensors:** For the development of anti-biofilm substances and surface treatments or properties of biofilm formation, it is highly essential to detect the physicochemical factors that activate the sensors [73]. During scanning electron microscopy (SEM), traditional plating, combination of biochemical methods and microscopy, surface microscopy, and epi-fluorescence microscopy, the application of nanotechnology is becoming more prevalent like total internal reflection aqueous fluorescence microscopy, quartz crystal microbalance (QCM), atomic force microscopy (AFM), and total internal reflection microscopy.

6.2.11 Application of Antimicrobial Nanodispersed Systems

Nano-sized materials can play an important role in any stage of food processing such as transporting, packaging, and labeling. During the packaging process, nanomaterials display strong abilities for microbiological control of the final food products (Table 6.2).

6.2.12 Nanotechnology in Food Packaging and Safe Delivery Systems

- Nanosensors and antimicrobial activators that has been used in packaging food items are capable of detecting food spoilage and releasing nanoantimicrobes to extend food shelf life [56].
- A desirable food packaging material should have gas and moisture permeability along with the strength and biodegradability of the material [80]. Nanocomposites, in this regard, could be considered as an active material for packaging and safe delivery of the food products [81].
- Using inorganic nanoparticles, a strong antibacterial activity can be achieved in low concentrations and more stability in extreme conditions. In this aspect many nanoparticles such as silver, copper, chitosan, and metal oxide nanoparticles like titanium oxide or zinc oxide have antibacterial property [81, 82].
- The incorporation of inorganic nanoparticles in polymers makes a more resistant packaging material with cost-effectiveness [83]. Further, use of inert chitin or

Table 6.2 Application of antimicrobial nanodispersed systems

Nanoparticles	Applications
AgNPs	Alters the surface properties of bacterial cells through degradation of several lipopolysaccharide [74] Ag + ions damage bacterial DNA
CuNPs	Blocks biochemical pathways Causes membrane disruption Causes DNA damage Complex formation with proteins [75, 76]
UV activated TiO_2	Has high potential to destroy both gram-negative and gram-positive bacteria [77]
Cyclodextrins Silica nanoparticles Clay Graphene Chitosan nanoparticles Starch nanocrystals	Positive impact on shelf life of vegetables and fruits by inhibiting senescence Presents strong antimicrobial characteristics [78] Biofilm formation may be prevented by killing the microbes adhered to the surface and through the application of specific covering materials [79]

chitosan into the polymer matrix makes it lighter but stronger, fire-resistant, and with better thermal properties [84, 85].
- PEG coating nanoparticles loaded with garlic essential oil can control the pests [86].
- Phytoglycogen octenyl succinate nanoparticles with ε-polylysine creates a stronger defense against free radical oxygen and metal ions that cause lipid oxidation and thereby significantly increases the shelf life of the product [87].
- Researchers are using silicate nanoparticles to provide a barrier to gasses (e.g., oxygen), or moisture in a plastic film used for packaging. This could reduce the possibility of food spoiling or drying out during delivery [88].

6.2.13 A List of Food Products Currently Containing Nanoproducts Include

- Canola Active Oil (Shemen, Haifa, Israel).
- Nanotea (Shenzhen Become Industry Trading Co. Guangdong, China).
- Fortified Fruit Juice (High Vive. com, USA).
- Nanoceuticals Slim Shake (assorted flavors, RBC Life sciences, Irving, USA).
- NanoSlim beverage (NanoSlim).
- Oat Nutritional Drink (assorted flavors, Toddler Health, Los Angeles, USA).
- 'Daily Vitamin Boost' fortified fruit juice (Jamba Juice Hawaii, USA) and nanocapsules containing tuna fish oil (a source of omega 3 fatty acids) in "Tip-Top" Up bread (Enfield, Australia) [89].
- A listing of nano-related food and beverage is provided by the Nanotech Project in its Nanotechnology Consumer Products Inventory [90].
- ε-Polylysine, a food-grade polypeptide, can be added to the oil droplets to help protect from oxidation.
- Polylysine is much smaller than the phytoglycogen octenyl succinate nanoparticles, allowing it to fill in the gaps between phytoglycogen octenyl succinate nanoparticles [91].
- **Nanoemulsions:** Nanoemulsions have been developed to decontaminate the food packaging equipment. A typical example is a nanomicelle-based product that contains natural glycerin, and it removes pesticide residues from fruits and vegetables, as well as the oil/dirt from cutlery. Nanoemulsions have recently received a lot of attention from the food industry due to their high clarity. These enable the addition of nanoemulsified bioactives and flavors to a beverage without a change in product appearance [92].
- The growth of *Salmonella typhimurium* colonies has been eliminated by treatment with nanoemulsion [93].
- Various types of nanoemulsion, including single-layer, double-layer, and triple-layer nanoemulsions, can be produced, depending on the polyelectrolytes, such as alginate and chitosan [94].

- Solid lipid nanoparticles have been reported for delivery of bioactives, such as lycopene and carotenoids [36, 95]. The major advantages of solid lipid nanoparticles include large-scale production without the use of organic solvents, high concentration of functional compounds in the system, long-term stability, and the ability to be spray dried into powder form.
- Further, polymeric nanoparticles, including vitamin E, itraconazole (an antimicrobial), and beta-carotene are reported as a colorant [96, 97].
- Weiss and his colleagues have demonstrated that the particles can also serve as carriers of antimicrobial components, with niacin-containing biopolymeric nanoparticles exhibiting much more potent activity against *E. coli* O157:H7 than particles without niacin [98].
- The discovery of antimicrobial properties of nanozinc oxide and nanomagnesium oxide at the University of Leeds may provide more affordable materials for such applications in food packaging [99].
- Recently, nanoscale processing has been used to make new solid state forms of food materials, such as cereals.
- Compared to conventional materials, these materials have new physicochemical properties, such as altered solubility, cohesiveness, stability, and reactivity [100].
- The vehicles are expanded micelles in the size of 30 nm and can be used in "clear" beverages without phase separation.
- They are coined fortifying nanovehicles. Their potential applications include lycopene, beta-carotene, CoQ10, omega-3 fatty acids, phytosterols, and isoflavones [101, 102].

6.2.14 The Limitations of Nanotechnology in Food Industry

In terms of safety issues associated with the using of nanomaterials, there are some important concerns that need to be considered. Physicochemical properties of nanomaterials in their nanostates are completely different from that are in their macrostate; therefore, more additional studies must be done to examine the risk of using its nano counterparts in foods. Moreover, the small-sized nanomaterials may accumulate within body organs and tissues [103]. Therefore, regulatory authorities must develop some standards for commercial use of nanoparticles containing products to ensure the quality, safety, and the environmental regulations.

Along with different merits of nanotechnology in food science like fast reproduction, the other issues such as ethical, regulatory, and policy points as well as public and environmental safety issues have also been raised.

- The destiny of nanocarriers in the gastrointestinal tract (GIT) varies depending on the conditions of GIT and their vulnerability to hydrolysis by digestive enzymes [104].
- Usually, the redundant application of emulsifiers and organic solvents cause many risks because of their toxicity for the arrangement of nanocarriers [105].

- Emulsifiers and solvents are categorized as hazardous, with safe doses prepared by the European Food Safety Authority (EFSA), and global organizations like WHO, and FDA [106].
- These organic solvents should be removed by a process called evaporation, but this method can bring undesirable residual solvents that cannot be removed in the final material, resulting in adverse consequences.
- Despite the current lack of risk management and specific regulations for nanotechnology, research in this field has witnessed important advances in terms of regulatory affairs.
- With information transparency, new specific nanotechnology regulations will guarantee the safe utilization of nanomaterials in the food industries [104].
- A large number of nanoparticles generally released in the environment for their degradation by microbial, chemical, and physical reactions [107].
- Unfortunately, up till now, there are no such specific protocols or guidelines to deal with the wastes. However, recently the scientists are trying to produce biodegradable plastics nanoparticles for packaging purposes that are lighter in weight and thermally more stable with improved barrier protection [108].

6.3 Impact on Society

Commercial applications of nanomaterials because of their unique properties will continue to impact the food industry and thereby on global health. Therefore, the safety of the food products containing the nanomaterials should be the most important concern for the public acceptance. A uniform international regulatory guideline for using nanotechnology in food industry is a must.

6.3.1 Positive Impact

In vivo, nanoparticles are generally attached to proteins, antibodies, and nucleic acids, and used as a probe for displaying and quantifying molecular reactions inside the body. Besides the imaging of the molecular and cellular changes that happen in vitro and in vivo in food items, fluorescent biological probes that use organic dyes and inert are able to interact with a variety of cellular reactions with sensitivity.

Nanoparticles-mediated targeted drug delivery is more effective due to their efficient bioavailability, minimal side effects, and decreased toxicity to other organs, and also inexpensive. It may have the potential to prevent the threat of antimicrobial resistance through foodborne transmission of antimicrobial residues, antimicrobial-resistant bacteria (ARBs) and antimicrobial-resistant genes (ARGs) particularly in livestock sector. Using of prebiotics and probiotics instead of antibiotics in farming sector also adds to this.

6.3.2 Negative Impact

Nanoparticles with their abnormal chemical and biological reactivity when enter in the body through skin, ingestion, inhalation, injection, or by implantation may cause some problems especially when they accumulate in the system. Toxicity of nanoparticles depends on their size, structure, surface properties, and ability to aggregate. If nanoparticles have poor solubility and/or biodegradability, they may cause cancer. Especially, the cosmetic products that contain the nanoparticles but do not require any clinical trials are the greater concern of experiencing the bad effects of the nanomaterials on health. The main health concerns so far reported from nanoparticles are erythema due to the damage of fibroblasts [109].

Nanoparticles-mediated cardiopulmonary morbidity and mortality have also been proposed. Further nanoparticles can also stimulate the neuron and affects the central nervous system. In the circulation system, nanoparticles can trigger an acute inflammation, as these particles are considered by the immune system as "foreign" materials. WHO has already notified the health implications of nanoparticles, but no regulatory policies have yet been circulated [110, 111]. It is very important and urgent for human health to use regulatory mechanisms and laws during the usage of nanomaterials in foods.

6.3.3 Message to the Community

Nanomaterials are being consumed in our daily lives in different ways, often without the consumer knowledge. A strong social supports and consumer awareness can further develop the vast use of nanotechnology. Our mission as educators is to instill and arouse students' curiosity in learning the future frontier science. The knowledge of Nanoscience could be integrated in the school science curricula as early as preschool through higher education. Some of the topics that could be included are origins of nanotechnology, associated challenges about nanotechnology for educational implementation, examination of currently available school activities, types of current product applications, about the ethical issues related to nanotechnology, recommendations for educational policy along with teaching approaches and practical implications of nanotechnology [4].

6.3.4 Public Acceptance

Unfortunately nanoscience, in developing countries, raises many ethical issues, like to accept genetically modified foods. Even the developed countries sometimes face many challenges like ethical, environmental, and equities that are obstructing the scientific advances of nanotechnology. In fact, acceptance of nanotechnology-based products depends on how it is being used as end products.

6.4 Conclusions

- The energy is produced from healthier food at the cellular level due to many pathways.
- The basic secret of nanotechnology in food industry is that the human cells and food components can easily interact, including processing, packaging, safety, and security of the food products.
- Packaging materials are nanocomposites, nanosensors, biodegradable nanocomposites for leakage-proof, gases-free, and pathogen-less food packaging.
- Nanocomposites act as barrier for exchange of gases or any kind of pathogens and toxins.
- Biodegradable nanocomposites packaging are of great potential to environment. Smart packaging provides good shelf life of food products.
- Nutritional supplements with the combination of nanotechnology deliver the drugs efficiently.
- Certain regulations have to be made by the food administration department of the countries to establish proper and safe commercialization of nanofood.
- The interaction of nanoparticles with cells may limit the use of nanofoods widely though we understand its benefits in providing rich nutritional value, quality packaging, and smart sensing to provide safe food to the consumers.

In conclusion, the expectations of nanoscience in the area of health and medicine are high; however, the safety of nanomedicine(s) should be explained well. Further, concurrent application of nanotechnology in other fields like diagnostics and development of molecular research tools should also be explored.

Nanotechnology endeavor has lots of commitment on the improvement in quality of human life. A large number of industries, including food, pharmaceutical, and textile, are utilizing nanotechnology for product improvement. In food industry, the application of nanotechnology is vivid as they improve not only quality of the food but also food safety on food storage and food delivery. Nutritional supplements like probiotics witnessed bacterial viability taking the advantages of nanotechnology. Nanomaterials also act as an excellent carrier to deliver bioactive substances to the target cells.

Although nanotechnology can bring lots of improvement on food qualities, the stability and bioavailability of this technology have some demerits from their toxic effects on the human/animal health, if they enter into the body system and accumulate there. Therefore, it warrants more investigations to minimize the detrimental side effects of nanoparticles in the living tissues. Compulsory testing of nanofood-products is required according to the FDA guidelines to increase the safety issues and minimize the environmental impact before they are released to the consumer market [39].

Acknowledgement We thank all our colleagues and secretaries for their help during the preparation of the manuscript by providing all the relevant information. Thanks to Ms. Bethany Pond for her editorial assistance.

References

1. Dingman, J. (2008). Nanotechnology: Its impact on food safety. (Guest commentary). *Journal of Environmental Health, 70*(6), 47–50.
2. Rai, M., Yadav, A., & Gade, A. (2009). Silver nanoparticles as a new generation of antimicrobials. *Biotechnology Advances, 27*, 76–83. https://doi.org/10.1016/j.biotechadv.2008.09.002
3. Gupta, A., Eral, H. B., Hatton, T. A., & Doyle, P. S. (2016). Nanoemulsions: Formation, properties and applications. *Soft Matter, 12*, 2826–2841. https://doi.org/10.1039/c5sm02958a
4. Chakraborty, A., Guha, S., & Patra, P. (2020). Nanotechnology and global applications: Bench to community. *Journal of Biology and Life Science*. (ISSN 2157-6076), *11*(2), 181–189.
5. Dasgupta, N., Ranjan, S., Mundekkad, D., Ramalingam, C., Shanker, R., & Kumar, A. (2015). Nanotechnology in agrofood: From field to plate. *Food Research International, 69*, 381–400. https://doi.org/10.1016/j.foodres.2015.01.005
6. Ezhilarasi, P. N., Karthik, P., Chhanwal, N., & Anandharamakrishnan, C. (2013). Nanoencapsulation techniques for food bioactive components: A review. *Food and Bioprocess Technology, 6*, 628–647. https://doi.org/10.1007/s11947-012-0944-0
7. Ajitha, B., Reddy, Y. A. K., & Reddy, P. S. (2014). Biosynthesis of silver nanoparticles using Plectranthusamboinicus leaf extract and its antimicrobial activity. *Spectro chimica Acta Part A: Molecular and Biomolecular Spectroscopy, 128*, 257–262.
8. Egger, S., Lehmann, R. P., Height, M. J., Loessner, M. J., & Schuppler, M. (2009). Antimicrobial properties of a novel silver-silica nanocomposite material. *Applied and Environmental Microbiology, 75*, 2973–2976.
9. Liu, S., Yuan, L., Yue, X., Zheng, Z., & Tang, Z. (2008). Recent advances in nanosensors for organophosphate pesticide detection. *Advanced Powder Technology, 19*, 419–441. https://doi.org/10.1016/S0921-8831(08)60910-3
10. Allahverdiyev, A. M., Abamor, E. S., Bagirova, M., & Rafailovich, M. (2014). Antimicrobial effects of TiO_2 and Ag_2O nanoparticles against drug-resistant bacteria and leishmania parasites. *Future Microbiology., 6*, 933–940.
11. Murthy, K. C., Monika, P., Jayaprakasha, G., & Patil, B. S. (2018). *Nanoencapsulation: An advanced nanotechnological approach to enhance the biological efficacy of curcumin advances in plant phenolics: From chemistry to human health* (pp. 383–405). ACS Publications.
12. Nasr, N. F. (2015). Applications of nanotechnology in food microbiology. *International Journal of Current Microbiology and Applied Sciences, 4*, 846–853.
13. Bugusu, B., Mejia, C., Magnuson, B., & Tafazoli, S. (2009). Global regulatory food policies on nanotechnology. *Food Technology, 63*(5), 24–29.
14. Semo, E., Kesselman, E., Danino, D., & Livney, Y. D. (2007). Casein micelle as a natural nano-capsular vehicle for nutraceuticals. *Food Hydrocolloids, 21*, 936–942.
15. Sekhon, B. S. (2010). Food nanotechnology – an overview. *Nanotechnology, Science and Applications, 3*(4), 1–15.
16. Fu, W. C., Opazo, M. A., Acuña, S. M., & Toledo, P. G. (2017). New route for self-assembly of α-lactalbumin nanotubes and their use as templates to grow silver nanotubes. *PLoS One, 12*(4), e0175680. https://doi.org/10.1371/journal.pone.0175680
17. Aguilera, J. M., & Stanley, D. W. (1999). *Microstructural principles of food processing and engineering* (2nd ed.). Springer-Verlag. Available from; http://www.knovel.com/web/portal/browse/display?_EXT_KNOVEL_DISPLAY_bookid=1164. Accessed 26 Feb 2010
18. Donsì, F., Annunziata, M., Sessa, M., & Ferrari, G. (2010). Nanoencapsulation of essential oils to enhance their antimicrobial activity in foods. *Journal of Biotechnology, 44*, 1908–1914.
19. Ravichandran, M., Hettiarachchy, N. S., Ganesh, V., Ricke, S. C., & Singh, S. (2011). Enhancement of antimicrobial activities of naturally occurring phenolic compounds by nanoscale delivery against Listeria monocytogenes, Escherichia coli O157: H7 and Salmonella typhimurium in broth and chicken meat system. *Journal of Food Safety, 31*(462), 471.

20. Chopra, M., Kaur, P., Bernela, M., & Thakur, R. (2014). Surfactant-assisted nisin loaded chitosan-carrageenan nanocapsule synthesis for controlling food pathogens. *Food Control, 37*, 158–164.
21. Handford, C. E., Dean, M., Henchion, M., Spence, M., Elliott, C. T., & Campbell, K. (2014). Implications of nanotechnology for the agri-food industry: Opportunities, benefits and risks. *Trends in Food Science and Technology, 40*, 226–241.
22. Rahmati, F. (2020). Microencapsulation of Lactobacillus acidophilus and Lactobacillus plantarum in Eudragit S100 and alginate chitosan under gastrointestinal and normal conditions. *Applied Nanoscience, 10*, 391–399.
23. Rahmati, F. (2017). Characterization of Lactobacillus, Bacillus and Saccharomyces isolated from Iranian traditional dairy products for potential sources of starter cultures. *AIMS Microbiology, 3*, 815–825.
24. Rahmati, F. (2018b). *Iran, the potential source of dairy starter microorganisms: Isolation, identification and characterization of LAB and yeast starter strain from traditional dairy products of Lorestan province*. GRIN Verlag.
25. Dulf, F. V., Pamfil, D., Baciu, A. D., & Pintea, A. (2013). Fatty acid composition of lipids in pot marigold (Calendula officinalis L.) seed genotypes. *Chemistry Central Journal, 7*, 8.
26. Vidhyalakshmi, R., Bhakyaraj, R., & Subhasree, R. S. (2009). Encapsulation "the future of probiotics" – A review. *Advances in Biology Research, 3–4*, 6–103.
27. Wang, X., Jiang, Y., Wang, Y. W., Huang, M. T., Ho, C. T., & Huang, Q. (2008). Enhancing antiinflammation activity of curcumin through O/W nanoemulsions. *Food Chemistry, 108*(2), 419–424.
28. Wang, X., Jiang, Y., Wang, Y. W., & Huang, Q. (2009). *Enhancing stability and oral bioavailability of polyphenols using nanoemulsions* (pp. 198–212). ACS Symposium Series, Vol. 1007: Micro/Nanoencapsulation of Active Food Ingredients, Chapter 13. https://doi.org/10.1021/bk-2009-1007.ch013
29. Relkin, P., Yung, J. M., Kalnin, D., & Ollivon, M. (2008). Structural behaviour of lipid droplets in protein-stabilized nano-emulsions and stability of α-tocopherol. *Food Biophysics, 3*(2), 163–168.
30. Brody, A. (2006). Nano and food packaging technologies converge. *Food Technology, 60*(3), 92–94.
31. IOM (Institute of Medicine). (2009). *Nanotechnology in food products: Workshop summary*. The National Academies Press. Available from: http://www.nap.edu/openbook.php?record_id=12633. Accessed 26 Feb 2010
32. Cientifica Report Nanotechnologies in the Food Industry, Published August 2006. Available: www.cientifica.com/www/details.php?id=47. Accessed 24 Oct 2006.
33. Nanotechnology and the Food Industry – The Pros and Cons of Nanofoods. Available from: http://www.azonano.com/news.asp?newsID=10016. (Posted February 19, 2009). Accessed 26 Feb 2010.
34. Pradhan, N., Singh, S., Ojha, N., Srivastava, A., Barla, A., Rai, V., et al. (2015). Facets of nanotechnology as seen in food processing, packaging, and preservation industry. *BioMed Research International*, 365672. https://doi.org/10.1155/2015/365672
35. Renton, A. (2006). *Welcome to the world of nano foods*. Available at: http://observer.guardian.co.uk/foodmonthly/futureoffood/story/0,1971266,00.html. Accessed 17 Jan 2008.
36. Weiss, J., Takhistov, P., & McClements, J. (2006). Functional materials in food nanotechnology. *Journal of Food Science, 71*, R107–R116. https://doi.org/10.1111/j.1750-3841.2006.00195.x
37. Sari, P., Mann, B., Kumar, R., Singh, R. R. B., Sharma, R., Bhardwaj, M., et al. (2015). Preparation and characterization of nanoemulsion encapsulating curcumin. *Food Hydrocolloids, 43*, 540–546. https://doi.org/10.1016/j.foodhyd.2014.07.011
38. Langer, R., & Peppas, N. A. (2003). Advances in biomaterials, drug delivery, and bionanotechnology. *AICHE Journal, 49*, 2990–3006. https://doi.org/10.1002/aic.690491202
39. Zhang, T., Lv, C., Chen, L., Bai, G., Zhao, G., & Chuanshan, X. C. (2014). Encapsulation of anthocyanin molecules within a ferritin nanocage increases their stability and cell

uptake efficiency. *Food Research International, 62*, 183–192. https://doi.org/10.1016/j.foodres.2014.02.041
40. Yang, R., Zhou, Z., Sun, G., Gao, Y., Xu, J., Strappe, P., et al. (2015). Synthesis of homogeneous protein-stabilized rutin nanodispersions by reversible assembly of soybean (Glycine max) seed ferritin. *RSC Advances, 5*, 31533–31540. https://doi.org/10.1039/C5RA03542B
41. Ozturk, A. B., Argin, S., Ozilgen, M., & McClements, D. J. (2015). Formation and stabilization of nanoemulsion-based vitamin E delivery systems using natural biopolymers: Whey protein isolate and gum. *Food Chemistry, 188*, 256–263. https://doi.org/10.1016/j.foodchem.2015.05.005
42. Ottaway P. B. Nanotechnology in supplements and foods – EU concerns. 2010. Available at: http://www.accessmylibrary.com/coms2/summary_0286-37130259_ITM. Accessed 26 Feb 2010.
43. Dekkers, S., Krystek, P., Peters, R. J., Lankveld, D. X., Bokkers, B. G., van Hoeven-Arentzen, P. H., et al. (2011). Presence and risks of nanosilica in food products. *Nanotoxicology, 5*, 393–405. https://doi.org/10.3109/17435390.2010.519836
44. Savolainen, K., Pylkkänen, L., Norppa, H., Falck, G., Lindberg, H., Tuomi, T., et al. (2010). Nanotechnologies, engineered nanomaterials and occupational health and safety – A review. *Safety Science, 6*, 1–7. https://doi.org/10.1016/j.ssci.2010.03.006
45. Estevinho, B. N., & Rocha, F. (2017). A key for the future of the flavors in food industry: Nanoencapsulation and microencapsulation. In *Nanotech appl in food* (pp. 1–19). Elsevier.
46. Yu, H., Park, J. Y., Kwon, C. W., Hong, S. C., Park, K. M., & Chang, P. S. (2018). An overview of nanotechnology in food science: Preparative methods, practical applications, and safety. *Journal of Chemistry, 2018*, e5427978–e5427973. https://doi.org/10.1155/2018/5427978
47. Estevinho, B. N., Rocha, F., Santos, L., & Alves, A. (2013). Microencapsulation with chitosan by spray drying for industry applications: A review. *Trends in Food Science & Technology, 31*, 138–155.
48. McClements, D. J. (2015). Nanoscale nutrient delivery systems for food applications: Improving bioactive dispersibility, stability, and bioavailability. *Journal of Food Science, 80*, N1602–N1611.
49. Bao, C., Jiang, P., Chai, J., Jiang, Y., Li, D., Bao, W., Liu, B., Liu, B., Norde, W., & Li, Y. (2019). The delivery of sensitive food bioactive ingredients: Absorption mechanisms, influencing factors, encapsulation techniques and evaluation models. *Food Research International, 120*, 130–140.
50. Stones, M. *Nanoscience to boost food safety, quality and shelf life*. Available from: http://www.meatprocess.com/Product-Categories/Ingredients-and-additives/Nanoscience-to-boost-food-safety-qualityand-shelf-life, http://www.foodhaccp.com/1news/060809i.html. Accessed 8 Jun 2009.
51. Chaudhry, Q., Scotter, M., Blackburn, J., et al. (2008). Applications and implications of nanotechnologies for the food sector. *Food Additives and Contaminants, 25*(3), 241–258.
52. Woodrow Wilson International Centre for Scholars. (2009). The Nanotechnology Consumer Inventory. Available at: www.nanotechproject. org/inventories/consumer/. Accessed 26 Feb 2010.
53. Sustech GMBH and Co. (2003). *Patent application EP20030748025, sweet containing calcium, Germany*; Sustech GMBH Co. 2004. International Patent Application: PCT/EP2003/010213 Coated chewing gum, Germany.
54. Sorrentino, A., Gorrasi, G., & Vittoria, V. (2007). Specialty biotech Thailand. *Trends in Food Science and Technology, 18*, 84–95.
55. Vartiainen, J., Rättö, M., & Paulussen, S. (2005). Antimicrobial activity of glucose oxidase-immobilized plasma-activated polypropylene films. *Packaging Technology & Science, 18*, 243–251.
56. Helmke, B. P., & Minerick, A. R. (2006). Designing a nano-interface in a microfluidic chip to probe living cells: Challenges and perspectives. *Proceedings of the National Academy*

of Sciences of the United States of America, 103, 6419–6424. https://doi.org/10.1073/pnas.0507304103
57. Bouwmeester, H., Dekkers, S., Noordam, M. Y., Hagens, W. I., Bulder, A. S., Heer, C., et al. (2009). Review of health safety aspects of nanotechnologies in food production. *Regulatory Toxicology and Pharmacology, 53*, 52–62. https://doi.org/10.1016/j.yrtph.2008.10.008
58. Jianrong, C., Yuqing, M., Nongyue, H., Xiaohua, W., & Sijiao, L. (2004). Nanotechnology and biosensors. *Biotechnology Advances, 22*, 505–518. https://doi.org/10.1016/j.biotechadv.2004.03.004
59. Subramanian, A. (2006). A mixed self-assembled monolayer-based surface Plasmon immunosensor for detection of E. Coli O157H7. *Biosensors & Bioelectronics, 7*, 998–1006. https://doi.org/10.1016/j.bios.2005.03.007
60. Inbaraj, B. S., & Chen, B. H. (2015). Nanomaterial-based sensors for detection of foodborne bacterial pathogens and toxins as well as pork adulteration in meat products. *Journal of Food and Drug Analysis, 24*, 15–28. https://doi.org/10.1016/j.jfda.2015.05.001
61. Nachay, K. (2007). Analyzing nanotechnology. *Food Technology, 1*, 34–36.
62. Wang, L., Chen, W., Xu, D., Shim, B. S., Zhu, Y., Sun, F., et al. (2009). Simple, rapid, sensitive, and versatile SWNT-paper sensor for environmental toxin detection competitive with ELISA. *Nano Letters, 9*, 4147–4152. https://doi.org/10.1021/nl902368r
63. Couch, L. M., Wien, M., Brown, J. L., & Davidson, P. (2016). Food nanotechnology: Proposed uses, safety concerns and regulations. *Agro Food Industry Hitech, 27*, 36–39.
64. Pinto, R. J. B., Daina, S., Sadocco, P., Neto, C. P., & Trindade, T. (2013). Antibacterial activity of nanocomposites of copper and cellulose. *BioMed Research International, 6*, 280512. https://doi.org/10.1155/2013/280512
65. Rico-Yuste, A., & Carrasco, S. (2019). Molecularly imprinted polymer-based hybrid materials for the development of optical sensors. *Polymers, 11*, 1173.
66. Durán, N., & Marcato, P. D. (2013). Nanobiotechnology perspectives. Role of nanotechnology in the food industry: A review. *International Journal of Food Science and Technology, 48*, 1127–1134.
67. Jackman, J. A., Ferhan, A. R., Yoon, B. K., Park, J. H., Zhdanov, V. P., & Cho, N. J. (2017). Indirect nanoplasmonic sensing platform for monitoring temperature-dependent protein adsorption. *Analytical Chemistry, 89*, 12976–12983.
68. Gonchar, M., Smutok, O., Karkovska, M., Stasyuk, N., & Gayda, G. (2017). Yeast-based biosensors for clinical diagnostics and food control. In *Biotechnology of yeasts and filamentous fungi* (pp. 391–412). Springer.
69. Cattò, C., & Cappitelli, F. (2019). Testing anti-biofilm polymeric surfaces: Where to start? *International Journal of Molecular Sciences, 20*, 3794.
70. Subramanian, A., Irudayaraj, J., & Ryan, T. (2006). A mixed self-assembled monolayer-based surface Plasmon immunosensor for detection of E. coli O157: H7. *Biosensors & Bioelectronics, 21*, 998–1006.
71. Szendrő, I., Erdélyi, K., Puskás, Z., Fabian, M., Adanyi, N., & Somogyi, K. (2010). Development and experiments with conductive oxide nanofilm coated planar waveguide sensors. *Nanopages, 7*, 17–24.
72. Vizzini, P., Braidot, M., Vidic, J., & Manzano, M. (2019). Electrochemical and optical biosensors for the detection of *campylobacter* and *listeria*: An update look. *Micromachines (Basel)., 10*(8), 500. https://doi.org/10.3390/mi10080500
73. Wolska, K. I., Grudniak, A. M., Kaminski, K., & Markowska, K. (2015). The potential of metal nanoparticles for inhibition of bacterial biofilms. In M. Rai & K. Kon (Eds.), *Nanotechnology in diagnosis* (pp. 119–132). Academic Press, Treatment and Prophylaxis of Infectious Diseases.
74. Slavin, Y. N., Asnis, J., Häfeli, U. O., & Bach, H. (2017). Metal nanoparticles: Understanding the mechanisms behind antibacterial activity. *Journal of Nanobiotechnology, 15*, 65.

75. Baptista, P. V., McCusker, M. P., Carvalho, A., Ferreira, D. A., Mohan, N. M., & Martins, M. (2018). Fernandes AR (2018) Nano-strategies to fight multidrug resistant bacteria "A Battle of the Titans". *Frontiers in Microbiology, 9*, 1441.
76. Yaqub, A., Malkani, N., Shabbir, A., Ditta, S. A., Tanvir, F., Ali, S., Naz, M., Kazmi, S. A., & Ullah, R. (2020). Novel biosynthesis of copper nanoparticles using Zingiber and allium sp. with synergic effect of doxycycline for anticancer and bactericidal activity. *Current Microbiology, 13*, 1–3.
77. Hwang, G. B., Patir, A., Allan, E., & Nair, S. P. (2017). Parkin IP Superhydrophobic and white light-activated bactericidal surface through a simple coating. *ACS Applied Materials & Interfaces, 9*, 29002–29009.
78. Hoseinzadeh, E., Makhdoumi, P., Taha, P., Hossini, H., Stelling, J., & Amjad, K. M. (2017). A review on nano-antimicrobials: Metal nanoparticles, methods and mechanisms. *Current Drug Metabolism, 18*, 120–128.
79. Gebreyohannes, G., Nyerere, A., Bii, C., & Sbhatu, D. B. (2019). Challenges of intervention, treatment, and antibiotic resistance of biofilm forming microorganisms. *Heliyon, 5*, e02192.
80. Bradley, E. L., Castle, L., & Chaudhry, Q. (2011). Applications of nanomaterials in food packaging with a consideration of opportunities for developing countries. *Trends in Food Science and Technology, 22*, 603–610. https://doi.org/10.1016/j.tifs.2011.01.002
81. Tan, H., Ma, R., Lin, C., Liu, Z., & Tang, T. (2013). Quaternized chitosan as an antimicrobial agent: Antimicrobial activity, mechanism of action and biomedical applications in orthopedics. *International Journal of Molecular Sciences, 14*, 1854–1869. https://doi.org/10.3390/ijms14011854
82. Sorrentino, A., Gorrasi, G., & Vittoria, V. (2007). Potential perspectives of bionanocomposites for food packaging applications. *Trends in Food Science and Technology, 18*, 84–95. https://doi.org/10.1021/acsami.7b04297
83. Duncan, T. V. (2011). Applications of nanotechnology in food packaging and food safety: Barrier materials, antimicrobials and sensors. *Journal of Colloid and Interface Science, 363*, 1–24. https://doi.org/10.1016/j.jcis.2011.07.017
84. Othman, S. H. (2014). Bio-nanocomposite materials for food packaging applications: Types of biopolymer and nano-sized filler. *Agriculture and Agricultural Science Procedia, 2*, 296–303. https://doi.org/10.1016/j.aaspro.2014.11.042
85. Wyser, Y., Adams, M., Avella, M., Carlander, D., Garcia, L., Pieper, G., Rennen, M., Schuermans, J., & Weiss, J. (2016). Outlook and challenges of nanotechnologies for food packaging. *Packaging Technology and Science, 29*, 615–648.
86. Yang, F. L., Li, X. G., Zhu, F., & Lei, C. L. (2009). Structural characterization of nanoparticles loaded with garlic essential oil and their insecticidal activity against Triboliumcastaneum (Herbst) (Coleoptera: Tenebrionidae). *Journal of Agricultural and Food Chemistry, 57*(21), 10156–10162.
87. Scheffler, S. L., Wang, X., Huang, L., Gonzales, F. S., & Yao, Y. (2010). Phytoglycogenoctenyl succinate, an amphilic carbohydrate nanoparticle, and epsilonpolylysine to improve lipid oxidative stability of emulsions. *Journal of Agricultural and Food Chemistry, 58*(1), 660–667.
88. Neethirajan, S., & Jayas, D. (2009). *Nanotechnology for food and bioprocessing industries*. 5th CIGR International Technical Symposium on Food Processing, Monitoring Technology in Bioprocesses and Food Quality Management, Potsdam, Germany. 8 p. Available from: www.docstoc.com/../Suresh-Neethirajan-Nanotechnology-for-Food-andBioprocessingIndustries. Accessed 26 Feb 2010.
89. Nanny, Nano, Boo, Boo Food?. Available from: http://towerofbabel.com/2008/08/28/nanny-nano-boo-boo-food/ (August 28, 2008). Accessed 26 Feb 2010.
90. Nanotechnology in Agriculture and Food. Available from: http://www.nanotechproject.org/inventories/agrifood/. Accessed 26 Feb 2010.
91. Nanoparticle protects oil in foods from oxidation, spoilage. Available from: www.purdue.edu/UNS/../091208YaoNanoparticles.html. (December 8, 2009). Accessed 26 Feb 2010.

92. Mustafa, I. F., & Hussein, M. Z. (2020). Synthesis and technology of nanoemulsion-based pesticide formulation. *Nanomaterials (Basel), 10*(8), 1608. https://doi.org/10.3390/nano10081608. PMID: 32824489; PMCID: PMC7466655.
93. Nanoemulsions. *Centre for biologic nanotechnology*. Available from: www.vitamincity.com/umichnanobio.htm. Accessed 26 Feb 2010.
94. Choi, A. J., Kim, C. J., Cho, Y. J., Hwang, J. K., & Kim, C. T. *Solubilization of capsaicin and its nanoemulsion formation in the sonication and self assembly processes*. Available from: http://www.nsti.org/BioNano2008/showabstract.html?absno=462. Accessed 26 Feb 2010.
95. Weiss, J., Decker, E. A., McClements, J., Kristbergsson, K., Helgason, T., & Awad, T. S. (2008). Solid lipid nanoparticles as delivery systems for bioactive food components. *Food Biophysics, 3*(2), 146–154.
96. Sabliov, C. M., & Astete, C. E. (2007). *Controlled release technologies for targeted nutrition: Encapsulation and controlled release of antioxidants and vitamins via polymeric nanoparticles*. Delivery and Controlled Release of Bioactives in Foods and Nutraceuticals Garti N. ed. Chapter 17. Available from: http://www.chipsbooks.com/delbioac.htm. Accessed 26 Feb 2010
97. Sabliov, C. *New and emerging food applications of polymeric nanoparticles for improved health*. IFT International Food Nanoscience Conference 2009 June 6. Available from: http://members.ift.org/IFT/Research/ConferencePapers/foodnanoscience2009.htm. Accessed 26 Feb 2010.
98. IOM (Institute of Medicine). (2009). *Nanotechnology in food products: Workshop summary*. The National Academies Press. Available from: http://www.nap.edu/openbook.php?record_id=12633. Accessed 26 Feb 2010.
99. Zhang, L., Jiang, Y., Ding, Y., Povey, M., & York, D. (2007). Investigation into the antibacterial behaviour of suspensions of ZnO nanoparticles (ZnOnanofluids). *Journal of Nanoparticle Research, 9*(3), 479–489.
100. Munro, I. C., Haighton, L. A., Lynch, B. S., & Tafazoli, S. (2009). Technological challenges of addressing new and more complex migrating products from novel food packaging materials. *Food Addit Contam: Part A: Chemistry, Analysis, Control, Exposure and Risk Assessment, 26*(12), 1534–1546.
101. Garti, N., & Aserin, A. (2007). Understanding and controlling the microstructure of complex foods. In M. D. Julian (Ed.), *Nanoscale liquid self-assembled dispersions in foods and the delivery of functional ingredients* (pp. 504–553). Woodhead Publishing Ltd..
102. Garti, N. (Ed.). (2008). *Delivery and controlled release of bioactives in foods and nutraceuticals*. Woodhead Publishing.
103. Cheng, Q., Li, C., Pavlinek, V., Saha, P., & Wang, H. (2006). Surface-modified antibacterial TiO2/AgC nanoparticles: Preparation and properties. *Applications of Surface Science, 252*, 4154–4160. https://doi.org/10.1016/j.apsusc.2005.06.022
104. McClements, D. J., & Xiao, H. (2017). Is nano safe in foods? Establishing the factors impacting the gastrointestinal fate and toxicity of organic and inorganic food-grade nanoparticles. *npj Science of Food, 1*, 6.
105. Elbagory, A. M., Cupido, C. N., Meyer, M., & Hussein, A. A. (2016). Large scale screening of southern African plant extracts for the green synthesis of gold nanoparticles using microtitre-plate method. *Molecules, 21*, E1498. https://doi.org/10.3390/molecules21111498
106. McClements, D. J., & Rao, J. (2011). Food-grade nanoemulsions: Formulation, fabrication, properties, performance, biological fate, and potential toxicity. *Critical Reviews in Food Science and Nutrition, 51*, 285–330.
107. Naseer, B., Srivastava, G., Qadri, O. S., Faridi, S. A., Islam, R. U., & Younis, K. (2018). Importance and health hazards of nanoparticles used in the food industry. *Nanotechnology Reviews, 7*, 623–641.
108. Pan, Z., Lee, W., Slutsky, L., Clark, R. A. F., Pernodet, N., & Rafailovich, M. H. (2009). Adverse effects of titanium dioxide nanoparticles on human dermal fibroblasts and how to protect cells. *Small, 5*(4), 511–520. https://doi.org/10.1002/smll.200800798

109. WHO: *Nanotechnology and human health: Scientific evidence and risk governance*. Report of the WHO expert meeting 10–11 December 2012, WHO Regional Office for Europe, 2013. http://apps.who.int/iris/bitstream/handle/10665/108626/e96927.pdf?sequence=1
110. WHO guidelines on protecting workers from potential risks of manufactured nanomaterials. Geneva: World Health Organization; 2017. License: CC BY-NC-SA 3.0 IGO. https://apps.who.int/iris/bitstream/handle/10665/259671/9789241550048-eng.pdf
111. Singh, T., Shukla, S., Kumar, P., Wahla, V., Bajpai, V. K., & Rather, I. A. (2017). Application of nanotechnology in food science: Perception and overview. *Frontiers in Microbiology, 8*, 1501. https://doi.org/10.3389/fmicb.2017.01501

Chapter 7
Application of CRISPR Technology and Nanomaterials to Advance Food Supply

Narayanaswamy Bharathan and Zhibek Turarbekova

7.1 Introduction

Food security and food production are critical issues facing global community. With growing population to feed, coupled with climate change, urbanization, and competition for natural assets, it is a greater challenge today to adapt to any new strategy that advances sustainable agriculture [1]. The world's ever-growing population and changing climate are putting serious pressure on global food security. According to a United Nations report, by 2030, the world's population and demand for food will grow to 8.5 billion and 11.6 billion tons, respectively. If recent trends continue, the number of people suffering from hunger could exceed 840 million by 2030. To prevent this, the UN has set "zero hunger" as a sustainable development goal. To achieve the goal of "zero hunger", the world needs to produce 15–20% more food than the yield predicted based on recent trends. In addition, the demand for cereals, root crops, tubers, legumes, sugar, vegetables, and oilseeds are growing compared to other food crops. While there is crop plant improvement over the years by domestication and selective plant breeding, the revolutionized crop enhancement and the dawn of high-yielding, diseases-resistant varieties have taken place during the Green Revolution in 1950s. In contrast, modern-day advances in genetics and genetic tools permit for a more effective and cautious way to manipulate genes and DNA to create crops with specific desired traits [2]. Genetic tools also allow for faster and more efficient plant breeding [3]. Marker-assisted selection (MAS) is a

N. Bharathan (✉)
Department of Biology, Kopchick College of Natural Sciences and Mathematics, Indiana University of Pennsylvania, Indiana, PA, USA
e-mail: bharathn@iup.edu

Z. Turarbekova
Department of Biotechnology and Microbiology, Eurasian National University, Astana, Kazakhstan

R. S. Chaughule, A. S. Lokur (eds.), *Applications of Nanotechnology in Microbiology*, https://doi.org/10.1007/978-3-031-49933-3_7

technique that uses biochemical or DNA markers to identify desirable traits in plants at early stage of development, allowing breeders to select only those plants with the desired characteristics to be included in the breeding program [4]. Plant genetic engineering involving tissue culture techniques have been established in wide range of cultivated and medicinal plant species using T-DNA transfer or biolistic gun [5–7]. While these techniques are promising, highly effective gene transfer system for engineering of crop plants, which is equally important and central to engineering crop plants, is not without a challenge. Traditional gene delivery methods are genotype dependent, extremely slow, and expensive with several limitations to realize the full potential of targeted DNA insertions using engineered nucleases [8].

7.2 Crop Improvement

Crop improvement to enhance genetic makeup of crops over centuries has been a gradual process through selective breeding and hybridization. The traditional farming practiced for thousands of years, selectively bred plants with desirable traits and selecting seeds from the best performing plants and saving them for next growing season. Overtime, this led to domestication of many important crop species, including wheat, corn, rice, and potatoes. In the twentieth century, advances in genetics and molecular biology revolutionized crop improvement. Researchers began to use new techniques such as mutagenesis, genetic engineering, and biomarker-assisted selection to create crops with desirable traits or to introduce genes from other organisms. Today, a variety of crop improvement strategies are used to develop crops that meet the challenges of modern agriculture, including climate change, emerging diseases, and pests. These strategies include genetic modification, genome editing, gene stacking, and precision breeding [7–10]. Overall, crop improvement over centuries has played a vital role in ensuring food security and sustaining human populations around the world. Crop improvement by modern genetic tools refers to the use of advance technologies to modify genetic makeup of crop plants to improve their performance, resistant to pests and diseases, and strategies that promote sustainable agriculture.

7.3 Genetic Management and Tools in Plant Breeding

The term genetic management of plant species refers to the implementation of strategies aimed at preserving and enhancing the genetic diversity. Genetic management of plant species is important for maintaining the health and productivity of plant populations in general and more specifically of the crop species, which in turn supports biodiversity and ecosystem services. This can be best achieved through

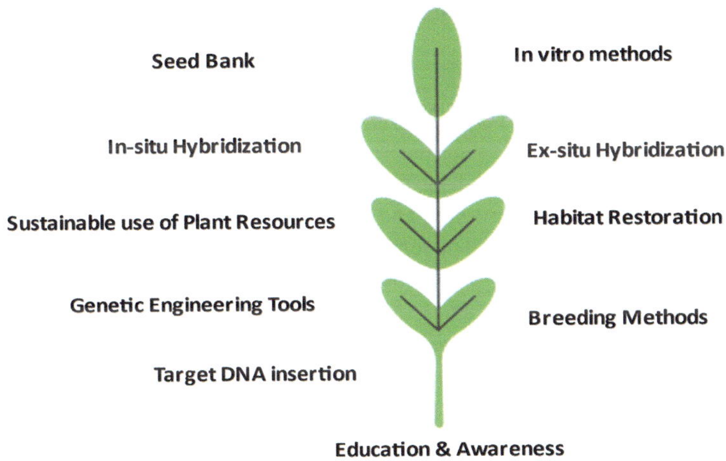

Fig. 7.1 Genetic management and tools in crop improvement

complementary strategies of various methods such as seed banking, in situ and ex situ conservation, hybridizations, In vitro methods, genetic engineering, and more importantly education and awareness (Fig. 7.1).

Seed banking and conservation involves the collection and storage of seeds from different plant populations to ensure their availability for future use. In this technique, seeds are stored in secured seed banks and can be recovered when crops are threatened due to unforeseen circumstances. Seed banks safeguard genetic diversity long-term for crop survival, protection against climate change and inevitable extinction, defense against natural and manmade disasters, and more importantly seed banks may be the only source of healthy seeds when there is a serious disease breakout that eliminates crops. Appropriately saved and deposited seeds can stay sustainable and viable for extended period, reducing the probability of crop loss that are critical for the existences of human beings and animals. Alternatively, in vitro tools provide additional backup collections as useful genetic resource and methods of propagation and conservation when seeds or vegetative material from genetically diverse wild populations are difficult to obtain. In vitro techniques are well developed for the collection, propagation, and cryopreservation of many species. In situ conservation involves protecting *wild* and *endangered* plant populations in their natural habitat while ex situ involves conservation involves growing plants outside their natural habitat such as botanical gardens or green houses. As more plant species are threatened with extinction preserving the biodiversity using a single method described may not be always adequate to guarantee the survival of a species. An effective integrated approach must be undertaken with complementary strategies that combine education and awareness of individuals and policy makers to modern day tools of molecular biology, plant genetic engineering, and molecular genetics for effective genetic management.

7.4 Plant Breeding Revolutions

Breeding new crop varieties that have delivered modern day highly productive crops can be classified into four major revolutions (Fig. 7.2). In selective conventional plant breeding favorable genes are incorporated along with several other traits. This form of breeding selection dates far back into human history. It is often slow, random, without any enforced crossing and often a genetic bottle neck on the diversity available in modern day crops. The recent challenges of climate change and extreme drought combined with rapid population growth and greater demand for plant-based products there is a greater demand to accelerate genetic improvement. Modern-day tools for crop productivity with the use of *mutation plant breeding* techniques offer new opportunities to introduce genetic variation into crops by inducing mutations in their genome, which can be further bred until a desired result has been achieved. It allows to (i) develop new crop varieties rich in nutrition that enrich human diet; (ii) further advance the effectiveness of agriculture output with very little or no environmental footprint; and (iii) more importantly speed up the efficiency of crop breeding and hence genetic yield potential to protect global food security. Mutation plant breeding can involve use of *mutagens*, such as chemicals or radiation, which can cause changes or alterations to the genetic material of plants. Mutation plant breeding using mutagens has been used successfully to improve agronomic traits, disease resistance, and tolerance in important crops including rice, wheat, and barley. Specifically, ethyl methane sulfonate (EMS) in rice [11] and N-methyl-N-nitrosourea

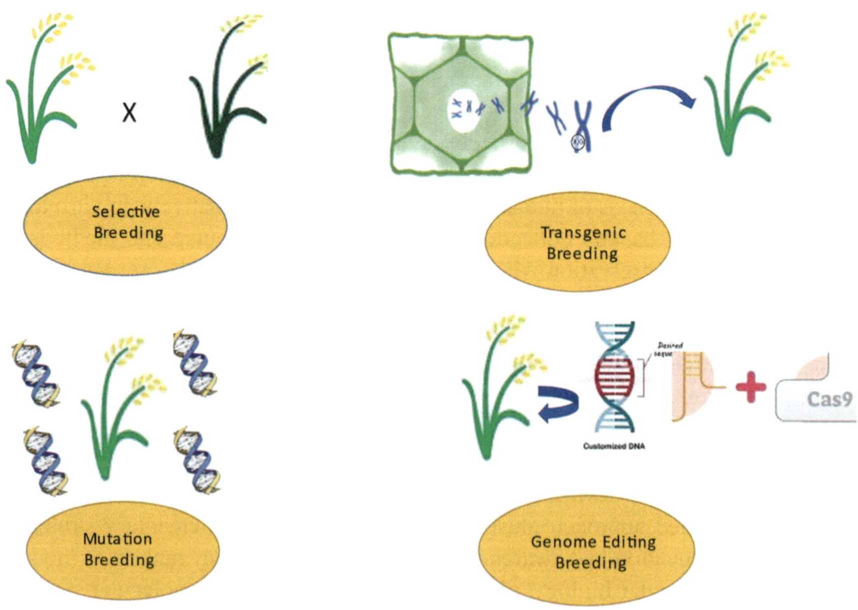

Fig. 7.2 Types of plant breeding for crop improvement

(MNU) are used to introduce genetic mutations in plants [12]. In all these cases mutated plants are then selected for desirable characteristics and are grown for several generations and further cultivated and bred to stabilize the desired traits and create new varieties. For many breeders' gamma radiation is the most preferred physical mutagen. Several crop varieties for food, herbal medicine, and ornamentals created by gamma irradiation mutagenesis for trait improvement have been released and widely cultivated. Combinational use of in vitro tissue culture and mutation breeding methods makes a significant contribution to improve new crops (Fig. 7.3). For a detailed review please see New Frontier of Plant Breeding Using Gamma Irradiation and Biotechnology [13].

Transgenic Plant Breeding is one other strategy of mutation breeding that helped to creating genome variability through genetic engineering that allowed breeders to introduce genetic variability into plants, which have accelerated the process of developing new and improved crop varieties. Unlike conventional cross-pollination breeding methods, plant genetic engineering facilitates direct delivery of gene coding desirable traits into the plant genome. This process of plant genetic engineering results in the integration of the DNA at random sites in the plant genome that involves inserting foreign genes into plants randomly to create new varieties with desirable traits. This technique unlike the others allows previously described allows for intraspecies gene transfer into the target plant genome and modifying the

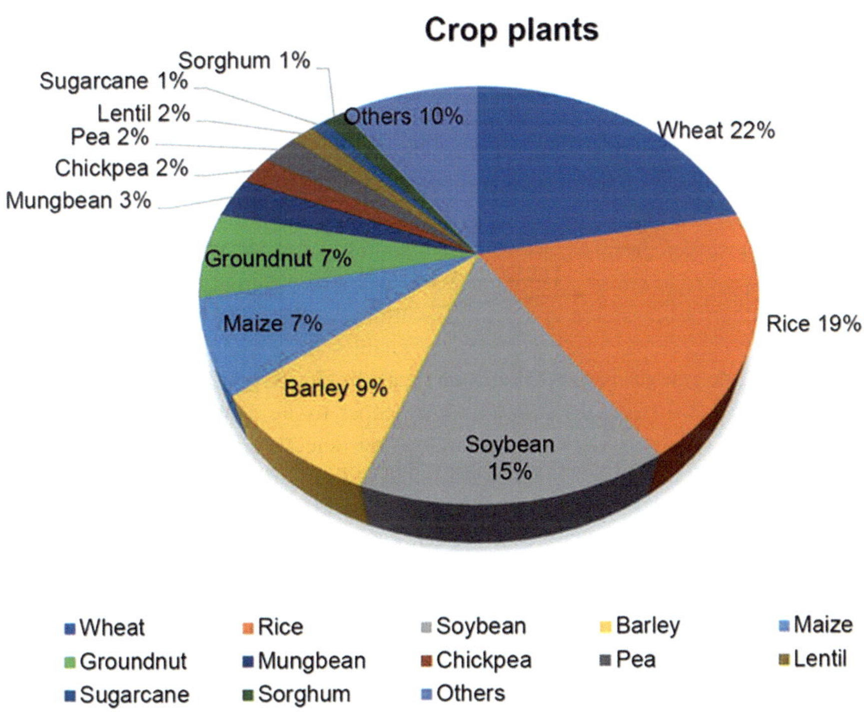

Fig. 7.3 Pie chart showing percentage of crop plants on FAO/IAES mutant variety database

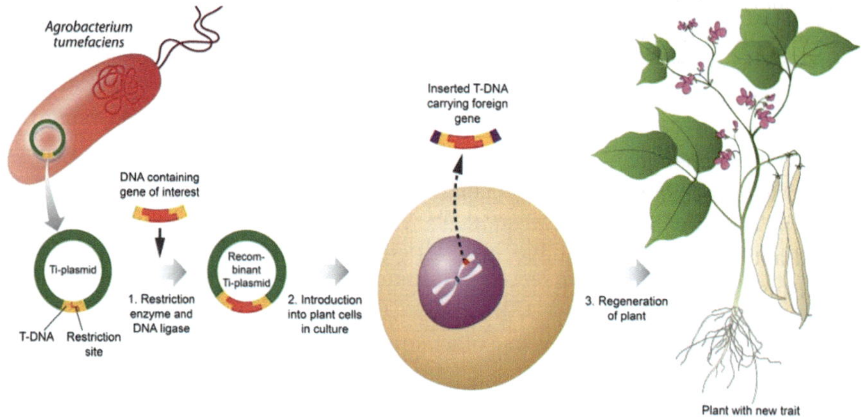

Fig. 7.4 Direct T-DNA transfer

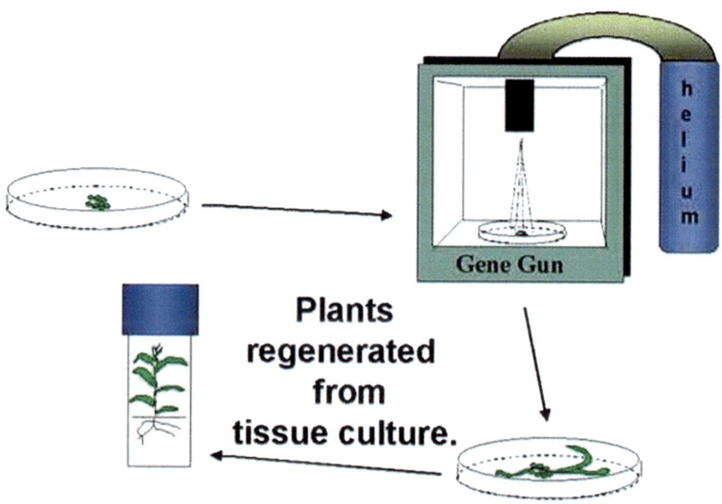

Fig. 7.5 Biolistic gene gun method to transform plant cells. (Image by P. Hain)

endogenous plant DNA by genome editing. The conventional transforming of plants is based on insertional mutagenesis due to T-DNA (Transfer DNA) (Fig. 7.4) [14–17] or biolistic bombardment (Fig. 7.5) [18]. The transformed plants are selected for the expression of selectable marker genes to eliminate untransformed cells. All such plants are regenerated under aseptic tissue culture conditions to induce growth and differentiation into whole plants. Direct gene transfer may also be achieved from protoplasts of plants [19, 20].

7.5 Genome Modification by DNA Repair-Based Methods

Recent improvements in methods in biotechnology coupled with genomic and proteomic traits of the crops and pathogen provide opportunity for targeted gene insertion at relatively high frequencies during DNA repair [21]. Targeted DNA insertions can be best achieved by using Site-Directed Nucleases (SDNs). These enzymes are capable of inducing double-stranded breaks (DSBs) at genomic targets with specific sequences [21, 22] (Fig. 7.6). The major SDN platforms that have been successfully used for gene targeting include Zinc Finger Nucleases (ZFN) Transcription activator-like effect or nucleases (TALENs), and Clustered Regularly Clustered Regularly Interspaced Short Palindrome Repeats (CRISPR)/Cas nuclease (Fig. 7.7). The interaction of the site directed engineered nucleases with plant cells result in site-specific DNA DSBs, which are repaired by endogenous systems that result in targeted genome modifications. The repair of the DSBs by host cell occurs via two pathways: (i) non-homologous end joining (NHEJ) or (ii) homologous directed repair (HDR). These pathways can be exploited to modify a gene or specific gene insertion, or gene repair [23–27] (Fig. 7.6).

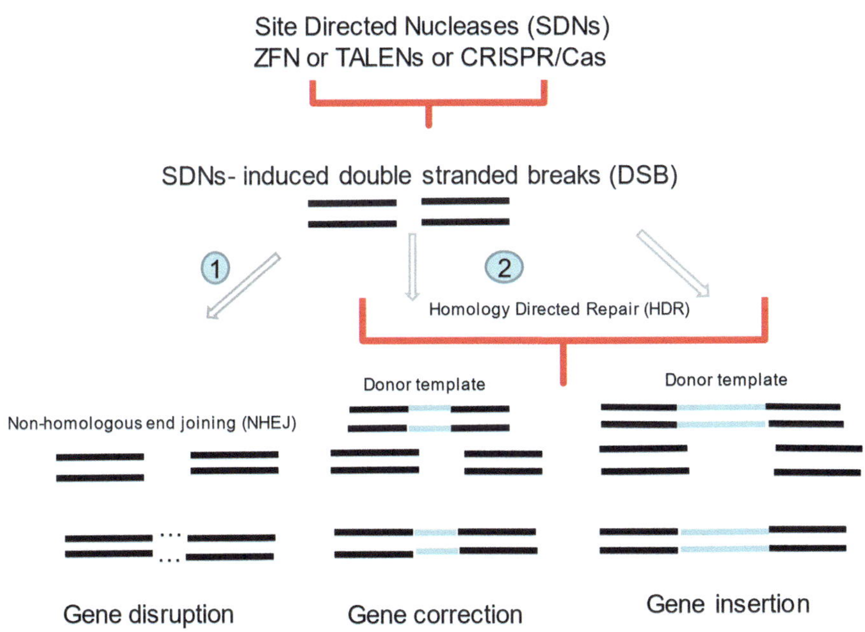

Fig. 7.6 DNA repair pathways in targeted DNA insertion using SDNs. Three different SDNs can be used to insert donor DNA fragments at desired genome targets. There are two known pathways for the repair of the DSBS in the host cell. The NHEJ pathway (1) is error-prone, and the broken ends are rejoined directly. This kind of repair is often associated with insertion or deletion (indels) of the bases. In the HDR (2) directed pathway there is always a donor template to guide the repair the DSBs

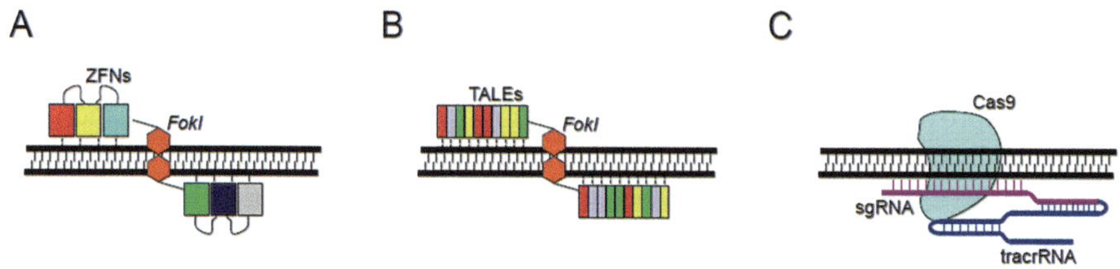

Fig. 7.7 Targeted DNA insertions in plants using ZFN, TALEN, and CRISPR platforms

The successful use of ZFN, a chimeric endonuclease in targeted DNA insertion, was reported in tobacco [28]. Subsequently, ZFN was used to disrupt herbicide tolerance gene within the corn metabolic gene inositol-1,3,4,5,6-pentakisphosphate 2-kinase (IPK1) [29]. ZFN is a chimeric engineered endonuclease that functions as dimmers (Fig. 7.7a). ZFN is created by the fusion of two monomers a nonspecific DNA binding domain and catalytic cleavage domain derived from the endonuclease *Fok*I. ZFNs are more flexible and programmable unlike previously known SDNs like meganucleases. Like ZFN the TALENs are also chimeric DNA cutting enzymes. The prospect of engineering transcription activator-like effectors for DNA binding was first established [30]. The TAL effect or nuclease (TALEN), which is based on the fusion of a Fok1 nuclease domain to the DNA-binding TALE repeats (Fig. 7.7b). High efficiency targeted gene insertion by HDR using TALE was shown in tobacco protoplasts. [31], tomato genome [32], and potato [33]. TALENs is an effective genome editing tool for rapid adoption because it is easy to program, target-specific, and efficient [34]. While molecular tools like ZFN and TALENs laid the footing for genome editing today, however, there are limits to this groundwork technologies primarily related to the requirements for protein engineering of ZFN and TALEN with target DNA. The design is complex, time-consuming, and limited to only certain parts of the plant genome. The discovery of CRISPR-Cas platform identified from bacteria was quickly adapted to the purpose of modifying eukaryotic genomes [35, 36] and wide range of applications [36, 37] including targeted gene insertion in plants (Table 7.1). CRISPR technology has allowed for the development for an RNA-guided genome editing tool that is simple, easy, and quick to implement (Fig. 7.7c). The CRISPR/Cas9 system consists of a single protein and a chimeric RNA. The RNA is composed of CRISPR RNA (crRNA), and a trans-activating crRNA (tracrRNA). In gene-editing applications, crRNA and tracrRNA have been fused into a single programmable guide RNA (gRNA). The recognition specificity of the gRNA can be easily changed by modifying the variable region of the gRNA to create gRNA library. The Cas9 protein is composed of two endonuclease domains the HNH and RuvC. The HNH domain helps to cleave the target strand and the RuvC the displaced non-target DNA strand to induce a DSB. The ribonucleoprotein complex (Cas9 protein and the CRISPR RNA (crRNA), and a trans-activating crRNA (tracrRNA) induce DSB only after it recognizes a protospacer adjacent motif (PAM), such as NGG (Fig. 7.8).

Table 7.1 Summary of plant gene modification by modern day molecular tools in crop species, their cargo delivery, and efficiency

Molecular Tool	Crop Species	Gene Delivery	Efficiency	Result	References
ZFN	Tobacco	T-DNA transfer	10%	Gene disrupted	38
	Corn	Biolistic bombardment	30%	Insertion	39
	Arabidopsis	Direct gene transfer	5.32%	Insertion	40
	Tobacco	T-DNA transfer	0%	Random insertion	28
	Potato	T-DNA transfer	Not known	Gene restoration	33
TALEN	Tobacco	Direct gene transfer	14%	Insertion	31
	Tomato	T-DNA transfer	7.20%	Gene activation	32
	Potato	T-DNA transfer	96.00%	Gene restoration	41
CRISPR-Cas	Rice	Direct gene transfer	4.7% -8.5%	Insertion	42
	Wheat	Biolistic bombardment	5.75%	Insertion	43
	Corn	Biolistic bombardment	18%	Insertion	44
	Tomato	T-DNA transfer	2.75% - 8.8%	Gene activation	31
	Arabidopsis	T-DNA transfer	6.3% -9.1%	Insertion	45
	Rice	Biolistic bombardment	25%	Insertion	46
	Tomato	T-DNA transfer	25.00%	Gene activation	26
	Arabidopsis	T-DNA transfer	6.3 -9.1%	Insertion	45

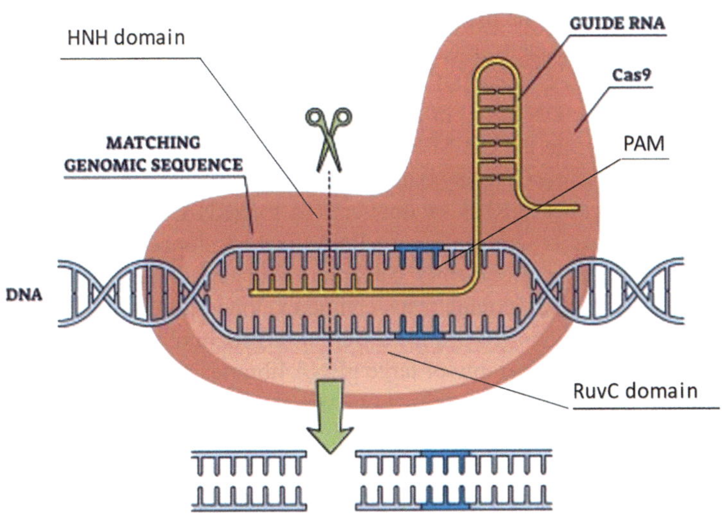

Fig. 7.8 CRISPR Cas platform

The chiefly understood Watson–Crick complementary base pairing with the target DNA sequence is the basis for gRNA-based cleavage, making complex protein engineering for each target unnecessary. Later it was shown that only a 20 nucleotide in the gRNA was needed to create a targeted DNA DSB in vitro. Soon CRISPR-Cas9 was demonstrated as a powerful RNA-guided site directed nuclease (SDN) for genome editing in human cells and plants. The repair of the DSB by host cell via non-homologous end joining (NHEJ) or homology directed repair (HDR) pathways can be utilized to create gene knockout or introduce a specific genetic modification through homologous recombination with a DNA donor (Fig. 7.6 and Table 7.1 [26–28, 31–33, 38–46].

7.6 Plant Editing Technologies and Their Applications in Crop Improvement

Ever since the introduction of CRISPR-Cas9 genome editing system in 2012, CRISPR platform has been widely adapted in plant genome modifications for its programmable versatility, site-specificity, simplicity, and cost-effectiveness, CRISPR Cas platform has moved beyond the proof-of-concept phase to create major new crops with desirable traits. Plant biotechnology is embarking on an era of editing technologies that enable target gene modifications thereby succeeding conventional methods of random mutagenesis as EMS or irradiation as previously described. The CRISPR/Cas9 technology has been effectively employed in model plants (*Nicotiana benthamiana*, *Arabidopsis thaliana*) and crops (rice, wheat, tomato, and potato) and the list is growing.

With continued advances in CRISPR Cas editing platform and development of tool kits like Multiplex Genome Editing (MGE) it is now possible to bioengineer two or more specific DNA loci in a plant genome with high precision simultaneously. To accommodate the CRISPR/Cas system for T-DNA-mediated plant transformation, designs using Gateway binary T-DNA vectors for co-expression of *CAS9* and guide RNA (either sgRNA or dual-crRNA:tracrRNA, With this tool kit and Golden Gate cloning several different gRNAs can be cloned into a single binary vector in a single delivery construct. This more efficient and less laborious to successive rounds of regular cloning steps can be exploited to insert different expression cassettes containing sgRNAs for different targets into a single construct, this allows the inexpensive assembly of large gRNA libraries so that the CRISPR/Cas9 system can be used for high-throughput functional genomics applications. It is obvious from the previous literature that multiple regions of the genome can be deleted or replaced using multiplex genome editing. MGE is used for multiple trait stacking, molecular farming, metabolic engineering, and regulatory pathway control. MGE has been used in various cultures to create climate-resilient and

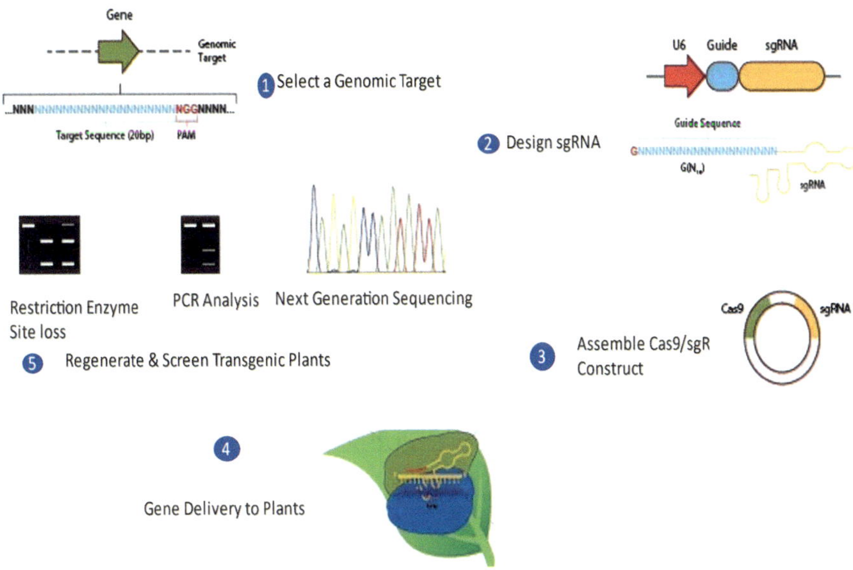

Fig. 7.9 Generalized workflow to develop CRISPR Cas 9 mutagenized plants (Sigma Aldrich, USA)

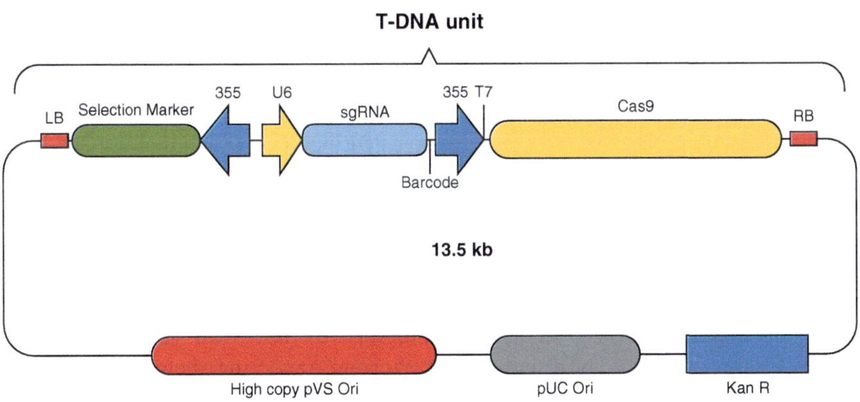

Fig. 7.10 Basic structure of CRISPR/Cas9 constructs for T-DNA-mediated Transformation (Sigma Aldrich, USA)

nutrient-enriched crop plants to combat climate change and ensure food security. A generalized workflow to generate mutagenized plant cell line is in plants shown in Fig. 7.9 along with several T-DNA transfer constructs that are commercially available from Sigma Aldrich, USA for biolistic bombardment (Fig. 7.10) or protoplast transformation (Fig. 7.11) [47–49].

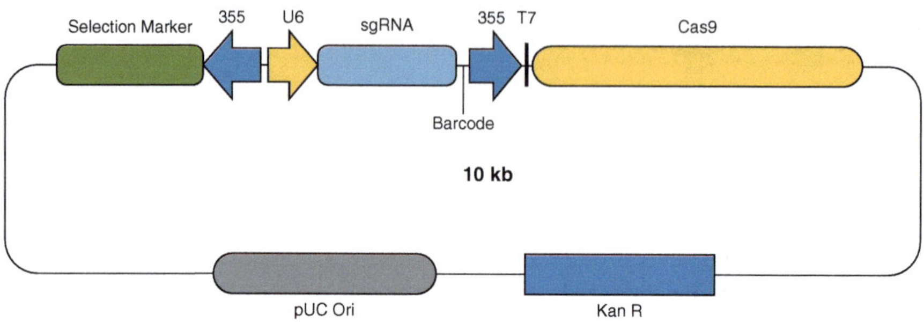

Fig. 7.11 Basic structure of CRISPR/Cas9 constructs for biolistic or protoplast transformation (Sigma Aldrich, USA)

7.7 CRISPR/Cas Genome Editing Platforms to Improve Agricultural Crops

CRISPR Cas technology has rapidly emerged in recent years as a powerful and reliable genome editing tool in various organisms [50, 51]. The targeted gene-editing tool has been adapted in various crop plants, including rice, tobacco, wheat, sorghum, corn, tomatoes, liverwort, and orange [49]. Unlike animals, direct delivery of RNA to cells in plants is technically difficult. In most cases, constructs expressing sgRNA and/or Cas9 are jointly transformed by T-DNA transfers into plant cells to form a functional CRISPR/Cas9 complex. Although sgRNA and Cas9 could be delivered by separate vectors, most studies have shown cassettes in a single binary vector to increase the efficiency of mutagenesis [52]. CRISPR was first shown to be a viable tool for plant genome development in 2013 [53]. Several other groups demonstrated proof of concept using the example of arabidopsis, tobacco, sorghum, rice, and wheat [23, 54]. In addition to the capabilities, groups have already started using CRISPR to produce cultures with the desired traits described above. DuPont Pioneer (now Corteva Agriscience) has solved the problem of stress resistance with the help of drought-resistant corn developed [55]. Cassava has been developed, which has increased protection against the disease of the brown streak of cassava, which belongs to the field of disease-resistant crops. And in order to increase yields, the flowering dates of soybeans were changed [56]. Traditional fruits and vegetables have also been improved with CRISPR. Tomatoes were edited by CRISPR [57]. Both tomatoes and potatoes have developed many targets and properties. Pathogenocarpy has been a particular target among many tomato diseases, which is an urgent problem for the industry, especially in conditions of heat stress [58]. Resistance to powdery mildew is another industry problem that has been solved [59]. Potatoes were edited to obtain a waxy phenotype [60]. Fruit engineering, still in its early stages, began in strawberries and apples [61, 62].

A comprehensive study of rice [23], sorghum [53], wheat [63], tomatoes [57], and arabidopsis [47] expand extensive knowledge about specific gene processing and editing. Prior to SSNs, RNAi technology was used to study gene function by suppressing target genes, which was not as beneficial as SSNs. Some of the characteristics discussed below include examples for certain garden and ornamental plants using the CRISPR system. The color and weight-to-size ratio of tomato fruits can be obtained by editing the PL and TBG4 genes [64]. The SlNPR1 and SlCBF1 genes corresponding to drought and cold resistance can also be modified [65], and the fruit ripening transcription factor RIN (ripening inhibitor) can be edited to preserve the tomato with the desired characteristics. The albino phenotype and flowering signs can be changed in cabbage by editing the FRI and PDS genes [66]. In addition, the biosynthesis of carotenoid pigment can be enhanced in wild cabbage by editing the BoaCRTISO (carotenoid isomerase) gene [67]. By editing the DcF3H and DcPDS, DcMYB113 genes, it is possible to increase the accumulation of acylated anthocyanins in carrot roots to obtain pigmented purple carrots [68, 69]. Cucumber mosaic virus (CMV-Z1) and zucchini yellow mosaic virus (ZYMV) are the two main rapidly infecting pathogens that can seriously damage crops. To overcome this, pathogenic resistance can be developed/enhanced by editing the elF4EF gene [70]. Drought resistance of an important cash crop—hot pepper (*Capsicum annuum* L. syn. Chile) was developed by editing the NAC72 gene [71]. In addition, to increase the duration of flowering, attractive petunia flowers can be edited using the PhACO1, 3 and 4 genes [72]. The color and accumulation of carotenoids in the Japanese ipomoea flower can be changed by changing the related gene InCCD4 (carotenoid cleavage dioxygenase) [73]. Comparably, flavonoid biosynthesis could be enriched in cherry seed flower by editing the F3H (flavanone-3-hydroxylase) gene [74]. Moreover, it was reported that the AhFAD2A andAhFAD2B genes encoding fatty acid desaturases in peanuts were edited [75]. TYLCV-IR gene (intergenic regions). It has been modified to overcome numerous viral diseases in *Nicotiana benthamiana*. Following gene modification, the plants showed resistance to geminiviridavirus, begomovirus, kurtovirus, bekurtovirus, eragrovirus, turnkurtovirus, and topokuvirus [54]. Resistance to *Phytophthora tropicalis* in cocoa was overcome by changing the TcNPR3 gene [76]. Similarly, several desirable phenotypic traits of ornamental flowers, including bloom induction, flower meristem initiation and organ development, as well as color, fragrance, and shelf life were identified [9]. The longevity of the flowers was induced in petunia by changing the group of genes of the hybrid 1-aminocyclopropane-1-carboxylate oxidase of petunia (PhACO, PhACO1, PhACO3, and PhACO4) [72]. Ulcers and diseases of Huanglong bin are the main factors in reducing the productivity of citrus plants; this problem was overcome by modifying the CsLOB1, CsWRKY22, and DMR6 genes using the CRISPR/Cas9 system [77–79].

Selective marker gene (SMG) systems are crucial and play an important role in the identification of transgenic cultures. Currently, scientists are considering SMGS that can affect human and animal health. Plants that have undergone the gene (GMP: genetically modified plants) usually contain an antibiotic resistance gene, so GM plants must survive and regenerate in an environment with antibiotics. Taking into account that plants that have not undergone gene transformation will not rejuvenate and will eventually die in toxic proximity. Despite the fact that antibiotics have a positive effect on human and animal health, the negative consequences of diarrhea associated with taking antibiotics and pseudomembranous colitis will increase the likelihood of subsequent diseases. Long-term consumption of these products (fruits and vegetables) can seriously affect human/animal health [80]. Thus, it is extremely important to exclude SMGS from transgenic cultures using CRISPR technology [81].

Soy storage protein genes have also been successfully edited to monitor the effectiveness of the CRISPR/Cas9 method using the hairy root transformation method mediated by *Agrobacterium rhizogenes* [82]. Acetolactate synthase (ALS) is involved in amino acid biosynthesis; numerous herbicides target this amino acid. These two plant enzymes (EPSPS, ALS, ACCase) and BFP genes ensure plant resistance to herbicides [83–85]. Hybrid rice is susceptible to bentazone and sulfonylurea, and the BEL gene has been mutated by radiation. In the production of hybrid rice, these mutants can be used to prevent contamination of batches of hybrid seeds [86]. Here, the BEL gene was edited using CRISPR-Cas9 and transformed into rice using T-DNA transfer [87]. The value of nutrients in vegetables and fruits was also increased by knocking out genes using the CRISPR CAS 9 system. Visually attractive flowers have a pleasant aroma due to the presence of anthocyanin, the expression of which is regulated by the MYB-bHLH-WD (MBW) complex [88, 89]. Gibberellin (GA) determines the height of the plant, and strigolactone (SL) affects the branching of the shoot, both of which can be modulated by modifying biosynthesis or transmitting a GA and/or SL signal [90].

Undesirable metabolites usually have a negative impact on crop yields and their quality; accumulation of these undesirable metabolites can be avoided by using GE. Cyanide intoxication, ataxia or partial paralysis, and goiter are caused by cyanide, which is contained in cassava [91]. Glucosinolates, which are formed from mustard and cabbage, also have a high content of toxins, were edited [92]. The FAD2 and FAD3 genes produce a high content of oleic acid and a low content of linolenic acid in soy; However, soybean oil promotes the accumulation of monounsaturated fats and reduces the content of linolenic acid in seeds [93].

The AtPDS3, AtFLS2, AtADH, ATFT, AtSPL4, and AtBRI1 genes are targets for *Arabidopsis* with a mutation rate (MRs) of 1.1–84.8% in the first generation [47]. The OsPDS and OsBADH2 genes were knocked out with MRs in 9.4% and 7.1% of cases [94, 95]. The DsRed2, DD20, and DD43 genes were targeted at sorghum with MRs 33%, 59%, and 76%, respectively. Similarly, the ZmIPK (13.1%), LIG1, MS26, MS45, and ALS1 genes were edited in maize with MRs less than 5% [96]. TaMLO-A, TaMLO-B, and TaMLO-Dare are three homeoalleles that provide resistance to powdery mildew and were modified with the same moderate mutation rate of 5.6% [63]. A mutation rate of 26–84% was observed in the BRI1, JAZ1, and GAI genes [97]. NtPDS and NtPDR6 have been mutated with MRs 81.8 and 87.5%, respectively [98]. The squamosa promoter binding protein-like 4 and the flowering locus T (FT) were mutated with mutation rate of 90%, resulting in late flowering [99]. Genome modifications at 46 target sites with an average of 85.4% mutations in monocotyledonous and dicotyledonous plants using either golden gate ligation or Gibson assembly [100]. Single, double, and triple mutants for CDKA2, CDKB1, and CDKB2 in rice were also obtained using sgRNA [101]. Forest pathosystemsproblems with diseases have been discussed in detail [102] using the CRISPR/Cas9 system. Zhang et al. also reported mutations in the genes of young albino seedlings (OsYSA) and OsROC5 with mutation rates of 65–66.7% [103]. Similarly, edited the OsERF922 gene, which encodes ERF transcription factors, for the development of resistance to rice blast disease have been reported [104] Transgenic poplar plants were modified, and phenotypic results showed a 51% increase in mutation rate [105]. El-Munadi et al. [106] explained the biosafety of genomically edited plants and the use of CRISPR/CAS technology, to increase yield, quality, and nutritional value, that was mutated with mutation rates of 81.8% and 87.5%, respectively [44].

Additionally, editing platform based on revolutionary *prime editing* CRISPR technique is a promising strategy [107] that will reduce off-target effects, and improve efficiency of targeted insertion in crop plants. Prime editing expands on the capabilities of traditional CRISPR-Cas9 gene editing by offering greater precision and versatility. While CRISPR-Cas9 acts like molecular scissors, cutting DNA at specific locations, prime editing functions more like a word processor, allowing researchers to directly edit specific DNA sequences without completely cutting the DNA strand. The prime editing system consists of two key components: a modified version of the Cas9 protein called "prime editor" and an RNA molecule known as the "prime editing guide RNA." The prime editor contains a nuclease domain capable of cutting one DNA strand and a reverse transcriptase domain capable of writing new genetic information into the DNA. To perform prime editing, the prime editing guide RNA directs the prime editor to the desired target site within the genome. The nuclease domain then creates a single-strand break in the DNA, exposing it to the reverse transcriptase domain. This domain uses the RNA template within the guide RNA as a blueprint to synthesize a new DNA sequence, effectively rewriting the genetic code. Prime editing allows researchers to insert, delete, or replace specific DNA sequences with high precision and efficiency [107]. It offers several advantages over traditional CRISPR gene-editing techniques, such as reducing off-target effects and enabling the introduction of multiple edits simultaneously [107]. The various applications of prime editing to bring about specific target sequence changes in plant biology is summarized in Fig. 7.12.

Numerous biotic and abiotic stresses, such as disease, salinity, drought, heavy metal toxicity, and climate change, can negatively affect plant growth, development and pose a serious challenge for crop production. They cause yield loss every year worldwide. High-precision gene modification by prime editing can contribute significantly to developing new crops that express stress tolerance [107]. This technology represents the best option for a sustainable and eco-friendly way to improve crop varieties. The cis-acting regulatory elements (CREs) play a crucial role in the response of plants to abiotic stress conditions. Cis regulation involves the interaction between specific DNA sequences, called cis regulatory elements or cis-elements, and transcription factors (proteins) that bind to these elements. These CRE are located within the promoter region of genes, which are involved in initiating gene expression [108]. Under abiotic stress conditions, various cis-elements within the promoter region of stress-responsive genes interact with specific

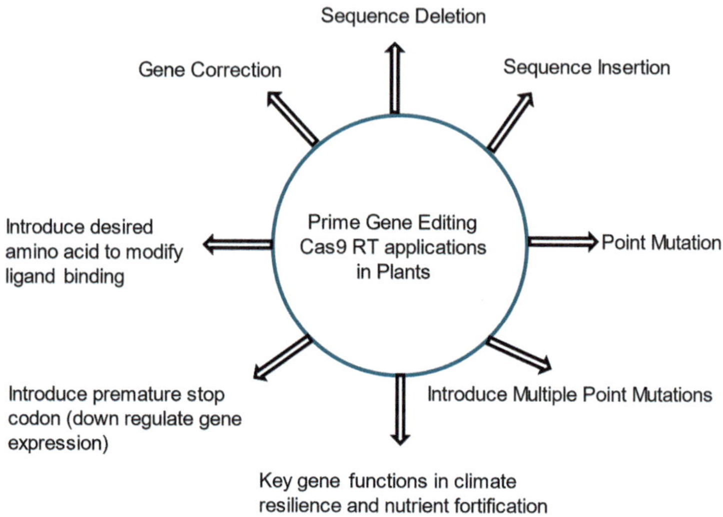

Fig. 7.12 Summary of possible genetic modification in plants mediated by prime editing. These could be best achieved by gene knockout, gene insertion, or protein engineering

transcription factors. This interaction leads to the activation or repression of gene expression, allowing plants to respond and adapt to the stress. Since prime editing allows generating base conversions, the CREs are ideal to create novel crop selections [107]. Furthermore, cis regulation allows for the fine-tuning of gene expression in response to different abiotic stress conditions. The presence of multiple cis-elements and their interactions with different transcription factors provide a complex regulatory network that not only determine the specificity and intensity of gene expression but have also been implicated in evolutionary processes underlying domestication and divergence of crop varieties [109–111]. Understanding the intricacies of cis-regulatory networks can contribute to the development of stress-tolerant crop varieties through genetic engineering and breeding strategies, thereby ensuring global food security in the face of increasingly unpredictable abiotic stress factors [107, 111].

Prime editing allows for the following:

(i) *Precision Crop Breeding*: Prime gene editing enables scientists to make highly precise modifications at the nucleotide level in crop genomes. This technology allows to edit both coding and noncoding DNA sequences, thus new prospects for precision crop breeding that can increase tolerance to both abiotic and biotic stresses. With targeted changes in specific genes, it is possible to enhance beneficial traits without disrupting the rest of the genome [107]. This precision offers considerable advantages over traditional breeding methods, which may involve crossing multiple generations to achieve desired characteristics.
(ii) *Disease Resistance*: One of the promising applications of prime editing could be developing crops for disease resistance. There are several reports of

genomics-based improvement and improved disease resistance. In one such study resistance gene distribution across 50 *Brassica napus* lines, a total of 1749 resistance gene analogs (RGAs) were identified, of which 996 are core and 753 are variable, with over predicted 15,000 single nucleotide polymorphisms (SNPs), 368 of which are not present in the reference genome [10]. Since many of the plant disease resistance genes are allelic in nature and differ only in single or a few nucleotides, they are known to give rise to missense mutations and certain alleles result in pseudogenes [112], leading to loss-of-resistance gene function and hence increased susceptibility. Prime editing could help restore the resistance gene function from pseudogenes and hence provide disease resistance in crop plants [113].

(iii) *Gene Regulation*: Systemic defense responses involve long-distance signaling molecules that travel throughout the plant to activate defense mechanisms in distant tissues. These signaling molecules, like salicylic acid and jasmonic acid, coordinate the expression of defense-related genes and prime the plant for future pathogen attacks. S gene regulation and host defense in plants involve a complex network of molecular mechanisms that enable plants to recognize pathogens and mount effective defense responses. Understanding these processes is crucial for developing strategies to enhance plant immunity and ultimately improve crop productivity. Prime gene editing could play a significant role in nonviral pathogens to manipulate operational conservation of the S genes across different plant species to create desired S gene mutants in several crop varieties of breeding value [113]. S genes are involved in the recognition and response to pathogens, allowing plants to mount an effective defense response. The regulation of S genes involves transcriptional regulation, post-transcriptional modifications, and signal transduction pathways. Transcriptional regulation is a key aspect of S gene regulation, where specific transcription factors interact with the promoter regions of S genes to either activate or repress their expression. These transcription factors can be induced by pathogen-derived signals or by internal signals triggered by stress conditions. Post-transcriptional modifications, such as alternative splicing and RNA editing, can also modulate the expression and function of S genes. The activation of S genes leads to the initiation of host defense responses in plants. These responses involve both local and systemic defense mechanisms. Local defense responses include the production of antimicrobial compounds, such as phytoalexins, that directly inhibit the growth of pathogens. Plants also produce cell wall reinforcement components, such as lignin and callose, to strengthen physical barriers. Additionally, plants can undergo programmed cell death at the site of infection, known as the hypersensitive response, to restrict the spread of pathogens. In our effort to identify novel proteins through high throughput proteomic analysis in Tomato mosaic virus (ToMV) infected and mock inoculated healthy plants several differentially expressed proteins were identified (Table 7.2 and Figs. 7.13 and 7.14) (Bharathan, 2023, unpublished results).

Table 7.2 Spot statistics showing the number of similar and differentially expressed protein spots in virus infected and healthy plant tissues

Spot statistics- twofold greater spot threshold	
Similar	503 (74.1%)
Decreased with ToMV infected	82 (12.1%)
Increased with ToMV infected	94 (13.8%)

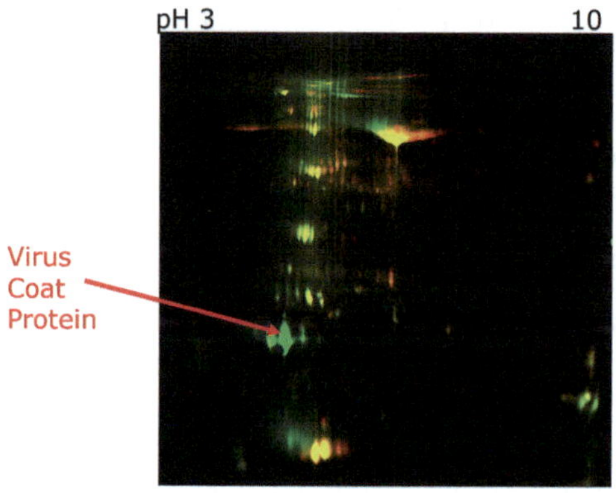

Fig. 7.13 2D-DIGE image analysis showing the spot pattern and spot distribution of proteins isolated from tomato mosaic virus (ToMV)-infected plant and uninfected plant. Proteins colored green are more abundant in the virus infected plants and the proteins colored red are more abundant in the uninfected plants. The proteins colored yellow are present at similar levels in both samples. The virus coat protein is only present in the virus-infected plant. The threshold for comparison is twofold or greater in abundance

In addition to plant-virus model system described above, my lab is actively involved with the genomics and proteomics of the soil-borne plant pathogenic fungus *Rhizoctonia solani*. It is a potential threat to food-biosecurity because it can kill or reduce the vigor of developing seedlings resulting in poor plant stand and productivity. The fungus is found in diverse soil habitats in temperate, neotropical, and tropical regions of the world and a pathogen of most cultivated and native species of plants, including staple foods like rice, wheat, potato, and various bio-energy crops such as conifers, corn, poplar, sorghum, soybean, and switch grass. *R. solani* also represents an important evolutionary link to beneficial and disease-causing fungi, suggesting that the fungus has novel genes for producing toxic molecules that kill and damage plant cells. The fungus also harbors double-stranded (ds) RNA viruses that influence the parasitic and saprobic activity of the fungus and could be potentially exploited for bioterrorism. However, our ability to detect and counter such threats is limited. My laboratory at Indiana University of Pennsylvania (IUP) has the largest collection of several genetically diverse isolates of *R. solani* from all over the world, including United States, Australia, Japan, and Poland. Since very few strategies are available for conventional resistance breeding of the crop plants infected with *R. solani*, understanding the biology of the fungus and molecular mechanisms of pathogenicity is paramount to address *Rhizoctonia* disease. With

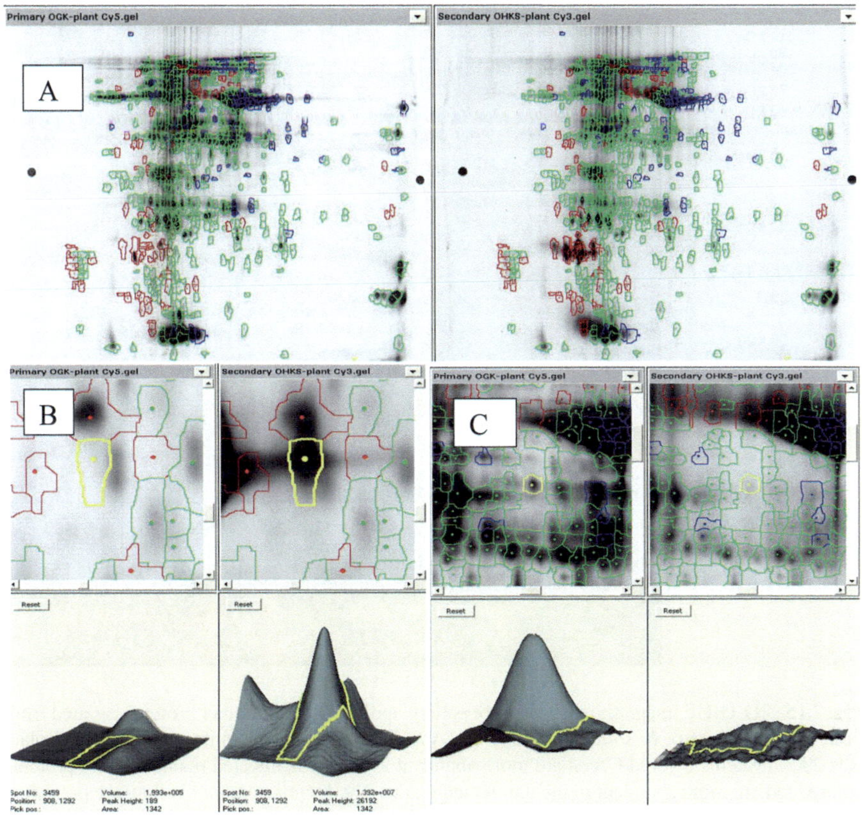

Fig. 7.14 2D-DIGE generated spot patterns from ToMV-infected plants and uninfected plants. (**a**) Global pattern of protein expression, (**b**) DeCyder analysis of representative spot upregulated in ToMV-infected and (**c**) representative spot downregulated in mock buffer inoculated (healthy plant)

that objective my laboratory in collaboration with 12 other scientists completed the genome sequencing of *R. solani* [114]. The genome of *R. solani* assembly is summarized below Potato is the world's fourth largest food crop, following corn, wheat, and rice. During the early stages of growth, potato seedlings are susceptible to attack by soil-borne *R. solani*, resulting in root rot, damping off, and stem cankers, triggering qualitative and quantitative damage to infected potato. Despite the threat of this pathogen in potato production, virtually little is known about the molecular mechanisms that regulate defense responses on potato when the fungus is interacting with the potato plant. With the genomic data generated in our lab has now applied mass spectrometry-based proteomics to identify specific proteins produced under potato infection and vegetative growth condition (Fig. 7.15). Preliminary data from such studies postulate significant relationships between the pathogen (disease-causing agent), host cell defense mechanisms, and their vulnerability. We observed on potato tubers the causal link between expression and non-expression of different types of proteins by decyder analysis (Fig. 7.16) and their role in host defense

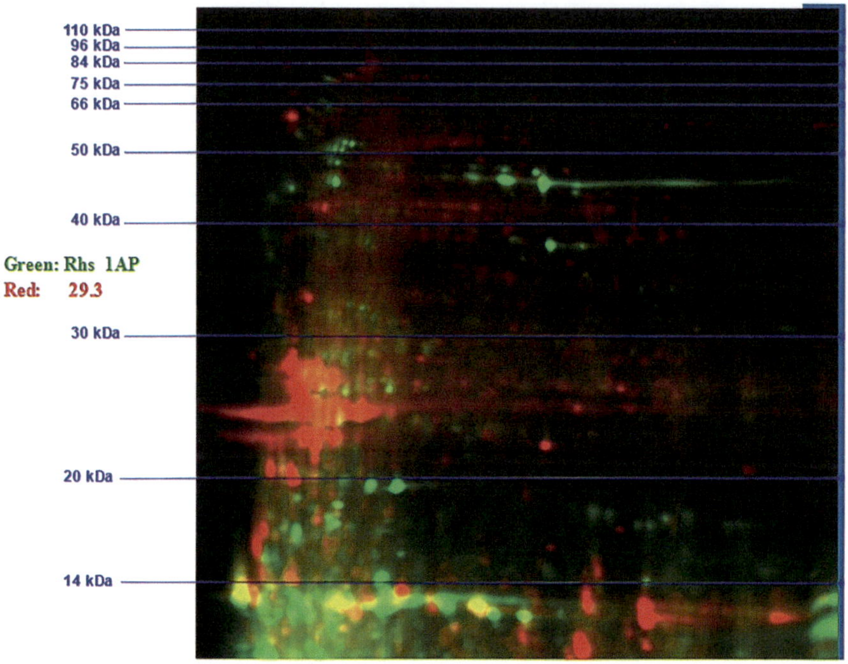

Fig. 7.15 2D-DIGE image showing the spot pattern and spot distribution of proteins isolated from virus infected (green). *R. solani* isolate (Rhs 1AP) and non-viral (red) infected *R. solani* isolate Rhs 29.3. Proteins colored Green are more abundant in the virus infected plants and the proteins colored red are more abundant in the uninfected plants. The proteins colored Yellow are present at similar levels in both samples (Bharathan 2023, unpublished results)

association genomics and molecular approaches will further be applied along with prime editing to better understand fungal-plant communication during critical phases of infection process.

7.8 *Rhizoctonia solani* Genome Assembly

Isolate Rhs 1AP

Genome size = 43,184,312 bp
N50 = 69,768 bp
Number of contigs = 19,310
Number of contigs > = 2000 bp = 1189
Length of contigs > = 2000 bp = 36,328,946 bp
Number of gene models = 14,214

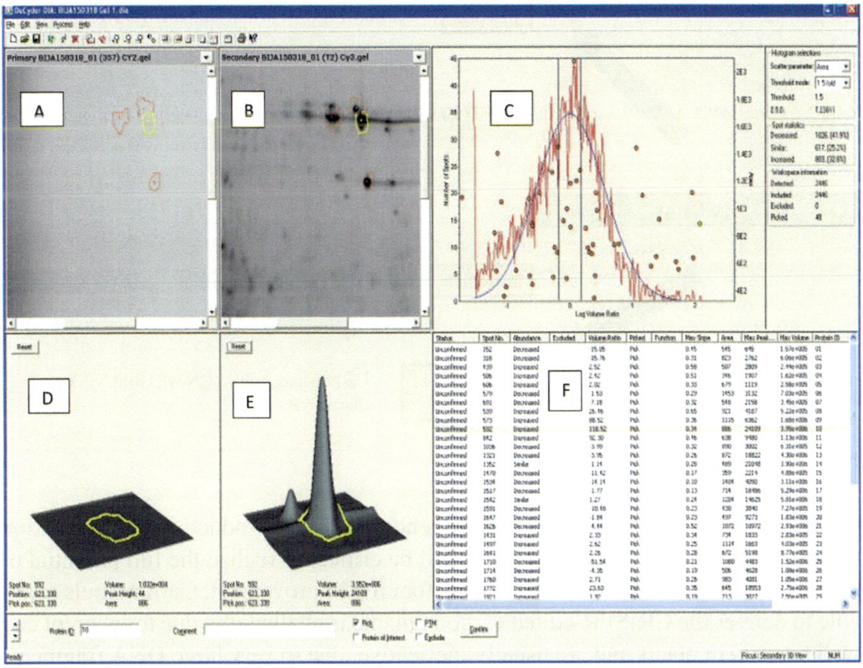

Fig. 7.16 DeCyder Analysis of protein spots from two isolates of *R.solani*Identification of In the above figure: (**a**) upper left hand panel control (non-viral infected); (**b**) represents viral infected *R. solani* isolate; (**c**) upper right hand panel indicates spot distribution. The spots within the threshold are the number of spots that increased and decreased; (**d**) lower left 3D view of characteristic spot of interest down regulated (**e**) upregulated; and (**f**) lower right is table view of spots of interest. Unique –Up and down regulated proteins due to viral infection in *R. solani* isolates. (Bharathan 2023, unpublished)

Isolate Rhs 29.3
Genome size = 46,582,548 bp
N50 = 31,048 bp
Number of contigs = 19,058
Number of contigs >= 2000 bp = 2177
Length of contigs >= 2000 bp = 40,855,208 bp
Number of gene models = 14,222

7.9 Advances in Gene Transfer Methods

A critical step to realize the full advantage of advanced gene-editing methods and precise application of gene modification in plants is to create transgene that is fast, scalable, and acceptable. A brief outline of gene transfer methods is described in Fig. 7.17.

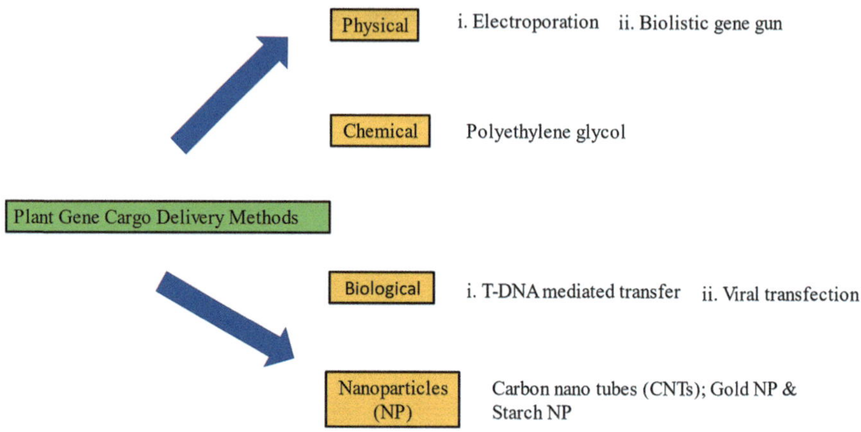

Fig. 7.17 Plant gene transfer methods

As more crop plants are modified by genome editing, productive transformation techniques and tissue culture methods will be critical to realize the full potential of CRISPR-mediated target genome editing for crop improvement. Current tools available to deliver the CRISPR-edited cargo in plants is challenging due to inherent cell wall barriers in plants and are usually ineffective, due to very large DNA fragments that need to be inserted along with selectable markers. It is, therefore, important to increase the efficiency of targeted gene insertion to reduce the labor for screening large plant populations without the selectable marker DNA fragments. To achieve these goals the frequency of targeted insertion with CRISPR applications needs to be improved. Increasing the efficiency of SDNs delivery will greatly enhance genome editing, including HDR applications. Editing plant genomes without introducing foreign DNA into cells may help to alleviate some of the regulatory fears associated to genetically modified plants. Off the shelf several transfected preassembled complexes of purified Cas9 protein and guide RNA have been introduced into plant protoplasts of *Arabidopsis thaliana*, tobacco, lettuce, and rice. In all these cases targeted mutagenesis in regenerated plants at frequencies of up to 46% were obtained. These targeted sites contained germline-transmissible small insertions or deletions that are identical to naturally occurring genetic variation (Fig. 7.18). New gene transfer technologies are being developed at an accelerating rate beyond the conventional plant transformation methods. Among the new delivery technologies carbon nanotubes [115], viral replicons [116], and de novo meristem induction [117] have emerged as promising solutions to overcome some of the hurdles of gene delivery.

Technological innovations such as these will aid to improve efficacy of targeted DNA insertion in plants. A recent report of using carbon nanotube-polymer hybrid modified with functional peptides to deliver DNA into intact plant mitochondria with almost 30 times higher efficiency than existing methods [118]; and *Baby boom* and *Wuschel* genes to improve makeover efficiency in recalcitrant monocot plants are exciting examples of this effort [119].

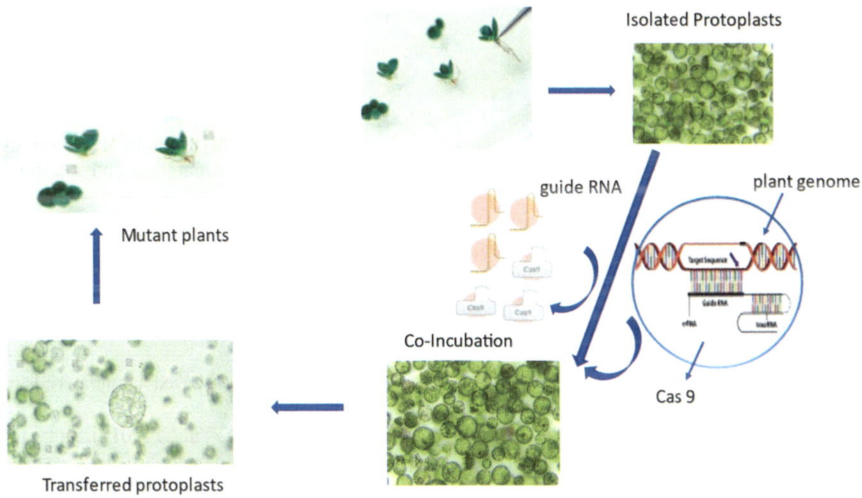

Fig. 7.18 Gene editing in plants without the use if foreign DNA

7.10 Using Nanotechnology for the Improvement of Agricultural Crops

Gene delivery using nanoparticles in plants is a rapidly growing field of research that has the potential to revolutionize agriculture. Nanoparticles are tiny particles that can carry genetic material into plant cells, allowing scientists to introduce new traits and characteristics into crops. There are several types of nanoparticles that can be used for gene delivery in plants, including liposomes, dendrimers, and carbon nanotubes as summarized in Fig. 7.19.

These nanoparticles are designed to protect the genetic material from degradation and deliver it directly to the plant cell. One of the major advantages of using nanoparticles for gene delivery in plants is that it allows for targeted delivery of genetic material to specific parts of the plant. This can help reduce the risk of unintended effects on other parts of the plant or the environment. Nanoparticle-mediated gene delivery in plants has already been used to introduce traits such as disease resistance, drought tolerance, and increased yield. However, there are still challenges that need to be overcome before this technology can be widely adopted in agriculture, including ensuring the safety and efficacy of the nanoparticles and minimizing any potential risks to human health or the environment. Overall, gene delivery using nanoparticles in plants has enormous potential to improve crop yields, reduce the use of pesticides and other chemicals, and help address global food security challenges.

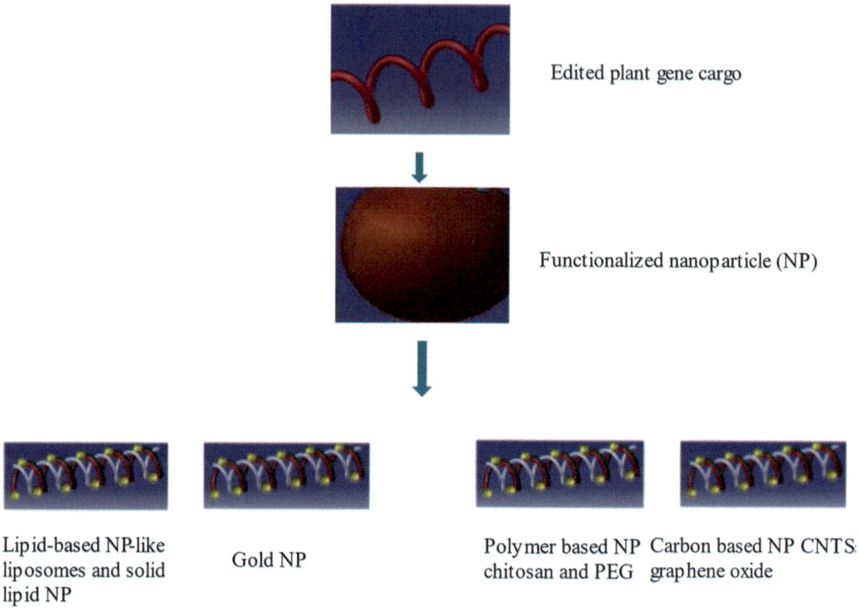

Fig. 7.19 Summary of types of plant gene cargo delivery in plants using different nanoparticles

Nanotechnology plays a prominent role in biology, medicine, and pharmaceuticals, including plant science. By increasing resistance to biotic and abiotic stress and improving crop yield/quality through gene editing, nanotechnology has similarities to CRISPR/Cas9. Nanofertilizers [120, 121] are used in horticultural plants, including vegetables and fruits, and are introduced into food crops to enhance growth, germination, and genetic manipulation [122, 123].

Photosynthesis is an energy conversion process in plants that converts light energy into chemical energy; however, it is ineffective in cloudy conditions and on sunlit plants during the rainy season. Consequently, cellular mechanisms may not be regulated. Gold nanoparticles can be useful for increasing the ability to capture light, thereby contributing to highly excited electron transfer in the chloroplast [124]. Environmental factors (abiotic stress) cause biochemical and physiological changes in plants, and they are more susceptible to stress. Even under stressful conditions, the use of metal nanoparticles can influence the physiology and level of antioxidant enzymes in plants [125] and reduce the levels of reactive oxygen species (ROS) in mitochondria and chloroplasts to protect the plant [126]. However, the use of these fertilizers with nanoparticles in the fields of agricultural crops not only increases soil fertility, but also significantly affects the pollution of water resources [127, 128]. Fertilizers containing microorganisms are labeled as biofertilizers that can activate the plant system and improve the absorption of nutrients from the soil [129]. Nanofertilizers have the same advantages as biofertilizers [130]. Moreover, metal nanoparticles have anti-pathogenic, antifungal, and antibacterial properties [131], so that they can survive in conditions of pathogenic exposure under soil.

7.11 The Role of Genetic Modification of Agricultural Crops Based on Nanoparticles

Nanoparticles act as a carrier for delivering the necessary materials to plant cells, animal cells, and specific organs for cancer therapy, treatment of genetic diseases and for obtaining the desired properties in plants [132–134] and explained a method in which gene transformation is performed using nanoparticles. Different types of nanoparticles have been used to deliver genetic material to plant cells using various platforms. These reports also describe in detail the advantages and disadvantages of using nanoparticles in gene transfer methods. Typically, mesoporous silica nanoparticles, carbon nanotubes, gold, and magnetic nanoparticles have been used to deliver plasmid DNA, double-stranded RNA, and siRNA to plant protoplasts or other intact cell lines [135, 136] and discussed in detail the mode of entry, uptake, and translocation [137] and reviewed in nanotechnology strategies for plant genetic engineering [138]. However, several groups have demonstrated gene suppression and gene editing in plants using nanoparticles; to achieve this task, the transformation of pollen based on magnetic nanoparticles was used. With this approach, a complex of vector-magnetic nanoparticles was associated with pollen that falls on the stigma of the desired plant flowers [115, 139]. Finally, plants produce the desired seeds by transferring a complex of vector-magnetic nanoparticles into the stigma of the flower. Later, these flowers turned into fruits, and the seeds were sifted on tablets with antibiotics. The rapid propagation protocol was used to obtain T0, T1, and T2 generations of transgenic plants; This breeding program is inexpensive for editing plant genomes and is used for various *Brassica* species [134]. Similarly, the double-stranded (ds) RNA was loaded into layered clay nanolayers with double hydroxide (LDH), which are non-degradable, non-toxic, and resistant to light washing. Moreover, when these complexes with RNA nanoparticles are sprayed onto plant leaves, they immediately attach to the leaf surface and are absorbed by plant viruses, inducing RNAi, eventually destroying the target plant pathogens, or endogenous mRNA can be minimized/eliminated [133]. Similarly, gene editing has also been demonstrated with a small NPs-CRISPR/Cas9 vector complex that has been micro-injected into leaves or any other parts of the plant that can be further propagated using tissue culture or other ease protocols [132]. The carbon dots-siRNA complex was used to suppress GFP in tobacco and tomato plants [140]. Demirer et al. (2021) recently demonstrated genome editing in plants using the CRISPR/Cas9 system along with nanoparticles and explained the regeneration and phenotypic/metabolic changes of genomically edited cultures [141].

Carbon nanotubes (CNTs) have shown great potential for gene delivery in plants due to their ability to penetrate the plant cell wall. Here are some examples of carbon nanotubes used in plant gene delivery:

1. Single-walled carbon nanotubes (SWCNTs) were used to deliver a green fluorescent protein gene into tobacco cells. The SWCNTs were functionalized with polyethyleneimine (PEI) to promote DNA binding and uptake [9, 141].
2. Multi-walled carbon nanotubes (MWCNTs) were used to deliver a luciferase gene into *Arabidopsis thaliana* plants. The MWCNTs were functionalized with chitosan to enhance their stability and biocompatibility [141].
3. A combination of SWCNTs and gold nanoparticles (AuNPs) was used to deliver a red fluorescent protein gene into onion epidermal cells. The SWCNTs were functionalized with PEI, while the AuNPs were coated with polyvinyl alcohol (PVA). Several studies have also demonstrated the potential of plant edited gene with nanoparticles for disease detection [142].
4. CNTs were used to deliver RNA interference (RNAi) molecules targeting a fungal pathogen into tomato plants. The CNTs were functionalized with cationic polymers to promote RNA binding and uptake. Overall, carbon nanotubes have shown promise as a tool for delivering genes and other biomolecules to plants, with potential applications in agriculture and biotechnology. However, further research is needed to optimize their properties and minimize any potential toxicity or environmental impact.

Many international agricultural companies have started to market CRISPR-edited crops for human consumption. Syngenta announced the acquisition of CRISPR IP (intellectual property) to begin its use in several of its cultures. This list includes many crops where editing has already been successful, such as corn, wheat, and rice, but also new plants such as sunflower [143]. Recently, one group managed to increase the yield of rice grains [144]. This opens the way for companies to target similar genes to overcome food shortages. Another group has developed mushrooms with reduced browning, which, according to the decision of the US Department of Agriculture, will not be regulated. This creates the basis for regulatory groups of companies when they develop new products (see further discussion below). One crop is already available on the market—pitted apples. Even more companies have begun to commit to the production of grain crops in the future DuPont is one of such companies; The company has announced the release of wax corn by 2020 [145] and is building up its IP technology portfolio, which is in principle open to the agricultural sector, including even competitors such as Bayer and BASF.Very recently waxy corn hybrids have been created by CRISPR–Cas9 editing of a *waxy* allele in 12 elite inbred maize lines. Field trials at 25 locations showed that CRISPR-waxy hybrids were agronomically superior to conventional hybrids, producing on average 5.5 bushels per acre higher yield [146].

7.12 Conclusion

Genome editing by CRISPR has an impact on the food supply chain at all levels—from starter cultures to crop improvement and animal husbandry. CRISPR is just one example of how microorganisms are gradually shaping our world. Advances in microbiome and metagenomic research are beginning to show what enormous potential lies ahead. This is especially true in the food industry, where bacteria and yeast are widely used in the fermentation and production of many foods and beverages. This also includes the possibility of using multiple CRISPR-based technologies in microbiomes that cover the farm (e.g. soil microbiome, livestock microbiome, and feed microbiome), production facilities (e.g. fermentation tanks, processing lines, food safety control points, and packaging media) and the consumer (e.g. oral and intestinal microbiomes), since along with CRISPR, microbes are easily broken down. The research is aimed at studying how soil microorganisms affect agricultural crops and their productivity. The microbiomes of livestock affect their health and growth. Finally, a person's gut health is largely influenced by their microbiome(s), which is often influenced by diet. Obviously, we are far from exhausting the possibilities of CRISPR in the food supply chain and must make significant progress in using CRISPR-based technologies to produce healthier and more environmentally friendly food.

References

1. Hunter, M. C., Smith, R. G., Schipanski, M. E., Atwood, L. W., & Mortensen, D. A. (2017). Agriculture in 2050: Recalibrating targets for sustainable intensification. *Bioscience, 67*, 385–390. https://doi.org/10.1093/biosci/bix010
2. Steinwand, M. A., & Ronald, P. C. (2020). Crop biotechnology and the future of food. *Nature Food, 1*, 273–283.
3. Mittler, R., & Blumwald, E. (2010). Genetic engineering for modern agriculture: Challenges and perspectives. *Annual Review of Plant Biology, 61*, 443–462.
4. Ben-Ari, G., & Lavi, U. (2012). Marker-assisted selection in plant breeding. In A. Altman & P. M. Hasegawa (Eds.), *Plant biotechnology and agriculture* (pp. 163–184). Academic. https://doi.org/10.1016/B978-0-12-381466-1.00011-0
5. Mathews, H., Narayanaswamy, B., Litz, R. E., Narayanan, K. R., Rao, P. S., & Bhatia, C. R. (1990). The promotion of *Agrobacterium* mediated transformation in *Atropa belladonna* by acetosyringone. *Journal of Plant Physiology, 136*, 404–409.
6. Kohli, A., et al. (2003). Transgene integration, organization, and interaction in plants. *Plant Molecular Biology, 52*, 247–258.
7. Altpeter, F., et al. (2016). Advancing crop transformation in the era of genome editing. *The Plant Cell, 28*, 1510–1520.
8. Anand, A., & Jones, T. J. (2018). Advancing agrobacterium-based crop transformation and genome modification technology for agricultural biotechnology. *Current Topics in Microbiology and Immunology, 418*, 489–507. https://doi.org/10.1007/82_2018_97
9. Giovannini, A., Laura, M., Nesi, B., Savona, M., & Cardi, T. (2021). Genes and genome editing tools for breeding desirable phenotypes in ornamentals. *Plant Cell Reports, 40*, 461–478. https://doi.org/10.1007/s00299-020-02632-x

10. Dolatabadian, A., Bayer, P. E., Tirnaz, S., Hurgobin, B., Edwards, D., & Batley, J. (2020). Characterization of disease resistance genes in the Brassica napus pangenome reveals significant structural variation. *Plant Biotechnology Journal, 4*, 969–982. https://doi.org/10.1111/pbi.13262
11. Feldman, A. B., Leung, H., Baraoidan, M., Elmido-Mabilangan, A., Canicosa, I., Quick, W. P., et al. (2017). Increasing leaf vein density *via* mutagenesis in rice results in an enhanced rate of photosynthesis, smaller cell sizes and can reduce interveinal mesophyll cell number. *Frontiers in Plant Science, 8*, 1–10. https://doi.org/10.3389/fpls.2017.01883
12. Feldman, A. B., Murchie, E. H., Leung, H., Baraoidan, M., Coe, R., Yu, S. M., et al. (2014). Increasing leaf vein density by mutagenesis: Laying the foundations for C4 rice. *PLoS One, 9*, e94947. https://doi.org/10.1371/journal.pone.0094947
13. Puripunyovanich, V., Lamai, M., Limtiyayothin, M., & Orpong, P. (2022). New frontier of plant breeding using gamma irradiation and biotechnology. https://doi.org/10.5772/intechopen.104667
14. Vergunst, A. C., Jansen, L. E., & Hooykaas, P. J. (1998). Site-specific integration of Agrobacterium T-DNA in *Arabidopsis thaliana* mediated by Cre recombinase. *Nucleic Acids Research, 26*, 2729–2734.
15. Louwerse, J. D., et al. (2007). Stable recombinase-mediated cassette exchange in Arabidopsis using Agrobacterium tumefaciens. *Plant Physiology, 145*, 1282–1293.
16. Vergunst, A. C., & Hooykaas, P. J. (1998). Cre/lox-mediated site-specific integration of Agrobacterium T-DNA in Arabidopsis thaliana by transient expression of cre. *Plant Molecular Biology, 38*, 393–406.
17. Li, Z., et al. (2009). Site-specific integration of transgenes in soybean via recombinase mediated DNA cassette exchange. *Plant Physiology, 151*, 1087–1095.
18. Srivastava, V., & Ow, D. W. (2002). Biolistic mediated site-specific integration in rice. *Molecular Breeding, 8*, 345–350.
19. Zhang, H. M., Yang, H., Rech, E. L., Golds, T. J., Davis, A. S., Mulligan, B. J., Cocking, E. C., & Davey, M. R. (1988). Transgenic rice plants produced by electroporation-mediated plasmid uptake into protoplasts. *Plant Cell Reports, 7*(6), 379–384.
20. Liang, Z., Chen, K., Li, T., Zhang, Y., Wang, Y., Zhao, Q., et al. (2017). Efficient DNA-free genome editing of bread wheat using CRISPR/Cas9 ribonucleoprotein complexes. *Nature Communications, 8*, 14261. https://doi.org/10.1038/ncomms14261
21. Puchta, H. (2005). The repair of double strand breaks in plants: Mechanisms and consequences for genome evolution. *Journal of Experimental Botany, 56*, 1–14.
22. Puchta, H., & Fuser, F. (2013). Gene targeting in plants: 25 years later. *The International Journal of Developmental Biology, 57*, 629–637.
23. Shan, Q., Wang, Y., Li, J., Zhang, Y., Chen, K., Liang, Z., Zhang, K., Liu, J., Xi, J. J., Qiu, J., et al. (2013). Targeted genome modification of crop plants using a CRISPR-Cas system. *Nature Biotechnology, 31*(8), 686–688. https://doi.org/10.1038/nbt.2650
24. Butt, H., et al. (2017). Efficient CRISPR/Cas9-mediated genome editing using a chimeric single-guide RNA molecule. *Frontiers in Plant Science*, 8.
25. Shi, J., et al. (2017). ARGOS8 variants generated by CRISPR-Cas9 improve maize grain yield under field drought stress conditions. *Plant Biotechnology Journal, 15*(2), 207–216.
26. Dahan-Meir, T., et al. (2018). Efficient in planta gene targeting in tomato using geminiviral replicons and the CRISPR/Cas9 system. *The Plant Journal, 95*, 5–16. https://doi.org/10.1111/tpj.13932
27. Hahn, F., et al. (2018). Homology-directed repair of a defective glabrous gene in Arabidopsis with Cas9-based gene targeting. *Frontiers in Plant Science, 9*, 424.
28. Wright, D. A., et al. (2005). High-frequency homologous recombination in plants mediated by zinc-finger nucleases. *The Plant Journal, 44*(4), 693–705.
29. Shukla, V. K., et al. (2009). Precise genome modification in the crop species Zea mays using zinc-finger nucleases. *Nature, 459*, 437–441.
30. Boch, J., et al. (2009). Breaking the code of DNA binding specificity of TAL-type III effectors. *Science, 326*, 1509–1512.

31. Zhang, Y., et al. (2013). Transcription activator-like effector nucleases enable efficient plant genome engineering. *Plant Physiology, 161*(1), 20–27.
32. Cermak, T., et al. (2015). High-frequency, precise modification of the tomato genome. *Genome Biology, 16*, 232.
33. Butler, N. M., et al. (2016). Gemini virus-mediated genome editing in potato (*Solanum tuberosum* L.) using sequence-specific nucleases. *Frontiers in Plant Science, 7*, 1045.
34. Gaj, T., Gersbach, C. A., & Barbas, C. F. (2013). ZFN, TALEN, and CRISPR/Cas-based methods for genome engineering. *Trends in Biotechnology, 31*, 397–405.
35. Doudna, J. A., & Charpentier, E. (2014). The new frontier of genome engineering with CRISPR-Cas9. *Science, 346*(6213), 1258096.
36. Zhu, H., Li, C., & Gao, C. (2020). Applications of CRISPR–Cas in agriculture and plant biotechnology. *Nature Reviews. Molecular Cell Biology, 21*, 661–677.
37. Doudna, J. A. (2020). The promise and challenge of therapeutic genome editing. *Nature, 578*, 229–236.
38. Cai, C. Q., et al. (2009). Targeted transgene integration in plant cells using designed zinc finger nucleases. *Plant Molecular Biology, 69*(6), 699–709.
39. Kumar, S., et al. (2015). A modular gene targeting system for sequential transgene stacking in plants. *Journal of Biotechnology, 207*, 12–20.
40. Qi, Y., et al. (2013). Increasing frequencies of site-specific mutagenesis and gene targeting in Arabidopsis by manipulating DNA repair pathways. *Genome Research, 23*(3), 547–554.
41. Forsyth, A., et al. (2016). Transcription activator-like effector nucleases (TALEN)-mediated targeted DNA insertion in potato plants. *Frontiers in Plant Science, 7*, 1572.
42. Wang, M., et al. (2017). Gene targeting by homology-directed repair in rice using a Geminivirus-based CRISPR/Cas9 system. *Molecular Plant, 10*(7), 1007–1010.
43. Gil-Humanes, J., et al. (2017). High-efficiency gene targeting in hexaploid wheat using DNA replicons and CRISPR/Cas9. *The Plant Journal, 89*(6), 1251–1262.
44. Gao, H., et al. (2020). Complex trait loci in maize enabled by CRISPR-Cas9 mediated gene insertion. *Frontiers in Plant Science, 11*, 535.
45. Miki, D., et al. (2018). CRISPR/Cas9-mediated gene targeting in Arabidopsis using sequential transformation. *Nature Communications, 9*(1), 1967.
46. Lu, Y., et al. (2020). Targeted, efficient sequence insertion and replacement in rice. *Nature Biotechnology*. https://doi.org/10.1038/s41587-020-0581-5
47. Li, J., Norville, J. E., Aach, J., McCormack, M., Zhang, D., Bush, J., Church, G. M., & Sheen, J. (2013). Multiplex and homologous recombination mediated genome editing in Arabidopsis and *Nicotiana benthamiana* using guide RNA and Cas9. *Nature Biotechnology, 31*(8), 688–691. https://doi.org/10.1038/nbt.2654
48. Miao, J., Guo, D., Zhang, J., Huang, Q., Qin, G., Zhang, X., Wan, J., Gu, H., & Qu, L. (2013). Targeted mutagenesis in rice using CRISPR-Cas system. *Cell Research, 23*(10), 1233–1236. https://doi.org/10.1038/cr.2013.123
49. Zhou, H., Liu, B., Weeks, D. P., Spalding, M. H., & Yang, B. (2014). Large chromosomal deletions and heritable small genetic changes induced by CRISPR/Cas9 in rice. *Nucleic Acids Research, 42*, 10903–10914.
50. Barrangou, R., Coбытй-Monvoisin, A.-C., Stahl, B., Chavichvily, I., Damange, F., et al. (2013). Genomic impact ofCRISPR immunization against bacteriophages. *Biochemical Society Transactions, 41*, 1383–1391.
51. Jinek, M., Chylinski, K., Fonfara, I., et al. (2012). A programmable dual-RNA-guided DNA endonuclease in adaptive bacterial immunity. *Science, 337*, 816–821.
52. Brazelton, V. A., Jr., Scott, Z., Wright, D. A., Wang, Y., Liu, J., Chen, K., Yang, G., & Lawrence-Dill, C. J. (2015). A quick guide to CRISPR sgRNA design tools. *GM Crops Food, 6*(4), 266–276.
53. JiangW, Z. H., Bi, H., Fromm, M., Yang, B., & Weeks, D. P. (2013). Demonstration of CRISPR/Cas9/sgRNAmediatedtargeted gene modification in Arabidopsis, tobacco, sorghum and rice. *Nucleic Acids Research, 41*, e188.

54. Ali, Z., Abulfaraj, A., Li, L., Ghosh, N., Piatek, M., Mahjoub, A., Aouida, M., Piatek, A., Baltes, N., Voytas, D., et al. (2015). Efficient virus-mediated genome editing in plants using the CRISPR/Cas9 system. *Molecular Plant, 8*(8), 1288–1291. https://doi.org/10.1016/j.molp.2015.02.011
55. Zhang, X., Liu, X., Zhang, D., Tang, H., Sun, B., Li, C., et al. (2017). Genome-wide identification of gene expression in contrasting maize inbred lines under field drought conditions reveals the significance of transcription factors in drought tolerance. *PLoS One, 12*(7), e0179477. https://doi.org/10.1371/journal.pone.0179477
56. Cai, Y., Chen, L., Zhang, Y., Yuan, S., Su, Q., Sun, S., Wu, C., Yao, W., Han, T., & Hou, W. (2020). Target base editing in soybean using a modified CRISPR/Cas9 system. *Plant Biotechnology Journal, 18*(10), 1996–1998. https://doi.org/10.1111/pbi.13386
57. Brooks, C., Nekrasov, V., Lippman, Z. B., & Van Eck, J. (2014). Efficient gene editing in tomato in the first generationusing the clustered regularly interspaced short palindromic repeats/CRISPR-associated9 system. *Plant Physiology, 166*, 1292–1297.
58. Karkute, S. G., Singh, A. K., Gupta, O. P., Singh, P. M., & Singh, B. (2017). CRISPR/Cas9 mediated genome engineeringfor improvement of horticultural crops. *Frontiers in Plant Science, 2017*(8), 1635.
59. Nekrasov, V., Wang, C., Win, J., Lanz, C., Weigel, D., & Kamoun, S. (2017). Rapid generation of a transgene-freepowdery mildew resistant tomato by genome deletion. *Scientific Reports, 7*, 482.
60. Andersson, M., Turesson, H., Nicolia, A., Fält, A. S., Samuelsson, M., & Hofvander, P. (2017). Efficient targeted multiallelic mutagenesis in tetraploid potato (Solanum tuberosum) by transient CRISPR-Cas9 expression in protoplasts. *Plant Cell Reports, 36*(1), 117–128. https://doi.org/10.1007/s00299-016-2062-3
61. Nishitani, C., Hirai, N., Komori, S., et al. (2016). Efficient genome editing in apple using a CRISPR/Cas9 system. *Scientific Reports, 6*, 31481. https://doi.org/10.1038/srep31481
62. Lobato-Gómez, M., Hewitt, S., Capell, T., et al. (2021). Transgenic and genome-edited fruits: Background, constraints, benefits, and commercial opportunities. *Horticulture Research, 8*, 166. https://doi.org/10.1038/s41438-021-00601-3
63. Wang, Y., Cheng, X., Shan, Q., Zhang, Y., Liu, J., Gao, C., et al. (2014). Simultaneous editing of three homoeoalleles in hexaploid bread wheat confers heritable resistance to powdery mildew. *Nature Biotechnology, 32*, 947–951. https://doi.org/10.1038/nbt.2969
64. Chen, K., Wang, Y., Zhang, R., Zhang, H., & Gao, C. (2019). Agriculture. *Annual Review of Plant Biology, 70*(1), 667–697.
65. Li, R., Liu, C., Zhao, R., et al. (2019). CRISPR/Cas9-mediated *SlNPR1* mutagenesis reduces tomato plant drought tolerance. *BMC Plant Biology, 19*, 38. https://doi.org/10.1186/s12870-018-1627-4
66. Murovec, J., Gucek, K., Bohanec, B., Avbelj, M., & Jerala, R. (2018). DNA-free genome editing ofBrassica oleracea and B. RapaProtoplasts using CRISPR-Cas9Ribonucleoprotein complexes. *Plant Biotechnology, 9*. https://doi.org/10.3389/fpls.2018.01594
67. Sun, B. O., et al. (2020). Color-related chlorophyll and carotenoid concentrations of Chinese kale can be altered through CRISPR/Cas9 targeted editing of the carotenoid isomerase gene. *Boa CRTISO, Horticulture Research, 7*, 16. https://doi.org/10.1038/s41438-020-00379-w
68. Klimek-Chodacka, M., Oleszkiewicz, T., Lowder, L. G., et al. (2018). Efficient CRISPR/Cas9-based genome editing in carrot cells. *Plant Cell Reports, 37*, 575–586. https://doi.org/10.1007/s00299-018-2252-2
69. Que, F., Hou, X. L., Wang, G. L., & Tan, G. F. (2019). Advances in research on the carrot, an important root vegetable in the Apiaceae family. *Horticulture Research, 6*, 69. https://doi.org/10.1038/s41438-019-0150-6
70. Chandrasekaran, J., Brumin, M., Wolf, D., Leibman, D., Klap, C., Pearlsman, M., et al. (2016). Development of broad virus resistance in non-transgenic cucumber using CRISPR/Cas9 technology. *Molecular Plant Pathology, 17*, 1140–1153. https://doi.org/10.1111/mpp.12375
71. Joshi, R. K. (2019). Genome editing in chili pepper using a CRISPR/Cas9 cytidine base editing system. *Asian Journal of Plant Science and Research, 9*, 445.

72. Xu, J., Kang, B. C., Naing, A. H., Bae, S. J., Kim, J. S., Kim, H., et al. (2020). CRISPR/Cas9 mediated editing of 1aminocyclopropane1carboxylate oxidase1enhances Petunia flower longevity. *Plant Biotechnology Journal, 18*, 287–297. https://doi.org/10.1111/pbi.13197
73. Watanabe, K., Oda-Yamamizo, C., Sage-Ono, K., Ohmiya, A., & Ono, M. (2018). Alteration of flower colour in Ipomocanil through CRISPR/Cas9-mediated mutagenesis of carotenoid cleavage dioxygenase 4. *Transgenic Research, 27*, 25–38. https://doi.org/10.1007/s11248-017-0051-0
74. Nishihara, M., Higuchi, A., Watanabe, A., et al. (2018). Application of the CRISPR/Cas9 system for modification of flower color in *Torenia fournieri*. *BMC Plant Biology, 18*, 331. https://doi.org/10.1186/s12870-018-1539-3
75. Yuan, M., Zhu, J., Gong, L., He, L., Lee, C., Han, S., et al. (2019). Mutagenesis of FAD2 genes in peanut with CRISPR/Cas9 based gene editing. *BMC Biotechnology, 19*, 24. https://doi.org/10.1186/s12896-019-0516-8
76. Fister, A. S., et al. (2018). Transient expression of CRISPR/Cas9 machinery targeting TcNPR3 enhances defense response in *Theobroma cacao*. *Frontiers in Plant Science*, 9. https://doi.org/10.3389/fpls.2018.00268
77. Peng, A., Chen, S., Lei, T., Xu, L., He, Y., Wu, L., et al. (2017). Engineering canker resistant plants through CRISPR/Cas9 targeted editing of the susceptibility gene CsLOB 1 promoter in citrus. *Plant Biotechnology Journal, 15*, 1509–1519. https://doi.org/10.1111/pbi.12733
78. Zhang, Z., et al. (2018). Simultaneous editing of two copies of Gh14-3-3d confers enhanced transgene-clean plant defense against *Verticillium dahliae* in allotetraploid upland cotton. *Frontiers in Plant Science, 9*, 842. https://doi.org/10.3389/fpls.2018.00842
79. Wang, L., Chen, S., Peng, A., Xie, Z., He, Y., & Zou, X. (2019). CRISPR/Cas9-mediated editing of CsWRKY22 reduces susceptibility to *Xanthomonas citri* subsp. citri in wanjincheng orange (*Citrus sinensis* (L.) Osbeck). *Plant Biotechnology Reports, 13*, 501–510. https://doi.org/10.1007/s11816-
80. Breyer, D., Kopertekh, L., & Reheul, D. (2014). Alternatives to antibiotic resistance marker genes for in vitro selection of genetically modified plants – Scientific developments, current use, operational access, and biosafety considerations. *Critical Reviews in Plant Sciences, 33*, 286–330. https://doi.org/10.1080/07352689.2013.870422
81. Yau, Y. Y., & Stewart, C. N. (2013). Less is more: Strategies to remove marker genes from transgenic plants. *BMC Biotechnology, 13*, 36. https://doi.org/10.1186/1472-6750-13-36
82. Li, C., Nguyen, V., Liu, J., Fu, W., Chen, C., Yu, K., et al. (2019). Mutagenesis of seed storage protein genes in Soybean using CRISPR/Cas9. *BMC Research Notes, 12*, 176. https://doi.org/10.1186/s13104-019-4207-2
83. Voytas, D. F., & Gao, C. (2014). Precision genome engineering and agriculture: Opportunities and regulatory challenges. *PLoS Biology, 12*, e1001877. https://doi.org/10.1371/journal.pbio.1001877
84. Sauer, N. J., Narváez-Vásquez, J., Mozoruk, J., Miller, R. B., Warburg, Z. J., & Woodward, M. J. (2016). Oligonucleotide-mediated genome editing provides precision and function to engineered nucleases and antibiotics in plants. *Plant Physiology, 170*, 1917–1928. https://doi.org/10.1104/pp.15.01696
85. Li, C., Zong, Y., Wang, Y., Jin, S., Zhang, D., Song, Q., et al. (2018). Expanded base editing in rice and wheat using a Cas9-adenosine deaminase fusion. *Genome Biology, 19*, 59. https://doi.org/10.1186/s13059-018-1443-z
86. Cantos, C., Francisco, Trijatmiko, K. R., Slamet-Loedin, I., & Chadha-Mohanty, P. K. (2014). Identification of "safe harbor" loci in indica rice genome by harnessing the property of zinc-finger nucleases to induce DNA damage and repair. *Frontiers in Plant Science, 2014*(5), 302. https://doi.org/10.3389/fpls.2014.00302
87. Xu, R., Li, H., Qin, R., Wang, L., Li, L., Wei, P., et al. (2014). Gene targeting using the Agrobacterium tumefaciens-mediated CRISPR-Cas system in rice. *Rice, 7*, 5. https://doi.org/10.1186/s12284-014-0005-6
88. Albert, N. W., et al. (2014). A conserved network of transcriptional activators and repressors regulates anthocyanin pigmentation in eudicots. *The Plant Cell, 26*(3), 962–980. https://doi.org/10.1105/tpc.113.122069

89. Lloyd, A., Brockman, A., Aguirre, L., Campbell, A., Bean, A., Cantero, A., et al. (2017). Advances in the MYB–bHLH–WD repeat (MBW) pigment regulatory model: Addition of a WRKY factor and co-option of an anthocyanin MYB for betalain regulation. *Plant & Cell Physiology, 58*, 1431–1441. https://doi.org/10.1093/pcp/pcx075
90. D'Halluin, K., Vanderstraeten, C., Stals, E., Cornelissen, M., & Ruiter, R. (2008). Homologous recombination: A basis for targeted genome optimization in crop species such as maize. *Plant Biotechnology Journal, 6*, 93–102. https://doi.org/10.1111/J.1467-7652.2007.0035.X
91. Padmaja, G. (1995). Cyanide detoxification in cassava for food and feed uses. *Critical Reviews in Food Science and Nutrition, 35*, 299–339. https://doi.org/10.1080/10408399509527703
92. Hannoufa, A., Pillai, B. V. S., & Chellamma, S. (2014). Genetic enhancement of *Brassica napus* seed quality. *Transgenic Research, 23*, 39–52. https://doi.org/10.1007/s11248-013-9742-3
93. Pham, A. T., Shannon, J. G., & Bilyeu, K. D. (2012). Combinations of mutant FAD2 and FAD3 genes to produce high oleic acid and low linolenic acid soybean oil. *Theoretical and Applied Genetics, 125*, 503–515. https://doi.org/10.1007/s00122-012-1849-z
94. Hinge, V., Patil, H., & Nadaf, A. (2015). Comparative characterization of aroma volatiles and related gene expression analysis at vegetative and mature stages in basmati and non-basmati rice (Oryza sativa L.) cultivars. *Applied Biochemistry and Biotechnology, 178*, 619–639. https://doi.org/10.1007/s12010-015-1898-2
95. Liang, Z., Zhang, K., Chen, K., & Gao, C. (2014). Targeted mutagenesis in Zea mays using TALENs and the CRISPR/Cas system. *Journal of Genetics and Genomics, 41*, 63–68. https://doi.org/10.1016/j.jgg.2013.12.001
96. Svitashev, S., Young, J. K., Schwartz, C., Gao, H., Falco, S. C., & Cigan, A. M. (2015). Targeted mutagenesis, precise gene editing, and site-specific gene insertion in maize using Cas9 and guide RNA. *Plant Physiology, 169*, 931–945. https://doi.org/10.1104/pp.15.00793
97. Feng, C., Su, H., Bai, H., Wang, R., Liu, Y., Guo, X., et al. (2018). High efficiency genome editing using a dmc1 promoter controlled CRISPR/Cas9 system in maize. *Plant Biotechnology Journal, 16*, 1848–1857. https://doi.org/10.1111/pbi.12920
98. Gao, J., Wang, G., Ma, S., Xie, X., Wu, X., & Xia, Q. (2015). CRISPR/Cas9-mediated targeted mutagenesis in *Nicotiana tabacum*. *Plant Molecular Biology, 87*, 99–110. https://doi.org/10.1007/s11103-014-0263-0
99. Hyun, Y., Kim, J., Cho, S. W., Choi, Y., Kim, J. S., & Coupland, G. (2015). Site-directed mutagenesis in Arabidopsis thaliana using dividing tissue-targeted RGEN of the CRISPR/Cas system to generate heritable null alleles. *Planta, 241*, 271–284. https://doi.org/10.1007/s00425-014-2180-5
100. Ma, X., Zhang, Q., Zhu, Q., Liu, W., Chen, Y., Qiu, R., et al. (2015). A robust CRISPR/Cas9 system for convenient, high-efficiency multiplex genome editing in monocot and dicot plants. *Molecular Plant, 8*, 1274–1284. https://doi.org/10.1016/j.molp.2015.04.007
101. Endo, M., Mikami, M., & Toki, S. (2015). Multigene knockout utilizing off-target mutations of the CRISPR/Cas9 system in rice. *Plant & Cell Physiology, 56*, 41–47.
102. Dort EN, Tanguay, P., and Hamelin, R. C. CRISPR/Cas9 gene editing: An unexplored frontier for forest pathology. Frontiers in Plant Science 2020; 11:1126. Doi:https://doi.org/10.3389/fpls.2020.01126.
103. Zhang, H., et al. (2014). The CRISPR/C as9 system produces specific and homozygous targeted gene editing in rice in one generation. *Plant Biotechnology Journal, 12*, 797–807. https://doi.org/10.1111/pbi.12200
104. Wang, F., Wang, C., Liu, P., Lei, C., Hao, W., & Gao, Y. (2016). Enhanced rice blast resistance by CRISPR/Cas9-targeted mutagenesis of the ERF transcription factor gene OsERF922. *PLoS One, 11*, e0154027. https://doi.org/10.1371/journal.pone.0154027
105. Fan, D., Liu, T., Li, C., Jiao, B., Li, S., Hou, Y., et al. (2015). Efficient CRISPR/Cas9-mediated targeted mutagenesis in Populus in the first generation. *Scientific Reports, 5*, 12217–12223. https://doi.org/10.1038/srep12217
106. El-Mounadi, K., Morales-Floriano, M. L., & Garcia-Ruiz, H. (2020). Principles, applications, and biosafety of plant genome editing using CRISPR-Cas9. *Frontiers in Plant Science, 11*, 56. https://doi.org/10.3389/fpls.2020.00056

107. Mahmudul Hassan, M. D., Yuan, G., Chen, J.-G., Tuskan, G. A., & Yang, X. (2020). Prime editing technology and its prospects for future applications in plant biology research. *BioDesign Research*. https://doi.org/10.34133/2020/9350905
108. Wolter, F., & Puchta, H. (2018). Application of CRISPR/Cas to understand cis- and trans-regulatory elements in plants. *Methods in Molecular Biology, 1830*, 23–40.
109. Koenig, D., et al. (2013). Comparative transcriptomics reveals patterns of selection in domesticated and wild tomato. *Proceedings of the National Academy of Sciences of the United States of America, 110*(28), E2655–E2662.
110. Gómez-Porras, J. L., Riaño-Pachón, D. M., Dreyer, I., et al. (2007). Genome-wide analysis of ABA-responsive elements ABRE and CE3 reveals divergent patterns in Arabidopsis and rice. *BMC Genomics, 8*, 260. https://doi.org/10.1186/1471-2164-8-260
111. Biłas, R., Szafran, K., Hnatuszko-Konka, K., & Kononowicz, A. (2016). Cis-regulatory elements used to control gene expression in plants. *Plant Cell, Tissue and Organ Culture (PCTOC), 127*(2), 269–287.
112. Yin, K., & Qiu, J. L. (2019). Genome editing for plant disease resis-tance: Applications and perspectives. *Philosophical Transactions of the Royal Society of London. Series B, Biological Sciences, 374*(1767), Article 20180322.
113. Van Schie, C. C. N., & Takken, F. L. W. (2014). Susceptibility genes 101: How to be a good host. *Annual Review of Phytopathology, 52*(1), 551–581.
114. Cubeta, M. A., Thomas, E., Dean, R. A., Jabaji, S., Neate, S., Tavantzis, S., Toda, T., Vilgalys, R., Narayanaswamy, B., Fedorova-Abrams, N., Pakala, S., Zafar, N., Joardar, V., Losada, L., & William, C. N. (2014). Genome sequences of the plant pathogenic fungus Rhizoctonia solani anastomosis group 3 strain Rhs1AP and a protoplast derived strain Rhs1AP-123E with a reduced nuclear complement. *Genome Announcements, 2*(5), e01072–e01014.
115. Demirer, G. S., et al. (2019). High aspect ratio nanomaterials enable delivery of functional genetic material without DNA integration in mature plants. *Nature Nanotechnology, 14*, 456–464.
116. Baltes, N. J., Gil-Humanes, J., Cermak, T., Atkins, P. A., & Voytas, P. F. (2014). DNA replicons for plant genome engineering. *The Plant Cell, 26*, 151–163.
117. Maher, F. M., et al. (2020). Plant gene editing through de novo induction of meristems. *Nature Biotechnology, 38*, 84–89.
118. Law, S. S. A., Liou, G., Nagai, Y., Giménez-Dejoz, J., Tateishi, A., Tsuchiya, K., Kodama, Y., Fujigaya, T., & Numata, K. (2022). Polymer-coated carbon nanotube hybrids with functional peptides for gene delivery into plant mitochondria. *Nature Communications, 13*, 2417. https://doi.org/10.1038/s41467-022-30185-y
119. Van, O. M., et al. (2016). DNA repair profiling reveals nonrandom outcomes at cas9-mediated breaks. *Molecular Cell, 63*, 633–646.
120. Subramanian, K. S., Manikandan, A., Thirunavukkarasu, M., & Rahale, C. S. (2015). Nano-fertilizers for balanced crop nutrition. In M. Rai, C. Ribeiro, L. Mattoso, & N. Duran (Eds.), *Nanotechnologies in food and agriculture*. Springer. https://doi.org/10.1007/978-3-319-14024-7_3
121. Elnahal, A. S. M., El-Saadony, M. T., Saad, A. M., et al. (2022). The use of microbial inoculants for biological control, plant growth promotion, and sustainable agriculture: A review. *European Journal of Plant Pathology, 162*, 759–792.
122. Wang, Y., Wang, S., Sun, J., Dai, H., Zhang, B., Xiang, W., Hu, Z., Li, P., Yang, J., & Zhang, W. (2021). Nanobubbles promote nutrient utilization and plant growth in rice by upregulating nutrient uptake genes and stimulating growth hormone production. *Science of the Total Environment, 800*, 149627.
123. Ma, Z., Ma, L., & Zhou, J. (2023). Applications of CRISPR/Cas genome editing in economically important fruit crops: recent advances and future directions. *Molecular Horticulture, 3*, 1. https://doi.org/10.1186/s43897-023-00049-0
124. Yuliya, V., Deryabin, A., Popov, V., Lev, D., & Moshkov, I. (2022). Priming with gold nanoparticles leads to changes in the photosynthetic apparatus and improves the cold tolerance of wheat. *Biochemistry*, 145–155.

125. Ahmad, A., Hashmi, S. S., Palma, J. M., & Corpas, F. J. (2022). Influence of metallic, metallic oxide, and organic nanoparticles on plant physiology. *Chemosphere, 290*, 133–329. https://doi.org/10.1016/j.chemosphere.2021.133329
126. Liu, J., Fu, C., Li, G., Khan, M. N., & Wu, H. (2021). ROS Homeostasis and Plant Salt Tolerance: Plant Nano biotechnology updates. *Sustainability, 13*(6), 3552. https://doi.org/10.3390/su13063552
127. Naderi, M., & Danesh-Shahraki, A. (2013). Nanofertilizers and their roles in sustainable agriculture. *International Journal of Agriculture Research and Crop Sciences, 5*(19), 2229–2232.
128. Mittal, D., Kaur, G., Singh, P., Yadav, K., & Ali, S. A. (2020). Nanoparticle-based sustainable agriculture and food science: Recent advances and future outlook. *Frontiers in Nanotechnology, 2*, 10. https://doi.org/10.3389/fnano.2020.579954
129. Yadav, A. N. (2021). Nanotechnology for agro-environmental. *Journal Of Applied Biology & Biotechnology, 9*(4), i–vi. https://doi.org/10.7324/JABB.2021.94edsustainability
130. Toksha, B., Sonawale, V. A. L. M., Vanarase, A., Bornare, D., Tonde, S., Hazra, C., Kundu, D., Satdive, A., Tayde, S., & Chatterjee, A. (2021). Nanofertilizers: A review on synthesis and impact of their use on crop yield and environment. *Environmental Technology & Innovation, 24*, 101986. https://doi.org/10.1016/j.eti.2021.101986
131. Seifi, T., & Kamali, K. R. (2021). Anti-pathogenic activity of graphene nanomaterials: A review. *Colloids and Surfaces B: Biointerfaces, 199*, 111509. ISSN: 0927-7765. https://doi.org/10.1016/j.colsurfb.2020.111509
132. Que, Q., Chilton, M. D., Elumalai, S., Zhong, H., Dong, S., & Shi, L. (2019). Repurposing macromolecule delivery tools for plant genetic modification in the era of precision genome engineering. In S. Kumar, P. Barone, & M. Smith (Eds.), *Transgenic plants. Methods in molecular biology* (Vol. 1864). Humana Press. https://doi.org/10.1007/978-1-4939-8778-8_1
133. Lv, J., Christie, P., & Zhang, S. (2019). Uptake, translocation, and transformation of metal-based nanoparticles in plants: Recent advances and methodological challenges. *Environmental Science. Nano, 6*, 41–59.
134. Ahmar, S., Mahmood, T., Fiaz, S., Mora-Poblete, F., Sohaib, M. S., Shafique, Sohaib, M. S., & Jung, K. H. (2021). Advantage of nanotechnology-based genome editing system and its application in crop improvement. https://doi.org/10.3389/fpls.2021.663849
135. Zhao, X., Meng, Z., et al. (2017). Pollen magnetofection for genetic modification with magnetic nanoparticles as gene carriers. *Nature Plants, 3*, 956–964. https://doi.org/10.1038/s41477-017-0063-z
136. Yuzhi, H., et al. (2013). Magnetic gold nanoparticles as a vehicle for fluorescein isothiocyanate and DNA delivery into plant cells. *Botany, 91*, 457–466.
137. Banerjee, K., Pramanik, P., Maity, A., Joshi, D. C., Wani, S. H., & Krishnan, P. (2019). Chapter 4: Methods of using nanomaterials to plant systems and their delivery to plants (mode of entry, uptake, translocation, accumulation, biotransformation and barriers). In M. Ghorbanpour & S. H. Wani (Eds.), *Advances in phytonanotechnology* (pp. 123–152). Academic. https://doi.org/10.1016/B978-0-12-815322-2.00005-5
138. Yan, Y., Zhu, X., Yu, Y., Li, C., Zhang, Z., & Wang, F. (2022 Feb). Nanotechnology strategies for plant genetic engineering. *Advanced Materials, 34*(7), e2106945. https://doi.org/10.1002/adma.202106945
139. Kwak, S.-Y., Lew, T. T. S., Sweeney, C. J., Koman, V. B., Wong, M. H., Bohmert-Tatarev, K., Snell, K. D., Seo, J. S., Chua, N.-H., & Strano, M. S. (2019). Chloroplast-selective gene delivery and expression in planta using chitosan-complexed single-walled carbon nanotube carriers. *Nature Nanotechnology, 14*, 447–455. https://doi.org/10.1038/s41565-019-0375-4
140. Schwartz, S. H., Hendrix, B., Hoffer, P., Sanders, R. A., & Zheng, W. (2020). Carbon dots for efficient small interfering RNA delivery and gene silencing in plants. *Plant Physiology, 184*, 647–657. https://doi.org/10.1104/pp.20.00733
141. Demirer, G. S., et al. (2021). Nanotechnology to advance CRISPR–Cas genetic engineering of plants. *Nature Nanotechnology*. https://doi.org/10.1038/s41565-021-00854-y

142. Kelich, P., Jeong, S., Navarro, N., Adams, J., Sun, X., Zhao, H., Landry, M. P., & Vuković, L. (2022). Discovery of DNA–carbon nanotube sensors for serotonin with machine learning and near-infrared fluorescence spectroscopy. *ACS Nano, 16*, 736–745. https://doi.org/10.1021/acsnano.1c08271
143. Maurer, A. (2017). Syngenta acquires non-exclusive license to use CRISPR-Cas9 gene-editing for agriculture applications. Press Release, November 3, 2017. https://www.ncbiotech.org/news/syngenta-acquires-non-exclusivelicense-use-crispr-cas9-gene-editing-agriculture-applications.
144. Miao, C., Xiao, L., Hua, K., Zou, C., Zhao, Y., et al. (2018). Mutations in a subfamily of abscisic acid receptor genespromote rice growth and productivity. *Proceedings. National Academy of Sciences. United States of America, 115*, 6058–6063. https://doi.org/10.1073/pnas.1804774115
145. Bomgardner, M. M. (2017). A new toolbox for better crops. *Chemical and Engineering News, 95*, 30–34.
146. Gao, H., Gadlage, M. J., Lafitte, H. R., et al. (2020). Superior field performance of waxy corn engineered using CRISPR–Cas9. *Nature Biotechnology, 38*, 579–581. https://doi.org/10.1038/s41587-020-0444-0

Chapter 8
Nanomaterial-Based Sensing Platforms for Food-Borne Pathogen Detection

Preeti K. Pathania, Navjeet Kaur Lotey, Dev Kashyap, and Priyanka Sabherwal

8.1 Introduction

Nanomaterial-based sensing involves fabrication and manipulation of materials and devices to understand the matter at nanoscale dimension, ranging between 1 and 100 nm [1]. A variety of nanomaterials include quantum dots, gold, silver, magnetic materials, metal oxides, and carbon-based materials display unique chemical, physical and surface-dependent properties because of their extremely small dimensions [2]. Nanomaterials, in conjunction with bioreceptors (antibodies and aptamers) are used to develop rapid, accurate, and cost-effective biosensors for pathogen monitoring [3]. There is a pressing need to consistently employ a reliable approach for collecting and analyse food-borne pathogens, as culture-dependent methods are not sufficiently sensitive, time-consuming and are more prone to give erroneous negative results. Additionally, many pathogens are not able to be cultured or produce false negative results. On the other hand, molecular-based methods are highly specific to species of pathogens, can provide phylogenetic information, and allow for the tracing of contamination sources [4]. However, these molecular methods require highly skilled labour

P. K. Pathania
Amity School of Biological Sciences, AMITY University, Mohali, Punjab, India

N. K. Lotey
National Centre for Nanoscience and Nanotechnology, University of Mumbai, Mumbai, Maharashtra, India

Somaiya Institute for Research & Consultancy, Somaiya Vidyavihar University, Mumbai, Maharashtra, India

D. Kashyap
CT University, Ludhiana, Punjab, India

P. Sabherwal (✉)
National Centre for Nanoscience and Nanotechnology, University of Mumbai, Mumbai, Maharashtra, India

R. S. Chaughule, A. S. Lokur (eds.), *Applications of Nanotechnology in Microbiology*, https://doi.org/10.1007/978-3-031-49933-3_8

and expensive instruments. Nano-based sensors, a possible alternative, can measure biological responses and convert the response signals for interpretation and analysis [5]. They can provide rapid, accurate, easy, and cost-effective results, can also be automated and have the potential for portability. Additionally, they obviate the need for skilled personnel and high-cost instrumentation [6]. Functional nanomaterials used as such or by hybridization can further add up the advantages for developing nano-biosensors including the increment of signals, retention of activity of biomolecules for a long-term period, and extension of investigating tools by using its unique plasmonic, optical and electronic properties of interest [7, 8]. Microfluidics lab-on-a-chip" is an advanced integrated nanosensors, which has been used recently for rapid and high-throughput detection of food-borne pathogens [9]. It offers solutions to detect pathogens from complex food sample matrices, a major challenge in diagnostics to separate these microbes from the matrix. Microfluidics-integrated nanosensors are gaining ever wider acceptance in order to increase the sample concentration, detection limits and sensitivity of biosensor before the onset of infection [10]. The prospects of these integrated technologies count on the triumph of evolving sophisticated technologies based on the novel bioreceptors/bio-recognition molecules (antibodies/aptamers) and advanced functional nanomaterials, their hybrid giving high specificity and sensitivity [11].

Food-borne infections have developed into an important health concern on a global basis because of the substantially increasing prevalence of these illnesses over the past 20 years. Every year, about one in ten people worldwide become sick after consuming contaminated food and, causing substantial mortality and illness [12]. Therefore, public health and national economy are seriously threatened by food-borne disease brought on by pathogenic microorganisms [13]. There are 31 pathogens that have been identified as the cause of food-borne illnesses, with the ones responsible for most food-borne disease outbreaks being *Salmonella*, *Campylobacter*, *Staphylococcus*, *Listeria*, and *Escherichia coli* [14]. The most common symptoms of these illnesses – which can even be life-threatening if not diagnosed timely – include diarrhoea, vomiting, nausea, abdominal cramps, joint aches, and fever (Table 8.1).

Thus, to protect public health, it is important to detect these pathogens to ensure a safe food supply and to minimize the occurrence of food-borne diseases. Here in this chapter, we will describe innovative approaches for the design of next-generation biosensors. Different sensing strategies based on antibodies, aptamers, DNA receptors, phage display peptides, etc., are presented. We also provide some creative breakthroughs in developing next-generation biosensor devices using novel nanomaterials, particularly rapid and sensitive platforms, which can be easily field deployable, for detecting harmful pathogens and toxins with high sensitivity.

8.2 Nanomaterials Design and Functionality for Detecting Pathogens

Nanomaterials play a pivotal role in advancing the diagnosis of pathogens, offering versatile tools to enhance the accuracy and efficiency of detection methods. They can be natural or artificially synthesized materials having external dimensions or at

Table 8.1 Representative bacterial pathogens, their symptoms, mode of transmission and incubation time [14]

Bacterial pathogen	Sign and symptoms	Food source/mode of transmission	Duration
Salmonella spp.	Ingestion of contaminated food or water	Eggs, meat, poultry, fruits, and vegetables	4–7 days
Campylobacter jejuni	Diarrhoea, abdominal cramps, fever, vomiting, bloody diarrhoea	Raw and uncooked poultry, unpasteurized milk, and contaminated water.	2–10 days
Shigella spp.	Gastrointestinal infections, watery diarrhoea, fever, fatigue, abdominal cramps, and malaise	Salads, milk, chicken, and shellfish	2–4 days
Listeria monocytogenes	Meningitis, gastroenteritis, and septicaemia.	Uncooked vegetables, meats, ingestion of contaminated food or water	1–15 days/ variable
Vibrio parahaemolyticus	Stomach-ache, diarrhoea, and vomiting	Sea food, such as fish, shrimp, and shellfish, undercooked/raw seafood	2–8 days
Clostridium botulinum	Vomiting, diarrhoea, blurred vision, double vision, difficulty in swallowing, muscle weakness can result in respiratory failure	Improperly canned foods, especially home-canned vegetables, fermented fish, baked potatoes in aluminium foil	Variable
Diarrheagenic *E. coli*	Watery or bloody diarrhoea, abdominal cramps, with or without fever	Water or food contaminated with human or animal faces.	3–10 days
Staphylococcus aureus	Sudden onset of severe nausea and vomiting, abdominal cramps, diarrhoea and fever	Unrefrigerated or improperly refrigerated meats, potato and egg salads, cream pastries	1–4 days
Yersinia enterocolitica	Diarrhoea, vomiting, abdominal pain	Raw or undercooked pork, unpasteurized milk or contaminated water	1–3 weeks
Brucella spp.	Profuse sweating and joint and muscle pain	Raw milk and soft cheeses made with unpasteurized goat or cow milk	Variable (1–2 months)

least one internal structural size within the range of 1–100 nm or less. Their microscopic structure alters bulk material behaviour and interactions with one another and the environment, giving rise to distinctive features that set them apart from their bulk counterparts. Changes in surface area and the emergence of the quantum effect due to discrete energy levels at the nanoscale are responsible for the startling changes in the properties of nanomaterials. They have very high surface area-to-volume ratios, meaning more surface area per unit of mass than conventional materials possess. This property enables nanomaterials to have more interaction sites, surface electrons, and surface energy, thereby improving the critical surface-dependent characteristics of the materials, such as plasmonic effect, hydrophobicity/hydrophilicity and charge transfer. One such instance is the difference between the bandgaps of nanoparticles and those of their bulk equivalents. Transforming bulk materials into nanomaterials has been shown to decrease band gaps, transforming nanomaterials into electrically better materials. As the material's band gap and conductivity can be controlled, it can be used to produce adjustable semi-conductors. Coupling of biomolecules or complex biological systems with electronic or optical devices is the general principle of various biosensors. The effective performance of biosensors requires transduction of the chemical signals generated by the biological components to electronic signals. Nanomaterials are categorized as zero-dimensional (0D) (for example, nanoparticles, quantum dots, carbon dots), one-dimensional (1D) (for example, nanotubes & nanorods), two-dimensional (2D) (for example, graphene, graphene oxide nanosheets, nanoplates, nanopores), and three dimensional (3D) (for example, nano-prisms, nanoflowers, bridged nanostructures, nanocomposites and complex hierarchical structures) [15]. Nanomaterials are also categorized as nanocomposites based on their atomic compositions, which exhibit much more varied optical or electrical characteristics because they can be produced using a variety of extremely complex atomic structures. Among the commonly utilized nanomaterials are gold nanoparticles, quantum dots, magnetic nanoparticles, carbon nanostructures and silica nanoparticles. These materials exhibit unique properties enabling precise tagging and visualization and isolation of pathogens from complex samples. Many researchers have integrated these materials into biosensors that employ techniques like surface-plasmon resonance, surface-enhanced Raman scattering, or electrochemical transduction to provide real-time and quantitative pathogen detection [16–18].

8.2.1 Gold Nanoparticles (AuNPs)

Gold is known for its biocompatibility and stability, making AuNPs suitable for various biological applications, including food-borne pathogen detection (Fig. 8.1). AuNPs offer high stability which allows development of long-lasting assays, making them reliable tools for pathogen detection. Moreover, the biocompatibility of AuNPs enables their usage in various sample types, including food matrices, without interfering with the accuracy of the detection process [19]. AuNPs can be easily functionalized with various biomolecules, antibodies, or ligands for specific pathogen recognition. This functionalization allows for the development of highly specific and sensitive assays for detecting food-borne pathogens. AuNPs have been used in immunosensors and lateral flow assays to detect these pathogens in food samples [20, 21]. Gold nanoparticles exhibit unique surface-plasmon resonance properties, which result in distinct colour changes when they interact with target pathogens. This optical phenomenon allows for simple and rapid visual detection in lateral flow assays or colorimetric-based methods. For instance, when AuNPs are functionalized with specific receptors or ligands that bind to the surface of food-borne pathogens, the binding interaction induces changes in the nanoparticles'

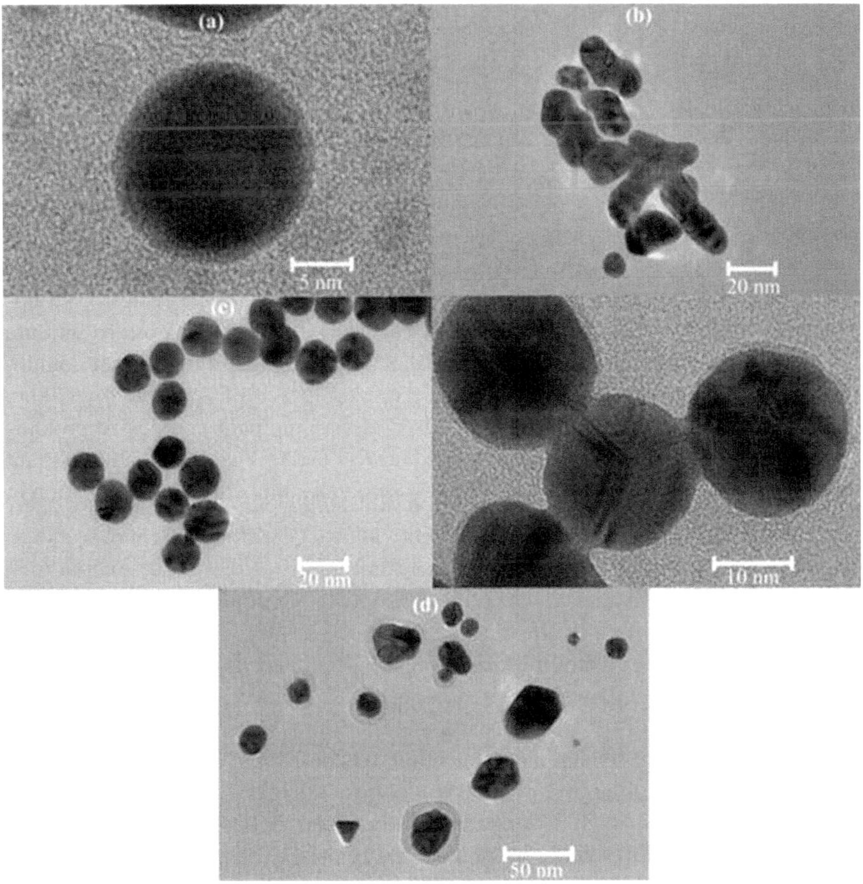

Fig. 8.1 TEM images of gold nanoparticles synthesized at 0.1% HAuCl4 at room temperature: (**a**) at 0.002 M α-cyclodextrin, pH 11.5 in the presence of NaCl after 54 min; (**b**) at 0.004 M β-cyclodextrin, pH 11.5, after 44 min and (**c**) 2 h 17 min; (**d**) at 0.002 M α-cycl [22]

surface-plasmon resonance, leading to observable colour shifts. This colour change serves as a direct and visible indication of the presence of the target [22].

8.2.2 *Quantum Dots (QDs)*

Quantum dots are semi-conducting nanocrystals having small dimensions (<10 nanometers in general) (Fig. 8.2). One of the unique properties of quantum dots (QDs) that distinguishes them from many other nanomaterials is their size-dependent tunable emission spectra. This property arises from the quantum confinement effect, which results in discrete energy levels for electrons within the QD structure.

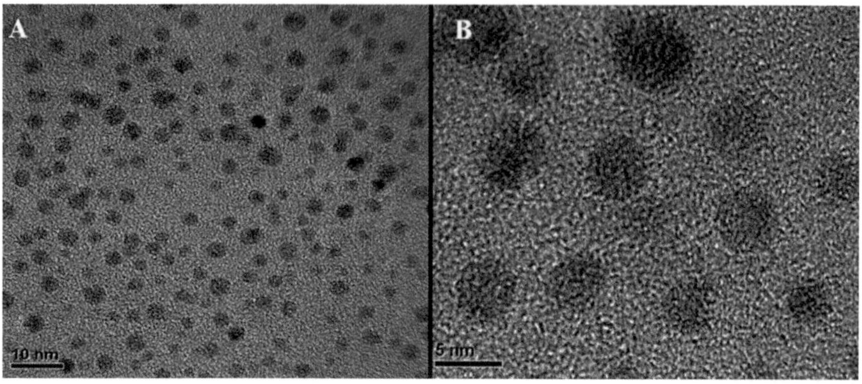

Fig. 8.2 TEM images of as-synthesized QDs. (**a**) QDs at low resolution. (**b**) QDs at high resolution, crystalline lattices can be seen [26]

Consequently, the bandgap and the emission wavelength of the QD can be precisely controlled by adjusting its size. This tunability allows QDs to emit light over a wide range of colours, from the visible to the near-infrared spectrum. They exhibit high molar absorption coefficient which could be about 10–100 times higher than organic dyes and are also resistant to chemical degradation [23–25]. They offer high sensitivity and multiplexing capabilities, enabling the simultaneous detection of multiple foodborne pathogens. QDs emit light at very specific and well-defined wavelengths based on their size. This property is particularly advantageous for applications such as biological imaging, sensing, and detection, where distinct and narrow emission bands are crucial. Their large surface area allows for attachment of multiple biomolecules, increasing the sensitivity of pathogen detection. Multiple types of QDs, each functionalized with different biomolecules, can be used to detect various pathogens simultaneously [24, 26]. QDs biocompatible coatings facilitate interaction with biological samples, making them suitable for biosensing applications. Their compatibility with optical readouts, such as fluorescence microscopy and flow cytometry, enables real-time and non-destructive analysis of biological samples.

8.2.3 Magnetic Nanoparticles (MNPs)

Magnetic nanoparticles (Fig. 8.3) exhibit unique magnetic properties, advantageous for biosensing applications in pathogen detection, including target capture and separation, magnetic relaxation-based sensing, signal amplification, real-time label-free detection, multiplexing capabilities, and compatibility with complex samples. These properties collectively make magnetic nanoparticles a powerful tool for developing sensitive, rapid, and versatile biosensing platforms for pathogen detection [28]. A variety of MNPs also exhibit super paramagnetism, a property where they become magnetized in the presence of an external magnetic field and lose their

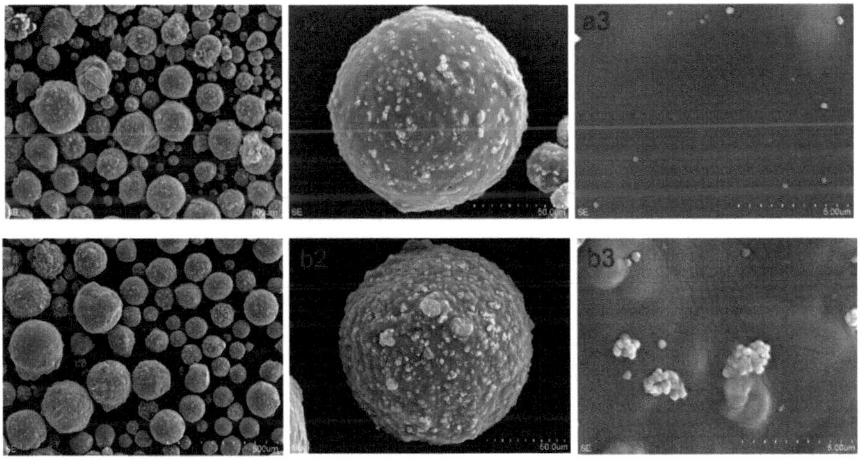

Fig. 8.3 SEM images of (**a**1–3) Fe_3O_4-chitosan spheres and (**b**1–3) Fe_3O_4-chitosan poly(acryloyloxyethyltrimethyl ammonium chloride) (PDAC) spheres [33]

magnetization once the field is removed. This property can be exploited for biosensing through techniques such as magnetic relaxation or magnetic relaxation switching. Changes in the relaxation behaviour of MNPs upon target binding can be detected, providing a quantitative readout for pathogen detection. MNPs can be easily manipulated and concentrated using external magnetic fields, facilitating efficient separation from complex food matrices and enhancing detection sensitivity. Thus, magnetic separation can efficiently remove large debris and impurities from samples, improving the compatibility of complex matrices like blood, serum, food and water samples [29–31].

MNPs can be easily functionalized with specific ligands such as antibodies or aptamers, that bind to target pathogens which enable the capture and concentration of pathogens from complex samples, facilitating their detection even at low concentrations. MNPs can then be manipulated using an external magnetic field to separate the captured targets from the sample matrix, enhancing the sensitivity and reducing interference from background components. This phenomenon has been demonstrated by multiple researchers and is commonly referred to as Immunomagnetic separation (IMS). Furthermore, IMS can be enhanced by complexing with additional signal amplification elements, such as enzymes or nanoparticles, further boosting the sensitivity of the assay [32, 33].

8.2.4 Carbon-Based Nanostructures

Carbon-based nanomaterials include fullerenes, carbon nanotubes, graphene and its derivatives, graphene oxide, nanodiamonds, and carbon-based quantum dots (Fig. 8.4). They are one of the most widely studied materials in the field of

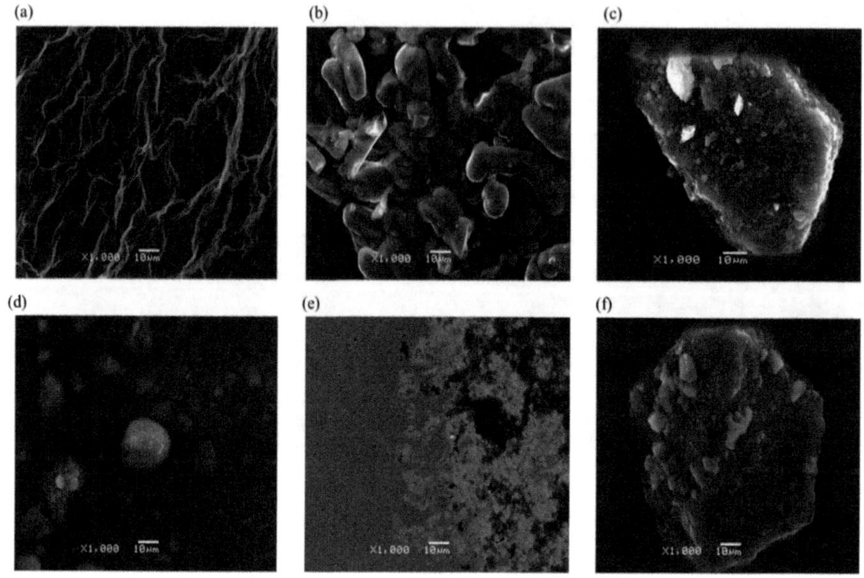

Fig. 8.4 SEM images of the surface morphology Carbon-based Nanomaterials (CBNs): (**a**) graphene oxide (GO) film; (**b**) amorphous carbon powder); (**c**) fluorescent nanodiamond), (**d**) plasma–chemically modified detonated nanodiamond particles (MDP1); (**e**) chemically modified detonated nanodiamond particles with hydroxyl functional groups (MDCHPOH); (**f**) pure detonated nanodiamond particles (DND) [34]

nanotechnology due to their low cost of mass production, low intrinsic toxicity and multifunctional surface functionalization that make them exceptionally well-suited for biosensing and diagnostic applications [34]. Their properties encompass unique electrical conductivity, fluorescence, biointerfacing, magnetism, biocompatibility, and enhanced Raman scattering, collectively enabling the development of rapid, sensitive, and multiplexed biosensing platforms for pathogen detection [35, 36]. For label-free real-time detection of pathogens, the changes in their electrical properties, such as conductivity or resistance, upon target binding can be exploited [37].

Carbon nanotubes (CNTs) are carbon-based nanostructures with a diameter of nanometers and a length of micrometres (where the length-to-diameter ratio exceeds 1000). CNTs have recently gained interest for use in electrochemical immunoassays due to their remarkable tensile strength, surface area, flexibility and other unique structural, mechanical, electrical as well as physicochemical properties reflecting increased signal currents. The possibility of functionalizing these smart nanomaterials directly with different biomolecules and their biocompatibility make them extremely attractive for electrochemical based sensing. For the detection of an electroactive analyte, bioreceptors are usually immobilized on the electrode and the corresponding redox responses can be measured [38].

Graphene, a monolayer of a hexagonal network of carbon atoms densely packed into a two-dimensional honeycomb crystal lattice demonstrated superior optical, chemical and electronic properties. These intrinsic properties of graphene make it a promising candidate for its use as a biosensing platform. Recently, graphene-based nanocomposite films have been used for constructing enhanced electrochemical sensors and biosensors, because of its synergistic effects of the electrochemical and catalytic activity [39]. Graphene's unique Raman scattering properties can be harnessed for biosensing applications. Graphene-enhanced Raman spectroscopy (GERS) can provide highly sensitive and selective detection by amplifying the Raman signals of target molecules. Additionally, Graphene's compatibility with biofilms can be exploited for pathogen detection. Graphene-based surfaces can mimic the extracellular matrix, promoting the adhesion of target pathogens and enabling their detection through various readout methods [40]. Graphene Quantum Dots (GQDs), derived out of graphene monolayer are characterized by atom-thin graphitized planes (usually 1 or 2 layers, not exceeding 2 nanometers in thickness) that exhibit unique quantum confinement effects. One of the outstanding features of GQDs is Photoluminescence (PL), emission of light in the form of fluorescence, after absorbing photons that can be exploited for biosensing applications. GQDs can be functionalized to specifically interact with pathogens, enabling sensitive and real-time detection.

Nanodiamonds or diamond nanoparticles (size <100 nm) made of a diamond core and outer layers of amorphous carbon possess a biocompatible surface that is amenable to functionalization with various biomolecules, such as antibodies, peptides, or aptamers. These functionalized nanodiamonds can specifically interact with target pathogens, enabling selective binding and enhancing the sensitivity and specificity of pathogen detection. Like graphene, nanodiamonds also mimic the extracellular matrix, promoting the adhesion of pathogens present in biofilms, which are often challenging to identify using conventional methods [41, 42].

8.2.5 Silica Nanoparticles

Silica nanoparticles possess several unique properties that make them well-suited for various applications (Fig. 8.5). Although, many of these properties may also be found in other types of nanoparticles to some extent, it is the combination of these properties, along with their biocompatibility and ease of functionalization. Silica nanoparticles possess a versatile surface chemistry providing an ease of functionalization making them suitable for a wide range of applications including target pathogen detection. They possess a mesoporous and well-defined controllable pore structure, which allows for efficient encapsulation, storage, and release of various molecules, such as drugs, fluorescent dyes, imaging agents, and other payloads. They can amplify Raman scattering signals of nearby molecules, enabling highly sensitive molecular detection and identification. Owing to the intrinsic properties of the core material, silica nanoparticles are highly stable and ensure robust protection to the encapsulated entity preserving its integrity [43, 44].

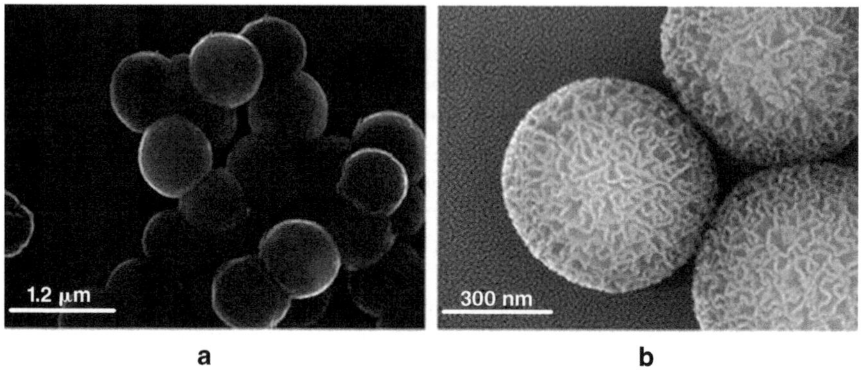

Fig. 8.5 SEM images of (SiO_2 mesoporous nanoparticles) KCC-1 particles after dispersion in ethanol: (**a**) low magnification and (**b**) high magnification, where the wrinkled surface is clearly visible [45]

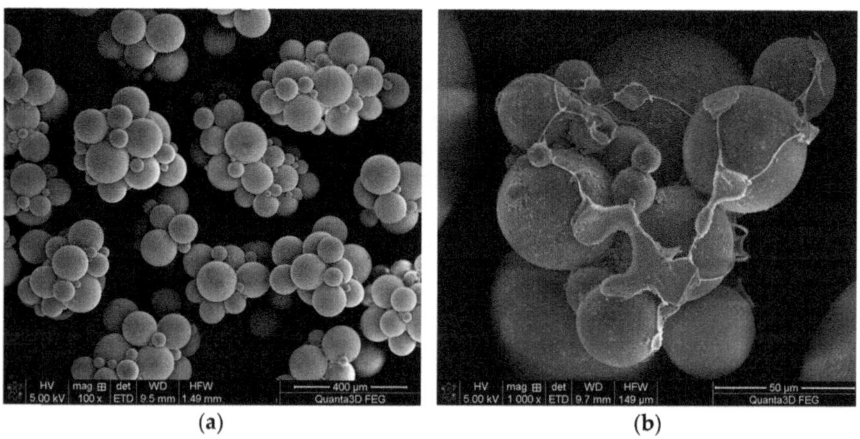

Fig. 8.6 SEM image of 9,10-bis(methacryloyloxymethyl)anthracene copolymer trimethylpropanetrimethacrylate (BMA-co-TRIM_C) microspheres synthesized in the presence of 0.5 wt.% (**a**) and 2.5 wt.% (**b**) of Poly(vinyl alcohol)(PVA) [51]

8.2.6 Organic Nanoparticles

Organic nanoparticles (Fig. 8.6) encompass a diverse group of nanomaterials derived from organic compounds and offer a range of characteristics that contribute to their effectiveness in pathogen detection. While they might not be an integral part of the sensing process, they have been widely used in enhancing the functionality of the diagnostic tool. Several types of organic nanomaterials have been proposed in this regard to be used as membranes, filters, surface functionalization tools for biosensing and pathogen detection devices [46].

Liposomes are spherical lipid-based vesicles composed of one or more lipid bilayers. They can encapsulate various payloads, including detection agents such as fluorescent dyes, antibodies, or nucleic-acid probes. Liposomes are used for target-specific binding to pathogens and can enhance the sensitivity of pathogen detection. They are often utilized in fluorescence-based assays and optical imaging techniques [47]. Solid lipid nanoparticles (SLNs) are composed of a solid lipid core stabilized by surfactants. SLNs have been explored for encapsulating hydrophobic detection agents, such as antimicrobial peptides or lipophilic dyes. These nanoparticles offer controlled release properties and can be designed for specific pathogen targeting, making them valuable for both detection and treatment strategies. Nanostructured lipid carriers (NLCs) are similar to SLNs but incorporate a mixture of solid and liquid lipids. This composition improves the encapsulation efficiency of hydrophobic agents and provides enhanced stability. NLCs are investigated for their potential in antimicrobial agents delivery and imaging of pathogens in food matrices. Micelles are lipid-based nanoparticles formed by the self-assembly of amphiphilic lipids in aqueous solutions. They have a hydrophobic core and a hydrophilic shell, making them suitable for encapsulating hydrophobic detection agents or antimicrobial agents. In addition to pure lipid-based nanoparticles, lipid coatings can be applied to other types of nanoparticles to enhance their stability, biocompatibility, and targeting capabilities. Lipid-coated nanoparticles are utilized to improve the interaction between the nanoparticle surface and the target pathogen, enhancing detection sensitivity [47, 48].

Polymeric nanoparticles are architectural moieties which can have different structural arrangements, such as linear, branched, crosslinked, or spherical, depending on the arrangement of monomer units and the polymerization process used. Most of them are biocompatible and biodegradable in nature making them a suitable choice in many biomedical applications. Some commonly used nanoparticles are polymeric micelles, nanogels, and dendrimers. Polymeric NPs have a large number of functional groups which enable simultaneous interaction with a variety of ligands, dyes and other molecular entities. They can also encapsulate detection agents and offer controlled release, making them suitable for targeted pathogen detection [49].

Although organic NPs may be regularly used as singular core materials in detection methods, but they can be combined with various techniques, such as polymerase chain reaction (PCR), fluorescence, surface-enhanced Raman spectroscopy (SERS), and electrochemical methods, to create sensitive and specific assays for the detection and diagnosis of food-borne pathogens in a variety of food matrices.

8.2.7 Upconversion Nanoparticles (UCNPs)

Photon upconversion is a recently reported phenomenon in which the absorption of multiple photons, typically infrared or lower-energy photons, leads to the emission of a single photon at a higher-energy level or shorter wavelength. This process results in anti-Stokes emission, which is contrary to the usual process where

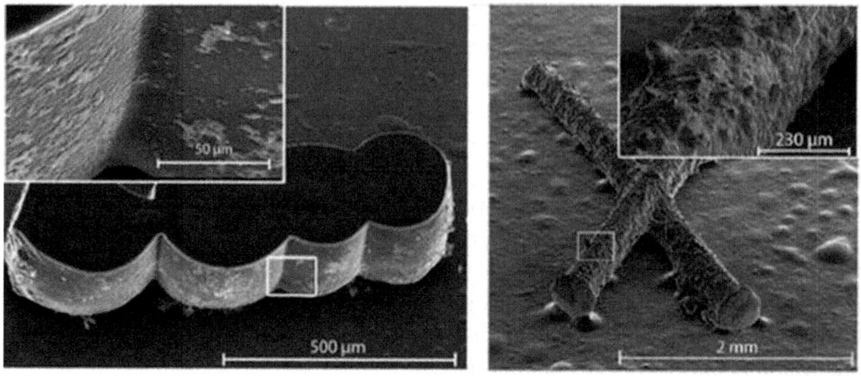

(a) (b)

Fig. 8.7 SEM image of 3D polymer microstructures obtained by NIR light-activated photopolymerization: (**a**) UCNPs concentration ~20 mg mL – 1 and (**b**) UCNPs concentration ~2 mg mL [51]

absorption of photons results in lower energy emissions. Upconversion nanoparticles (Fig. 8.7) exhibit a unique optical phenomenon, also referred as – upconversion luminescence. Upconversion is achieved through specific nanoparticles composed of rare-earth elements like erbium, ytterbium, and thulium, that have the unique ability to convert lower-energy photons into higher-energy ones. These nanoparticles have applications in various fields, including bioimaging and biosensing [52] and can be exploited in multiplexing of food pathogen diagnosis.

UCNPs exhibit several unique properties which enhance their suitability for detecting pathogens in food samples. Lanthanide-doped UCNPs consist of a host material doped with specific lanthanide ions wherein different lanthanide dopants can lead to various emission colours, allowing for multicolour imaging and detection. This multicolour emission capability enables the simultaneous detection of multiple targets or pathogens within a single assay [52–54]. UCNPs can be excited using near-infrared (NIR) light, which has deeper tissue penetration and reduced scattering. This feature allows for non-invasive detection in complex biological samples, making them suitable for in vivo imaging and diagnostics. UCNPs exhibit reduced photobleaching compared to organic fluorophores, ensuring consistent and long-lasting emission for prolonged detection periods [54]. They are usually more photostable compared to conventional materials, making them suitable for repeated imaging and prolonged detection processes without a significant loss of signal intensity. UCNPs can be hybridized with other nanomaterials such as silica or polymer, to enhance stability, biocompatibility, and allows for functionalization of the surface for specific and selective targeting of pathogens. Some sensitizers and activators can also be integrated to enhance energy absorption and transfer to activators, which then participate in the upconversion process [55, 56]. Due to these exclusive properties, upconversion nanoparticles have gained traction as promising tools for pathogen detection. They offer the potential to develop highly sensitive, specific, and multiplexed assays for identifying foodborne pathogens, enhancing food safety measures, and contributing to the early diagnosis of infections.

8.2.8 Metal-Organic Frameworks (MOFs)

MOFs are composed of metal ions or clusters linked together by organic ligands, forming a porous and crystalline framework with a high surface area. MOF nanostructures have a crystalline and porous structure, with well-defined and uniform pores (Fig. 8.8). This structure provides ample space for guest molecules to enter and be trapped within the nanoparticle, which is advantageous for molecular capture and biosensing applications. This structure allows MOFs to adsorb and store gases, liquids, or other molecules within their pores. These nanostructures have an exceptionally high surface area, which means they can interact with many target molecules. This property is crucial for effective capture and detection of pathogens present in low concentrations [57–59]. MOF nanoparticles can be designed to have specific optical and electronic properties based on the choice of metal ions and ligands making them useful for applications such as sensors, detectors, and optoelectronic devices. They can be tailored by choosing specific metal ions and ligands, allowing researchers to design nanoparticles with the desired pore size and surface chemistry. This tunability enables MOFs to selectively capture specific pathogens while excluding interfering substances with high sensitivity. MOF nanoparticles are generally stable and can withstand harsh conditions. This makes them suitable for repeated use, reducing the cost and waste associated with single-use detection methods [60].

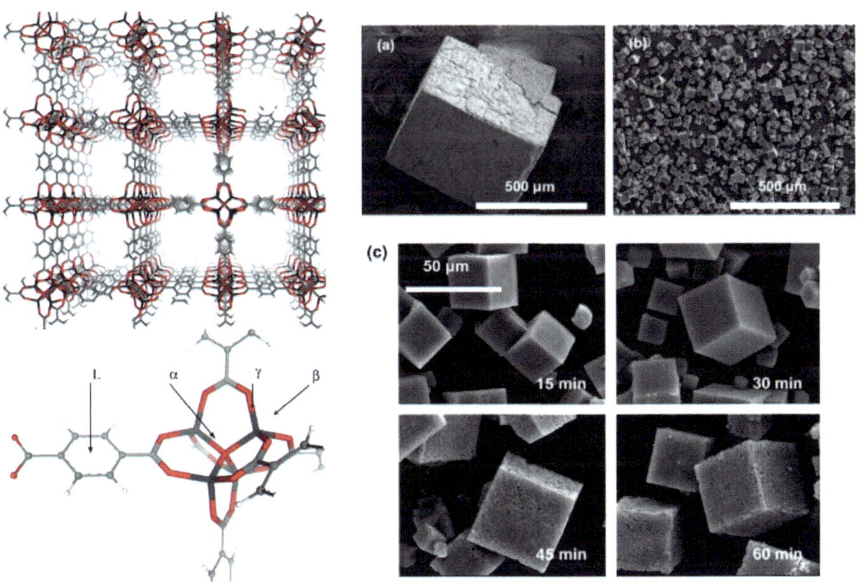

Fig. 8.8 SEM images of Metal-Organic Framework crystals synthesized using (**a**) solvothermal method, (**b**) microwave, and (**c**) degradation of MOF crystals as a result of microwave irradiation for 15, 30, 45, and 60 min [62]

MOF nanoparticles can be used to capture and concentrate target pathogens which can then be detected using techniques such as fluorescence, colorimetry, or other spectroscopic methods. The intensity of the signal can be correlated to the concentration of pathogens in the original food sample, allowing for quantification. MOF nanoparticles can serve as a scaffold for further functionalization. Additional functional groups or molecules can be attached to the MOF surface, expanding their capabilities, and enabling more complex applications. MOF nanoparticles have garnered significant interest in various scientific and industrial fields, and their potential applications continue to expand as researchers explore new ways to harness their distinctive characteristics [52, 61]. Some of the commonly used nanomaterials used in pathogen detection have been discussed in this section and certain examples have been tabulated in Table 8.2.

Table 8.2 Nanomaterials used in food pathogen detection

Nanomaterial	Target pathogen	Food sample	Method	Reference
Carbon QDs	*S. typhimurium*	Eggshell and tap water	Aptasensor	[63]
Carbon QDs & Gold NPs	AFB1, ochratoxin (fungal toxins)	–	Aptasensor	[36]
GO-Gold nanocomposite	AFB1	–	Immunosensor	[64]
GO-modified iron oxide-chitosan hybrid nanocomposite	*E. coli O157:H7*	–	Electrochemical sensor	[65]
Gold Nanorods	*Salmonella*	–	IMS, thermal ablation	[22]
Gold NPs	*Bacillus spores*	Food sample	Surface-enhanced Raman scattering (SERS)	[66]
Gold NPs	*Norovirus, Adenovirus, Parvovirus, Simian rotovirus, Coronavirus, Sendai virus and Herpes virus*	Food and water samples	–	[16]
Gold NPs	*E. coli O157: H7 antibodies*	–	HRP enzymatic reaction	[67]
Gold NPs	*Norovirus*	Lettuce	Concanavalin A detection	[68]
Gold NPs	*L. monocytogenes*	Spiked blueberries	Biosensor	[55]
Gold NPs	*S. aureus* and *S. typhimurium*	Spiked pork	Aptasensor, SERS	[70]
Gold NPs - magnetic	*S. choleraesuis*	Whole milk	–	[71]
Iron oxide with Gold Nanorods	Multiplex detection of food borne pathogens	–	IMS and NIR spectroscopy	[19]
Magnetic NPs	*L. monocytogenes*	Artificially contaminated milk	Immuno-magnetic separation (IMS) followed by qPCR	[32]

(continued)

Table 8.2 (continued)

Nanomaterial	Target pathogen	Food sample	Method	Reference
Magnetic NPs	*E. coli*	Freshly ground beef		[75]
Magnetic NPs	*Mycobacterium avium* spp. *paratuberculosis*	Contaminated whole milk	Conjugation-induced magnetic particle agglomeration	[72]
Magnetic NPs	*Brucella antibodies*	Blood serum of infected cows		[73]
Magnetic NPs	*E. coli*	2% milk, spinach extract	Fluorescent staining, FTIR	[74]
Magnetic NPs	*Bacillus anthracis*	Starch and baking soda	Magnetic, optical detection	[75]
Magnetic NPs	*E. coli O157:H7*	Contaminated food matrices	Magnetic, optical detection	[76]
Magnetic NPs	*Escherichia coli O157:H7, Salmonella enterica, Vibrio cholerae and Campylobacter jejuni*	Food samples	Oligonucleotide microarray assay	[77]
MWCNTs	*Salmonella enteritidis, Salmonella typhimurium*	–	Aptasensor	[37]
TiO2 nanowires	*L. monocytogenes*	–	Conductance resistance	[78]
Silica coated magnetic NPs and Gold NPs	*S. enterica serovar Typhimurium* and *S. aureus*	Spinach and peanut butter	Raman spectrometry and SERS	[79]
PAMAM dendrimers	AFB1	Contaminated peanuts sample	Aptasensor	[80]
Poly [pyrrole-co-3-carboxyl-pyrrole] copolymer	*S. typhimurium*		Aptasensor, Impedimetric biosensor	[81]
Polyaniline	*E. coli O157:H7*		Immunosensor	[82]
Polyaniline	*E. coli O157:H8*	Lettuce, alfalfa sprouts, and straw-berries	Electrochemical sandwich assay	[83]
Polymer based	*S. enterica, B. cereus*, and *E. coli*	Food samples	Chromatic polymer	[18]
Silica NPs	*E. coli*		Chemiluminiscence assay	[84]
Silica NPs	*M. avium* subsp. Paratuberculosis, c-DNA SARS CoV2	Cultured samples	Electrochemical, qPCR	[85]
SWCNTs	*Anti-Escherichia coli antibodies*	–	Aptasensor	[86]
QDs and UCNPs	*Staphylococcus aureus*, and *Salmonella typhimurium*		DNA aptamer-based, Spectophotometric	[26]
UCNPs	*Shigella*	Contaminated chicken	Aptamer-based FRET	[56]

Nanomaterials discussed in the section demonstrate versatile applications due to their remarkable properties and are opening a novel opportunity for developing a new generation of biosensor technologies [87]. Nanomaterial-based biosensors are progressing towards single-molecule biosensors and high output biosensor arrays, which can enhance the mechanical, electrochemical, optical, and magnetic characteristics of biosensors [88–90]. For various diagnosis uses, functionalization of nanoparticles with macromolecules (antigens) using various surface chemistries has also been established. Biosensing is suitable for early diagnosis, detection, and prevention of life-threatening diseases via use of combination of nanostructures.

8.3 Biosensors Concepts and Applications

A biosensor, in general, employs a biological identification moiety detecting the existence of an analyte (the species to be detected), a biorecognition/bioreceptor element, a physicochemical transducer, a signal amplification device, and a detection system (Fig. 8.9). Biosensors endure critical role across numerous fields including clinical diagnosis, prognosis, disease progression, drug discovery, food control and environmental monitoring [91]. Recent technological developments in biosensor-based pathogen diagnosis, have proven to be a viable replacement for established techniques such as culture-based methods, analytical and bioanalytical techniques [92].

Enzymes, antibodies, whole cells, aptamers, and peptides are some of the most versatile biorecognition elements that have been widely researched and are being effectively used in biosensor development. The precise interaction between biorecognition element and target analyte are highly specific and the presence of sensitive interface helps to achieve analysis at very low concentration, thus helping to overcome issues related to cross-reactivity and sensitivity. The desired detection sensitivity is achieved by overcoming the issues related to reproducibility, preconcentration, and pre-cleaning steps of the sample, and monitoring real time of target analyte with quick recovery time resulting in reusability are some other advantages of biosensors [93].

One of the major requirements in developing a biosensor for infectious diseases is the need for a sensitive analytical device that can easily go down to very low detection levels without significant changes in selectivity [94]. Most infectious diseases spread rapidly through a community before any symptoms are identified; a biosensor that can easily detect low levels of antigen at the onset of infection is highly in-demand. Besides, a biosensor that is relatively cheap, robust, responds rapidly and provides high-throughput is highly desired for field applications. Conventional techniques for detecting food-borne pathogens have several drawbacks, including lengthy detection times, low sensitivity, and poor selectivity

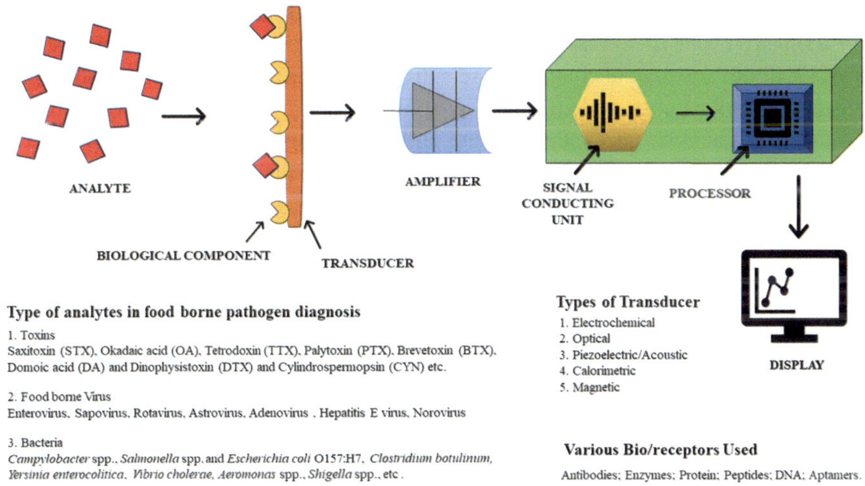

Fig. 8.9 Biosensor construct and design for different types of food borne pathogen analytes using a variety of bioreceptors employed on potential transducer approaches

[95]. The development of nanomaterial-based biosensors for the detection of foodborne pathogens offers many advantages viz. sensitivity, selectivity, rapidity, precision, and ease based on the exceptional characteristics of nanomaterials. However, the type of biosensor used will depend on the properties of the analyte (i.e., size, structure, concentration, etc.) and the matrix (i.e., air, liquid) in which the analyte is found. In fact, the sensitivity of the biosensor is highly dependent on the surface preparation/structure and a great deal of effort has been focused on understanding and tailoring the interfacial properties using various nanocomposites.

Biosensors can be categorized based on type of bioreceptor molecules viz. enzymatic sensors, nucleic-acid sensors, immunosensors, aptamer-based sensors, and so on. In addition, various techniques can be introduced for the analysis of biosensing reactions between sensing molecules and target materials, including electrochemical, fluorescence, and optical property analysis, surface-plasmon resonance measurements, and surface-enhanced Raman spectroscopy (SERS). Molecular/Nucleic-acid-based methods include simple polymerase chain reaction (PCR), real-time PCR, Loop-mediated isothermal amplification (LAMP), Nucleic-acid sequence-based amplification and oligonucleotide DNA microarray which are sensitive, specific and have shorter response times than the culture-based detection methods. Nevertheless, each of these methods has one or more limitations such as high cost or a requirement for lengthy special training. Biosensor-based methods include optical, and electrochemical approaches which are time-effective, labour saving and have higher sensitivity in comparison to culture-based methods. One of the most important criteria in creating a biosensor for foodborne illness is the need for a sensitive analytical instrument that can readily go down to very low detection levels without substantial changes in selectivity. In addition, a biosensor with high-throughput, robustness, and affordability is highly sought for field uses. In reality, a lot of work has gone into understanding and modifying the interfacial characteristics because the sensitivity of the biosensor is heavily reliant on surface preparation and structure. In order to solve these issues without losing accuracy or finesse, novel sensing strategies are required. To surmount current limitations, the most common food-borne bacterial pathogens must be controlled and identified using a fast, sensitive, particular, cost-effective, and reliable method.

8.3.1 Recognition Elements (Bioreceptors): Key Players in Biosensing

8.3.1.1 Antibody

Antibodies are class of immunoglobulins which are formed after the reaction of external molecules by the cells of immune system [96]. Entry of these external (recognized non-self) molecules results in their molecular recognition as foreign entities by the host's immune system ultimately leading to the generation of a specific class of immunoglobulin molecules (known as antibodies), which interact and bind with these foreign molecules in a specific manner ultimately leading to the clearance of these molecules from the body of the host [97]. These antibodies are produced by a special class of lymphocytes (B-cells) of immune system, which usually circulates in the blood and lymph throughout the body. Invasion of the foreign molecule into the blood or lymph leads to the encounter of these specific antigens, leading to the removal from the body. Antibodies exhibit vital functions in the identification of diverse molecules and pathogens, rendering them indispensable assets in contemporary biomedical research and clinical diagnostics. As glycoproteins derived from B cells, antibodies demonstrate exceptional specificity in binding to their target antigens. Antibodies are broadly divided into two categories, monoclonal and polyclonal type. Monoclonal antibodies (Fig. 8.10) are typically derived from a single clone of immune cell (typically B cells), engineered to produce identical antibodies. They are obtained by fusing spleen cells from an immunized mouse with human or mouse myeloma cells (malignant self-perpetuating antibody-producing cells) and selecting and cloning the hybrid cells (hybridomas) that produced the desired antibody [98]. On the other hand, polyclonal antibodies (Fig. 8.10) are derived from multiple B-cell clones that have differentiated into antibody-producing plasma cells in response to an immunogen injected inside a rabbit. In contrast to a monoclonal antibody, which recognizes a single epitope, a polyclonal antibody against a single molecular species of antigen recognizes more than one epitope on the target molecule [99].

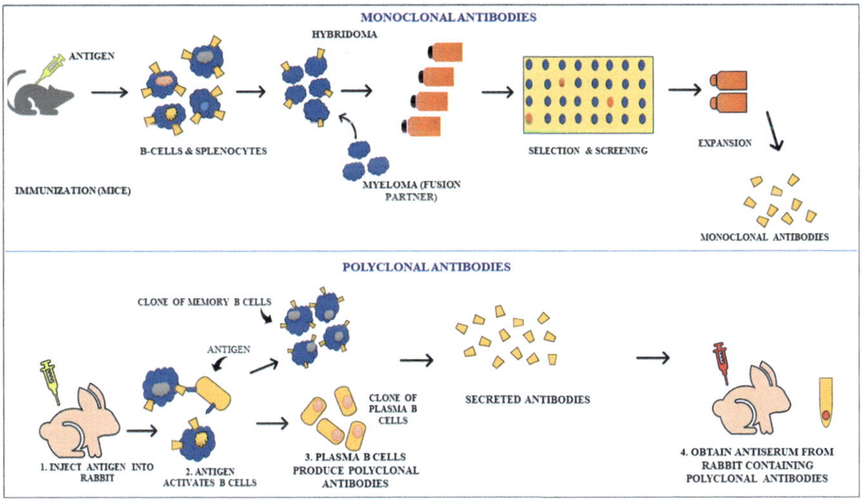

Fig. 8.10 Process flow of generation of monoclonal and polyclonal antibodies

Antibodies due to their high specificity, show remarkable properties such as well-characterized protein structure and a relatively uniform response, that make them easy to purify, label and detect with high reproducibility and predictability via general procedures. Thus, the ability of these antibody molecules to bind specifically to their target antigen can be utilized to develop bio-probes, which can be used for detecting potential food pathogens. Polysaccharides and low molecular weight antigens such as peptides or small molecules are usually antigenic but non-immunogenic and are unable to activate the immune system alone to raise antibodies [100]. Such small compounds are also called as haptens which can be immunogenic unless they are conjugated with the carrier immunogenic protein using various available functional groups (–NH_2 or –COOH). In diagnostic applications, antibodies are extensively utilized in serological methods for the identification and quantification of specific antigens, such as proteins, peptides, and small molecules. Techniques like enzyme-linked immunosorbent assay (ELISA), immunofluorescence assay (IFA), chemiluminescence immunoassay (CLIA), and lateral flow immunoassay (LFIA), rely on the binding affinity of antibodies to achieve sensitive and specific detection [101]. In recent times, progress in antibody engineering has resulted in the creation of recombinant antibodies, such as single-chain variable fragments (scFv) and nanobodies. These condensed and adaptable antibody formats demonstrate exceptional stability and simplified production processes, augmenting their value in point-of-care testing and biosensor applications [102]. Antibodies are commonly used as bioreceptors in biosensors due to their high specificity for antigens [103]. Nanomaterials can be modified with antibodies to create highly selective nanoprobes. When the target antigen binds to the antibody on the nanomaterial surface, it triggers a detectable signal. This signal can be based on changes in electronic, optical, fluorescence, [104] or other physical properties of the nanomaterial. Mutreja and coworkers have developed a novel OmpD-based immunosensor using rGO functionalized screen-printed electrodes in an electrochemical format. The developed nano-biosensor exhibited high sensitivity with a limit of detection (LOD) of 10 CFU mL^{-1} [6]. Shrinivasan and group have reported two label-free fluorescent aptasensor methods for the detection of *S. typhimurium.* The lowest detection limit achieved with one approach was in the range of 733 CFU/mL and while the alternative approach has LOD of 464 CFU/mL [105].

8.3.1.2 Aptamer

Aptamers, an alternative to antibodies are single-stranded DNA or RNA molecules, typically less than 100 nucleotides, that can specifically bind to target molecules with high affinity. They are a valuable class of recognition elements in biosensor design due to their unique ability to specifically bind with target molecules. They are generated by evolving from a library of oligonucleotides via a process known as the "Systematic Evolution of Ligands by Exponential Enrichment" (SELEX) [106, 107]. The process of generation of aptamers is attributed to the ability of these small oligos to take unique 3D structures by iterative rounds of SELEX, which can possibly interact with the target in a highly specific manner. These selective interactions between aptamers with a diverse range of targets, irrespective of their chemical nature, are exploited for different food pathogens detection, which is possible due to their high affinity, ease of selection and specificity towards their targets. Aptamers, or chemical antibodies, offer various advantages compared to antibodies, such as their capacity for in vitro generation, ease of functionalization, suitability for large-scale synthesis without compromising quality, storage without the need for a cold chain, and the ability to undergo target-induced structural switching in an electrochemical DNA aptasensor for swift diagnosis [108].

SELEX process is based on the iteration of successive rounds of ssDNA molecules followed by the elution and amplification of bound molecules (Fig. 8.11). The technique consists of four steps: exposure, binding, selection of binders, amplification, and then separation of amplicon to single strands. This iterative process is continued till the pool is enriched for sequences specific towards the target. High-binding aptamers are screened iteratively from the oligonucleotide library, which is made up of a random base sequence containing primer binding sites [11]. A wide range of SELEX versions have been reported (Table 8.3), mainly based on whole cells [109], capillary electrophoresis CE-SELEX [110], microfluidics [111], FACS [112], microtitre plates [11, 107, 113] and magnetic beads [114], have been developed. Aptamers in conjunction with nanomaterials have been known to enhance the specificity, sensitivity and performance of the biosensor that recognize a wide range of targets from small molecules viz. toxins to food pathogens. Recently, Pathania and colleagues have demonstrated that functionalization of anti-Vi aptamers with 2D MoS-rGO nanosheets increased the efficacy of electrochemical sensors with LOD of 100 pg mL^{-1} for the detection of *Salmonella typhi* [4].

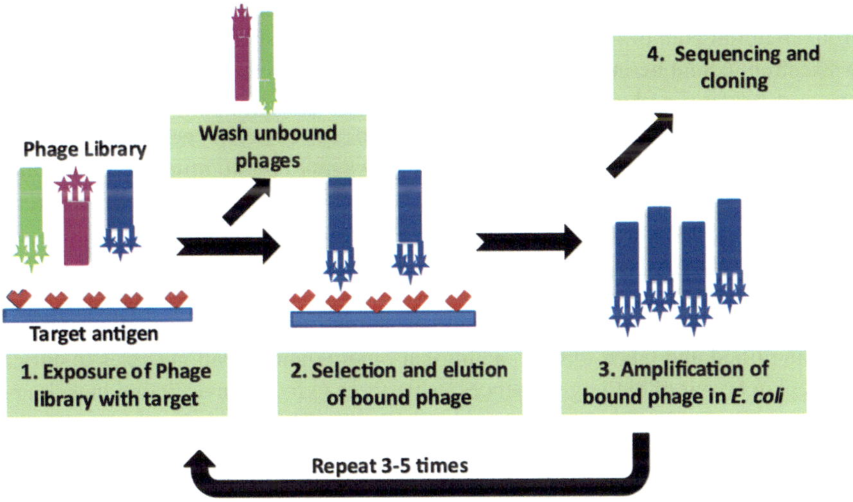

Fig. 8.11 Schematic representation of bio-panning for selection of phage display peptides

Table 8.3 Different types of SELEX

S.No.	Type	Description
1.	Classical SELEX	Traditional SELEX process involving iterative rounds of selection and amplification to enrich aptamers
2.	Counter-SELEX	Selection performed on whole cells as targets, enabling the identification of cell-specific aptamers
3.	Cell-SELEX	Selection performed on whole cells as targets, enabling the identification of cell-specific aptamers
4.	Photo SELEX	Utilizes light irradiation to introduce photo-cleavable linkers, allowing targeted aptamer release
5.	Capillary-SELEX	Selection carried out in capillary tubes to enhance partitioning efficiency, increasing aptamer yield
6.	Microfluidic-SELEX	Uses microfluidic devices for precise control over reaction conditions and improved aptamer yield
7.	Magnetic-SELEX	Incorporates magnetic beads for target immobilization, simplifying the aptamer selection process

8.3.1.3 Phage Display Peptides

Phage display is a selection technique in which a peptide is displayed on filamentous phage by fusing the peptide of interest to gene III of filamentous phage [115]. Phage display peptides have been well-documented tools for the identification or screening of peptide bioreceptors against a target molecule and are widely used for many diseases [116, 117]. Existing literature has proven the use of bacteriophage in several detection platforms that can be excellent alternatives to antibodies. The advantages of using phages as biorecognition elements include exceptional stability, precise host identification, ease of replication [118]. This in vitro technique selects phage clones expressing the peptides with the highest affinity to a specific target from an initial pool of random peptides library. Phage display technology describes a selection technique in which a library of peptide or protein variants is expressed on the outside of a phage virion, while the genetic material encoding each variant resides on the inside. This creates a physical linkage between each variant protein sequence and the DNA encoding it, which allows rapid partitioning based on binding affinity to a given target molecule (antibodies, enzymes, cell-surface receptors, etc.) by an in vitro selection process called panning (Fig. 8.11).

In its simplest form, panning is carried out by incubating a library of phage-displayed peptides on a plate (or bead) coated with the target, washing away the unbound phage, and eluting the specifically bound phage. The eluted phages are then amplified and taken through additional binding/amplification cycles to enrich the pool in favour of binding sequences. After three to four rounds, individual clones are characterized by DNA sequencing and ELISA. The targeted and high-affinity peptides discovered by the phage display technology have been shown to be a better alternative for antibody replacement [119].

8.3.1.4 Molecularly Imprinted Polymers (MIPs)

MIPs are interesting class of materials that can be employed as another class of bioreceptor analogues. MIPs are synthetic polymers, like antibodies or other natural receptors and are designed to identify and selectively detect specific molecules of interest. These are polymers that specifically adsorb and recognize targets or structural mimics. MIPs contribute to the creation of stable "solid-state-like" synthetic-detection elements in sensors [120, 121]. They are generated by polymerizing monomers around the target molecule in a process known as molecular imprinting. After polymerization the template was removed, and the cavities with corresponding forms to the template are left inside the polymer matrix. When exposed to the target of interest molecule, these spaces enable MIPs to selectively bind [122]. MIPs have shown promising approaches for the detection of food borne pathogen in recent years. Based on molecularly imprinted polymer (MIP)-coated quantum dot (QDs@MIP), a novel class-specific artificial receptor has been developed by Sun and coworkers for the detection of acylated homoserine lactone molecules produced by Gram-negative bacteria in fluorescent-based format within 2 to 18 nM. The developed sensor exhibited high sensitivity, good stability, and fast response (30 s) towards the target molecules due to the successful formation of surface imprints [123]. In another format, Percin and his coworkers have presented a microcontact imprinted chip-based surface-plasmon resonance sensor for the detection of Salmonella Paratyphi with a limit of detection (LOD) of 1.4×10^6 CFU mL^{-1} [124]. Wang and group have developed a simple and rapid ultrasensitive method for the detection of Salmonella based on MXene/polypyrrole (PPy)-based bacterial imprinted polymer (MPBIP) sensor, which showed a LOD 23 CFU mL-1 [125].

8.3.2 Sensor Design/Transducer Aspects of Biosensors

8.3.2.1 Electrochemical Sensors

An electrochemical biosensor is based on the recognition of electrical signal produced by the interaction of the biorecognition element/molecule, which selectively reacts with the biomolecule or converts the product depicting the concentration of analyte, which is proportional to an electrical signal. Several strategies have been employed to measure electrochemical changes produced by a recognition of biomolecular components, and these can be categorized into conductometric, impedimetric, potentiometric and amperometric. Alterations with regards to current can be observed between electrodes due to oxidation and reduction reactions, and these changes can be utilized for the quantification of analyte using electrochemical sensors [126]. Electrochemical sensors offer various merits like fast response, high sensitivity, more stable output, lesser interferences. The known formats of these electrochemical-based sensors are as follows:

8.3.2.2 Conductometric Transducer

These are based on measuring the change in the conductance of the system due to the presence of the analyte in the sample used. The interaction of the analyte and recognition molecule affects the ionic concentration of the solution changing the flow of current or the conductivity of the solution. A general conductometric transducer is based on metal electrodes separated by a certain distance. Applying an alternating current across the separated electrodes leads to the flow of current, as during a biorecognition event, the conductivity properties change due to the change in ions, which can be measured via a conductometric transducer based on ohm's law [127].

8.3.2.3 Amperometric Transducers

These transducers measure alteration in current produced due to the consequence of the presence of oxidative or reductive species at a constant applied potential. Typically, the potential is fixed at a value within the working electrode, and the reference electrode is usually Ag/AgCl. The current is measured by immobilizing the bioreceptor molecule on the working electrode modified with mostly gold, carbon, or platinum nanostructures. These electrodes function by the production of a current when the potential is applied between two electrodes, the magnitude of current being proportional to the substrate concentration [128]. The current produced with respect to the applied potential is a direct measure of the rate of electron transfer and is directly proportional to the reaction produced between bioreceptor molecule and analyte. Biosensors based on such transducers offers advantages like ultra-high sensitivity upto ~10^{-9} M concentration and also have ability to perform measurements on turbid/opaque solutions, over the conventional techniques. The performance of the electrochemical sensor is highly dependent on the employment of electrode materials of the electrodes, mainly working and reference [129]. Some other important parameters that can influence the performance of sensor are the nature of the electrolyte and its concentrations are properties. Other parameters such as reproducibility, stability and detection limit are also crucial for efficient electron transfer processes.

8.3.2.4 Potentiometric Transducers

Potentiometry is based on the measurement of accumulation of change in the potential of an electrochemical cell. The potentiometric biosensor consists of two electrodes: a reference electrode and an ion selective electrode (ISE). The latter is made up of a membrane that selectively interacts ion of interest, leading to the stock of a charge potential as compared to the reference electrode. The former electrode remains unaffected by the concentration of the analyte. The potential change is measured and compared after analyte exposure when zero or no current flows between the electrodes through a low resistance voltammeter. The response in case of potentiometric biosensor is directed by the Nernst equation, as the applied potential is proportional to the logarithm of the concentration of the analyte [93].

8.3.2.5 Impedimetric Transducers

This type of transducer is the most used approach in food pathogen detection. It measures the changes in the capacitance and resistance that occurred due to the biomolecular interactions at the interface, measured on the basis of electrochemical impedance spectroscopy (EIS) [107]. The technique involves typical three-electrode system, working, reference and counter electrodes to which a small electrical field is applied, which makes the current flow in a biosensor. The in and out-of-phase responses are determined via capacitance and resistance-based components by varying the frequency range. The event of biomolecular recognition at the interface leads to the increase in impedance properties of the electrodes that can be measured by an impedimetric analyser [6].

Kaur et al., reported a newly designed Bridged Rebar Graphene (BRG) nanostructured aptasensor for label-free impedimetric sensing of pathogenic bacteria *E. coli* O78:K80:H11 (Fig. 8.12). Based on chemically-enhanced multi-walled carbon nanotubes (MWCNTs) unwinding and subsequent crosslinking with terephthalaldehyde (TPA) to form 3D hierarchical BRG nanostructures provides a unique chemical function leading to improved electrical properties and stable bio-interfaces. Bacterial-aptamer (DNA) biological interactions were recorded on nanostructured BRG electrodes using specific anti-*E. coli* DNA aptamers (Kd ~14 nM), screened by SELEX method using phenylboronic acid, a chemical moiety specifically

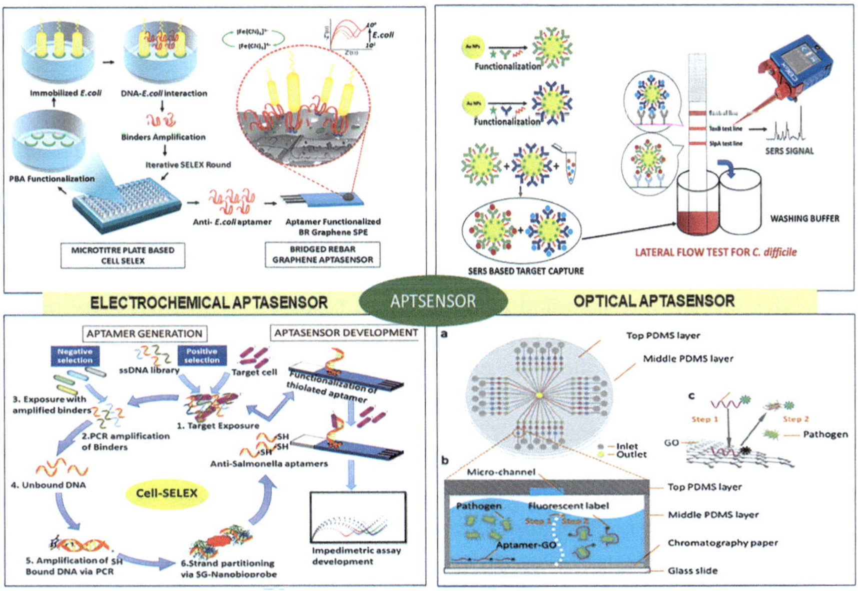

Fig. 8.12 Left (top): Aptamer generation through microtitre plate SELEX for E. coli and impedimetric sensing on bridged rebar graphene nanostructured screen-printed electrode (SPE). Reprinted from Biosensors Bioelectronics, 98, Kaur et al., Bridged Rebar Graphene functionalized aptasensor for pathogenic E. coli O78:K80:H11detection, 486–493 Copyright (2017), with permission from Elsevier. Left (bottom): Aptamer generation using cell-selex for the label-free detection of S. typhimurium). Reprinted from Microchimica Acta, 184, Pathania et al., Selective identification of specific aptamers for the detection of non-typhoidal salmonellosis in an apta-impedimetric sensing format,1499–1508, Copyright (2017), with permission from Springer. Right (top): Duplex detection of SlpA and ToxB by SERS-based LFA. Reproduced from Ref. [130] with permission from the Royal Society of Chemistry. Right (bottom): Schematic of a PDMS/paper hybrid microfluidic system for one-step multiplex pathogen detection using aptamer-functionalized GO biosensors (not to scale). (a) Layout of the microfluidic biochip, (b) and (c) demonstrate the principle of a one-step 'turn-on' detection approach based on GO, aptamer, and pathogen interactions. Step 1: When the aptamer is adsorbed on the GO surface, its fluorescence is quenched. Step 2: If the target pathogen is present, the target pathogen induces release of the aptamer from the GO, thereby restoring its fluorescence for detection. Reproduced from Ref. [131] with permission from the Royal Society of Chemistry

binding with diol groups present in polysaccharides of bacterial membrane. The developed nanostructured aptasensor showed a low detection limit and a sensitivity of ~10^1 cfu/mL for *E. coli* O78:K80:H11 in water, juice and milk samples, with a dynamic response range of 10^1–10^6 cfu/mL [11]. In another study, Pathania and colleagues' work demonstrated an intriguing aptamer-based electrochemical approach for precise detection of *S.* Typhimurium (Fig. 8.12). The work is based on the selection of species-specific aptamers through counter cell-SELEX against Salmonella Typhimurium with little cross-reactivity among related genera and species. Gold nanoparticles (AuNPs) are electrochemically deposited on screen-printed carbon electrodes and are crosslinked to thiol-modified aptamers on the platform. The aptasensor could detect Salmonella Typhimurium in an impedimetric format with a linear range of 10 CFU mL^{-1} to 10^5 CFU mL^{-1} and LOD of up to ~10 CFU mL^{-1} in spiked egg and water samples [107].

8.3.3 *Optical Biosensors*

Optical sensors utilize photons for transduction of signal in order to gather information about the bio interface and analyte. These types of biosensors measure the changes in the light and are known for their high sensitivity, specificity, smaller size and cost-effectiveness. Optical biosensors can be broadly divided into two general modes: label-free and label-based assays. Briefly, a label-free method employs direct detection of the signal generated, produced by the interaction of the target material on the transducer. In contrast, label-based sensing involves the use of a label and the optical signal is then generated by a colorimetric, fluorescent or luminescent method. The principle for the working of an optical bio-transducer relies on the detector molecules used (nanoparticle/enzyme system) that converts the signal produced by the analyte to get reduced or oxidized on the working areas of the functionalized electrode [132]. Some of the widely studied examples of optical transducers include surface-plasmon resonance (SPR), evanescent wave (EW), fluorescence and chemiluminescence devices. These optical biosensors typically measure properties like luminescence, absorbance, fluorescence or reflectance changes upon the binding of the specific bioreceptor with analyte. These emissions detectable by optical transducers mainly occur in the visible, ultraviolet (UV), or near-infrared (NIR) spectral regions and measures both the catalytic as well as affinity reactions [133].

Hassanain*et al.*, reported detection of two specific biomarkers, surface layer protein A (SlpA) and toxin B (ToxB) present in *Clostridium difficile*, using a surface-enhanced Raman scattering-based lateral flow assay (SERS-based LFA) (Fig. 8.12). The developed SERS-based LFA platform enabled rapid dual detection of SlpA and ToxB within 20 min on separate test lines using dual LF test strips. Scanning the test line using a handheld Raman spectrometer enabled sensitive quantitative detection of both biomarkers at the lowest observable concentration of 0.01 pg μL-1. This SERS-based LFA enables rapid, selective, sensitive, and efficient CDI at the point of care (POC) with minimal sample backlog by using a handheld device instead of a benchtop device and paves the way for cost-effective clinical evaluation [130].

In another study, Zuo*etal.*, developed a polydimethylsiloxane (PDMS)/paper/glass hybrid microfluidic system integrated with aptamer-functionalized graphene oxide (GO) nanobiosensors for easy one-step multiplexed detection of pathogens (Fig. 8.12). The paper substrate used in this hybrid microfluidic system facilitates the integration of aptamer biosensors on microfluidic biochips, avoiding complex surface treatments and aptamer probe immobilization in PDMS or all-glass microfluidic systems. Lactobacillus acidophilus is used as a bacterial model to develop microfluidic sensing platform with a detection limit of 11.0 CFU mL^{-1}. The group

Table 8.4 Various transducer approaches used for food pathogen detection

Biosensor/transducer	Target food pathogen	LOD	Reference
QCM	*Campylobacter jejuni*	150 CFU mL^{-1}	[134]
Electrochemical	*Salmonella typhimurium*	10 CFU mL^{-1}	[107]
Dual mode lateral flow assay (colour mode and SERS mode)	*Campylobacter jejuni*	50 CFU mL^{-1}	[135]
Fluorescent	*Vibrio parahaemolyticus*	10 CFU mL^{-1}	[136]
Fluorescent	*Shigella* spp.	102 CFU mL^{-1}	[137]
Fluorescent	*Pseudomonas aeruginosa*	102 CFU mL^{-1}	[138]
Electrochemical	*Salmonella typhimurium*	1 CFU mL^{-1}	[139]
FRET	*Salmonella typhi*	1 pg mL^{-1}	[140]
Chemiluminescence	*E. coli O157:H7*	143 CFU mL^{-1}	[141]
Electrochemical	*Salmonella typhimurium*	10 CFU mL^{-1}	[6]
Electronic sensor	*Escherichia coli*	10^2 CFU mL^{-1}	[142]
Fibre optic surface-plasmon resonance (SPR)	*Escherichia coli*	94 CFU mL^{-1}	[143]
Fluorescent	*Salmonella typhimurium*	10 CFU mL^{-1}	[144]
Photoelectrochemical	*Escherichia coli*	2 CFU mL^{-1}	[145]
Electrochemical	*Shigella* spp. (Shiga toxin)	44.5 pg mL^{-1}	[5]
Electrochemical	*Salmonella typhimurium* (Vi Antigen)	100 pg mL^{-1}	[4]
SERS	*Staphylococcus aureus*, *Escherichia coli* O157:H7, *Salmonella typhimurium*	8, 10, and 10 cells mL^{-1}	[146]
Electrochemical	*Escherichia coli* O157: H7	10 CFU mL^{-1}	[7]
Microfluidic immunosensor	*Salmonella* sp.	93 CFU mL^{-1}	[147]
Optical	*Escherichia coli*	3 CFU mL^{-1}	[148]
Electrochemical	*Escherichia coli* O157: H7	10 CFU mL^{-1}	[11]

successfully extended this method to the simultaneous detection of two infectious agents (Staphylococcus aureus and Salmonella enterica). This microfluidic sensing platform has a great potential for rapid detection of various other bacterial and viral pathogens [131].

To summarize, the most commonly used biosensing methodologies for detecting food pathogens are shown in Table 8.4, and described in detail.

8.4 Conclusion

Foodborne infections have developed into an important health concern on a global basis because of the substantially increasing prevalence of these illnesses over a period. There is a pressing need to consistently employ a reliable approach for collecting and analyse foodborne pathogens, as culture-dependent methods are not sufficiently sensitive, time-consuming and are more prone to give erroneous negative results. Nano-based biosensors, a possible alternative, relying on variously modified nanomaterial types, design, functionality in conjunction with specific bioreceptors, aided improved sensitivity and specificity. To get around the drawbacks of traditional methods, experts have created a variety of biosensors which are mainly based upon optical and electrochemical-based approaches. Both have their own pros and cons. Optical methods are simple, inexpensive, and easy to use compared to electrochemical methods, which are highly sensitive and can easily detect even a single cell. However, it is relatively expensive, and sometimes requires specialized personnel or fabrication facility to develop sensors. Nonetheless, biosensor as analytical tool has a huge potential in food pathogen diagnosis, due to ease of use, potential miniaturization, and real-time analytical capabilities, which can provide rapid and effective diagnostics. Biosensing technology can explored in building total human health surveillance system, targeting potential food pathogens, which are of a major concern of morbidity across globe.

References

1. Baig, N., Kammakakam, I., & Falath, W. (2021). Nanomaterials: A review of synthesis methods, properties, recent progress, and challenges. *Materials Advances, 2*, 1821–1871.
2. Yaqoob, A. A., Ahmad, H., Parveen, T., Ahmad, A., Oves, M., Ismail, I. M., Qari, H. A., Umar, K., & Mohamad Ibrahim, M. N. (2020). Recent advances in metal decorated nanomaterials and their various biological applications: A review. *Frontiers in Chemistry, 19*(8), 341.
3. Zamora-Galvez, A., Morales-Narváez, E., Mayorga-Martinez, C. C., & Merkoçi, A. (2017). Nanomaterials connected to antibodies and molecularly imprinted polymers as bio/receptors for bio/sensor applications. *Applied Materials Today, 9*, 387–401.
4. Pathania, P. K., Saini, J. K., Vij, S., Tewari, R., Sabherwal, P., Rishi, P., & Suri, C. R. (2018). Aptamer functionalized MoS2-rGO nanocomposite based biosensor for the detection of Vi antigen. *Biosensors and Bioelectronics, 122*, 121–126.
5. Kaur, H., Shorie, M., & Sabherwal, P. (2020). Biolayer interferometry-SELEX for Shiga toxin antigenic-peptide aptamers & detection via chitosan-WSe_2aptasensor. *Biosensors and Bioelectronics, 167*, 112498.
6. Mutreja, R., Jariyal, M., Pathania, P., Sharma, A., Sahoo, D. K., & Suri, C. R. (2016). Novel surface antigen based impedimetric immunosensor for detection of Salmonella typhimurium in water and juice samples. *Biosensors and Bioelectronics, 85*, 707–713.
7. Kaur, H., Shorie, M., & Sabherwal, P. (2020). Electrochemical aptasensor using boron-carbon nanorods decorated by nickel nanoparticles for detection of E. coli O157:H7. *Microchimica Acta, 187*(8), 461.
8. Xu, C., Zhang, Y. S., Begin, S., & Thanh, N. T. (2022). Introduction to advanced functional nanomaterials for biomedical applications. *Nanoscale, 14*(20), 7441–7443.
9. Lonchamps, P. L., He, Y., Wang, K., & Lu, X. (2022). Detection of pathogens in foods using microfluidic "lab-on-chip": A mini review. *Journal of Agriculture and Food Research, 20*, 100430.
10. Radhakrishnan, S., Mathew, M., & Rout, C. S. (2022). Microfluidic sensors based on two-dimensional materials for chemical and biological assessments. *Materials Advances, 3*(4), 1874–1904.
11. Kaur, H., Shorie, M., Sharma, M., Ganguli, A. K., & Sabherwal, P. (2017). Bridged Rebar Graphene functionalized aptasensor for pathogenic E. coli O78: K80: H11 detection. *Biosensors and Bioelectronics, 98*, 486–493.

12. https://www.who.int/activities/estimating-the-burden-of-foodborne-diseases
13. Scharff, R. L., Besser, J., Sharp, D. J., Jones, T. F., Peter, G. S., & Hedberg, C. W. (2016). An economic evaluation of PulseNet: A network for foodborne disease surveillance. *American Journal of Preventive Medicine, 50*(5), 566–573.
14. Adley, C. C., & Ryan, M. P. (2016). The nature and extent of foodborne disease. In *Antimicrobial food packaging* (pp. 1–10). Academic.
15. Chouhan, R. S., Horvat, M., Ahmed, J., Alhokbany, N., Alshehri, S. M., & Gandhi, S. (2021). Magnetic nanoparticles – A multifunctional potential agent for diagnosis and therapy. *Cancers, 13*(9), 2213.
16. Duncan, T. V. (2011). Applications of nanotechnology in food packaging and food safety: Barrier materials, antimicrobials and sensors. *Journal of Colloid and Interface Science, 363*, 1–24.
17. Bobrinetskiy, I., Radovic, M., Rizzotto, F., Vizzini, P., Jaric, S., Pavlovic, Z., Radonic, V., Nikolic, M. V., & Vidic, J. (2021). Advances in nanomaterials-based electrochemical biosensors for foodborne pathogen detection. *Nanomaterials, 11*. https://doi.org/10.3390/nano11102700
18. Velusamy, V., Arshak, K., Korostynska, O., Oliwa, K., & Adley, C. (2010). An overview of foodborne pathogen detection: In the perspective of biosensors. *Biotechnology Advances, 28*, 232–254.
19. Wang, C., & Irudayaraj, J. (2008). Gold nanorod probes for the detection of multiple pathogens. *Small, 4*, 2204–2208.
20. Ding, X., Li, D., & Jiang, J. (2020). Gold-based inorganic nanohybrids for nanomedicine applications. *Theranostics, 10*, 8061–8079.
21. Zhang, N., Chen, H., Liu, A. Y., Shen, J. J., Shah, V., Zhang, C., Hong, J., & Ding, Y. (2016). Gold conjugate-based liposomes with hybrid cluster bomb structure for liver cancer therapy. *Biomaterials, 74*, 280–291.
22. Wang, S., Singh, A. K., Senapati, D., Neely, A., Yu, H., & Ray, P. C. (2010). Rapid colorimetric identification and targeted photothermal lysis of Salmonella bacteria by using bioconjugated oval-shaped gold nanoparticles. *Chemistry – A European Journal, 16*, 5600–5606.
23. Hahn, M. A., Tabb, J. S., & Krauss, T. D. (2005). Detection of single bacterial pathogens with semiconductor quantum dots. *Analytical Chemistry, 77*, 4861–4869.
24. Hahn, M. A., Keng, P. C., & Krauss, T. D. (2008). Flow cytometric analysis to detect pathogens in bacterial cell mixtures using semiconductor quantum dots. *Analytical Chemistry, 80*, 864–872.
25. Du, H., Wang, X., Yang, Q., & Wu, W. (2021). Quantum dot: Lightning invisible foodborne pathogens. *Trends in Food Science and Technology, 110*, 1–12.
26. Kurt, H., Yüce, M., Hussain, B., & Budak, H. (2016). Dual-excitation upconverting nanoparticle and quantum dot aptasensor for multiplexed food pathogen detection. *Biosensors and Bioelectronics, 81*, 280–286.
27. Dager, A., Baliyan, A., Kurosu, S., Toru, M., & Tachibana, M. (2020). Ultrafast synthesis of carbon quantum dots from fenugreek seeds using microwave plasma enhanced decomposition: application of c-QDs to grow fluorescent protein crystals, *10*, 12333.
28. Mornet, S., Vasseur, S., Grasset, F., & Duguet, E. (2004). Magnetic nanoparticle design for medical diagnosis and therapy. *Journal of Materials Chemistry, 14*, 2161–2175.
29. Kumar, H., Kuča, K., Bhatia, S. K., Saini, K., Kaushal, A., Verma, R., Bhalla, T. C., & Kumar, D. (2020). Applications of nanotechnology in biosensor-based detection of foodborne pathogens. *Sensors (Switzerland), 20*. https://doi.org/10.3390/s20071966
30. Mocan, T., Matea, C. T., Pop, T., Mosteanu, O., Buzoianu, A. D., Puia, C., Iancu, C., & Mocan, L. (2017). Development of nanoparticle-based optical sensors for pathogenic bacterial detection. *Journal of Nanobiotechnology, 15*, 25. https://doi.org/10.1186/s12951-017-0260-y
31. Reddy, B. L., Jatav, H. S., Rajput, V. D., et al. (2022). Nanomaterials based monitoring of food- and water-borne pathogens. *Journal of Nanomaterials, 2022*, 1. https://doi.org/10.1155/2022/9543532

32. Yang, H., Qu, L., Wimbrow, A. N., Jiang, X., & Sun, Y. (2007). Rapid detection of Listeria monocytogenes by nanoparticle-based immunomagnetic separation and real-time PCR. *International Journal of Food Microbiology, 118*, 132–138.
33. Wu, K., Cheeran, M. C. J., Wang, J. P., Saha, R., Su, D., Krishna, V. D., & Liu, J. (2020). Magnetic-nanosensor-based virus and pathogen detection strategies before and during covid-19. *ACS Applied Nano Materials, 3*, 9560–9580.
34. Govan, J. (2020). Recent advances in magnetic nanoparticles and nanocomposites for the remediation of water resources. *Magnetochemistry, 6*. https://doi.org/10.3390/MAGNETOCHEMISTRY6040049
35. Mitura, K., Kornacka, J., Kopczyńska, E., et al. (2021). Active carbon-based nanomaterials in food packaging. *Coatings, 11*, 1–31.
36. Wang, B., Chen, Y., Wu, Y., Weng, B., Liu, Y., Lu, Z., Li, C. M., & Yu, C. (2016). Aptamer induced assembly of fluorescent nitrogen-doped carbon dots on gold nanoparticles for sensitive detection of AFB1. *Biosensors and Bioelectronics, 78*, 23–30.
37. Hasan, M. R., Pulingam, T., Appaturi, J. N., Zifruddin, A. N., Teh, S. J., Lim, T. W., Ibrahim, F., Leo, B. F., & Thong, K. L. (2018). Carbon nanotube-based aptasensor for sensitive electrochemical detection of whole-cell Salmonella. *Analytical Biochemistry, 554*, 34–43.
38. Sharma, P., Bhalla, V., Tuteja, S., Kukkar, M., & Suri, C. R. (2012). Rapid extraction and quantitative detection of the herbicide diuron in surface water by a hapten-functionalized carbon nanotubes based electrochemical analyzer. *The Analyst, 137*(10), 2495–2502.
39. Sharma, P., Tuteja, S. K., Bhalla, V., Shekhawat, G., Dravid, V. P., & Suri, C. R. (2013). Bio-functionalized graphene–graphene oxide nanocomposite based electrochemical immunosensing. *Biosensors and Bioelectronics, 39*(1), 99–105.
40. Mansuriya, B. D., & Altintas, Z. (2020). Graphene quantum dot-based electrochemical immunosensors for biomedical applications. *Materials, 13*, 96.
41. Motlag, M., Liu, X., Nurmalasari, N. P. D., Jin, S., Nian, Q., Park, C., Jin, L., Huang, L., Liu, J., & Cheng, G. J. (2020). Molecular-scale nanodiamond with high-density color centers fabricated from graphite by laser shocking. *Cell Reports Physical Science, 1*, 100054. https://doi.org/10.1016/j.xcrp.2020.100054
42. Martel-Estrada, S.-A. (2018). Recent Progress in biomedical applications of nanodiamonds. *Nanoscience and Nanotechnology, 8*, 11–24.
43. Karaman, D. Ş., Pamukçu, A., Karakaplan, M. B., Kocaoglu, O., & Rosenholm, J. M. (2021). Recent advances in the use of mesoporous silica nanoparticles for the diagnosis of bacterial infections. *International Journal of Nanomedicine, 16*, 6575–6591.
44. Chitra, K., & Annadurai, G. (2013). Fluorescent silica nanoparticles in the detection and control of the growth of pathogen. *Journal of Nanotechnology, 2013*, 1. https://doi.org/10.1155/2013/509628
45. Luisetto, I., Tuti, S., Marconi, E., Veroli, A., Buzzin, A., de Cesare, G., Natali, S., Verotti, M., Giovine, E., & Belfiore, N. P. (2018). An interdisciplinary approach to the nanomanipulation of SiO2 nanoparticles: Design, fabricationand feasibility. *Applied Sciences (Switzerland), 8*. https://doi.org/10.3390/app8122645
46. Lotey, N. K., Pednekar, S., & Chaughule, R. (2022). Hybrid nanoparticles in biomedical applications. *Nanomaterials for Cancer Detection Using Imaging Techniques and Their Clinical Applications*. https://doi.org/10.1007/978-3-031-09636-5_13
47. van den Hurk, R., & Evoy, S. (2015). A review of membrane-based biosensors for pathogen detection. *Sensors (Switzerland), 15*, 14045–14078.
48. Teixeira, M. C., Carbone, C., Sousa, M. C., Espina, M., Garcia, M. L., Sanchez-Lopez, E., & Souto, E. B. (2020). Nanomedicines for the delivery of antimicrobial peptides (Amps). *Nanomaterials, 10*. https://doi.org/10.3390/nano10030560
49. Dong, S., Kong, G., Qutob, D., Yu, X., Tang, J., Kang, J., Dai, T., Wang, H., Gijzen, M., & Wang, Y. (2012). The NLP toxin family in Phytophthora sojae includes rapidly evolving groups that lack necrosis-inducing activity. *Molecular Plant-Microbe Interactions, 25*, 896. https://doi.org/10.1094/MPMI

50. Hu, J., Tang, F., Wang, L., Tang, M., Jiang, Y. Z., & Liu, C. (2021). Nanozyme sensor based-on platinum-decorated polymer nanosphere for rapid and sensitive detection of Salmonella typhimurium with the naked eye. *Sensors and Actuators B: Chemical, 346*, 130560. https://doi.org/10.1016/j.snb.2021.130560
51. Maciejewska, M., & Józwicki, M. (2023). Porous polymers based on 9,10-Bis(methacryloyloxymethyl)anthracene – Towards synthesis and characterization. *Materials, 16*. https://doi.org/10.3390/ma16072610
52. Jiang, W., Yi, J., Li, X., He, F., Niu, N., & Chen, L. (2022). A comprehensive review on upconversion nanomaterials-based fluorescent sensor for environment, biology. *Food and Medicine Applications. Biosensors (Basel), 12*. https://doi.org/10.3390/bios12111036
53. Yin, M., Jing, C., Li, H., Deng, Q., & Wang, S. (2020). Surface chemistry modified upconversion nanoparticles as fluorescent sensor array for discrimination of foodborne pathogenic bacteria. *Journal of Nanobiotechnology, 18*, 41. https://doi.org/10.1186/s12951-020-00596-4
54. Huang, X., Sheng, W., Chen, H., Zhang, B., Huang, N., & Wang, S. (2022). Upconversion nanoparticles-based fluorescence immunoassay for the sensitive detection of 2-amino-3-methylimidazo [4,5-f] quinoline (Iq) in heat processed meat. *Sensors, 22*. https://doi.org/10.3390/s22010008
55. González-Béjar, M., Francés-Soriano, L., & Pérez-Prieto, J. (2016). Upconversion nanoparticles for bioimaging and regenerative medicine. *Frontiers in Bioengineering and Biotechnology, 4*. https://doi.org/10.3389/fbioe.2016.00047
56. Chen, M., Yan, Z., Han, L., Zhou, D., Wang, Y., Pan, L., & Tu, K. (2022). Upconversion fluorescence nanoprobe-based FRET for the sensitive determination of Shigella, *12*. https://doi.org/10.3390/bios12100795
57. Huang, J., Li, W., Bai, X., Xiao, F., & Xu, H. (2023). Metal–organic framework detectives meet infectious pathogens: A powerful tool against pandemics. *Coordination Chemistry Reviews, 488*, 215160. https://doi.org/10.1016/j.ccr.2023.215160
58. Cheng, W., Tang, X., Zhang, Y., Wu, D., & Yang, W. (2021). Applications of metal-organic framework (MOF)-based sensors for food safety: Enhancing mechanisms and recent advances. *Trends in Food Science and Technology, 112*, 268–282.
59. Sun, Z., Peng, Y., Wang, M., Lin, Y., Jalalah, M., Alsareii, S. A., Harraz, F. A., Yang, J., & Li, G. (2020). Electrochemical deposition of Cu metal-organic framework films for the dual analysis of pathogens. *Analytical Chemistry, 93*, 8994. https://doi.org/10.1021/acs.analchem.1c01763
60. Wang, K., Du, L., Zhang, L., Luo, X., Yang, T., & Zeng, H. (2022). On-site marine pathogen (Vibrio parahaemolyticus) rapid colorimetric determination based on modified-free aptamer and metal-organic frameworks with simple washing step. *Sensors and Actuators B: Chemical, 372*, 132695. https://doi.org/10.1016/j.snb.2022.132695
61. Hitabatuma, A., Wang, P., Su, X., & Ma, M. (2022). Metal-organic frameworks-based sensors for food safety. *Food, 11*. https://doi.org/10.3390/foods11030382
62. Livesey, T. C., Mahmoud, L. A. M., Katsikogianni, M. G., & Nayak, S. (2023). Metal–organic frameworks and their biodegradable composites for controlled delivery of antimicrobial drugs. *Pharmaceutics, 15*. https://doi.org/10.3390/pharmaceutics15010274
63. Wang, R., Xu, Y., Zhang, T., & Jiang, Y. (2015). Rapid and sensitive detection of salmonella typhimurium using aptamer-conjugated carbon dots as fluorescence probe. *Analytical Methods, 7*, 1701–1706.
64. Srivastava, S., Abraham, S., Singh, C., Ali, M. A., Srivastava, A., Sumana, G., & Malhotra, B. D. (2015). Protein conjugated carboxylatedgold@reduced graphene oxide for aflatoxin B1 detection. *RSC Advances, 5*, 5406–5414.
65. Tiwari, I., Singh, M., Pandey, C. M., & Sumana, G. (2015). Electrochemical genosensor based on graphene oxide modified iron oxide-chitosan hybrid nanocomposite for pathogen detection. *Sensors and Actuators B: Chemical, 206*, 276–283.
66. He, L., Liu, Y., Lin, M., Mustapha, A., & Wang, Y. (2008). Detecting single Bacillus spores by surface enhanced Raman spectroscopy. *Sensing and Instrumentation for Food Quality and Safety, 2*, 247–225.

67. Lin, Y. H., Chen, S. H., Chuang, Y. C., Lu, Y. C., Shen, T. Y., Chang, C. A., & Lin, C. S. (2008). Disposable amperometricimmunosensing strips fabricated by Au nanoparticles-modified screen-printed carbon electrodes for the detection of foodborne pathogen Escherichia coli O157:H7. *Biosensors and Bioelectronics, 23*, 1832–1837.
68. Hong, S. A., Kwon, J., Kim, D., & Yang, S. (2015). A rapid, sensitive and selective electrochemical biosensor with concanavalin A for the preemptive detection of norovirus. *Biosensors and Bioelectronics, 64*, 338–344.
69. Davis, D., Guo, X., Musavi, L., Lin, C. S., Chen, S. H., & Wu, V. C. H. (2013). Gold nanoparticle-modified carbon electrode biosensor for the detection of listeria monocytogenes. *Industrial Biotechnology, 9*, 31–36.
70. Zhang, H., Ma, X., Liu, Y., Duan, N., Wu, S., Wang, Z., & Xu, B. (2015). Gold nanoparticles enhanced SERS aptasensor for the simultaneous detection of Salmonella typhimurium and Staphylococcus aureus. *Biosensors and Bioelectronics, 74*, 872–877.
71. Xia, S., Yu, Z., Liu, D., Xu, C., & Lai, W. (2016). Developing a novel immunochromatographic test strip with gold magnetic bifunctional nanobeads (GMBN) for efficient detection of Salmonella choleraesuis in milk. *Food Control, 59*, 507–512.
72. Kaittanis, C., Naser, S. A., & Perez, J. M. (2007). One-step, nanoparticle-mediated bacterial detection with magnetic relaxation. *Nano Letters, 7*, 380–383.
73. Fornara, A., Johansson, P., Petersson, K., Gustafsson, S., Jian, Q., Olsson, E., Iiver, D., Krozer, A., Muhammed, M., & Johansson, C. (2008). Tailored magnetic nanoparticles for direct and sensitive detection of biomolecules in biological samples. *Nano Letters, 8*, 3423–3428.
74. El-Boubbou, K., Gruden, C., & Huang, X. (2007). Magnetic glyco-nanoparticles: A unique tool for rapid pathogen detection, decontamination, and strain differentiation. *Journal of the American Chemical Society, 129*, 13392–13393.
75. Wang, D. B., Tian, B., Zhang, Z. P., Wang, X. Y., Fleming, J., Bi, L. J., Yang, R. F., & Zhang, X. E. (2015). Detection of Bacillus anthracis spores by super-paramagnetic lateral-flow immunoassays based on "Road Closure". *Biosensors and Bioelectronics, 67*, 608–614.
76. Suaifan, G. A. R. Y., Alhogail, S., & Zourob, M. (2017). Paper-based magnetic nanoparticle-peptide probe for rapid and quantitative colorimetric detection of Escherichia coli O157:H7. *Biosensors and Bioelectronics, 92*, 702–708.
77. Liu, H., Dong, H., Chen, Z., Lin, L., Chen, H., Li, S., & Deng, Y. (2017). Magnetic nanoparticles enhanced microarray detection of multiple foodborne pathogens. *Journal of Biomedical Nanotechnology, 13*, 1333–1343.
78. Wang, R., Ruan, C., Kanayeva, D., Lassiter, K., & Li, Y. (2008). TiO 2 nanowire bundle microelectrode based impedance immunosensor for rapid and sensitive detection of Listeria monocytogenes. *Nano Letters, 8*, 2625–2631.
79. Wang, Y., Ravindranath, S., & Irudayaraj, J. (2011). Separation and detection of multiple pathogens in a food matrix by magnetic SERS nanoprobes. *Analytical and Bioanalytical Chemistry, 399*, 1271–1278.
80. Castillo, G., Spinella, K., Poturnayová, A., Šnejdárková, M., Mosiello, L., & Hianik, T. (2015). Detection of aflatoxin B1 by aptamer-based biosensor using PAMAM dendrimers as immobilization platform. *Food Control, 52*, 9–18.
81. Sheikhzadeh, E., Chamsaz, M., Turner, A. P. F., Jager, E. W. H., & Beni, V. (2016). Label-free impedimetric biosensor for Salmonella Typhimurium detection based on poly [pyrrole-co-3-carboxyl-pyrrole] copolymer supported aptamer. *Biosensors and Bioelectronics, 80*, 194–200.
82. Muhammad-Tahir, Z., & Alocilja, E. C. (2003). Fabrication of a disposable biosensor for Escherichia coli O157:H7 detection. *IEEE Sensors Journal, 3*, 345–351.
83. Muhammad-Tahir, Z., & Alocilja, E. C. (2004). A disposable biosensor for pathogen detection in fresh produce samples. *Biosystems Engineering, 88*, 145–151.
84. Mathew, F. P., & Alocilja, E. C. (2005). Porous silicon-based biosensor for pathogen detection. In *Biosensors and bioelectronics* (pp. 1656–1661). Elsevier.

85. Nunez-Bajo, E., Silva, A., Collins, P., et al. *Disposable silicon-based all-in-one micro-qPCR for rapid on-site detection of pathogens*. https://doi.org/10.1038/s41467-020-19911-6
86. Kara Yamada, Chong-Tai Kim, Jong-Hoon Kim, Jae-Hyun Chung, Hyeon Gyu Lee, Soojin Jun (2014) Single walled carbon nanotube-based junction biosensor for detection of Escherichia coli. PLoS One 9:do:https://doi.org/10.1371/journal.pone.0105767, e105767.
87. Lavilla, M., Moros, M., Puertas, S., Grazú, V., Pérez, M. D., Calvo, M., de la Fuente, J. M., & Sánchez, L. (2012). Specific peptides as alternative to antibody ligands for biomagnetic separation of Clostridium tyrobutyricum spores. *Analytical and Bioanalytical Chemistry, 402*, 3219–3226. Welch, E. C., Powell, J. M., Clevinger, T. B., Fairman, A. E., Shukla, A. (2021). Advances in biosensors and diagnostic technologies using nanostructures and nanomaterials. *Advanced Functional Materials, 31*(44): 2104126.
88. Kerman, K., Endo, T., Tsukamoto, M., Chikae, M., Takamura, Y., & Tamiya, E. (2007). Quantum dot-based immunosensor for the detection of prostate-specific antigen using fluorescence microscopy. *Talanta, 71*(4), 1494–1499.
89. Ahuja, T., & Kumar, D. (2009). Recent progress in the development of nano-structured conducting polymers/nanocomposites for sensor applications. *Sensors and Actuators B: Chemical, 136*(1), 275–286.
90. Mani, V., Chikkaveeraiah, B. V., Patel, V., Gutkind, J. S., & Rusling, J. F. (2009). Ultrasensitive immunosensor for cancer biomarker proteins using gold nanoparticle film electrodes and multienzyme-particle amplification. *ACS Nano, 3*(3), 585–594.
91. Tetyana, P., Shumbula, P. M., & Njengele-Tetyana, Z. (2021). Biosensors: Design, development and applications. In *Nanopores*. IntechOpen.
92. Sin, M., Mack, K. E., Wong, P. K., & Liao, J. C. (2014). Advances and challenges in biosensor-based diagnosis of infectious diseases. *Expert Review of Molecular Diagnostics, 14*(2), 225–244.
93. Thakur, M. S., & Ragavan, K. V. (2013). Biosensors in food processing. *Journal of Food Science and Technology, 50*, 625–641.
94. Sin, M. L., Mach, K. E., Wong, P. K., & Liao, J. C. (2014). Advances and challenges in biosensor-based diagnosis of infectious diseases. *Expert Review of Molecular Diagnostics, 14*(2), 225–244.
95. Kabiraz, M. P., Majumdar, P. R., Mahmud, M. C., Bhowmik, S., & Ali, A. (2023). Conventional and advanced detection techniques of foodborne pathogens: A comprehensive review. *Heliyon, 9*, e15482.
96. Duan, L. (2016). Janeway's immunobiology. *The Yale Journal of Biology and Medicine, 89*(3), 424.
97. Finlay, B. B., & McFadden, G. (2006). Anti-immunology: Evasion of the host immune system by bacterial and viral pathogens. *Cell, 124*(4), 767–782.
98. Singh, S., Tank, N. K., Dwiwedi, P., Charan, J., Kaur, R., Sidhu, P., & Chugh, V. K. (2018). Monoclonal antibodies: A review. *Current Clinical Pharmacology, 13*(2), 85–99.
99. Nakazawa, M., Mukumoto, M., & Miyatake, K. (2010). Production and purification of polyclonal antibodies. *Immunoelectron Microscopy: Methods and Protocols*, 63–74.
100. Sabherwal, P., Shorie, M., Pathania, P., Chaudhary, S., Bhasin, K. K., Bhalla, V., & Suri, C. R. (2014). Hybrid aptamer-antibody linked fluorescence resonance energy transfer-based detection of trinitrotoluene. *Analytical Chemistry, 86*(15), 7200–7204.
101. Gong, F., Wei, H. X., Li, Q., Liu, L., & Li, B. (2021). Evaluation and comparison of serological methods for COVID-19 diagnosis. *Frontiers in Molecular Biosciences, 8*, 682405.
102. Fodey, T., Leonard, P., O'Mahony, J., O'Kennedy, R., & Danaher, M. (2011). Developments in the production of biological and synthetic binders for immunoassay and sensor-based detection of small molecules. *TrAC Trends in Analytical Chemistry, 30*(2), 254–269.
103. Mutreja, R., Kumar, P., Semwal, A., Jain, S., Dhyani, R., Agarwal, R., Chand, U., Baba, S. A., Navani, N. K., & Kumar, P. (2019). Aptamer: A futuristic approach in diagnosis rivaling antibodies. *Aptamers: Biotechnological Applications of a Next Generation Tool*, 37–57.

104. Yi, J., Wu, P., Li, G., Xiao, W., Li, L., He, Y., He, Y., Ding, P., & Chen, C. (2019). A composite prepared from carboxymethyl chitosan and aptamer-modified gold nanoparticles for the colorimetric determination of Salmonella typhimurium. *Microchimica Acta, 186*, 1–8.
105. Srinivasan, S., Ranganathan, V., DeRosa, M. C., & Murari, B. M. (2018). Label-free aptasensors based on fluorescent screening assays for the detection of Salmonella typhimurium. *Analytical Biochemistry, 559*, 17–23.
106. Tuerk, C., & Gold, L. (1990). Systematic evolution of ligands by exponential enrichment: RNA ligands to bacteriophage T4 DNA polymerase. *Science, 249*(4968), 505–510.
107. Pathania, P., Sharma, A., Kumar, B., Rishi, P., & Raman, S. C. (2017). Selective identification of specific aptamers for the detection of non-typhoidal salmonellosis in an apta-impedimetric sensing format. *Microchimica Acta, 184*, 1499–1508.
108. Reid, R., Chatterjee, B., Das, S. J., Ghosh, S., & Sharma, T. K. (2020). Application of aptamers as molecular recognition elements in lateral flow assays. *Analytical Biochemistry, 593*, 113574.
109. Sefah, K., Shangguan, D., Xiong, X., O'Donoghue, M. B., & Tan, W. (2010). Development of DNA aptamers using cell-SELEX. *Nature Protocols, 5*(6), 1169–1185.
110. Cella, L. N., Sanchez, P., Zhong, W., Myung, N. V., Chen, W., & Mulchandani, A. (2010). Nano aptasensor for protective antigen toxin of anthrax. *Analytical Chemistry, 82*(5), 2042–2047.
111. Fraser, L. A., Cheung, Y. W., Kinghorn, A. B., Guo, W., Shiu, S. C., Jinata, C., Liu, M., Bhuyan, S., Nan, L., Shum, H. C., & Tanner, J. A. (2019). Microfluidic technology for nucleic acid aptamer evolution and application. *Advanced Biosystems, 3*(5), 1900012.
112. Mayer, G., Ahmed, M. S., Dolf, A., Endl, E., Knolle, P. A., & Famulok, M. (2010). Fluorescence-activated cell sorting for aptamer SELEX with cell mixtures. *Nature Protocols, 5*(12), 1993–2004.
113. Shorie, M., Bhalla, V., Pathania, P., & Suri, C. R. (2014). Nanobioprobe mediated DNA aptamers for explosive detection. *Chemical Communications, 50*(9), 1080–1082.
114. Wang, M., Wang, Q., Li, X., Lu, L., Du, S., & Zhang, H. (2020). Selection and identification of diethylstilbestrol-specific aptamers based on magnetic-bead SELEX. *Microchemical Journal, 159*, 105354.
115. Wu, C. H., Liu, I. J., Lu, R. M., & Wu, H. C. (2016). Advancement and applications of peptide phage display technology in biomedical science. *Journal of Biomedical Science, 23*(1), 1–4.
116. Fairlie, W. D., Spurck, T. P., McCoubrie, J. E., Gilson, P. R., Miller, S. K., McFadden, G. I., Malby, R., Crabb, B. S., & Hodder, A. N. (2008). Inhibition of malaria parasite development by a cyclic peptide that targets the vital parasite protein SERA5. *Infection and Immunity, 76*(9), 4332–4344.
117. Mukai, Y., Okamoto, T., Kawamura, M., Shibata, H., Sugita, T., Imai, S., Abe, Y., Nagano, K., Nomura, T., Kamada, H., & Tsutsumi, Y. (2006). Optimization of anti-tumor necrosis factor-α single chain Fv displayed on phages for creation of functional antibodies. *Die Pharmazie-An International Journal of Pharmaceutical Sciences, 61*(10), 889–890.
118. Anand, T., Virmani, N., Bera, B. C., Vaid, R. K., Vashisth, M., Bardajatya, P., Kumar, A., & Tripathi, B. N. (2021). Phage display technique as a tool for diagnosis and antibody selection for coronaviruses. *Current Microbiology, 78*(4), 1124–1134.
119. Petrenko, V. A. (2008). Evolution of phage display: From bioactive peptides to bioselective nanomaterials. *Expert Opinion on Drug Delivery, 5*(8), 825–836.
120. Bazin, I., Tria, S. A., Hayat, A., & Marty, J. L. (2017). New biorecognition molecules in biosensors for the detection of toxins. *Biosensors and Bioelectronics, 87*, 285–298.
121. Rico-Yuste, A., & Carrasco, S. (2019). Molecularly imprinted polymer-based hybrid materials for the development of optical sensors. *Polymers, 11*(7), 1173.
122. Wang, O., Jia, X., Liu, J., Sun, M., & Wu, J. (2022). Rapid and simple preparation of an MXene/polypyrrole-based bacteria imprinted sensor for ultrasensitive Salmonella detection. *Journal of Electroanalytical Chemistry, 918*, 116513.

123. de Dieu, H. J., Ji, J., Pi, F., Karangwa, E., Sun, J., Guo, W., Cui, F., Shao, J., Ntakirutimana, C., & Sun, X. (2018). A class-specific artificial receptor-based on molecularly imprinted polymer-coated quantum dot centers for the detection of signaling molecules, N-acyl-homoserine lactones present in gram-negative bacteria. *Analytica Chimica Acta, 1031*, 134–144.
124. Perçin, I., Idil, N., Bakhshpour, M., Yılmaz, E., Mattiasson, B., & Denizli, A. (2017). Microcontact imprinted plasmonic nanosensors: Powerful tools in the detection of Salmonella paratyphi. *Sensors, 17*(6), 1375.
125. Wang, Y., Wang, M., Yu, H., Wang, G., Ma, P., Pang, S., Jiao, Y., & Liu, A. (2022). Screening of peptide selectively recognizing prostate-specific antigen and its application in detecting total prostate-specific antigen. *Sensors and Actuators B: Chemical, 367*, 132009.
126. Pejcic, B., & De Marco, R. (2006). Impedance spectroscopy: Over 35 years of electrochemical sensor optimization. *Electrochimica Acta, 51*(28), 6217–6229.
127. Heimlich, H. J., & Patrick, E. A. (1990). The Heimlich maneuver: Best technique for saving any choking victim's life. *Postgraduate Medicine, 87*(6), 38–53.
128. Grieshaber, D., MacKenzie, R., Vörös, J., & Reimhult, E. (2008). Electrochemical biosensors-sensor principles and architectures. *Sensors, 8*(3), 1400–1458.
129. Nayak, P., Kurra, N., Xia, C., & Alshareef, H. N. (2016). Highly efficient laser scribed graphene electrodes for on-chip electrochemical sensing applications. *Advanced Electronic Materials, 2*(10), 1600185.
130. Hassanain, W. A., Spoors, J., Johnson, C. L., Faulds, K., Keegan, N., & Graham, D. (2021). Rapid ultra-sensitive diagnosis of clostridium difficile infection using a SERS-based lateral flow assay. *The Analyst, 146*(14), 4495–4505.
131. Zuo, P., Li, X., Dominguez, D. C., & Ye, B. C. (2013). A PDMS/paper/glass hybrid microfluidic biochip integrated with aptamer-functionalized graphene oxide nano-biosensors for one-step multiplexed pathogen detection. *Lab on a Chip, 13*(19), 3921–3928.
132. Ligler, F. S., & Taitt, C. R. (Eds.). (2002). *Optical biosensors: Present & future*. Gulf Professional Publishing.
133. Damborský, P., Švitel, J., & Katrlík, J. (2016). Optical biosensors. *Essays in Biochemistry, 60*(1), 91–100.
134. Masdor, N. A., Altintas, Z., & Tothill, I. E. (2016). Sensitive detection of Campylobacter jejuni using nanoparticles enhanced QCM sensor. *Biosensors and Bioelectronics, 78*, 328–336.
135. He, D., Wu, Z., Cui, B., Xu, E., & Jin, Z. (2019). Establishment of a dual mode immunochromatographic assay for Campylobacter jejuni detection. *Food Chemistry, 289*, 708–713.
136. Shen, M., Duan, N., Wu, S., Zou, Y., & Wang, Z. (2019). Polydimethylsiloxane gold nanoparticle composite film as structure for aptamer-based detection of vibrio parahaemolyticus by surface-enhanced Raman spectroscopy. *Food Analytical Methods, 12*, 595–603.
137. Elahi, N., Kamali, M., Baghersad, M. H., & Amini, B. (2019). A fluorescence nano-biosensors immobilization on iron (MNPs) and gold (AuNPs) nanoparticles for detection of *Shigella* spp. *Materials Science & Engineering: C, 105*, 110113.
138. Gao, R., Zhong, Z. T., Gao, X. M., & Jia, L. (2018). Graphene oxide quantum dots assisted construction of fluorescent aptasensor for rapid detection of Pseudomonas aeruginosa in food samples. *Journal of Agricultural and Food Chemistry, 66*(41), 10898–10905.
139. Ranjbar, S., Shahrokhian, S., & Nurmohammadi, F. (2018). Nanoporous gold as a suitable substrate for preparation of a new sensitive electrochemical aptasensor for detection of Salmonella typhimurium. *Sensors and Actuators B: Chemical, 255*, 1536–1544.
140. Kamal, Z., Ghobadi, M. Z., Mohseni, S. M., & Ghourchian, H. (2021). High-performance porphyrin-like graphene quantum dots for immuno-sensing of Salmonella typhi. *Biosensors and Bioelectronics, 188*, 113334.
141. Liu, D. Q., Li, T. C., Huang, W. C., Ma, Z., Zhang, W., Zhang, R., et al. (2019). Electrochemiluminescent detection of Escherichia coli O157:H7 based on Ru(bpy)3 2+/ZnO nanorod arrays. *Nanotechnology, 30*(2), 025501.

142. Benserhir, Y., Salaün, A. C., Dutertre, S., Loréal, O., Geneste, F., Pichon, L., & Jolivet-Gougeon, A. (2020). *Development of electronic nano-sensors for the specific in situ detection of Escherichia coli*. European Congress of Clinical Microbiology & Infectious Diseases.
143. Kaushik, S., Tiwari, U. K., Pal, S. S., & Sinha, R. K. (2019). Rapid detection of Escherichia coli using fiber optic surface plasmon resonance immunosensor based on biofunctionalized Molybdenum disulfide (MoS2) nanosheets. *Biosensors and Bioelectronics, 126*, 501–509.
144. Singh, P., Gupta, R., Sinha, M., Kumar, R., & Bhalla, V. (2016). MoS_2 based digital response platform for aptamer based fluorescent detection of pathogens. *Microchimica Acta, 183*, 1501–1506.
145. Ge, R., Lin, X., Dai, H., Wei, J., Jiao, T., Chen, Q., Oyama, M., Chen, Q., & Chen, X. (2022). Photoelectrochemical sensors with near-infrared-responsive reduced graphene oxide and MoS2 for quantification of Escherichia coli O157: H7. *ACS Applied Materials & Interfaces, 14*(36), 41649–41658.
146. Shen, W., Wang, C., Zheng, S., Jiang, B., Li, J., Pang, Y., Wang, C., Hao, R., & Xiao, R. (2022). Ultrasensitive multichannel immunochromatographic assay for rapid detection of foodborne bacteria based on two-dimensional film-like SERS labels. *Journal of Hazardous Materials, 437*, 129347.
147. Guo, Y., Li, J., Islam, M. S., Yan, T., Zhou, Y., Liang, L., Connerton, I. F., Deng, K., & Li, J. (2021). Application of a novel phage vB_SalS-LPSTLL for the biological control of Salmonella in foods. *Food Research International, 147*, 110492.
148. Gupta, A., Garg, M., Singh, S., Deep, A., & Sharma, A. L. (2020). Highly sensitive optical detection of Escherichia coli using terbium-based metal–organic framework. *ACS Applied Materials & Interfaces, 12*(42), 48198–48205.

Chapter 9
Dental Microbial Biofilms: Control and Treatment Through Nanotechnology Approaches

Heeresh Shetty, Varsha Shukla, Anushree S. Lokur, and Ramesh S. Chaughule

9.1 Introduction

Bacterial infections have been a persistent challenge to human life since its existence. Throughout history, numerous attempts have been made to control these infections, but the effectiveness of these measures has been limited. Traditional medicines from ancient Chinese, Roman, and Egyptian cultures used herbs, copper, and mercury salts to treat infections with little success and many side effects. In the late eighteenth century, there was a shift towards using probiotics – beneficial bacteria – to combat infections. Elie Metchnikoff's work with lactic acid bacteria earned him a Nobel Prize in 1908 for his ground-breaking research in enhancing the health and longevity of Bulgarian peasant populations. However, the development of probiotics was halted because they could not match the efficacy of penicillin, discovered by Alexander Fleming in 1923, who also received a Nobel Prize in the field of infection control. Since then, many new antibiotics have been developed and have been widely used to combat biofilm-associated bacterial infections across the human body, making antibiotics the primary solution for several decades [1]. The overuse of antibiotics, not only in the prevention and treatment of human infections but also in veterinary medicine and farming, has resulted in the acquisition of mutations that have led to the development of bacterial strains and species that are

H. Shetty (✉)
Department of Conservative Dentistry and Endodontics, Nair Hospital Dental College, Mumbai, India

V. Shukla · A. S. Lokur
Department of Microbiology, Ramnarain Ruia Autonomous College, Mumbai, India

R. S. Chaughule
Department of Chemistry, Ramnarain Ruia Autonomous College, Mumbai, India

R. S. Chaughule, A. S. Lokur (eds.), *Applications of Nanotechnology in Microbiology*, https://doi.org/10.1007/978-3-031-49933-3_9

intrinsically resistant to all known antibiotics. Despite the urgent need for new antibiotics to combat these resistant strains, the development of such antibiotics is hindered by the high costs involved and regulatory scrutiny. Moreover, the efficacy of new antibiotics is often short-lived, as bacteria can rapidly adapt and become resistant to them. Therefore, the long-term sustainability of antibiotic use as a solution to bacterial infections remains questionable [2]. The challenge of intrinsic antimicrobial resistance is not the only obstacle in the effective control of bacterial infections. As far back as 1684, Van Leeuwenhoek observed that the vinegar he used to rinse his teeth only killed the bacteria on the outer layer of dental plaque, highlighting a persistent issue in infection control [3]. The main issue in the effective control of bacterial infections is the inadequate penetration of antimicrobial agents into the "syntrophic consortium of microorganisms embedded within an extracellular matrix," which we commonly refer to as biofilms, which are the primary mode of growth for most bacteria causing infections. Therefore, the development of effective strategies to penetrate biofilms and treat infections caused by these complex microbial communities remains a critical area of research in infection control.

The current methods of controlling biofilms rely heavily on chemical biocides, antiseptic solutions, and antibiotics, which produce harmful by-products and are limited in their efficacy against resistant and slow-growing bacteria. Moreover, biofilms are notorious hot spots for the emergence of resistant mutants, genetic transfer of mobile elements, and antimicrobial tolerance.

Recently, nanotechnology and nanomedicine have emerged as promising tools for the diagnosis and treatment of various diseases. Nanomaterials possess unique surface properties and dimensions that enable them to overcome physiological barriers. Nanoparticles (NPs) have been extensively studied in the last decade as drug carriers in several different fields, including oncology, immunotherapy, neuroscience and dentistry.

The application of nanotechnology to the prevention and eradication of oral biofilms is an emerging area of research that holds great promise. Nanoparticles (NPs) have several advantages over conventional approaches in biofilm control, such as targeted delivery, enhanced antimicrobial activity, and reduced toxicity [4, 5].

The unique surface properties and dimensions of NPs enable them to penetrate the complex biofilm matrix, making them suitable carriers for antimicrobial agents. Furthermore, NPs can be engineered to deliver multiple payloads simultaneously, including drugs and imaging agents, providing a platform for simultaneous diagnosis and treatment of oral biofilms [6].

The use of NPs in controlling oral biofilms has the potential to revolutionize the way we approach oral hygiene and dentistry. Targeted delivery of antimicrobial agents could significantly improve their efficacy while reducing their adverse effects on host tissues and the environment. Additionally, the ability to simultaneously diagnose and treat oral biofilms could lead to more efficient and cost-effective treatment options.

This chapter provides a comprehensive overview of microorganisms living as oral biofilms, their formation, and characteristics, followed by an appraisal of the strategies used to treat bacterial biofilm communities and their associated drawbacks. Furthermore, alternative therapies and effective methods for delivering antibiotics into the biofilm will be discussed, along with the means for localized antibacterial action at a surface. Overall, this chapter emphasizes the importance of developing novel and effective strategies for treating biofilms in the oral cavity, given their propensity to cause dental caries, periodontal disease, and other oral health issues. Understanding the specificities of oral biofilms is crucial to develop effective treatment options that can overcome the inherent resistance and persistence of these microbial communities.

9.2 The Impact of Oral Biofilm: A Comprehensive Guide

9.2.1 The Human Oral Microbiome

The term microbiome was coined by the Human Microbiome Project (HMP). Scholars in the field maintain that a comprehensive understanding of human health and disease is impossible without a full comprehension of the integrated microbiome/human system [7].

The oral microbiome encompasses a vast number of microorganisms that inhabit the oral cavity. These microbes comprise predominantly bacteria, which can be classified as either commensal or pathogenic, along with fungi, viruses, and yeast. Notable examples of microbial species that reside in the oral cavity include *Streptococcus* spp. and *Corynebacteria* spp. [8] (Fig. 9.1).

The presence of these living organisms plays a crucial role in maintaining oral hygiene by impeding the attachment of pathogenic microorganisms to the mucosal surface [9]. In scientific literature, microorganisms residing within the human oral

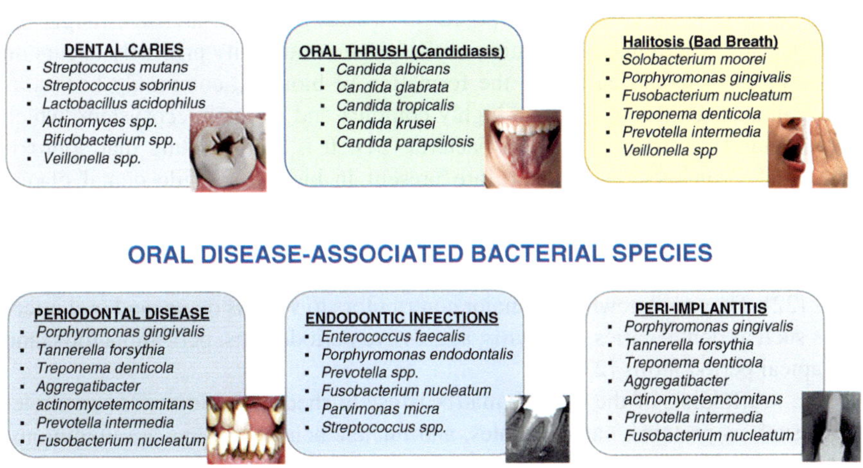

Fig. 9.1 Bacteria related to oral diseases

cavity have been variably denoted as the oral microflora, oral microbiota, or more recently as the oral microbiome. The term microbiome was originally introduced by Joshua Lederberg to describe the "ecological community" of commensal, symbiotic, and pathogenic microorganisms that literally share our body space and have been all but ignored as determinants of health and disease [10]. The human oral cavity comprises various distinct microbial habitats, such as teeth, gingival sulcus, attached gingiva, tongue, cheek, lip, hard palate, and soft palate. In addition, the tonsils, pharynx, oesophagus, Eustachian tube, middle ear, trachea, lungs, nasal passages, and sinuses are contiguous with the oral cavity. The human oral microbiome is defined as the collection of microorganisms that inhabit the human oral cavity and its contiguous extensions, terminating at the distal oesophagus. Research has demonstrated that distinct microbial communities colonize different oral structures and tissues [11, 12].

The human oral cavity serves as a major gateway to the body, where food is ingested and mixed with saliva for digestion, and air passes through the nose and mouth to reach the lungs via the trachea. As contiguous epithelial surfaces are present, microorganisms from one area of the oral cavity can easily spread to neighbouring sites. Consequently, the oral cavity harbours a range of microorganisms that can lead to various infectious diseases such as caries (tooth decay), periodontitis (gum disease), endodontic (root canal) infections, alveolar osteitis (dry socket), and tonsillitis [13]. Recent studies have provided evidence that oral bacteria are associated with several systemic diseases such as cardiovascular disease, stroke, preterm birth, diabetes, and pneumonia [14–18].

9.2.2 Composition of the Oral Microbiome

The human oral cavity harbours one of the most diverse microbiomes in the body [19]. The oral microbiome is composed of viruses, protozoa, archaea, fungi, and bacteria [20]. Microbial colonization of the human oral cavity primarily occurs on the surfaces of teeth, leading to the formation of biofilms, commonly known as dental plaque. Dental plaque is a highly complex and dynamic ecosystem, which predominantly consists of biofilm bacteria [21]. It is worth noting that approximately 95% of bacteria in nature are present in biofilms. While dental plaque research has been ongoing for more than a century, it is only recently that it has been recognized as a microbiological ecosystem and a biofilm. Biofilms are organized communities of microbes that form self-made patterns of extracellular polysaccharides [22]. They are known to be major contributors to various dental and oral infections such as dental caries, gingivitis, aggressive periodontitis, peri-implantitis, and periapical periodontitis [23].

The scaffolding of the biofilm matrix is established by biological macromolecules such as proteins, carbohydrates, and nucleic acids. Biofilms are ubiquitous, and they can be found on almost all moist surfaces, including the oral cavity and

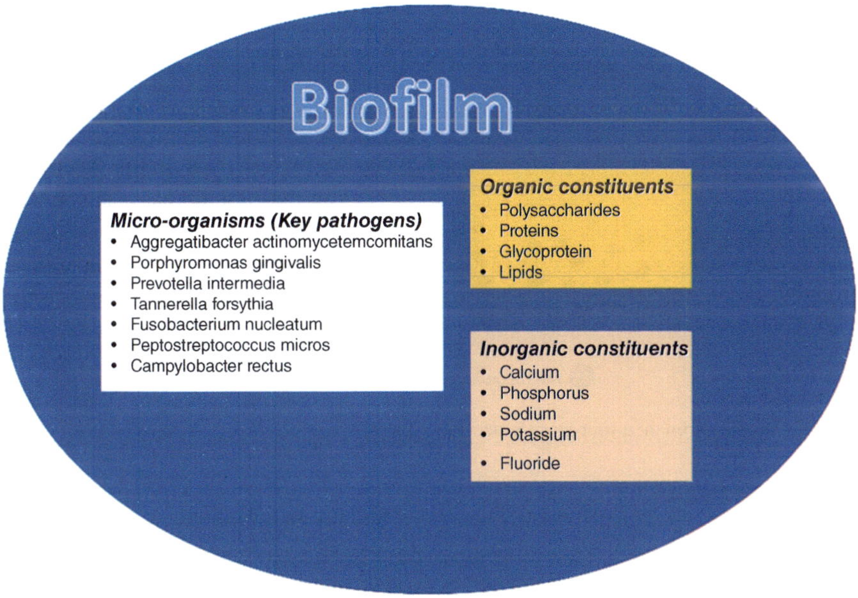

Fig. 9.2 Composition of biofilm

natural wet environments such as ponds and lakes. Since oral biofilms are primarily propagated in the mouth using saliva ingredients as the initial source, food components are quickly cleared [24]. The complex microbial network works in a synchronized manner to acquire nutrients, sugars, and amino acids from salivary ingredients, including epithelial mucin, through the activity of glycosidic enzymes such as sialidase, N-acetylglucosaminidases, β-galactosidase, mannosidases, α-fucosidase, exoproteolytic, and endo-proteolytic activities [25] (Fig. 9.2). There is a cross-feeding relationship among the species within the biofilm, which allows for the exchange of metabolites and other important nutrients. Oral biofilms can grow under various conditions and environments. In vitro studies have shown that glycans decomposition occurs sequentially, and the liberated sugars are rapidly transported [26]. *S. oralis* is an example of a bacterium that has a high capability to deglycosylate both N- and O-linked glycans and is considered a model microorganism [27].

9.2.3 Formation of Biofilm

The process of biofilm formation follows a four-stage sequence, as illustrated in Fig. 9.3. Firstly, bacterial adhesion to a surface occurs, followed by the formation of microcolonies. Next, the biofilm matures and, lastly, cells detach from the biofilm to establish new colonies elsewhere.

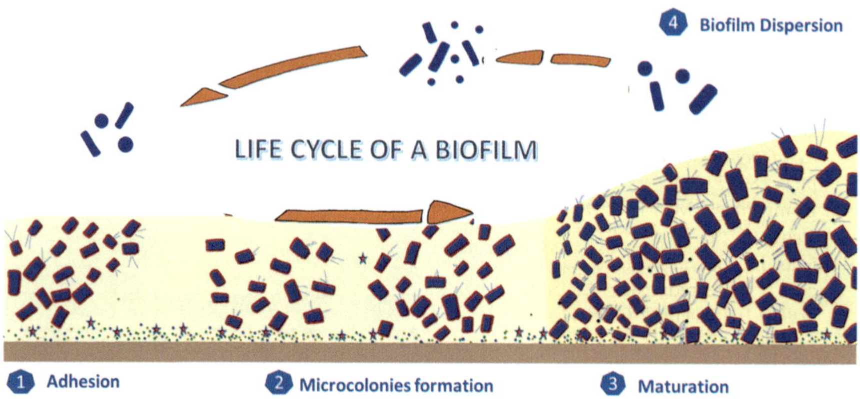

Fig. 9.3 Biofilm growth: a four-stage process

9.2.3.1 Adhesion

Cell attachment to a substrate is a crucial step in biofilm formation, facilitated by various physical, chemical, and biological interactions [28]. The adhesion process is significantly influenced by the physicochemical properties of the bacterial cell surface, as well as the chemical composition and physical properties of the substrate [29]. The forces responsible for bacterial adhesion include Lifshitz-van der Waals, Lewis acid-base, and electrostatic interactions [30, 31]. The environmental conditions, such as texture, surface charge, pH, temperature, nutrient composition, and organic matter, also affect the adhesion of cells to abiotic surfaces [32–35]. The conditioning layer, consisting of organic and inorganic particles, plays a significant role in biofilm development, as it alters the substratum and provides anchorage and nutrients for bacterial growth [36]. The conversion from reversible to irreversible attachment results from weak interactions between bacteria and the substrate that transform into permanent bonding with the extracellular polymeric substances (EPS) [37]. The irreversible attachment makes biofilms challenging to remove using shear forces or chemical breakdown, as enzymes, surfactants, detergents, sanitizers, or heat may be required [38–40]. Some biofilm-forming species require motility for growth or cell attachment to a substrate, while others use biofilm-associated proteins (Bap) or microbial surface components recognizing adhesive matrix molecules (MSCRAMM) to mediate initial cell attachment to an abiotic or biotic surface via ionic or hydrophobic interactions [41, 42].

9.2.3.2 Microcolony Formation and Biofilm Maturation

During microcolony formation and biofilm formation, cells begin to replicate and adhere irreversibly to surfaces due to the secretion of EPS that constitutes the biofilm matrix [43]. The EPS enables the attachment of bacteria to a surface and to each other [44]. Biofilms mature to form an enclosed biofilm structure where the cells grow in a sessile form in heterogeneous complex-enclosed microcolonies, contained in an EPS matrix, with an efficient network of water channels. This matrix forms a scaffold to stabilize this 3D biofilm structure [45].

*P. aeruginosa and S. aureus*biofilms develop a mushroom-like structure comprising of microcolonies in the stalk and cap, with structural integrity provided by Pel and Psl polysaccharides and eDNA [46, 47]. In *S. aureus* biofilms, the polymer poly-N-acetyl-β-(1-6)-glucosamine (PNAG) (also known as polysaccharide intercellular adhesin (PIA)) is predominant in the *S.aureus* EPS matrix and plays an important role in *S. aureus* biofilm morphology. A weak PNAG-producing *S.aureus* strain will have a simple morphology, similar to that of a young biofilm, whereas a strong PNAG-producing *S.aureus* strain will form a compact biofilm with mushroom-like colonies with channels. PNAG facilitates biofilm formation and helps in providing resistance to antimicrobials [48].

9.2.3.3 Dispersion

Biofilms are complex and dynamic structures that consist of sessile cells that can detach from the biofilm matrix and revert back to their planktonic state through a regulated or passive dispersion process triggered by environmental factors such as nutrient availability, oxygen depletion, fluid shear, and abrasion [49]. This dispersion process is driven by the expression of genes necessary for detachment, such as those involved in endogenous enzymatic degradation, extracellular polymeric substance (EPS) release, and surface binding proteins release [50].

One example of this process can be observed in *Paeruginosa* biofilms, where a rapid increase in nutrients such as succinate, glutamate, and glucose can trigger a significant reduction in surface-associated biofilm biomass. This reduction in nutrient levels can lead to increased flagella expression and decreased pilus expression, allowing for twitching motility and dispersal of the biofilm cells [51]. Similarly, in *S aureus* biofilms, depletion of glucose levels can reactivate the agr QS system, leading to matrix-degrading enzyme production and surfactant production, ultimately resulting in biofilm breakdown and dispersal [52].

Detachment and dispersal of biofilm cells enable colonization when nutrients are limited, allowing bacteria to survive and expand their community. In *P. aeruginosa*, detachment is triggered in the centre of the caps, leading to the formation of a fluid-filled cavity within the cap, which enables planktonic cells to be released and form new biofilms elsewhere [53].

Overall, biofilm dispersal is a complex and regulated process that allows bacteria to adapt and survive in changing environments by re-entering the planktonic state and colonizing new areas.

9.2.4 Qurom Sensing

Bacteria were once thought to be capable of only simple processes and single-celled life, but they are now recognized for their ability to function collectively in multicellular groups [54]. This coordinated behaviour includes activities such as bioluminescence, virulence factor production, secondary metabolite production, competence for DNA uptake, and biofilm formation [55]. These processes cannot be carried out effectively by a single bacterium acting alone; instead, success requires population-wide coordination of individual cells. To achieve this, bacteria use quorum-sensing, a cell-to-cell communication process that involves the production, release, accumulation, and group-wide detection of extracellular signalling molecules known as autoinducers [56].

Gram-negative quorum-sensing bacteria use small molecules as autoinducers, which are detected by cytoplasmic transcription factors or transmembrane two-component histidine sensor kinases [57]. Gram-positive bacteria, on the other hand, typically use oligopeptides as autoinducers, which are detected by transmembrane two-component histidine sensor kinases [58].Autoinducer-receptor complexes direct the expression of quorum-sensing-dependent target genes, often including the gene encoding the autoinducer synthase, which increases the extracellular autoinducer concentration as the bacteria enter into quorum-sensing mode. This feed-forward autoinduction loop synchronizes behaviours across the bacterial population [59].

Bacteria integrate information encoded in several quorum-sensing autoinducers to control gene expression, enabling communication not only within a species and genus but also between species and with bacteria in the microbiota [57].

Apart from the autoinduction loop described above, quorum-sensing circuits also often contain multiple feedbacks and feedforward regulatory loops that precisely modulate the response. These loops can adjust the input-output range and dynamics, reduce noise, and commit cells to the individual or group lifestyle programme [60–62]. Quorum-sensing circuits can interact with global regulators, such as the alternative sigma factor RpoN, RNA-binding proteins Hfq and CsrA, and the nucleoid protein Fis, to further refine the control of quorum-sensing-dependent gene expression [63–65].

Upon reaching a critical threshold of group or colony size, this process triggers a cascade of diverse cellular responses, such as the synthesis of various enzymes, mucus, competence for genetic exchange, reproduction, antibiotic production, and the emission of a sequence of signals [56] (Fig. 9.4). This discovery of bacterial quorum-sensing has revealed the sophisticated social lives of these microorganisms and has broad implications for fields ranging from medicine to ecology.

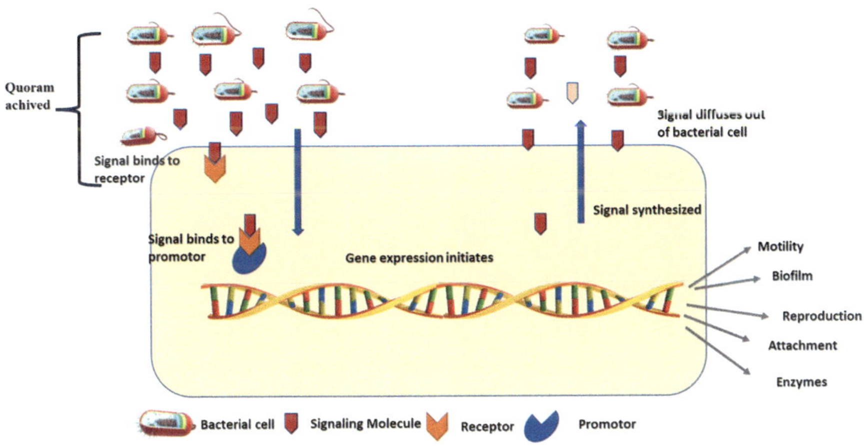

Fig. 9.4 General mechanism of quorum-sensing in bacteria

9.2.5 Clinical Implications of Biofilm

"The oral biofilm" represents an intricate microbial community within the human body, characterized by its complexity. It has been estimated that over 700 species contribute to the formation of dental plaque biofilm, which is categorized into colour-coded complexes based on their sequential colonization and impact on oral health. Notably, oral diseases such as dental caries, gingivitis, periodontitis, and peri-implantitis are commonly associated with biofilm formation. Periodontitis characterized by chronic and destructive inflammation of the tissues, results in the degradation of tooth attachment apparatus and can lead to tooth loss and complete edentulousness in severe cases [66].

The initiation of a dental biofilm occurs in stagnant areas of the teeth where bacteria find protection, such as occlusal fissures, interproximal spaces between adjacent teeth, and along the gingival margin. If left undisturbed due to insufficient dental hygiene practices, the supragingival biofilm gradually extends along the tooth root into the periodontal pocket, forming a subgingival biofilm. Biofilms on tooth surfaces contribute to dental caries, while supra- and subgingival biofilms along and beneath the gingival margin are implicated in the development of periodontal diseases. Moreover, the placement of dental implants may lead to the development of peri-implantitis [67].

Microbial plaque or biofilm forms on both hard and soft tissues, initially supragingivally and then subgingivally. This subgingival biofilm predominantly consists of bacterial pathogens from the orange and red complexes, which are strongly associated with periodontitis. Examples of such pathogens include *Aggregatibacter actinomycetemcomitans*, *Porphyromonas gingivalis,* and *Treponema denticola.* The formation of this subgingival biofilm, accompanied by a significant bacterial load in proximity to blood vessels, combined with inflammation and the release of inflammatory mediators within the periodontium, is linked to various chronic diseases. Notably, oral health is implicated in the aetiology of diverse conditions, including cardiovascular diseases such as endocarditis and stroke, as well as systemic inflammatory diseases like rheumatoid arthritis [68].

9.3 Dental Plaque as Biofilm

Dental plaque, an intricate biofilm comprised of diverse microbial species, exhibits adherence to both living and non-living surfaces. Recent advancements in microscopic and molecular techniques have enabled the exploration of dental plaque as a biofilm, confirming its characteristic features [69]. The development of this microbial community involves intricate interactions of cooperation and competition among a highly diverse range of organisms.

Defined as a matrix of bacterial and salivary polymers, dental plaque forms a coating on the tooth surface [70]. When plaque becomes calcified, it is known as calculus or tartar. Within this microbial community, a wide array of Gram-positive and Gram-negative bacteria coexist, encompassing facultatively anaerobic and obligately anaerobic species. The composition of plaque exhibits variation across distinct sites on the tooth surface, influenced by specific local biological properties [71].

Plaque bacteria primarily rely on endogenous nutrients present in saliva, specifically proteins and glycoproteins, to fuel their metabolic activities. However, the ecology of the gingival crevice, a critical factor in periodontal disease, is predominantly shaped by the characteristics of gingival crevicular fluid (GCF). GCF, a fluid resembling serum, envelops the tooth root, and its flow increases during periods of inflammation associated with periodontal disease [72]. This heightened flow of GCF not only provides components of the host's immune defences, such as IgG and neutrophils, but also supplies a broader range of potential nutrients for bacterial growth, including peptides, proteins, and glycoproteins. These endogenous nutrients serve as nourishment for numerous bacteria that do not ferment sugars. As a result, the gingival crevice harbours a higher proportion of Gram-negative species, particularly obligately anaerobic bacteria, notably in sites affected by periodontal disease [73]

9.3.1 Formation of Dental Plaque Biofilms

"Dental biofilm" formation initiates immediately after tooth surface cleaning, as an organic film known as the pellicle adsorbs and serves as ligands for bacterial receptors. Pioneer bacteria establish direct contact with the pellicle surface, while later colonizing bacteria adhere either to the already attached bacteria through co-adhesion or to different components of the extracellular matrix, as described previously in this chapter. The early colonizing microbiota is predominantly comprised of *Streptococcal* spp. If the dental biofilm remains undisturbed, it undergoes microbial succession and maturation, leading to a shift in bacterial composition from gram-positive cocci dominance to a microbiota containing cocci, filamentous organisms, spirals, spirochetes, and increased numbers of Gram-negative bacteria. This acquisition of bacteria and maturation of the biofilm microbiota contribute to the development of gingivitis.

Coaggregation, the adhesion of genetically distinct bacteria to each other, plays a crucial role in dental biofilm formation and is a widespread phenomenon among oral bacteria. Coaggregation mediates the colonization of late colonizers, which may not directly coaggregate with early colonizing species but can coaggregate with strains of the Gram-negative bacterium, *Fusobacterium nucleatum. F. nucleatum* has been reported to coaggregate with both late and early colonizers of dental hard tissues.

Bacteria within the biofilm may detach through either a passive or an active process mediated by the bacteria themselves. This dispersal from the biofilm enables bacteria to colonize new sites within the oral cavity [74]. The close proximity of bacteria in the oral biofilm allows for synergistic and antagonistic interactions. Nutritional cooperation is a significant feature of oral biofilms, leading to the formation of food chains and nutritional webs. Bacteria cooperate in metabolizing salivary molecules, and the metabolic capacity of a bacterium has recently been proposed as a determining factor for its spatial and temporal positioning within the oral biofilm [74]. Upon initial surface adhesion, bacteria start producing an extracellular matrix composed of polysaccharides, lipids, proteins, and extracellular DNA. This matrix is crucial for biofilm architecture, integrity, and its characteristic properties, including enhanced tolerance to antimicrobial compounds. "The significance of polysaccharides in the formation of oral biofilms and their association with dental caries has been acknowledged since early research." Bacterial glucosyltransferases, which produce extracellular polysaccharides in both soluble and insoluble forms, have been linked to the adhesion of *S. mutans* to tooth surfaces and the development of caries. It is believed that glucosyltransferases play a role in glycosylating proteins that are crucial for adhesion and the formation of biofilms [75].

9.3.2 Dynamics of Dental Plaque Biofilm Growth and Host Inhibition

The growth of dental plaque biofilm is primarily governed by bacterial growth, although attachment capacity plays a key role in the diversity of microbial species present. Plaque doubling times are fast during early development and slow down as the biofilm matures. Growth rates in mixed populations are complex and do not follow those observed in suspended cultures. This is consistent with the structure of bacterial biofilms, which contain areas of high and low bacterial biomass interlaced with aqueous channels of different size, which provide nutrients and facilitate the movement of metabolic waste products within the colony (Fig. 9.3).

Physiological cooperation within biofilms may play a role in the bacterial blooms observed in periodontal plaque. Bacterial blooms, where specific species or groups of species grow at rapidly accelerated rates, have been reported in vitro and are believed to occur in vivo. Gingival crevicular fluid is the main nutritional component in this ecosystem, accounting for the predominance of *Asaccharolytic* spp. Interspecies cooperation in the metabolism of alpha-gingipain in digested proteins has also been suggested. Although this cysteine protease is considered an important virulence factor for *P. gingivalis*, it may have evolved simply to provide nutrition in this environment [76].

Supragingival plaque growth is subject to much more intraoral abrasion and the flow characteristic of saliva, which restricts its net accumulation. Saliva contains a range of host defence components, including secretory immunoglobulin A (IgA), lactoferrin, lysozyme, and peroxidases, which display a wide spectrum of antimicrobial activity and serve to limit both colonization and spread of the supragingival bacterial biofilm. Saliva also contains antimicrobial proteins, such as histatins, which have antifungal as well as antibacterial activity. These components work synergistically to limit bacterial growth [77].

In contrast, subgingival plaque, situated in a more protected location, is not subjected to the same level of intraoral abrasion or the presence of salivary host defence components. The growth of subgingival plaque is primarily limited by physical space and the innate defence mechanisms of the host. Gingival crevicular fluid serves as a plentiful source of nutrients, suggesting that nutritional constraints do not significantly impede bacterial biofilm growth. However, the subgingival space available for bacterial colonization is restricted in periodontally healthy individuals. As subgingival plaque accumulates, there is a continual expansion of available space through reduced epithelial cell attachment levels and increased pocket depth. The intact epithelial cell barrier maintained by the host's innate defence system limits the spread of subgingival plaque [77].

Studies involving patients with leukocyte adhesion deficiencies have confirmed the crucial role of these cells in exiting the vascular compartment, entering gingival tissue, and eliminating potential microbial invaders. Recently, the molecular components responsible for the controlled migration of these cells from the vasculature have been identified, including low levels of E-selectin and intracellular adhesion molecule. Additionally, the presence of monocyte chemotaxis protein 1 (MCP-1) and interleukin 8 (IL-8) has been detected. Interestingly, IL-8 exhibits a gradient of expression, with the highest levels observed in the area closest to the bacterial-epithelial cell interface, gradually decreasing deeper into the periodontal tissue [78].

9.3.3 Role of Dental Plaque Biofilm in Oral Health

The dental plaque biofilm is a complex microbial community primarily consisting of commensal and non-pathogenic microorganisms in a healthy state. Contrary to their seemingly inactive nature, commensal members actively engage in a continuous cross-talk with host tissues, such as the gingiva. This harmonious relationship between commensal bacteria and the host is mutually beneficial, as the host provides a surface for colonization while the bacteria offer "colonization resistance" against harmful pathogens [79]. This symbiotic association becomes particularly apparent when considering the disruption of normal flora due to conditions like antibiotic-induced oral lesions, which can lead to the overgrowth of opportunistic pathogens. Research suggests that specific commensal bacterial species, including *Veillonella*, *S. salivarius*, *S. sanguinis*, and *Atopobiumparvulum*, may serve as indicators of a healthy biofilm, although further investigations are required for confirmation. Intriguingly, *S. salivarius* has demonstrated the ability to inhibit quorum-sensing and biofilm formation of *S. mutans*, providing supporting evidence for its protective role against dental caries [80].

Furthermore, commensal bacteria within the dental plaque contribute significantly to the development of a well-balanced immune system by continually exposing the host's innate immune system to a diverse range of bacterial antigens. Commensals initiate signalling cascades that promote tolerance, whereas pathogenic bacteria elicit robust inflammatory responses from the host. Consequently, oral epithelial cells exhibit a low-level production of pro-inflammatory cytokines, leading to the expression of E-selectin in vascular endothelial tissues and the absence of an interleukin-8 chemokine gradient. As a result, commensal bacteria activate a host innate immune response that strategically positions neutrophils alongside subgingival plaque bacteria and the junctional epithelium [81].

9.3.3.1 The Role of Dental Plaque in Dental Caries: Insights and Challenges

The ecological plaque hypothesis posits that changes in the local environment surrounding the dental plaque biofilm contribute to the development of dental diseases, including dental caries and periodontal diseases [79]. Increased consumption of dietary sugars provides an opportunity for acidogenic and aciduric bacteria within the dental plaque biofilm, such as *S. mutans* and L.*acidophilus*, to create a persistent acidic environment [82]. This shift in balance towards demineralization of the tooth surface is facilitated by the biofilm mode of growth, which exhibits greater tolerance to acidic stress compared to planktonic bacteria [83]. Mature *S. mutans* biofilms down-regulate the glycolytic pathway to enhance acid tolerance. Phenotypic changes have been observed in the dental plaque biofilm community during both health and disease states. Additionally, variations in sensitivity to host antimicrobial peptides and differential gene expression related to glucan and fructan production have been documented between planktonic and biofilm bacteria. These ecological changes within the dental plaque biofilm may contribute to the development of dental caries [79]. However, it is important to note that the presence of a single species, such as *S.mutans*, may not be sufficient for caries initiation. Multiple cariogenic species, including *S.mutans*, *S. mitis*, *Rothia*, *Actinomyces*, *Lactobacillus bifidobacterium*, and even fungal species like *Candida*, could collectively contribute to the cariogenic potential of the biofilm. Recent studies have identified *Candida albicans* as a significant contributor to occlusal caries, forming structures with streptococcal species within the supragingival plaque [84]. Similarly, the discovery of *Scardoviawiggsia*e as a potential cariogenic bacterium challenges the traditional notion of a single species being solely responsible for dental caries [85]. Advancements in understanding the molecular microbiology of dental plaque biofilm have led to clinical benefits, such as the use of xylitol to selectively inhibit the growth of *S. mutans*. However, there is ongoing debate regarding the effectiveness of xylitol and sorbitol in cariogenic biofilm models [86]. The use of probiotic bacteria, such as *Lactobacillu*s *rhamnosus* LB21, as a milk supplement has also been proposed, although its efficacy in controlling the colonization of cariogenic bacteria in adolescents remains uncertain [87]. Other strategies, including the use of proteases produced by early dental plaque biofilm colonizers to inhibit *S. mutans* colonization, have been explored. Despite these advances, the complete understanding of the dental plaque biofilm remains elusive, posing a significant challenge in addressing the global epidemic of dental caries.

9.3.3.2 Dental Plaque Biofilm in Periodontal Disease: Insights into Microbial Interactions and Pathogenesis

Periodontal disease is characterized by inflammation of the periodontal tissues in response to Gram-negative pathogenic bacteria, including *P. gingivalis* and spirochetes such as *T. denticola*, present in dental plaque biofilms [88]. The ecological plaque hypothesis suggests that inflammation of the periodontal tissues leads to an increased secretion of gingival crevicular fluid, resulting in a rise in local pH above the neutral value. Even a minor pH increase provides a favourable environment for periodontopathic bacteria like *P. gingivalis* to dominate the dental plaque, overriding other microorganisms [89]. *P. gingivalis*, a hemin-dependent bacterium, acquires hemin from gingival crevicular fluid, aided by secretory protease/hemagglutinins [90]. The rise in local hemin concentration during periodontitis enhances the competitive advantage of *P. gingivalis* and other red-complex bacteria over commensal species. Intriguingly, *P. gingivalis* can alter its lipopolysaccharide (LPS) structure, shifting from penta-acylated to tetra-acylated lipid A structures depending on the hemin concentration [91]. This LPS structural change allows *P. gingivalis* to evade the human immune system by "paralysing" the local cytokine network, facilitating invasion of the gingival tissue [92]. Additionally, other virulence factors, such as type IV fimbriae, contribute to periodontal disease. Notably, smoking has been found to enhance the activity of these virulent factors and alter the microbial composition of dental plaque biofilm, favouring colonization by periodontal pathogens [93]. Studies examining the in vivo development of dental plaque have shown differences in microbiota between healthy and periodontally diseased individuals, with rapid plaque formation observed in the latter group [94]. While the association of red-complex bacteria with periodontitis is well-established, the specific roles and mechanisms of each bacterium in pathogenesis remain incompletely understood. Moreover, emerging evidence suggests the involvement of other bacterial species, such as *Selenomonas,* complicating our understanding of periodontal disease pathogenesis [95].

9.3.4 Conventional Approaches for Eliminating Dental Plaque: Mechanisms and Limitations

9.3.4.1 Mechanical Plaque Control

Since the 1960s, when Loe and his colleagues established the significant role of dental plaque in causing gingivitis, mechanical plaque control has remained the cornerstone of periodontal therapy [96]. Mechanical removal involves the scaling and polishing of teeth, either above or below the gum line, or a combination of both approaches [97]. However, despite extensive efforts in mechanical plaque control, the prevalence of gingivitis suggests that relying solely on mechanical means is not entirely effective. This challenge has prompted scientists and clinicians to explore chemical antimicrobial agents that can help hinder biofilm formation on tooth surfaces [98].

9.3.4.2 Challenges of Mechanical Plaque Control

Mechanical plaque control techniques, such as brushing, have limitations when it comes to accessing interproximal plaque in pre-molars and molars, necessitating the use of additional tools like dental floss [99]. Research has shown that microorganisms responsible for gingivitis and periodontitis can accumulate on various soft tissue surfaces in the oral cavity, serving as a source for bacterial colonization on teeth [100]. Another constraint of mechanical plaque control procedures is their focus solely on the hard surfaces of the mouth [101]. To address these limitations, agents with antiplaque and antigingivitis properties can be incorporated into dentifrices or mouth rinses. Utilizing these agents in dentifrices is particularly advantageous, as they are typically used in conjunction with toothbrushing for routine oral hygiene [102]. By including chemical antiplaque agents in mouth rinses or dentifrices, it becomes possible to reach the soft tissue surfaces, improving biofilm control and delaying microbial accumulation on teeth [103].

9.3.4.3 Antiplaque and Antigingivitis Agents

The incorporation of antibacterial agents, such as the use of a dentifrice containing triclosan/copolymer and chlorhexidine, in oral care products has shown significant improvements in gingival health, preventing the onset of periodontitis and reducing further tissue destruction [104]. Conventional dentifrices (toothpastes) are complex combinations of abrasives, surfactants, and humectants that enhance the physical removal of plaque bacteria from tooth surfaces. The addition of chemical agents to dentifrices has demonstrated its effectiveness, as seen with the inclusion of fluoride, which has played a crucial role in reducing dental caries [105].

Triclosan: Triclosan is a phenolic agent with broad-spectrum antibacterial activity, effective against both Gram-positive and Gram-negative bacteria, and has low toxicity [106]. While triclosan can be formulated in conventional dentifrices, its main limitation is its inability to bind to oral surfaces for an extended period, resulting in a lack of sustained antiplaque activity [107]. The efficacy of triclosan/copolymer dentifrice in reducing plaque and improving gingival health has been evaluated in numerous randomized, controlled clinical trials, with results summarized in various reviews [108, 109]. The benefits of triclosan/copolymer dentifrice were initially attributed solely to the antibacterial activity of triclosan. However, several studies have shown that triclosan may also exert a significant anti-inflammatory effect [106].

Chlorhexidine: Chlorhexidine, a cationic agent with both hydrophilic and hydrophobic properties, is a leading antimicrobial agent used to combat supragingival and mucosal plaques [110]. Its effectiveness is attributed to the di-cationic nature of chlorhexidine, which enables reversible binding with negatively charged molecules in the oral cavity, resulting in prolonged activity after application. However, long-term use of chlorhexidine is limited due to its tendency to stain teeth, resulting from interactions between chlorhexidine on tooth surfaces and tannins in the diet [111]. The ultrastructural changes in oral biofilm caused by chlorhexidine treatment have clinical significance. Studies have shown that these alterations affect only a small portion of the oral biofilm and do not cause its complete disintegration [112]. These findings suggest that chlorhexidine may be insufficient in effectively combating oral biofilm.

9.4 Biofilm Formation in Endodontic Infections: Insights into Microbial Diversity and Ecological Niches

The oral cavity harbours a diverse array of bacterial species, with advanced DNA sequencing techniques revealing an even greater number than previously identified. Within the field of endodontics, the microbial diversity associated with various forms of apical periodontitis has been extensively investigated [113]. These studies have demonstrated the presence of over 400 different microbial species, primarily belonging to the *Phylum Firmicutes, Bacteroidetes, Actinobacteria, Fusobacteria, Proteobacteria, Spirochaetes, and Synergistetes* [114]. In addition to bacteria, occasional findings of archaea and fungi have been reported in intraradicular infections, with fungi being more prevalent in root canal treated teeth with post-treatment complications [115–117].

The microbial composition of root canal infections undergoes significant changes as the disease progresses. Initially, the root canal is colonized by aerobes and facultative anaerobes, but the introduction of treatment procedures, root canal irrigants, and materials can alter the ecological conditions, resulting in phenotypic shifts driven by genetic population changes [118]. Endodontic infections can be categorized as primary or secondary, with primary infections associated with pulp inflammation and subsequent invasion by microbial species, leading to apical periodontitis. Secondary infections, including reinfection, remnant infection, or recurrent infection, occur in previously treated teeth [119]. Primary infections are polymicrobial, predominantly involving *Bacteroides, Porphyromonas, Prevotella, Fusobacterium, Treponema, Peptostreptococcus, Eubacterium, and Campylobacter* spp. [120].

Persistent microbial colonization within the root canal system after treatment is considered a major cause of treatment failure [121]. The microbial composition in primary infections may differ after root canal treatment, with a shift in species propagation and abundance. Gram-positive bacteria dominate secondary infections, which are characterized by their ability to survive harsh conditions such as a wide pH range and nutrient limitations [122]. Specific species such as *E. faecalis, Streptococci, Lactobacilli, Actinomyces, and Candida* have been associated with post-treatment apical periodontitis [123, 124]. Mixed infections are more prevalent than single-organism isolates, with variations in microbial profiles attributed to differences in investigative approaches and culture techniques [125, 126].

Biofilm formation plays a crucial role in the pathogenesis of endodontic infections. Organisms initially penetrate the pulp, attach to the root canal surfaces, and form biofilms, which can subsequently lead to pulpal tissue destruction. These biofilms can be located within the root canal, on the root surface near the root apex, or at the bacteria/inflammatory interface zone [127]. The presence of anatomical complexities, such as deltas and isthmuses, provides protected niches that shield adherent bacteria from mechanical cleaning procedures [128]. Biofilm formation may occur in various sites within the root canal system, facilitated by host-derived proteins and bacterial adhesive substances.

9.4.1 Endodontic Biofilms

Various categories of biofilms are found in endodontic infections. These include intracanal microbial biofilms, extraradicular biofilms, periapical biofilms, and biomaterial-centred biofilms.

1. *Intracanal Microbial Biofilms*: Intracanal biofilms are formed on the dentin of the infected tooth's root canal. Previous studies identified these biofilms using transmission electron microscopy. The biofilms primarily consist of loose collections of filaments, spirochetes, cocci, and rods. Similar to dental plaque found on tooth surfaces, bacterial condensations in the form of palisade structures were observed. The extracellular matrix, derived from bacterial origin, was also detected within these biofilms [114].
2. *Extraradicular Biofilms*: Extraradicular biofilms develop on the root surface adjacent to the apex of endodontically infected teeth. These biofilms exhibited structure less smooth biofilms with varying bacterial species and degrees of extracellular matrix using scanning electron microscopy. Polymerase chain reaction (PCR)-based 16s rRNA gene assay identified *F. nucleatum, P. gingivalis,* and *Tannerella forsythea* as associated species with extraradicular biofilms [120].
3. *Periapical Biofilms*: Periapical microbial biofilms are isolated biofilms found in the periapical region of endodontically infected teeth, which can manifest even in the absence of root canal infections. *Actinomyces* spp. and *Propionibacterium propionicus* are implicated in periapical lesions when these bacteria overcome host defence mechanisms. The aggregation of Actinomyces cells, facilitated by factors such as pH, ionic strength, and cell concentration, contributes to biofilm formation [123].
4. *Foreign Body-Centred Biofilms*: Foreign body-centred biofilms occur when bacteria adhere to artificial biomaterial surfaces, often associated with prosthesis and implant-supported prosthesis complications. These infections are characterized by opportunistic invasion by nosocomial organisms. Gram-positive facultative anaerobes colonize and form an extracellular polymeric matrix surrounding Gutta-Percha, with the serum playing a significant role in biofilm formation. Extraradicular microbial biofilms and biomaterial-centred biofilms have been associated with refractory periapical disease [123].

Understanding the various types of biofilms encountered in endodontic infections is crucial for developing effective treatment strategies. Intraradicular and extraradicular biofilms can form on root canal obturating materials, depending on whether the obturating material remains within the root canal space or extends beyond the root apex. Different bacterial species exhibit varying abilities to form biofilms on Gutta-Percha (GP) points, with *E. faecalis*, Str. *sanguinis*, *Streptococcus intermedius*, *Streptococcuspyogenes*, and *S. aureus* demonstrate biofilm formation, while *F. nucleatum*, *Propionibacterium acnes*, *Po. gingivalis* and *Pr. intermedia* do not form biofilm on Gutta-Percha (GP) points.

9.4.2 Counteracting Endodontic Biofilms

Endodontic biofilms pose a formidable challenge in clinical practice due to their complex structure and inherent resistance to antimicrobial agents. The effective management of endodontic biofilms requires targeted strategies aimed at disrupting their architecture and eliminating the bacteria within. Various approaches have been investigated for countering endodontic biofilms, including mechanical debridement, antimicrobial irrigants, biofilm-modifying agents, advanced disinfection techniques, and biofilm-targeting nanoparticles [129].

Mechanical debridement techniques, such as instrumentation with rotary and hand files, aim to physically remove the biofilm from the root canal system. However, the intricate anatomy of root canals often hinders complete biofilm removal, necessitating adjunctive measures [129]. Antimicrobial irrigants, such as sodium hypochlorite (NaOCl) and chlorhexidine (CHX), are commonly employed to complement mechanical debridement [130]. NaOCl exhibits potent tissue-dissolving properties and antimicrobial activity, while CHX offers broad-spectrum antimicrobial action and sustained substantivity [131].

To enhance the efficacy of antimicrobial irrigants, biofilm-modifying agents can be employed. These agents, such as ethylene diaminetetra acetic acid (EDTA), serve to disrupt the biofilm matrix, facilitating the penetration of antimicrobials into the biofilm structure [132]. Additionally, advanced disinfection techniques, including laser irradiation and photoactivated disinfection, have shown promise in targeting endodontic biofilms. These methods employ light energy to selectively destroy biofilm bacteria and alter the biofilm matrix, enhancing the effectiveness of antimicrobial agents [133].

Furthermore, the use of nanoparticles specifically designed to target biofilms has gained attention. Nanoparticles can possess unique properties that enable them to penetrate biofilms, interact with bacterial cells, and disrupt the biofilm architecture. Silver nanoparticles, for instance, have demonstrated antimicrobial efficacy against endodontic biofilms, while chitosan nanoparticles exhibit biofilm-disrupting and antibacterial activities [134, 135]. A comprehensive examination of the role of nanoparticles in addressing biofilms will be provided in the subsequent section of this chapter, offering a detailed analysis of their potential and applications.

Overall, countering endodontic biofilms requires a multifaceted approach that combines mechanical debridement, antimicrobial irrigants, biofilm-modifying agents, advanced disinfection techniques, and targeted nanoparticles. These strategies, when employed in a synergistic manner, hold promise for effectively combating the resilience of endodontic biofilms and improving the success of endodontic treatment.

9.5 Biofilm-Enabled Antibiotic Resistance

Biofilms, consisting primarily of water and solutes, allow for the relatively easy diffusion of antibiotics into the biofilm matrix [136]. However, effectively treating biofilm infections requires the use of antibiotics that are both sensitive to the target bacteria and capable of penetrating deep into the biofilm structure [137]. In the context of oral health care, antimicrobial-containing products designed to control plaque biofilms must meet regulatory guidelines by delivering clinically and microbiologically significant benefits while preserving the natural microbial ecology of the oral cavity [138].

The formation of an exopolysaccharide matrix, or glycocalyx, is a defining characteristic of biofilms and serves multiple functions, including impeding the access of antibiotics to the bacterial cells within the community [139]. Although antibiotics possess physical mobility within biofilms, this alone does not guarantee their penetration into the depths of the biofilm. Various factors, such as chemical reactions or sequestration through binding, can significantly hinder the delivery of antibiotics as they diffuse into the biofilm [140].

Currently, antibiotic treatment represents the primary and most effective non-mechanical approach for controlling microbial infections in the oral cavity. However, completely eradicating biofilm infections with antibiotics alone proves to be exceedingly challenging [141]. An essential objective in oral infection treatment is the discovery of anti-infective agents specifically targeting microorganisms embedded within biofilms, which often exhibit heightened resistance to antibiotics and conventional treatment approaches [142].

9.5.1 Mechanisms of Drug Resistance in Biofilms

The formation of biofilms invariably results in a substantial increase in resistance to antimicrobial agents, including commonly used antibiotics such as amoxicillin, doxycycline, and metronidazole. This heightened resistance is observed not only in dental products like toothpaste and mouth rinses containing chlorhexidine but also in a wide range of antimicrobial agents [143]. Resistance in biofilms can arise through various mechanisms, including the production of inactivating enzymes, which have been found to be present in relatively high quantities [143]. Additionally, cells within biofilms can acquire resistance through mutations affecting drug targets, the presence of efflux pumps that actively remove antibiotics from cells, or the production of modifying enzymes, among other factors.

The structure of a biofilm itself can impede the penetration of antimicrobial agents. Some charged inhibitors may bind to oppositely charged polymers that constitute the biofilm matrix, thereby reducing their effectiveness in reaching bacterial cells [144]. The resistance exhibited by biofilms can be attributed to a combination of several mechanisms working in concert.

The following section provides a detailed discussion of these mechanisms and their contribution to drug resistance in biofilms.

9.5.1.1 Impaired Cell Permeability of Antimicrobial Agents in Biofilm Structures

The permeability of antimicrobial agents in biofilms is influenced by various mechanisms, including factors such as molecular size of the drug (e.g., chlorhexidine), hydrophobicity of the antimicrobial, electrostatic interactions, absorption, interaction with biofilm components, and neutralization by enzymes [139]. Studies have shown that certain antimicrobials, like ampicillin, can readily penetrate biofilms formed by beta-lactamase negative mutants [139]. Conversely, the binding of positively charged aminoglycosides to negatively charged biofilm matrix polymers has been observed to retard the penetration of these drugs [140].

The bacterial exopolysaccharide matrix serves as a significant barrier for drug molecules, influencing the rate of their transport to the deeper layers of the biofilm [142]. For antimicrobial drugs to effectively inactivate bacterial cells, they must diffuse through the biofilm matrix [145]. However, the resistance provided by the matrix can be overcome by longer exposure times or higher concentrations of antimicrobial agents [146]. These approaches may compensate for the reduced permeability and enhance the efficacy of antimicrobials against biofilm-encased bacteria.

9.5.1.2 Efflux Pump-Mediated Multidrug Resistance

Efflux pumps play a significant role in the resistance of biofilm-embedded bacteria to antimicrobial penetration [144]. These pumps are integral membrane proteins that utilize metabolic energy to actively expel drugs across the bacterial membrane against the concentration gradient. This efflux system is commonly referred to as a multidrug resistance (MDR) transporter. The MDR transporter serves a vital function for bacteria, enabling their survival in complex environments and facilitating biofilm formation [142].

Efflux pumps contribute to antimicrobial resistance by efficiently removing antimicrobial agents from the bacterial cell, thereby reducing their intracellular concentration and effectiveness. This mechanism allows biofilm-embedded bacteria to evade the action of antimicrobial drugs, leading to persistent infections and treatment challenges.

The presence and activity of efflux pumps in biofilm-associated bacteria highlight the adaptive nature of these microbial communities and their ability to develop sophisticated defence mechanisms. Understanding the role of efflux pumps in biofilms is crucial for the development of strategies to combat antimicrobial resistance and improve the effectiveness of antimicrobial therapies targeting biofilm infections.

9.5.1.3 Quorum-Sensing Molecules: Regulators of Biofilm Formation and Virulence

As mentioned earlier, many bacteria engage in cell-cell communication, and the formation of biofilms is heavily reliant on quorum-sensing activity. Quorum-sensing refers to a system that enables bacteria to sense the presence of a critical concentration of bacteria in a confined space and respond by activating specific genes that contribute to the production of enzymes or toxins [145]. Quorum-sensing molecules serve as crucial regulators in both bacteria and fungi. They allow bacteria to coordinate gene expression in response to their population density, serving as a decision-making process that controls the production of virulence factors and subsequent infections [141].

Quorum-sensing represents an intricate mechanism by which bacteria regulate their behaviour collectively. By monitoring their population density, bacteria can adjust their gene expression patterns to ensure optimal coordination of virulence factors and other essential processes within a biofilm community. Understanding the role of quorum-sensing in biofilm formation and its impact on the pathogenicity of bacteria is vital for developing strategies to disrupt quorum-sensing signalling and potentially mitigate biofilm-associated infections.

9.5.1.4 Dynamic Alterations in Outer Membrane Protein Profile

In the presence of antimicrobial agents, biofilm cells may undergo alterations in their membrane protein composition, leading to a reduction in cell permeability. Porin channels, which are utilized by several antibiotics to penetrate Gram-negative bacteria, play a crucial role in this process. Consequently, decreased expression of porins results in antibiotic resistance in Gram-negative bacteria [142]. The modulation of membrane protein composition represents a significant adaptive response of biofilm cells to counter the detrimental effects of antimicrobial agents. By reducing the expression of porins, biofilm cells can limit the entry of antibiotics into the bacterial cell, effectively minimizing their antimicrobial effects. This alteration in membrane protein composition and subsequent decrease in permeability contribute to the development of antibiotic resistance in Gram-negative bacteria.

Strategies aimed at targeting these alterations in membrane protein expression offer valuable avenues for enhancing the overall effectiveness of antimicrobial therapies.

9.5.1.5 Role of Slow Growth and Stress Response in Biofilms

Bacterial populations residing in non-growing states within biofilms exhibit enhanced survival capabilities and reduced susceptibility to antimicrobial challenges compared to biofilms consisting of uniformly growing bacteria at an intermediate rate [140]. Notably, mature biofilms often harbour bacteria exhibiting slow growth rates. This phenomenon is attributed to the presence of nutrient limitations within the biofilm, which has been proposed as a physiological change contributing to the inherent resistance of biofilms against antimicrobial agents [139].

However, it is important to note that the slow growth rate of certain cells within the biofilm is not solely a result of nutrient limitation. Rather, it is primarily driven by a general stress response triggered by the growth within a biofilm community [139]. Bacterial cells employ a stress response strategy to protect themselves from the detrimental effects of environmental changes, including alterations in pH and exposure to various chemical agents [142].

The combination of slow growth and stress response mechanisms confers significant advantages to bacteria residing within biofilms, enabling them to withstand antimicrobial challenges and enhance their survival. Understanding the complex interplay between slow growth, stress response, and antimicrobial resistance in biofilms is critical for developing effective strategies to combat biofilm-associated infections.

9.6 Approaches in Targeting Biofilms

As previously discussed, the application of mechanical methods alone for biofilm removal presents significant limitations in the treatment of antibiotic-resistant oral bacterial infections. To address this challenge, various anti-biofilm strategies have been investigated, and their summary is presented in Table 9.1. The treatment of biofilm-associated diseases can be broadly categorized into two main groups: restrictive and regenerative methods [147].

Table 9.1 Strategies for combating biofilm formation: a comprehensive overview of diverse anti-biofilm approaches

Anti-biofilm stategies	Method	Counteraction effect
Enhancement of salivary flow postprandially [144].	Sugar-free chewing gums	Inhibition effect encompass host defences, augmented buffering capacity, removal of fermentable substrates, and facilitation of remineralization processes
Suppression of Plaque Acid Production [144].	Microbial fermentation of dietary sugars and acidogenic activity	Reduces pH within the dental biofilm
Bacteriophage-based therapies [148]	Viruses targeting Bacteria	Expeditious eradication of the bacterial cells
Enhancedantimicrobial activity through electrical stimulation [143]	Electricity-Based Biological Response	Preventingbiofilm formation and improving antimicrobial activity
Enhancement of antimicrobial transport through ultrasound [149]	Ultrasound	Enhanced oxygen and nutrient delivery to cellular components
Enzymes for biofilm disruption [150]	Deoxyribonuclease I, Lysostaphin, α-Amylases	Biofilm structure disruption
Advancements in the design of inhibitors and antiplaque agents [144].	Mouthwashes, Gels	Augmented retention of agents
Photodynamic therapy (PDT) [151]	Photodynamic Agents	Elimination of biofilm structures
Innovative Approaches [152]	Nanotechnology	Extended release of agents and biofilm disruption

9.6.1 Enhancement of Salivary Flow Postprandially

Stimulation of salivary flow plays a vital role in oral health as it aids in the introduction of host defence components, enhances buffering capacity, eliminates fermentable substrates, promotes remineralization, and expedites the restoration of plaque pH to resting levels [144]. Among the notable approaches that leverage this principle, sugar-free gums serve as exemplary examples.

9.6.2 Suppression of Plaque Acid Production

This approach focuses on attenuating the production of acid resulting from microbial fermentation of dietary sugars and the subsequent decrease in pH within dental biofilms. The acidic environment promotes demineralization of the tooth's hard tissues. Furthermore, the low pH conditions favour the proliferation of acidogenic and acid-tolerating bacteria while inhibiting the growth of beneficial species. Several strategies have been employed to mitigate microbial acid production, such as the use of fluoride-containing products and other metabolic inhibitors like xylitol [144].

Fluoride not only restores enamel chemistry but also inhibits key enzymes involved in glycolysis and intracellular pH regulation. By reducing the fall in pH resulting from sugar metabolism, fluoride and xylitol prevent the favourable conditions for the growth of acid-tolerant cariogenic species. Xylitol specifically interferes with sugar transport in Streptococcus mutans, preventing their metabolism into acid and the subsequent generation of low pH that facilitates plaque formation.

9.6.3 Bacteriophage-Based Therapies

Bacteriophages, also known as phages, are viruses that infect and replicate within bacterial cells, leading to either lysogeny or host cell lysis [141]. Phages possess a unique ability to selectively replicate at the site of infection and tend to accumulate in areas where their target bacteria are present [153]. These viruses consist of an outer protein capsid that encloses genetic material and employ specific interactions with bacterial surface receptors, such as lipopolysaccharides, teichoic acids, proteins, or flagella, to enter the host cell [148].

Phage therapy offers a targeted approach due to the narrow host range of phages and their ability to self-replicate, which permits the administration of low dosages. One advantageous aspect of phage therapy is that lytic phages lack virulence genes, yet they can rapidly destroy bacterial cells through lysis [154]. This specificity and mode of action make phages an attractive alternative in combating bacterial infections, as they can selectively target and eliminate pathogenic bacteria without affecting the normal flora or harbouring virulence factors.

9.6.4 *Enhanced Antimicrobial Activity Through Electrical Stimulation*

The bioelectric effect encompasses a remarkable phenomenon wherein the bactericidal effectiveness of biocides against bacterial biofilms is enhanced through the concurrent application of Direct Current (DC). This bioelectric approach not only impedes biofilm formation but also boosts the activity of antimicrobial agents against established biofilms. One example of this approach involves the utilization of iontophoresis, employing a low-power current source to facilitate the controlled release of antimicrobial silver ions from a specialized device.

The bioelectric effect signifies the synergistic application of antibiotics and a weak electric field to effectively eradicate bacteria [143]. This innovative approach holds great promise for combating biofilm-related challenges by harnessing the potential of bioelectricity to enhance the efficacy of antimicrobial agents.

9.6.5 *Enhancement of Antimicrobial Transport Through Ultrasound*

Ultrasound has been utilized to augment the transport of antibiotics through biofilms, employing various strategies. For instance, the application of pulsed ultrasound for 24 hours was found to significantly enhance the efficacy of gentamicin in eradicating *E. coli* biofilms on polyethylene discs implanted subcutaneously into rabbits, while preserving the integrity of the skin [155].

Furthermore, low-frequency ultrasound at 70 kHz and low acoustic intensity demonstrated the ability to enhance the growth rate of *S.* epidermis, *P. aeruginosa*, and *E. coli* cultivated on polyethylene surfaces. It was postulated that ultrasound promotes the transportation of oxygen and nutrients to the cells, thereby facilitating their growth [149]. These findings suggest that ultrasound holds potential as a valuable tool for improving antibiotic penetration through biofilms. By harnessing the beneficial effects of ultrasound, such as increased transport of antimicrobials and improved nutrient availability, more effective treatment strategies can be developed to combat biofilm-associated infections.

9.6.6 *Enzymes for Biofilm Disruption*

The biofilm matrix, consisting of DNA, proteins, and extracellular polysaccharides, plays a crucial role in maintaining the integrity and resilience of biofilms. However, targeted degradation of the biofilm matrix can result in the disruption of its structural integrity. Various enzymes have been identified and utilized for their biofilm-disrupting properties, including deoxyribonuclease I, lysostaphin, amylases, and lyases [150].

Deoxyribonuclease I (DNase I), an endonuclease enzyme, acts by cleaving the DNA molecules present within the biofilm matrix. By targeting the DNA component, DNase I can degrade the biofilm matrix, leading to the destabilization and disintegration of the biofilm structure. Lysostaphin, on the other hand, is an enzyme specifically effective against Staphylococcus species, as it hydrolyses the peptidoglycan layer present in their cell walls, contributing to the disruption of the biofilm structure [156].

Amylases, a group of enzymes that break down complex carbohydrates, have also shown potential in biofilm disruption. By degrading the extracellular polysaccharides within the biofilm matrix, amylases can compromise the structural integrity and stability of the biofilm. Lyases, a class of enzymes that catalyse the breaking of chemical bonds, have been explored for their ability to degrade specific components of the biofilm matrix, such as polysaccharides or proteins, leading to the disruption of biofilm structure [157]. The utilization of these biofilm-disrupting enzymes offers a promising avenue for combating biofilm-associated infections.

9.6.7 Advancements in the Design of Inhibitors and Antiplaque Agents

Inhibitors and antiplaque agents have demonstrated heightened effectiveness in combating surface-associated microorganisms, especially when coupled with innovative delivery systems that specifically target particular bacteria and enhance agent retention within the oral cavity. The utilization of inhibitors and antiplaque agents is particularly advantageous in addressing surface-associated microorganisms, which predominantly reside on oral surfaces. By directly targeting these microbial populations, the efficacy of these agents is significantly enhanced. Furthermore, the development of targeted delivery systems has revolutionized the field by allowing precise delivery of inhibitors and antiplaque agents to specific bacterial species or sites of infection within the oral cavity. This targeted approach improves the retention and concentration of therapeutic agents, thereby maximizing their effectiveness [144].

9.6.8 Photodynamic Therapy (PDT)

Photodynamic approaches have emerged as a potential solution to combat resistance and disintegrate biofilms. Photodynamic therapy comprises three fundamental components: light, a photosensitizer (PS) chemical molecule, and oxygen [151]. Lee et al. (2016) investigated the beneficial effects of photodynamic therapy (PDT) in reducing biofilms formed by *Smutans*, a major cariogenic bacterium. They employed erythrosine and a halogen curing unit, commonly used in dental clinics, and observed a substantial decrease in *S. mutans* biofilm formation in vitro with the simple PDT technique. This study demonstrated the potential of PDT for biofilm control, particularly in caries lesion prevention [158].

Phenothiazine dyes (e.g., methylene blue) and toluidine blue, as well as porphyrin derivatives, fullerenes, and cyanine derivatives, are extensively studied photosensitizers (PSs). PSs can be excited by absorbing specific energy from light, generating reactive oxygen species. Excitation occurs when the light wavelength range overlaps with the PS's absorbance spectrum. After excitation, PSs typically form a long-lived triplet excited state, from which energy can be transferred to biomolecules or molecular oxygen, depending on the reaction type. Two primary types of reactions are involved: electron transfer from the triplet state PS to a substrate (e.g., unsaturated membrane phospholipids or amino lipids), leading to the production of lipid-derived radicals or hydroxyl radicals derived from water. These radicals can combine or react with other biomolecules and oxygen, resulting in lipid peroxidation or the generation of reactive oxygen species that induce cellular damage and death. The second reaction type involves energy transfer from the triplet state PS to the ground state (triplet) molecular oxygen, producing excited singlet oxygen—a highly reactive species capable of oxidizing biomolecules in the cell, such as proteins, nucleic acids, and lipids. This oxidative process leads to cellular damage and death. The two major types of cellular damage include DNA damage and the destruction of cellular membranes and organelles [150].

In a study focusing on the oral pathogen *Actinomyces viscosus,* biofilms were exposed to laser light at 666 nm in the presence of methylene blue. Confocal microscopy revealed that a single photomechanical wave increased the penetration of methylene blue by 75% and enhanced the photo destruction of the biofilm [143]. Suci et al. investigated the targeted delivery of a photosensitizer to *Aggregatibacter actinomycetemcomitans* biofilms. Previous research had shown a decrease in the susceptibility of *A. actinomycetemcomitans* biofilms to antimicrobials as the biofilm matured [145].

9.6.9 Innovative Approaches

The task of devising new therapeutic strategies to combat oral biofilms poses a significant challenge due to the rapid clearance of topically applied antibacterial agents by saliva, resulting in insufficient concentrations for sustained periods. There is an imperative to improve the availability and retention of antibacterial agents at dental surfaces and within biofilms. Extensive research into the assembly of biofilm matrices and alterations in the biofilm microenvironment has paved the way for the development of numerous innovative nanotechnology-based approaches, which will be comprehensively explored in the subsequent section of this chapter.

9.7 Nanoparticles in the Fight Against Oral Biofilm

9.7.1 Understanding the Interactions Between Nanoparticles and Biofilm

The mechanism of nanoparticles (NPs) interaction with biofilms involves several stages. Firstly, NPs are transferred in the vicinity of the biofilm. Subsequently, they attach to the biofilm surface and migrate within the biofilm matrix (Fig. 9.5). The implementation of each stage is influenced by various factors, including the physicochemical characteristics of NPs, extracellular polymeric matrix (EPM), and the surrounding environment [159]. Upon reaching the biofilm boundary, the physicochemical characteristics of the EPM play a crucial role in the initial attachment of NPs to the biofilm surface and their subsequent movement within the matrix. Electrostatic interactions primarily govern the interaction between NPs and biofilms. The zeta potential of NPs and the charge of the biofilm matrix are key factors in this interaction [160–162]. Biofilm matrices of most bacteria possess a negative charge due to the presence of uronic acid or metal-bound pyruvate, containing carboxylic acid and residual phosphate or occasionally sulphate groups [163]. This negatively charged matrix can interact with positively charged metal ions and organic compounds through electrostatic forces [164].

NPs that have successfully associated with the EPM on the biofilm surface can penetrate into the deeper regions of the biofilm at varying rates. The penetration and movement of NPs within the biofilm are primarily attributed to diffusion [165]. The diffusion of NPs inside the biofilm may depend on factors such as the size of the biofilm's pores, the presence of water channels, the charge of NPs and EPM, the hydrophobicity of the environment, and the chemical gradient within the matrix [162, 165–167]. The pore spaces of the EPM containing water can exhibit different ion concentrations, allowing ions and organic molecules to diffuse, penetrate the biofilm, and distribute through these pore spaces. The intervals between EPM pores

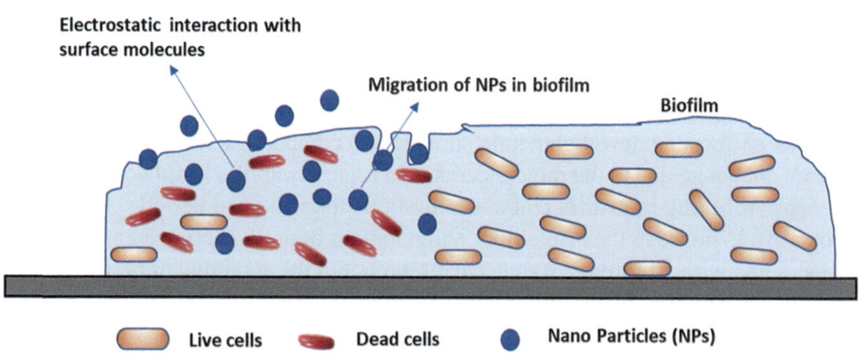

Fig. 9.5 Interaction between metal oxide nanoparticles (NPs) and biofilm

are considered particularly crucial in this process, although the specific nanoscale characteristics are not yet fully characterized and understood [168].

Therefore, the penetration and migration of NPs within the biofilm are mainly influenced by the charge and size of the particles, as well as the composition and structure of the EPM. However, many aspects of this interaction remain to be determined and further investigated.

9.7.2 *Nanoparticle-Based Metal Formulations for Microbial Control*

Metals have been utilized throughout history for their antimicrobial properties, with silver, copper, gold, titanium, and zinc being of particular interest due to their distinct properties and ranges of activity [169]. In the realm of oral care, zinc citrate or acetate in powdered micron-sized form has been incorporated into toothpastes to regulate the formation of dental plaque [170]. Titanium dioxide powder is commonly employed as a toothpaste whitener. In the field of dentistry, metallic nanoparticles have also been explored as coatings for dental implants to enhance their antimicrobial efficacy.

Among nanoparticulate metals, silver and copper have garnered significant attention for their antimicrobial properties [171, 172]. These metals have been coated onto or incorporated into various base materials, including PMMA and hydrogels [173]. Studies have demonstrated an inverse relationship between nanoparticle size and antimicrobial activity, with particles in the size range of 1–10 nm exhibiting the highest biocidal activity against bacteria [174]. It has been observed that smaller silver nanoparticles, particularly when oxidized, possess greater toxicity compared to larger particles [175]. At the nanoscale, Ag1 ions are known to be released from the surface [176]. Sotiriou et al. (2010) proposed that the antimicrobial activity of small nanosilver particles (less than 10 nm) is primarily attributed to Ag1 ions, whereas for larger particles (greater than 15 nm), both Ag1 ions and particles contribute comparably to the antibacterial activity, with the release of Ag1 ions being proportional to the exposed nanosilver surface area [177].

Certain nanoparticles, due to their small size, offer additional advantages in the field of biomedicine by improving biocompatibility [178]. Furthermore, bacteria are less likely to develop resistance to metal nanoparticles compared to conventional antibiotics with narrower spectrums [179]. This is believed to be because metals can act on a broad range of microbial targets, necessitating multiple mutations for microorganisms to resist their antimicrobial activity. The shape of nanoparticles may also impact their activity. For instance, in the case of Escherichia coli, truncated triangular silver nanoplates with a {111} lattice pattern as the basal plane have demonstrated superior biocidal activity compared to spherical and rod-shaped nanoparticles. The observed differences in activity are attributed to the proportion of active facets present in nanoparticles of different shapes.

Nanoparticulate metals and metal oxides, particularly those capable of generating reactive oxygen species under UV light, such as titanium dioxide (TiO_2) and zinc oxide (ZnO), are increasingly utilized in antimicrobial formulations [180]. Silver metal nanoparticles (5–40 nm) have been reported to inactivate various microorganisms, including HIV-1. The high reactivity of nanotitanium dioxide and nanosilicon dioxide (SiO_2) is extensively exploited for their bactericidal properties in filters and coatings on substrates like polymers, ceramics, glasses, and alumina [181].

Studies have demonstrated significant activity of metal and metal oxide nanoparticles, as well as their compound clusters, against fungal and bacterial pathogens such as methicillin-resistant *S. aureus* (MRSA) and *E. Coli* [182]. These nanoparticles have also shown efficacy in inactivating viruses, including severe acute respiratory syndrome, H1N1 swine flu, and H5N1 bird flu. For instance, new broad-spectrum materials in the size range of 5–60 nm have been shown to reduce virus levels by 80% to 100% through direct or indirect contact.

Various nanoparticle preparations based on nickel (Ni, NiO), zirconium (ZrO_2), copper (Cu, CuO, and Cu_2O), titanium (Ti, TiO_2), zinc (ZnO), aluminium (Al_2O_3), silicon nitride (Si3N4), silver (Ag), and tungsten carbide (WC) have been compared for their antimicrobial potential. Significant activity against bacterial pathogens, including MRSA and *Pseudomonas aeruginosa*, has been demonstrated for silver (Ag), zinc oxide (ZnO), titanium dioxide (TiO_2) in the presence of UV light, silicon dioxide (SiO_2), copper (Cu), copper (II) oxide (Cu_2O), and copper (II) oxide (CuO). The minimum bactericidal concentrations (MBCs) for these nanoparticles were found to be in the range of 0.15 mg/mL. In comparison, traditional antibiotics are effective at concentrations 1000-fold lower. However, nickel oxide (NiO), nickel (Ni), aluminium oxide (Al_2O_3), titanium dioxide (TiO_2) in the absence of UV light, silicon nitride (Si_3N_4), tungsten carbide (WC), and zirconium oxide (ZrO_2) lacked antimicrobial activity at the concentrations tested. Furthermore, oral pathogens such as *S. intermedius, P. gingivalis, F. nucleatum, P. intermedia,* and *A. actinomycetemcomitans* were found to be susceptible to silver (Ag) and copper oxide (CuO) nanoparticles under anaerobic conditions, with MBC values ranging from 0.025 to 2.5 mg/mL [183].

9.7.3 Nanoparticle-Based Photodynamic Therapy

As discussed previously, Photodynamic therapy (PDT) has emerged as a promising approach for combating oral biofilms, which are responsible for various oral infections. The combination of PDT with nanoparticles offers enhanced therapeutic potential by improving the targeted delivery of photosensitizers and enhancing their antimicrobial efficacy.

R.P. Allaker and C.W.I. Douglas (2015) emphasized the need for non-conventional therapeutics in the treatment of oral infections. They highlighted the potential of PDT as an alternative approach to antibiotics and demonstrated its effectiveness in eradicating oral biofilms. PDT involves the use of a photosensitizer, light of specific wavelength, and molecular oxygen to generate reactive oxygen species (ROS) that can damage and kill bacteria [184]. A.J. MacRobert et al. (1989) discussed the ideal photo properties required for an effective photosensitizer. They emphasized the importance of high absorption in the therapeutic window of light, selective accumulation in the target tissues, and efficient ROS generation. Nanoparticles can serve as carriers for photosensitizers, facilitating their delivery to the target sites and improving their photodynamic efficacy [185]. T.C. Pagonis et al. (2010) investigated nanoparticle-based endodontic antimicrobial PDT. They encapsulated a photosensitizer within nanoparticles to improve its stability and targeted delivery. The nanoparticles efficiently penetrated into the biofilm matrix, enhancing the photodynamic effect and leading to significant biofilm reduction. This study highlighted the potential of nanoparticles in enhancing the efficacy of PDT against oral biofilms [186]. S. Wood et al. (2006) explored the use of erythrosine, a photosensitizer, in combination with PDT for the treatment of oral plaque biofilms. They demonstrated that erythrosine-mediated PDT effectively reduced biofilm viability and disrupted the biofilm structure. The addition of nanoparticles could further enhance the delivery and retention of erythrosine within the biofilm, improving its antimicrobial activity [187].

The incorporation of nanoparticles in PDT offers several advantages in combating biofilms. Nanoparticles can enhance the stability, solubility, and bioavailability of photosensitizers, allowing for their controlled release and improved targeting. Furthermore, nanoparticles can facilitate the penetration of photosensitizers into the biofilm matrix, overcoming the limited diffusion and accessibility of traditional PDT agents. These properties contribute to enhanced antimicrobial efficacy and improved biofilm eradication. This approach holds great potential for the development of effective therapies against oral infections. Further research and optimization of nanoparticle-based PDT systems are needed to maximize their antimicrobial efficacy, minimize side effects, and ensure their safe and efficient translation into clinical practice.

9.8 Biocompatibility Considerations of Nano Antimicrobials

The increasing interest in utilizing nano antimicrobials to combat oral infections stems from their potential to enhance antimicrobial efficacy and enable targeted delivery. However, it is imperative to thoroughly evaluate the biocompatibility of these nanoparticles within the oral cavity to ensure their safe and effective utilization.

The potential toxicity of nanoparticles has been highlighted, emphasizing the need to evaluate their safety [188, 189]. Understanding the kinetic properties of nanoparticles in the body is important to assess any potential adverse effects [189]. It has been emphasized that comprehensive evaluation of the functionality and toxicity of nanoparticles is necessary, particularly on the central nervous system [190].

To assess the biocompatibility of nano antimicrobials within the oral cavity, several factors need to be considered. Studies have investigated the reactivity and biocompatibility of bioactive glass nanoparticles, indicating their promising properties [191]. The antifungal activity of silver nanoparticles against Candida spp. has also been explored, suggesting their potential for oral antimicrobial applications [192].

Coating dental implants with nanoparticles has been discussed as a strategy to enhance osseointegration and provide antimicrobial benefits [193]. The selection of biocompatible materials is crucial for successful integration and minimizing adverse reactions.

Considering the toxic potential of nanoparticles at the nano level is essential [194]. Surface properties, such as hydrophilic modification using nanoparticles, can impact biocompatibility [195]. Surfactant-coated nanoparticles have shown anti-adherent and antifungal activities, suggesting their potential as safe and effective antimicrobial agents [196].

Assessing the biocompatibility of nano antimicrobials within the oral cavity requires thorough evaluation of their toxicity, reactivity, and potential adverse effects. Understanding the potential toxic effects and reactivity of nanoparticles is essential for their safe and efficient use. Further research should focus on conducting in-depth studies to assess the long-term effects and potential interactions of nano antimicrobials with oral tissues and microorganisms. This knowledge will contribute to the development of biocompatible nano antimicrobial agents for effective oral infection control.

9.9 Final Considerations

Self-performed and professionally administered control of oral biofilms remains the primary approach for preventing oral biofilm-associated diseases. The exploration of nanotechnology for the treatment and control of microbial biofilms in the oral cavity shows great potential. The use of nanoscale antimicrobials offers several advantages, including enhanced biocidal activity, anti-adhesive properties, and targeted delivery. These nanoparticles can also work in synergy with active molecules, leading to more efficient utilization of existing antimicrobials. Additionally, incorporating nanoparticles into prosthetic device coatings, topical agents, and dental materials holds promise for combating oral infections. Future research will focus on identifying nanoparticles with optimal antimicrobial activity and minimal toxicity to the host. It is crucial to consider the biocompatibility of nanoparticles, as their surface characteristics play a vital role in determining their suitability for oral applications. Various strategies, such as altering aggregation behaviour, applying surface coatings, and modifying oxidative state and charge density, are being explored to enhance biocompatibility. With further advancements in nanotechnology, significant progress can be expected in the development of effective and safe nanoantimicrobial approaches to combat oral biofilms and improve oral health.

References

1. Busscher, H. J., van der Mei, H. C., Subbiahdoss, G., Jutte, P. C., van den Dungen, J. J. A. M., Zaat, S. A. J., Schultz, M. J., & Grainger, D. W. (2012). Biomaterial-associated infection: Locating the finish line in the race for the surface. *Science Translational Medicine, 4*(153), 153rv110.
2. Flemming, H.-C., Wingender, J., Szewzyk, U., Steinberg, P., Rice, S. A., & Kjelleberg, S. (2016). Biofilms: An emergent form of bacterial life. *Nature Reviews. Microbiology, 14*(9), 563–575.
3. Valm, A. M., Welch, J. L. M., Rieken, C. W., Hasegawa, Y., Sogin, M. L., Oldenbourg, R., Dewhirst, F. E., & Borisy, G. G. (2011). Systems-level analysis of microbial community organization through combinatorial labeling and spectral imaging. *Proceedings of the National Academy of Sciences, 108*(10), 4152–4157.
4. Yan, L., Shen, J., Wang, J., Yang, X., Dong, S., & Lu, S. (2020). Nanoparticle-based drug delivery system: A patient-friendly chemotherapy for oncology. *Dose Response, 18*, 1559325820936161.
5. Colombo, E., Feyen, P., Antognazza, M. R., Lanzani, G., & Benfenati, F. (2016). Nanoparticles: A challenging vehicle for neural stimulation. *Frontiers in Neuroscience, 10*, 105.
6. Allaker, R. P., & Memarzadeh, K. (2014). Nanoparticles and the control of oral infections. *International Journal of Antimicrobial Agents, 43*(2), 95–104. https://doi.org/10.1016/j.ijantimicag.2013.11.002. Epub 2013 Dec 1. PMID: 24388116.
7. Dewhirst, F. E., Chen, T., Izard, J., Paster, B. J., Tanner, A. C., Yu, W.-H., Lakshmanan, A., & Wade, W. G. (2010). The human oral microbiome. *Journal of Bacteriology, 192*(19), 5002–5017.
8. Mosaddad, S. A., Tahmasebi, E., Yazdanian, A., Rezvani, M. B., Seifalian, A., Yazdanian, M., & Tebyanian, H. (2019). Oral microbial biofilms: An update. *European Journal of Clinical Microbiology & Infectious Diseases, 38*(11), 2005–2019. https://doi.org/10.1007/s10096-019-03641-9. Epub 2019 Aug 1. PMID: 31372904.
9. Štšepetova, J., Truu, J., Runnel, R., Nõmmela, R., Saag, M., Olak, J., Nõlvak, H., Preem, J.-K., Oopkaup, K., & Krjutškov, K. (2019). Impact of polyols on oral microbiome of Estonian schoolchildren. *BMC Oral Health, 19*(1), 60.
10. Lederberg, J., & McCray, A. T. (2001). Ome SweetOmics--A genealogical treasury of words. *The Scientist, 15*(7), 8–8.
11. Aas, J. A., Paster, B. J., Stokes, L. N., Olsen, I., & Dewhirst, F. E. (2005). Defining the normal bacterial flora of the oral cavity. *Journal of Clinical Microbiology, 43*(11), 5721–5732.
12. Mager, D. L., Ximenez-Fyvie, L. A., Haffajee, A. D., & Socransky, S. S. (2003). Distribution of selected bacterial species on intraoral surfaces. *Journal of Clinical Periodontology, 30*(7), 644–654.
13. Rashid, S. A., & Sorchee, S. (2021). Role of the bacteria population inside the mouth: Oral bacteria. *International Journal of Medical Sciences, 4*(3), 17–30.
14. Seymour, G. J., Ford, P. J., Cullinan, M. P., Leishman, S., & Yamazaki, K. (2007). Relationship between periodontal infections and systemic disease. *Clinical Microbiology and Infection, 13*(Suppl 4), 3–10.
15. Joshipura, K. J., Rimm, E. B., Douglass, C. W., Trichopoulos, D., Ascherio, A., & Willett, W. C. (1996). Poor oral health and coronary heart disease. *Journal of Dental Research, 75*(11), 1631–1636.
16. Offenbacher, S., Jared, H. L., O'Reilly, P. G., Wells, S. R., Salvi, G. E., Lawrence, H. P., Socransky, S. S., & Beck, J. D. (1998). Potential pathogenic mechanisms of periodontitis associated pregnancy complications. *Annals of Periodontology, 3*(1), 233–250.
17. Genco, R. J., Grossi, S. G., Ho, A., Nishimura, F., & Murayama, Y. (2005). A proposed model linking inflammation to obesity, diabetes, and periodontal infections. *Journal of Periodontology, 76*(11-s), 2075–2084.

18. Awano, S., Ansai, T., Takata, Y., Soh, I., Akifusa, S., Hamasaki, T., Yoshida, A., Sonoki, K., Fujisawa, K., & Takehara, T. (2008). Oral health and mortality risk from pneumonia in the elderly. *Journal of Dental Research, 87*(4), 334–339.
19. Yamashita, Y., & Takeshita, T. (2017). The oral microbiome and human health. *Journal of Oral Science, 59*(2), 201–206.
20. Olsen, I. (2016). The oral microbiome in health and disease. In *Oral infections and general health* (pp. 97–114). Springer.
21. Filoche, S., Wong, L., & Sissons, C. (2010). Oral biofilms: Emerging concepts in microbial ecology. *Journal of Dental Research, 89*(1), 8–18.
22. Kuang, X., Chen, V., & Xu, X. (2018). Novel approaches to the control of oral microbial biofilms. *BioMed Research International, 2018*, 6498932.
23. Besinis, A., De Peralta, T., & Handy, R. D. (2014). Inhibition of biofilm formation and antibacterial properties of a silver nano-coating on human dentine. *Nanotoxicology, 8*(7), 745–754.
24. Jakubovics, N. S., & Palmer, R. J. (Eds.). (2013). *Oral microbial ecology: Current research and new perspectives*. Horizon Scientific Press.
25. Rowland, I., Gibson, G., Heinken, A., Scott, K., Swann, J., Thiele, I., & Tuohy, K. (2018). Gut microbiota functions: Metabolism of nutrients and other food components. *European Journal of Nutrition, 57*(1), 1–2.
26. Darrene, L.-N., & Cecile, B. (2016). Experimental models of oral biofilms developed on inert substrates: A review of the literature. *BioMed Research International, 2016*, 7461047.
27. Beighton, D., Mughal, S. R., & Do, T. (2013). Bacterial catabolism of salivary substrates. In *Oral microbial ecology: Current research and new perspectives* (pp. 37–46). Caister Academic Press.
28. Renner, L. D., & Weibel, D. B. (2011). Physicochemical regulation of biofilm formation. *MRS Bulletin, 36*, 347–355.
29. Bayoudh, S., Othmane, A., Bettaieb, F., Bakhrouf, A., Ouada, H. B., & Ponsonnet, L. (2006). Quantification of the adhesion free energy between bacteria and hydrophobic and hydrophilic substrata. *Materials Science and Engineering: C, 26*, 300–305.
30. Marais, F., Mehtar, S., & Chalkley, L. (2010). Antimicrobial efficacy of copper touch surfaces in reducing environmental bioburden in a South African community healthcare facility. *The Journal of Hospital Infection, 74*, 80–82.
31. Habimana, O., Semião, A. J. C., & Casey, E. (2014). The role of cell-surface interactions in bacterial initial adhesion and consequent biofilm formation on nanofiltration/reverse osmosis membranes. *Journal of Membrane Science, 454*, 82–96.
32. Donlan, R. M. (2002). Biofilms: Microbial life on surfaces. *Emerging Infectious Diseases, 8*, 881–890.
33. Ferreira, C., Pereira, A. M., Melo, L. F., & Simões, M. (2010). Advances in industrial biofilm control with micro-nanotechnology. *Current Research, Technology and Education Topics in Applied Microbiology and Microbial Biotechnology, 2*, 845–854.
34. Nilsson, R. E., Ross, T., & Bowman, J. P. (2011). Variability in biofilm production by Listeria monocytogenes correlated to strain origin and growth conditions. *International Journal of Food Microbiology, 150*, 14–24.
35. Gerstel, U., & Romling, U. (2001). Oxygen tension and nutrient starvation are major signals that regulate agfD promoter activity and expression of the multicellular morphotype in Salmonella typhimurium. *Environmental Microbiology, 3*, 638–648.
36. Garrett, T. R., Bhakoo, M., & Zhang, Z. (2008). Bacterial adhesion and biofilms on surfaces. *Progress in Natural Science, 18*, 1049–1056.
37. Stoodley, P., Sauer, K., Davies, D. G., & Costerton, J. W. (2002). Biofilms as complex differentiated communities. *Annual Review of Microbiology, 56*, 187–209.
38. Sinde, E., & Carballo, J. (2000). Attachment of Salmonella spp. and Listeria monocytogenes to stainless steel, rubber and polytetrafluoroethylene: The influence of free energy and the effect of commercial sanitizers. *Food Microbiology, 17*, 439–447.

39. Augustin, M., Ali-Vehmas, T., & Atroshi, F. (2004). Assessment of enzymatic cleaning agents and disinfectants against bacterial biofilms. *Journal of Pharmacy & Pharmaceutical Sciences, 7*, 55–64.
40. Maukonen, J., Mättö, J., Wirtanen, G., Raaska, L., Mattila-Sandholm, T., & Saarela, M. (2003). Methodologies for the characterization of microbes in industrial environments: A review. *Journal of Industrial Microbiology & Biotechnology, 30*, 327–356.
41. Latasa, C., Solano, C., Penadés, J. R., & Lasaa, I. (2006). Biofilm-associated proteins. *Comptes Rendus Biologies, 329*, 849–857.
42. Heilmann, C. (2011). Adhesion mechanisms of staphylococci. *Advances in Experimental Medicine and Biology, 715*, 105–123.
43. Flemming, H. C., & Wingender, J. (2010). The biofilm matrix. *Nature Reviews. Microbiology, 8*, 623–633.
44. Steinberg, N., & Kolodkin-Gal, I. (2015). The matrix reloaded: Probing the extracellular matrix synchronizes bacterial communities. *Journal of Bacteriology, 197*, 02516–02514.
45. Dunne, W. M. (2002). Bacterial adhesion: Seen any good biofilms lately? *Clinical Microbiology Reviews, 15*, 155–166.
46. Yang, L., Nilsson, M., Gjermansen, M., Givskov, M., & Tolker-Nielsen, T. (2009). Pyoverdine and PQS mediated subpopulation interactions involved in Pseudomonas aeruginosa biofilm formation. *Molecular Microbiology, 74*, 1380–1392.
47. Barken, K. B., Pamp, S. J., Yang, L., Gjermansen, M., Bertrand, J. J., Klausen, M., Givskov, M., Whitchurch, C. B., Engel, J. N., & Tolker-Nielsen, T. (2008). Roles of type IV pili, flagellum-mediated motility and extracellular DNA in the formation of mature multicellular structures in Pseudomonas aeruginosa biofilms. *Environmental Microbiology, 10*, 2331–2343.
48. Jefferson, K. K. (2004). What drives bacteria to produce a biofilm? *FEMS Microbiology Letters, 236*, 163–173.
49. Kaplan, J. B. (2010). Biofilm dispersal: Mechanisms, clinical implications, and potential therapeutic uses. *Journal of Dental Research, 89*, 205–218.
50. Kaplan, J. B., Ragunath, C., Ramasubbu, N., & Fine, D. H. (2003). Detachment of Actinobacillus actinomycetemcomitans biofilm cells by an endogenous β-hexosaminidase activity. *Journal of Bacteriology, 185*, 4693–4698.
51. Sauer, K., Cullen, M. C., Rickard, A. H., Zeef, L. A. H., Davies, D. G., & Gilbert, P. (2004). Characterization of nutrient-induced dispersion in Pseudomonas aeruginosa PAO1 biofilm. *Journal of Bacteriology, 186*, 7312–7326.
52. Boles, B. R., & Horswill, A. R. (2008). Agr-mediated dispersal of Staphylococcus aureus biofilms. *PLoS Pathogens, 4*, e1000052.
53. Harmsen, M., Yang, L., Pamp, S. J., & Tolker-Nielsen, T. (2010). An update on Pseudomonas aeruginosa biofilm formation, tolerance, and dispersal. *FEMS Immunology and Medical Microbiology, 59*, 253–268.
54. Bassler, B. L., & Losick, R. (2006). Bacterially speaking. *Cell, 125*, 237–246.
55. Mukherjee, S., & Bassler, B. L. (2019). Bacterial quorum sensing in complex and dynamically changing environments. *Nature Reviews Microbiology*. https://doi.org/10.1038/s41579-019-0186
56. Miller, M. B., & Bassler, B. L. (2001). Quorum sensing in bacteria. *Annual Review of Microbiology, 55*, 165–199.
57. Papenfort, K., & Bassler, B. L. (2016). Quorum sensing signal-response systems in Gram-negative bacteria. *Nature Reviews. Microbiology, 14*, 576–588.
58. Novick, R. P., & Geisinger, E. (2008). Quorum sensing in staphylococci. *Annual Review of Genetics, 42*, 541–564.
59. Waters, C. M., & Bassler, B. L. (2005). Quorum sensing: Cell-to-cell communication in bacteria. *Annual Review of Cell and Developmental Biology, 21*, 319–346.
60. Svenningsen, S. L., Tu, K. C., & Bassler, B. L. (2009). Gene dosage compensation calibrates four regulatory RNAs to control Vibrio cholerae quorum sensing. *The EMBO Journal, 28*, 429–439.

61. Lenz, D. H., et al. (2004). The small RNA chaperone Hfq and multiple small RNAs control quorum sensing in Vibrio harveyi and Vibrio cholerae. *Cell, 118*, 69–82.
62. Feng, L., et al. (2015). A Qrr noncoding RNA deploys four different regulatory mechanisms to optimize quorum sensing dynamics. *Cell, 160*, 228–240.
63. Lenz, D. H., et al. (2005). CsrA and three redundant small RNAs regulate quorum sensing in Vibrio cholerae. *Molecular Microbiology, 58*, 1186–1202.
64. Lenz, D. H., & Bassler, B. L. (2007). The small nucleoid protein Fis is involved in Vibrio cholerae quorum sensing. *Molecular Microbiology, 63*, 859–871.
65. Thompson, L. S., Webb, J. S., Rice, S. A., & Kjelleberg, S. (2003). The alternative sigma factor RpoN regulates the quorum sensing gene rhlI in Pseudomonas aeruginosa. *FEMS Microbiology Letters, 220*, 187–195.
66. Darveau, R. P., Tanner, A., & Page, R. C. (1997). The microbial challenge in periodontitis. *Periodontology 2000, 14*(1), 12–32.
67. Larsen, T., & Fiehn, N. E. (2017). Dental biofilm infections–an update. *APMIS, 125*(4), 376–384.
68. Donlan, R. M. (2001). Biofilm formation: A clinically relevant microbiological process. *Clinical Infectious Diseases, 33*(8), 1387–1392.
69. Seneviratne, C. J., Zhang, C. F., & Samaranayake, L. P. (2011). Dental plaque biofilm in oral health and disease. *Chinese Journal of Dental Research, 14*(2), 87–94.
70. Marsh, P. D., & Martin, M. V. (Eds.). (1992). *Oral microbiology* (3rd ed.). Chapman and Hall.
71. Theilade, E., Theilade, J., & Mikkelsen, L. (1982). Microbiological studies on early dento-gingival plaque on teeth and Mylar strips in humans. *Journal of Periodontal Research, 17*(1), 12–25.
72. Kuramitsu, H. K., He, X., Lux, R., Anderson, M. H., & Shi, W. (2007). Interspecies interactions within oral microbial communities. *Microbiology and Molecular Biology Reviews, 71*(4), 653–670.
73. Ebersole, J. L., et al. (2017). The periodontal war: Microbes and immunity. *Periodontology 2000, 75*(1), 52–115.
74. Miller, D. P., Fitzsimonds, Z. R., & Lamont, R. J. (2019). Metabolic signaling and spatial interactions in the oral polymicrobial community. *Journal of Dental Research, 98*(12), 1308–1314.
75. Valen, H., & Scheie, A. A. (2018). Biofilms and their properties. *European Journal of Oral Sciences, 126*, 13–18.
76. Christersson, L. A., Zambon, J. J., & Genco, R. J. (1991). Dental bacterial plaques. Nature and role in periodontal disease. *Journal of Clinical Periodontology, 18*(6), 441–446.
77. Kawasaki, M., Ikeda, Y., Ikeda, E., Takahashi, M., Tanaka, D., Nakajima, Y., Arakawa, S., Izumi, Y., & Miyake, S. (2021). Oral infectious bacteria in dental plaque and saliva as risk factors in patients with esophageal cancer. *Cancer, 127*, 512–519.
78. Das, J., et al. (2020). Leukocyte adhesion defect: Where do we stand circa 2019? *Genes & Diseases, 7*(1), 107–114.
79. Marsh, P. D., Moter, A., & Devine, D. A. (2011). Dental plaque biofilms: Communities, conflict, and control. *Periodontology 2000, 55*, 16.
80. Tamura, S., et al. (2009). Inhibiting effects of Streptococcus salivarius on competence-stimulating peptide-dependent biofilm formation by Streptococcus mutans. *Oral Microbiology and Immunology, 24*(2), 152–161.
81. Dixon, D. R., Bainbridge, B. W., & Darveau, R. P. (2004). Modulation of the innate immune response within the periodontium. *Periodontology 2000, 35*(1), 53–74.
82. Marsh, P. D. (2010). Microbiology of dental plaque biofilms and their role in oral health and caries. *Dental Clinics, 54*(3), 441–454.
83. Welin-Neilands, J., & Svensater, G. (2007). Acid tolerance of biofilm cells of Streptococcus mutans. *Applied and Environmental Microbiology, 73*(17), 5633–5638.
84. Klinke, T., et al. (2011). Dental caries in rats associated with Candida albicans. *Caries Research, 45*(2), 100–106.

85. Tanner, A. C., et al. (2011). Cultivable anaerobic microbiota of severe early childhood caries. *Journal of Clinical Microbiology, 49*(4), 1464–1474.
86. Giertsen, E., Arthur, R. A., & Guggenheim, B. (2011). Effects of xylitol on survival of mutans streptococci in mixed-six-species in vitro biofilms modelling supragingival plaque. *Caries Research, 45*(1), 31 39.
87. Lexner, M. O., et al. (2010). Microbiological profiles in saliva and supragingival plaque from caries-active adolescents before and after a short-term daily intake of milk supplemented with probiotic bacteria-A pilot study. *Oral Health & Preventive Dentistry, 8*(4), 383–388.
88. Pihlstrom, B. L., Michalowicz, B. S., & Johnson, N. W. (2005). Periodontal diseases. *The Lancet, 366*(9499), 1809–1820.
89. Marsh, P. D. (1994). Microbial ecology of dental plaque and its significance in health and disease. *Advances in Dental Research, 8*(2), 263–271.
90. Genco, C. A. (1995). Regulation of hemin and iron transport in Porphyromonas gingivalis. *Advances in Dental Research, 9*(1), 41–47.
91. Al-Qutub, M. N., et al. (2006). Hemin-dependent modulation of the lipid A structure of Porphyromonas gingivalis lipopolysaccharide. *Infection and Immunity, 74*(8), 4474–4485.
92. Darveau, R. P. (2010). Periodontitis: A polymicrobial disruption of host homeostasis. *Nature Reviews Microbiology, 8*(7), 481–490.
93. Shchipkova, A. Y., Nagaraja, H. N., & Kumar, P. S. (2010). Subgingival microbial profiles of smokers with periodontitis. *Journal of Dental Research, 89*(11), 1247–1253.
94. Haffajee, A. D., & Socransky, S. S. (2001). Relationship of cigarette smoking to attachment level profiles. *Journal of Clinical Periodontology, 28*(4), 283–295.
95. Drescher, J., et al. (2010). Molecular epidemiology and spatial distribution of Selenomonas spp. in subgingival biofilms. *European Journal of Oral Sciences, 118*(5), 466–474.
96. Loe, H., Theilade, E., & Jensen, S. B. (1965). Experimental Gingivitis in man. *Journal of Periodontology, 36*(3), 177–187.
97. Needleman, I., et al. (2005). A systematic review of professional mechanical plaque removal for prevention of periodontal diseases. *Journal of Clinical Periodontology, 32*(6), 229–282.
98. Teles, R. P., & Teles, F. R. F. (2009). Antimicrobial agents used in the control of periodontal biofilms: Effective adjuncts to mechanical plaque control? *Brazilian Oral Research, 23*(1), 39–48.
99. Axelsson, P., Albandar, J. M., & Rams, T. E. (2002). Prevention and control of periodontal diseases in developing and industrialized nations. *Periodontology 2000, 29*, 235–246.
100. Socransky, S. S., & Haffajee, A. D. (2005). Periodontal microbial ecology. *Periodontology 2000, 38*, 135–187.
101. Kerr, W. J., Kelly, J., & Geddes, D. A. (1991). The areas of various surfaces in the human mouth from nine years to adulthood. *Journal of Dental Research, 70*(12), 1528–1530.
102. Ciancio, S., & Panagakos, F. S. (2010). Superior management of plaque and gingivitis through the use of a triclosan/copolymer dentifrice. *Journal of Clinical Dentistry, 21*(4), 93–95.
103. Sekino, S., Ramberg, P., Uzel, N. G., Socransky, S., & Lindhe, J. (2004). The effect of a chlorhexidine regimen on de novo plaque formation. *Journal of Clinical Periodontology, 31*(8), 609–614.
104. Davies, R. M., Ellwood, R. P., & Davies, G. M. (2004). The effectiveness of toothpaste containing Triclosan and polyvinyl-methyl ether maleic acid copolymer in improving plaque control and gingival health. A systematic review. *Journal of Clinical Periodontology, 31*, 102.
105. Bratthall, D., Hansel-Petersson, G., & Sundberg, H. (1996). Reasons for the caries decline: What do the experts think? *European Journal of Oral Sciences, 104*(4), 416–422.
106. Blinkhorn, A., et al. (2009). Is there a role for triclosan/copolymer toothpaste in the management of periodontal disease. *British Dental Journal, 207*, 117–125.
107. Davies, R. M. (2007). The clinical efficacy of triclosan/copolymer and other common therapeutic approaches to periodontal health. *Clinical Microbiology and Infection, 13*(Suppl. 4), 25–29.

108. Biesbrock, A. R., Bartizek, R. D., & Gerlach, R. W. (2007). Oral hygiene regimens, plaque control, and gingival health: A two-month clinical trial with antimicrobial agents. *Journal of Clinical Dentistry, 18*, 103–107.
109. Gunsolley, J. C. (2006). A meta-analysis of six-month studies of antiplaque and antigingivitis agents. *Journal of the American Dental Association, 137*(12), 1649–1657.
110. Strydonck, D. A., Slot, D. E., Velden, U., & Weijden, F. (2013). Chlorhexidine mouthwash reduces plaque and gingivitis. *Evidence-Based Dentistry, 14*, 17–18.
111. Dutt, P., Rathore, P. R., & Khurana, D. (2014). Chlorhexidine – An antiseptic in periodontics. *Journal of Dental and Medical Sciences, 13*, 85–88.
112. Vitkov, L., Hermann, A., Krautgartner, W. D., et al. (2005). Chlorhexidine-induced ultrastructural alterations in oral biofilm. *Microscopy Research and Technique, 68*(2), 85–89.
113. Siqueira, J. F., Jr., & Rôças, I. N. (2005). Exploiting molecular methods to explore endodontic infections: Part 2-Redefining the endodontic microbiota. *Journal of Endodontia, 31*, 488–498.
114. Saito, D., Leonardo Rde, T., Rodrigues, J. L., Tsai, S. M., Höfling, J. F., & Gonçalves, R. B. (2006). Identification of bacteria in endodontic infections by sequence analysis of 16S rDNA clone libraries. *Journal of Medical Microbiology, 55*, 101–107.
115. Waltimo, T. M., Sirén, E. K., Torkko, H. L., Olsen, I., & Haapasalo, M. P. (1997). Fungi in therapy-resistant apical periodontitis. *International Endodontic Journal, 30*, 96–101.
116. Siqueira, J. F., Jr., & Sen, B. H. (2004). Fungi in endodontic infections. *Oral Surgery, Oral Medicine, Oral Pathology, Oral Radiology, and Endodontology, 97*, 632–641.
117. Vianna, M. E., Conrads, G., Gomes, B. P., & Horz, H. P. (2006). Identification and quantification of archaea involved in primary endodontic infections. *Journal of Clinical Microbiology, 44*, 1274–1282.
118. Antunes, H. S., Rocas, I. N., Alves, F. R., & Siqueira, J. F., Jr. (2015). Total and specific bacterial levels in the apical root canal system of teeth with post-treatment apical periodontitis. *Journal of Endodontia, 41*, 1037–1042.
119. Haapasalo, M., Udnæs, T., & Endal, U. (2003). Persistent, recurrent, and acquired infection of the root canal system post-treatment. *Endodontic Topics, 6*, 29–56.
120. Rocas, I. N., & Siqueira, J. F., Jr. (2010). Identification of bacteria enduring endodontic treatment procedures by a combined reverse transcriptase-polymerase chain reaction and reverse-capture checkerboard approach. *Journal of Endodontia, 36*, 45–52.
121. Alves, F. R., Andrade-Junior, C. V., Marceliano-Alves, M. F., Perez, A. R., Rocas, I. N., Versiani, M. A., Sousa-Neto, M. D., Provenzano, J. C., & Siqueira, J. F., Jr. (2016). Adjunctive steps for disinfection of the mandibular molar root canal system: A correlative bacteriologic, micro-computed tomography, and cryopulverization approach. *Journal of Endodontia, 42*, 1667–1672.
122. Lins, R. X., de Oliveira, A. A., Hirata Junior, R., Wilson, M. J., Lewis, M. A., Williams, D. W., & Fidel, R. A. (2013). Antimicrobial resistance and virulence traits of Enterococcus faecalis from primary endodontic infections. *Journal of Dentistry, 41*, 779–786.
123. Wang, J., Jiang, Y., Chen, W., Zhu, C., & Liang, J. (2012). Bacterial flora and extraradicular biofilm associated with the apical segment of teeth with post-treatment apical periodontitis. *Journal of Endodontia, 38*, 954–959.
124. Love, R. M. (2001). Enterococcus faecalis—A mechanism for its role in endodontic failure. *International Endodontic Journal, 34*, 399–405.
125. Guo, H., Gao, C., Zhang, C., Zheng, S., & Yue, L. (2014). Morphology of bacterial flora in root canals associated with apical abscesses. *Chinese Medical Journal, 127*, 3254–3258.
126. Siqueira, J. F., Jr. (2003). Microbial causes of endodontic flare-ups. *International Endodontic Journal, 36*, 453–463.
127. Langeland, K. (1987). Tissue response to dental caries. *Endodontics & Dental Traumatology, 3*, 149–171.
128. Leonardo, M. R., Rossi, M. A., Silva, L. A., Ito, I. Y., & Bonifácio, K. C. (2002). EM evaluation of bacterial biofilm and microorganisms on the apical external root surface of human teeth. *Journal of Endodontia, 28*, 815–818.

129. Kishen, A. (2010). Advanced therapeutic options for endodontic biofilms. *Endodontic Topics, 22*, 99–123.
130. Zehnder, M. (2006). Root canal irrigants. *Journal of Endodontia, 32*, 389–398.
131. Haapasalo, M., Shen, Y., Wang, Z., & Gao, Y. (2014). Irrigation in endodontics. *British Dental Journal, 216*, 299–303.
132. Ozdemir, H. O., Buzoglu, H. D., Calt, S., Stabholz, A., & Steinberg, D. (2010). Effect of ethylene diaminetetra acetic acid and sodium hypochlorite irrigation on Enterococcus faecalis biofilm colonization in young and old human root canal dentin: In vitro study. *Journal of Endodontia, 36*, 842–846.
133. Neelakantan, P., Cheng, C. Q., Mohanraj, R., Sriraman, P., Subbarao, C., & Sharma, S. (2015). Antibiofilm activity of three irrigation protocols activated by ultrasonic, diode laser or Er:YAG laser in vitro. *International Endodontic Journal, 48*, 602–610.
134. Skóra, B., Krajewska, U., Nowak, A., et al. (2021). Noncytotoxic silver nanoparticles as a new antimicrobial strategy. *Scientific Reports, 11*, 13451.
135. Kishen, A., Shi, Z., Shrestha, A., & Neoh, K. G. (2008). An investigation on the antibacterial and antibiofilm efficacy of cationic nanoparticulates for root canal disinfection. *Journal of Endodontia, 34*, 1515–1520.
136. Hasan, A., & Palmer, R. M. (2014). A clinical guide to periodontology: Pathology of periodontal disease. *British Dental Journal, 2*, 457.
137. Marsh, P. D., Head, D. A., & Devine, D. A. (2014). Prospects of oral disease control in the future: An opinion. *Journal of Microbiology, 6*, 26176.
138. Høiby, N., Bjarnsholt, T., Givskov, M., et al. (2010). Antibiotic resistance of bacterial biofilms. *International Journal of Antimicrobial Agents, 35*, 322–332.
139. Mah, T. F. C., & Toole, G. A. (2001). Mechanisms of biofilm resistance to antimicrobial agents. *Trends in Microbiology, 9*, 34–39.
140. Stewart, P. S. (2002). Mechanism of antibiotic resistance in bacterial biofilms. *International Journal of Medical Microbiology, 292*, 107–113.
141. Wu, H., Moser, C., Wang, H. Z., et al. (2015). Strategies for combating bacterial biofilm infections. *International Journal of Oral Science, 7*, 1–7.
142. Kouidhi, B., Qurashi, Y. M., & Chaieb, K. (2015). Drug resistance of bacterial dental biofilm and the potential use of natural compounds as an alternative for prevention and treatment. *Microbial Pathogenesis, 80*, 39–49.
143. Smith, A. W. (2005). Biofilms and antibiotic therapy: Is there a role for combating bacterial resistance by the use of novel drug delivery systems. *Advanced Drug Delivery Reviews, 57*, 1539–1550.
144. Marsh, P. D. (2009). Dental plaque as a biofilm: The significance of pH in oral health and caries. *Caries Research, 30*(2), 76–87.
145. Suci, P. A., Kang, S., Gmur, R., et al. (2010). Targeted delivery of a photosensitizer to Aggregatibacter actinomycetemcomitans biofilm. *Antimicrobial Agents and Chemotherapy, 54*(6), 2489–2496.
146. Taraszkiewicz, A., Fila, G., Grinholc, M., et al. (2013). Innovative strategies to overcome biofilm resistance. *BioMed Research International, 2013*, 1–13.
147. Sbordone, L., & Bortolaia, C. (2003). Oral microbial biofilms and plaque-related diseases: Microbial communities and their role in the shift from oral health to disease. *Clinical Oral Investigations, 7*(4), 181–188.
148. Cortes, M. E., Bonilla, J. C., & Sinisterra, R. D. (2011). Biofilm formation, control, and novel strategies for eradication. In A. Méndez-Vilas (Ed.), *Science against microbial pathogens: Communicating current research and technological advances. Formatex microbiology series no. 3* (pp. 896–905). Formatex Research Center.
149. Raheja, I., Kholi, K., & Drabu, S. (2013). Periodontal drug delivery system containing antimicrobial agents. *International Journal of Pharmacy and Pharmaceutical Sciences, 5*(3), 11–16.

150. Upadhyay, A., Pal, D., & Kumar, A. (2023). Combinatorial enzyme therapy: A promising neoteric approach for bacterial biofilm disruption. *Process Biochemistry, 129*, 56–66.
151. Horev, B., Klein, M. I., Hwang, G., et al. (2015). pH-activated nanoparticles for controlled topical delivery of farnesol to disrupt oral biofilm virulence. *ACS Nano, 9*(3), 2390–2394.
152. Rediske, A. M., Roeder, B. L., Schaalje, G. B., et al. (2000). Pulsed ultrasound enhances the killing of Escherichia coli biofilms by aminoglycoside antibiotics in vivo. *Antimicrobial Agents and Chemotherapy, 44*(3), 771–772.
153. Kostakioti, M., Hadjifrangiskou, M., & Hultgren, S. J. (2013). Bacterial biofilms: Development, dispersal, and therapeutic strategies in the dawn of the postantibiotic era. *Cold Spring Harbor Perspectives in Medicine, 3*(4), a010306.
154. Harini, G., & Kaarthikeyan, G. (2014). Advanced drug delivery systems in treating periodontal diseases – A review. *Journal of Dental and Medical Sciences, 13*(1), 27–32.
155. Pitt, W. G., & Ross, S. A. (2003). Ultrasound increases the rate of bacterial cell growth. *Biotechnology Progress, 19*(3), 1038–1044.
156. Watters, C. M., Burton, T., Kirui, D. K., & Millenbaugh, N. J. (2016). Enzymatic degradation of in vitro Staphylococcus aureus biofilms supplemented with human plasma. *Infection and Drug Resistance, 9*, 71–78.
157. Yu, M., & Chua, S. L. (2020). Demolishing the great wall of biofilms in Gram-negative bacteria: To disrupt or disperse? *Medicinal Research Reviews, 40*(3), 1103–1116.
158. Lee, Y. H., Park, H. W., Lee, J. H., Seo, H. W., & Lee, S. Y. (2012). Photodynamic therapy on Streptococcus mutans biofilms using erythrosine and dental halogen curing unit. *International Journal of Oral Science, 4*(4), 196–201.
159. Kohanski, M. A., Dwyer, D. J., Hayete, B., Lawrence, C. A., & Collins, J. J. (2007). A common mechanism of cellular death induced by bactericidal antibiotics. *Cell, 130*(5), 797–810.
160. Liang, Y., Hilal, N., Langston, P., & Starov, V. (2007). Interaction forces between colloidal particles in liquid: Theory and experiment. *Advances in Colloid and Interface Science, 134–135*, 151–166.
161. Ikuma, K., Madden, A. S., Decho, A. W., & Lau, B. L. T. (2014). Deposition of nanoparticles onto polysaccharide-coated surfaces: Implications for nanoparticle-biofilm interactions. *Environmental Science. Nano, 1*(2), 117–122.
162. Sahle-Demessie, E., & Tadesse, H. (2011). Kinetics and equilibrium adsorption of nano- particles on synthetic biofilm. *Surface Science, 605*(11–12), 1177–1184.
163. Hajipour, M. J., Fromm, K. M., Ashkarran, A. A., et al. (2012). Antibacterial properties of nanoparticles. *Trends in Biotechnology, 30*(9), 499–511.
164. Selvakumar, R., Aravindh, S., Ashok, A. M., & Balachandran, Y. L. (2014). A facile synthesis of silver nanoparticles with SERS and antimicrobial activity using Bacillus subtilis exopolysaccharides. *Journal of Experimental Nanoscience, 9*(10), 1075–1087.
165. Peulen, T.-O., & Wilkinson, K. J. (2011). Diffusion of nanoparticles in a biofilm. *Environmental Science & Technology, 45*(9), 3367–3373.
166. Stewart, P. S. (2003). Diffusion in biofilms. *Journal of Bacteriology, 185*(5), 1485–1491.
167. Habimana, O., Steenkeste, K., Fontaine-Aupart, M. P., Bellon-Fontaine, M. N., Kulakauskas, S., & Briandet, R. (2011). Diffusion of nanoparticles in biofilms is altered by bacterial cell wall hydrophobicity. *Applied and Environmental Microbiology, 77*(10), 367–368.
168. Ikuma, K., Decho, A. W., & Lau, B. L. T. (2015). When nanoparticles meet biofilms: Interactions guiding the environmental fate and accumulation of nanoparticles. *Frontiers in Microbiology, 6*, 591.
169. Cushing, B. L., Kolesnichenko, V. L., & O'Connor, C. J. (2004). Recent advances in the liquid phase syntheses of inorganic nanoparticles. *Chemical Reviews, 104*(9), 3893–3946.
170. Giersten, E. (2004). Effects of mouth rinses with triclosan, zinc ions, copolymer, and sodium lauryl sulphate combined with fluoride on acid formation by dental plaque in vivo. *Caries Research, 38*(5), 430–435.
171. Sondi, I., & Salopek-Sondi, B. (2004). Silver nanoparticles as an antimicrobial agent: A case study on Escherichia coli as a model for gram-negative bacteria. *Journal of Colloid and Interface Science, 275*, 177–182.

172. Cioffi, N., Torsi, L., Ditaranto, N., Sabbatini, L., Zambonin, P. G., Tantillo, G., et al. (2005). Copper nanoparticle/polymer composites with antifungal and bacteriostatic properties. *Chemistry of Materials, 17*, 5255–5262.
173. Li, Z., Lee, D., Sheng, X., Cohen, R. E., & Rubner, M. F. (2006). Two-level antibacterial coating with both release-killing and contact-killing capabilities. *Langmuir, 22*, 9820–9823.
174. Verran, J., Sandoval, G., Allen, N. S., Edge, M., & Stratton, J. (2007). Variables affecting the antibacterial properties of nano and pigmentary titania particles in suspension. *Dyes and Pigments, 73*, 298–304.
175. Lok, C. N., Ho, C. M., Chen, R., He, Q. Y., Yu, W. Y., Sun, H., et al. (2007). Silver nanoparticles: Partial oxidation and antibacterial activities. *Journal of Biological Inorganic Chemistry, 12*, 527–534.
176. Benn, T. M., & Westerhoff, P. (2008). Nanoparticle silver released into water from commercially available sock fabrics. *Environmental Science & Technology, 42*, 4133–4139.
177. Sotiriou, G. A., & Pratsinis, S. E. (2010). Antibacterial activity of nanosilver ions and particles. *Environmental Science & Technology, 44*, 5649–5654.
178. Kim, J. S., Kuk, E., Yu, K. N., Kim, J. H., Park, S. J., Lee, H. J., et al. (2007). Antimicrobial effects of silver nanoparticles. *Nanomedicine, 3*, 95–101.
179. Pal, S., Tak, Y. K., & Song, J. M. (2007). Does the antibacterial activity of silver nanoparticles depend on the shape of the nanoparticle? A study of the gram-negative bacterium Escherichia coli. *Applied and Environmental Microbiology, 73*, 1712–1720.
180. Elechiguerra, J. L., Burt, J. L., Morones, J. R., Camacho-Bragado, A., Gao, X., Lara, H. H., et al. (2005). Interaction of silver nanoparticles with HIV-1. *Journal of Nanobiotechnology, 3*, 1–10.
181. Han, J., Chen, L., Duan, S., Yang, Q. X., Yang, M., Gao, C., et al. (2005). Efficient and quick inactivation of SARS coronavirus and other microbes exposed to the surfaces of some metal catalysts. *Biomedical and Environmental Sciences, 18*, 176–180.
182. Ren, G., Hu, D., Cheng, E. W. C., Vargas-Reus, M. A., Reip, P., & Allaker, R. P. (2009). Characterisation of copper oxide nanoparticles for antimicrobial applications. *International Journal of Antimicrobial Agents, 33*, 587–590.
183. Vargas-Reus, M. A., Memarzadeh, K., Huang, J., Ren, G. G., & Allaker, R. P. (2012). Antimicrobial activity of nanoparticulate metal oxides against peri-implantitis pathogens. *International Journal of Antimicrobial Agents, 40*, 135–139.
184. Allaker, R. P., & Douglas, C. W. I. (2015). Non-conventional therapeutics for oral infections. *Virulence, 6*(3), 196–207.
185. MacRobert, A. J., Bown, S. G., & Phillips, D. (1989). What are the ideal photoproperties for a sensitizer? *Ciba Foundation Symposium, 146*, 4–12.
186. Pagonis, T. C., Chen, J., Fontana, C. R., Devalapally, H., Ruggiero, K., Song, X., et al. (2010). Nanoparticle-based endodontic antimicrobial photodynamic therapy. *Journal of Endodontia, 36*, 322–328.
187. Wood, S., Metcalf, D., Devine, D., & Robinson, C. (2006). Erythrosine is a potential photosensitizer for the photodynamic therapy of oral plaque biofilms. *The Journal of Antimicrobial Chemotherapy, 57*, 680–684.
188. Seetharam, R. N., & Sridhar, K. R. (2006). Nanotoxicity: Threat posed by nanoparticles. *Current Science, 93*, 769–770.
189. Hagens, W. I., Oomen, A. G., de Jong, W. H., Cassee, F. R., & Sips, A. J. (2007). What do we (need to) know about the kinetic properties of nanoparticles in the body? *Regulatory Toxicology and Pharmacology, 49*, 217–229.
190. Yang, Z., Liu, Z. W., Allaker, R. P., Reip, P., Oxford, J., Ahmad, J., et al. (2010). A review of nanoparticle functionality and toxicity on the central nervous system. *Journal of the Royal Society Interface, 7*, 411–422.
191. Macković, M., Hoppe, A., Detsch, R., Mohn, D., Stark, W. J., Spiecker, E., et al. (2012). Bioactive glass (type 45S5) nanoparticles: In vitro reactivity on nanoscale and biocompatibility. *Journal of Nanoparticle Research, 14*, 966.

192. Panacek, A., Kolar, M., Vecerova, R., Prucek, R., Soukupova, J., Krystof, V., et al. (2009). Antifungal activity of silver nanoparticles against Candida spp. *Biomaterials, 30*, 6333–6340.
193. Parnia, F., Yazdani, J., Javaherzadeh, V., & Dizaj, S. M. (2017). Overview of nanoparticle coating of dental implants for enhanced osseointegration and antimicrobial purposes. *Journal of Pharmacy & Pharmaceutical Sciences, 20*, 148–160.
194. Nel, A., Xia, T., Madler, L., & Li, N. (2006). Toxic potential of materials at the nanolevel. *Science, 311*, 622–627.
195. Luo, M. L., Zhao, J. Q., Tang, W., & Pu, C.-S. (2005). Hydrophilic modification of poly(ether sulfone) ultrafiltration membrane surface by self-assembly of nanoparticles. *Applied Surface Science, 49*, 76–84.
196. McCarron, P. A., Donnelly, R. F., Marouf, W., & Calvert, D. E. (2007). Anti-adherent and anti-fungal activities of surfactant-coated poly(ethylcyanoacrylate) nanoparticles. *International Journal of Pharmaceutics, 340*, 182–190.

Chapter 10
Recent Advances in Antifungal Nanomaterials for Combating Biofilm Infection Caused by *Candida albicans*

Amruta Shelar, Pravin N. Didwal, and Rajendra Patil

10.1 Introduction

The world health organization (WHO) identifies antimicrobial resistance as a top 10 global health threat [1]. In recent years, pathogenic microbes have developed resistance to antimicrobials, putting their effectiveness at risk [2]. Due to the long drug treatment courses, toxicity issues, and drug-drug interactions, the current management of severe infections caused by pathogens is challenging [3]. *C. albicans*, one of the most harmful yeast-like fungi, cause serious health issues and has shown resistance to many antifungal medications [4]. As one of the most important dimorphic-fungal pathogens, morphogenesis between yeast and filamentous forms of *C. albicans* plays a crucial role in initiating biofilm formation. It is known that biofilms can provide protection against host immune defences, drug treatments and thus promote the growth of this fungal pathogen [5]. As a result, classical therapies, such as removing infected lesions or administering antifungal agents, cannot eliminate *C. albicans*, and high drug dosages usually result in severe side effects. After treatment also, the inhibition rate for *C. albicans* infections remains above 30% due to the absence of effective anti-biofilm strategies. Several niches of the body are colonized by biofilm infection of *C. albicans*, including the gastrointestinal tract, female reproductive tract, oral cavity, and skin. Infections caused by *C. albicans*

A. Shelar
Department of Technology, Savitribai Phule Pune University, Pune, India

P. N. Didwal (✉)
Department of Materials, University of Oxford, Oxford, UK
e-mail: pravin.didwal@materials.ox.ac.uk

R. Patil (✉)
Department of Biotechnology, Savitribai Phule Pune University, Pune, India
e-mail: rpatil@unipune.ac.in

R. S. Chaughule, A. S. Lokur (eds.), *Applications of Nanotechnology in Microbiology*, https://doi.org/10.1007/978-3-031-49933-3_10

biofilm formation account for 40% of clinically significant bloodstream infections, which are the leading cause of hospital-acquired infections [6].

It is important to note that this organism forms biofilms on the host tissues and on implanted medical devices such as heart valves and catheters etc. The biofilms of *C. albicans* consist of yeast, pseudohyphals, and hyphal cells surrounded by extracellular polymeric matrix (EPM). In order to form biofilm, *C. albicans* undergo a multi-step process that involves adhesion, proliferation, maturation, and dispersion [7]. A mature biofilm with EPM is effective in protecting fungal cells from adverse environmental conditions, immune attacks (particularly neutrophils), and the development of reactive oxygen species (ROS), which are highly detrimental to the organisms, as well as protecting them from antifungal agents [8]. Various molecular mechanisms and cellular factors have been associated with drug resistance in *C. albicans*, including changes in the expression and function of drug efflux pump proteins, membrane fluidity, membrane lipid composition (sterols and fatty acids), the ability to elude host defences, adhesion, biofilm formation on deep-seated host tissues, hyphal morphogenesis, and the production of enzymes that damage tissues. In order to treat the infection effectively, it is crucial to disrupt the biofilm and inhibit hyphal morphogenesis in *C. albicans* [9]. Innovative approaches to inhibiting and disrupting biofilm formation are deemed necessary due to the issue of drug resistance and the critical role that biofilms play in *C. albicans* pathogenesis.

Medical nanotechnology has received considerable interest and attention in recent years. Due to their large surface-to-volume ratios and ability to control their physicochemical properties, nanoscale materials have been adopted for a wide range of antimicrobial applications. Antimicrobial drug resistance can be controlled with NPs that contain antimicrobial NPs or combine nanomaterials with existing drugs [10]. Several NPs, including silver, gold, copper, zinc oxide, iron oxide, titanium oxide, and zirconium oxide, as well as combinations of NPs, including graphene oxide coated with chitosan/guanidine, and graphene oxide coated with chitosan and iron oxide, showed potential antimicrobial properties against biofilm-forming pathogens [11–13].

Antimicrobial nanomaterial applications in medicine have resulted in the development of a new field that can provide novel treatment options for *C. albicans* biofilm-causing infections. As shown in Fig. 10.1, developing antimicrobial nanomaterials is seen as a promising strategy to control or treat *C. albicans* biofilm-causing infections on indwelling medical devices [14]. NPs loaded with drugs could overcome limitations associated with conventional antibiotic treatments such as toxicity, inadequate delivery, or enzymatic degradation. The bioactivity, biocompatibility, low toxicities, and anti-inflammatory and anti-immunogenic properties of hydroxyapatite, chitosan, collagen, silica, and titanium dioxide make these NPs ideal for incorporation of antimicrobials [15]. Recent developments have combined NPs with novel physical approaches for irreversible thermal damage to cell surfaces and eradication of biofilm-forming *C. albicans*. A noninvasive, on-demand treatment for external *C. Albicans* biofilm infections could possibly be developed based on these promising developments. These potential effects happen due to the unique

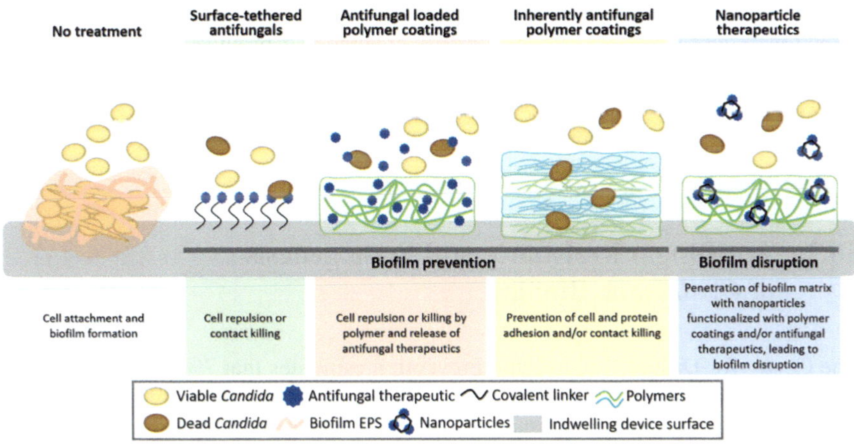

Fig. 10.1 The use of nanomaterials to combat surface-associated *C. albicans* biofilms. The use of NPs and direct surface functionalization with antifungal NPs may enhance penetration of dense biofilm matrix and the potential for targeting pathogenic fungal cells. As a result of these strategies, it may be possible to prevent the formation of biofilms by preventing the initial attachment of fungi to surfaces and eradicating existing biofilms as well. (This is a reprinted image of Ref. [17] with permission)

properties and phenomena that appear at the nanometer scale that cannot be found in bulk materials, such as physicochemical and biological properties [16].

Compared to biomolecules of various antimicrobials, nanomaterials have a much higher surface area-to-volume ratio, which enhances their chemical reactivity, solubility and bioactivity. Furthermore, NPs can damage cell membranes and deoxyribonucleic acid (DNA) irreversibly because they are small enough to penetrate microbial cell walls [15]. Moreover, they have long plasma half-lives and high surface-to-volume ratios, which facilitate the high loading of drugs. A variety of surface-engineered NPs such as metal NPs, polymer NPs, metal-polymer composites, biologically active NPs, ROS-releasing NPs, and stimuli-responsive smart NPs are being proven to successfully preventing or controlling *C. albicans* biofilm-related infections on medical devices [15]. Therefore, the use of nanomaterials in drug delivery systems has become a hot topic because of their ability to carry, protect, and stabilize therapeutic payloads. Nanomaterials, such as metal, metal-oxide NPs, and polymeric NPs, have been shown to be antibacterial against many pathogenic bacteriain laboratory tests.

Researchers have paid more attention to the prevention of bacterial biofilm infections and have paid less attention to the effects of different nanomaterials on fungal biofilm-forming pathogens like *C. albicans*. Now, researchers are more interested in investigating the antifungal properties of various organic and inorganic nanomaterials to prevent biofilm infection caused by *C. albicans*. Up till now, there are no more studies that investigated the potential to prevent the growth of biofilm-forming *C. albicans*. However, much research remains to be done to determine their exact mechanism of action. The purpose of this chapter is to summarize current strategies of nanotechnology-based approaches in the control or eradication of biofilm-related infections caused by *C. albicans*. In this chapter, we provide an overview of a biofilm-forming pathogen *C. albicans,* with drug resistance issues, nano-enabled strategies for tackling *C. albicans* biofilm infections for antifungal purposes, and a nanomaterial-based anti-biofilm mechanism in *C. albicans* biofilms.

10.2 Antifungal Resistance and Pathogenesis of *C. albicans* Biofilm

The Centre for Disease Control and Prevention estimates that 7% of all blood samples are from patients suffering from *C. albicans* which is resistant to drug fluconazole [18]. However, azole resistance has persisted for the past 20 years, while echinocandin resistance and other drug resistance are a major concern. Therefore, because of the ever-increasing population, this situation can be lethal. It is important to address antifungal drug resistance immediately since it has arisen as a result of its inappropriate use. As a result of overuse of antibiotics, the normal microflora of humans is affected, which makes it possible for *C. albicans* to grow in a favourable environment [19]. It is challenging to fight the problem of drug resistance despite a vast array of antifungal drugs. Under normal conditions, *C. albicans* colonizes mucosal surfaces without causing any symptoms; however, any disruption in the host environment or conditions of immune dysfunction can cause it to proliferate and invade virtually anywhere in the body. The virulence factors of this highly adaptable fungal species allow it to transition from commensal to pathogen. The ability to switch morphologies and form biofilms are both central to the pathogenesis of *C. albicans* [5]. Approximately 80 percent of *C. albicans* infections are caused by biofilms formed on host or abiotic surfaces, including indwelling medical devices that carry a high death and morbidity rate. As a result, *Candida* biofilms remain relatively resistant to antimicrobial therapy since *C. albicans* biofilms are inherently tolerable to antimicrobial therapy. Since biofilm-associated infections can be treated with novel antifungal therapies, their demand has been increasing several-fold in recent years [9]. The formation of biofilms is due to sessile *C. albicans* attached to either abiotic or biotic surfaces and encapsulated in an exopolymeric matrix, resulting in new phenotypic characteristics and intrinsic resistance to host immune response and antimicrobial drugs. Biofilms of *C. albicans* are complex associations of hyphal cells that are associated with both human and abiotic tissues. Several factors involved in the formation and maturation of biofilms have been discovered, due to the challenges linked with biofilm-associated diseases [19].

Biofilm formation by *C. albicans* is still a mystery both in terms of pathogenesis and virulence mechanisms. To understand the aetiology of biofilm formation, it is necessary to comprehend *C. albicans* biofilm formation by analysing its transcriptional regulatory network. According to Nobile et al., transcription regulators form a complex and interconnected network with thousands of genes in controlling

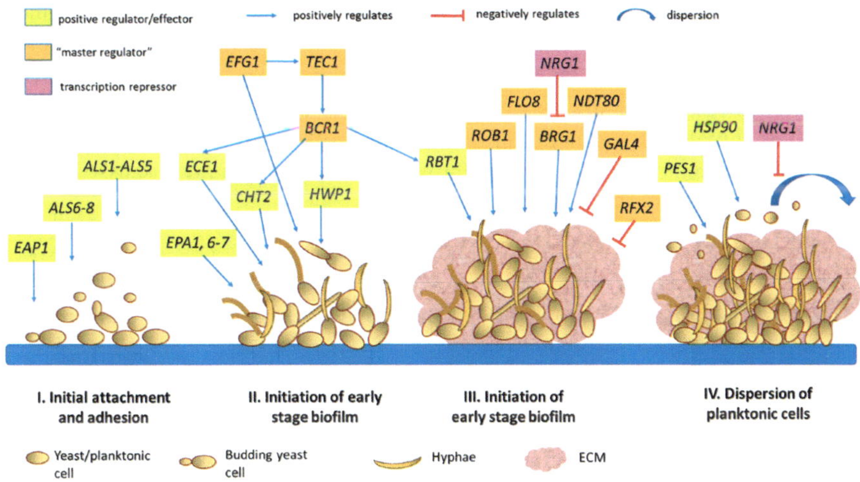

Fig. 10.2 An overview of the involvement of different transcription factors, master regulators, and effectors in the complex biofilm regulatory circuitry of *C. albicans* according to their distinct stages of development. (This is a reprinted image of Ref. [21] with permission)

biofilm formation, in which most of the target genes are regulated by two or more transcription factors [20]. As shown in Fig. 10.2, different transcriptional factors are involved in biofilm formation at their different stages. A positive regulator is responsible for binding specific proteins (activators) to initiate transcription. A master regulator gene is an important gene in a gene regulation hierarchy, particularly when regulating cell fate or differentiation. A transcriptional repressor inhibits transcription factors and RNA polymerase from accessing gene promoters, thus providing an alternative mechanism for genetically controlling autophagy [20]. *C. albicans* cells initiate biofilm formation by switching on positive regulators, Als 1–8 and Eap 1 at the time of attachment and adhesion. In the early stages of biofilm formation, Efg1, Tec1, Bcr1, Hwp1, Ece1, Cht2, Epa 1,6–7 factors are active to start morphogenesis. In next stage of biofilm formation Rbt1, Rob1, Brg1, Ndt80, Flo8 are active to take biofilm formation in next stage by forming extracellular matrix secretion (ECM). In this process Nrg1, Gal4 and Rfx2 act as negative regulators for biofilm formation. In final stage of biofilm, Pes1 and Hsp90 act as positive regulator and Nrg1 as a negative regulator for dispersion of planktonic cells [20].

The development of biofilms occurs in sequential steps over a period of 24–48 h. In adherence step, the initial attachment of one yeast cell to the substrate forms the foundation for the subsequent adherence of yeast cells. Following this adherence step, initiation phase is the proliferation phase in which the hyphal cells grow and grow into filamentous structure [19]. Next step is the maturation step, in after the formation of hyphae (hyphae assembly), an extracellular matrix (ECM) is accumulated during the maturation of the biofilm. Finally, in dispersal step, non-adhering yeast cells separate from the biofilm and disperse into the surrounding environment for a suitable attachment site [22]. *C. albicans* biofilm formation consists of initial yeast cell adherence (0–2 h), then germination and microcolonies are formed (2–4 h), filamentation (4–6 h), monolayer development (6–8 h), proliferation (8–24 h), and maturation (24–48 h) [19]. As biofilm-associated yeast cells spread throughout the host cells and tissues, they can cause invasive disease or Candidaemia as well as initiate the formation of new biofilms. A variety of factors have been reported to promote *C. albicans* biofilm pathogenesis, which are discussed in the following section.

10.2.1 Quorum Sensing (QS) Mechanism

Recent studies have shown that *C. albicans* regulates its morphogenic shift through changes in cell density. Observations have shown that *C. albicans* inoculated around 10^7 cells/mL under conditions that favour hyphal morphogenesis (pH 7.5, 37 °C) to form hyphae. Morphogenesis appears to be controlled by cell density in a similar way to how bacterial cells regulate their activities via QS [23]. Therefore, *C. albicans* are capable of quorum sensing, which result in to the biofilm formation as a strongest virulence factor in *C. albicans*. A density-dependent cell-cell communication mechanism allows auto-inducers (signalling molecules) to be released as a response to an increasing density, thereby enhancing or repressing certain genes or factors activation. QS has been shown to indirectly regulate virulent genes in *C. albicans*, leading to the emergence of multi-drug resistant strains. Therefore, alternative strategies need to be found to target and restrain QS [24].

10.2.2 Formation of Extracellular Matrix (ECM)

In biofilms of *C. albicans*, microbial communities are embedded in a 3D extracellular matrix. An important feature of biofilms is the extracellular matrix, which forms a dense and extensive structure that protects adherent cells against the host immune system and antifungal agents. Several pioneer works have shown that biofilms of *Candida* species can increase in matrices when highly dynamic flow environments affect the biofilm, and the quantity varies widely depending on the strain and species. Biofilms possess a unique virulence due to their complex matrix of extracellular polymeric substances. The extracellular matrix of mature biofilms is composed mainly of proteins (55%), carbohydrates (25%), lipids (15%), and DNA (5%). Despite the fact that protein comprises the largest proportion of the matrix, only a few things are known about each protein's function. In the extracellular matrix of *C. albicans*, approximately 12,000 residues of α-1,2-branched α-1,6mannan make up one abundant high-molecular-weight component. The polysaccharide residues in the cell wall of *C. albicans* are nearly ten times greater than the mannans. The biofilm matrix also contains nucleic acid and lipids which are less abundant macromolecules. Among the lipids found in *C. albicans* biofilms are phospholipids (primarily phosphatidylcholine and phosphatidylethanolamine), sphingolipids, and eicosanoids [25].

10.2.3 Extracellular DNA (e-DNA) and Genetic Factors

e-DNA inside the extracellular matrix contributes significantly to the stability of *C. albicans* biofilms. As e-DNA is present in *C. albicans* biofilms, the biofilm matrix becomes thinner when the biofilm-forming microbes are treated with DNAase enzyme and the respective drugs. It has been found that each of these genetic factors acts together and interacts with distinct genes in order to regulate and generate biofilms; therefore, a new understanding of the process of biofilm formation has been gained [26].

10.2.4 Efflux Pumps (EP) Regulation

An EP helps to regulate *C. albicans* internal environment by expelling toxic substances, quorum sensing molecules (auto-inducers), biofilm formation molecules, and virulence factors. *C. albicans* is highly resistant to antifungal drugs due to overexpression of EP, which lead to sequestration by pumping out the antifungal drugs. A normal antifungal treatment for planktonic cells increases the expression of EP to prevent the accumulation of antifungal drugs within the cell. There are two major classes of EP that control drug exportation in *C. albicans*, the Cdr1, 2 (Adenosine triphosphate (ATP) binding cassette transporter superfamily) and the Mdr1 (Major facilitator transporter superfamily). Furthermore, initially up-regulated biofilm genes remain up-regulated throughout the growth of the biofilm, even if no antifungal drugs are used. The EP is rapidly regulated during the biofilm formation process [27].

10.2.5 Persister Cell and Stress Response

Resistance is also greatly influenced by persister cells, which are inconsequential yeast cells with low metabolic activity that are raised as phenotypic variants, but not mutants. Cells within biofilms possess a high level of resistance. However, they regain their active metabolism in a stressful situation and re-establish as biofilms. Drug sequestration by persister cells is caused by some virulent traits, such as hyphal growth and not by efflux pumps or cell membrane structure. Persistent cells were discovered after amphotericin B treatment on *C. albicans*. Due to their ability to remain dormant, these cells are regarded as persisters [28].

The above are all key virulence factors that are involved in the pathogenesis of *C. albicans* biofilm infections and contributing to drug resistance. Antifungal drugs are available in a wide variety, but there is a persistent problem of drug resistance, making it difficult to overcome *C. albicans* biofilm infections. In addition, polyenes and azoles are ineffective against *C albicans* biofilms, thereby limiting treatment options. Therefore, new antifungal therapies with high efficiency against biofilm mode of growth must be developed urgently.

10.3 Nano-enabled Strategies for Developing Therapeutics Against *C. albicans* Biofilm Infection

NPs have the potential to inhibit virulence factors in biofilm-forming *C. albicans*; they inhibit morphogenesis, surface adhesion, and penetrating ECM to target pathogenic fungal cells, thus making them an attractive treatment approach for combating biofilm infections caused by *C. albicans* [11]. Metal, metal oxide, and polymeric NPs are a few examples of inorganic NPs that have been proposed for their excellent antifungal activity against biofilm-forming *C. albicans*. Potential inorganic NPs for biofilm inhibition are discussed as follows:

10.3.1 Metal NPs for Combating Biofilm Formation of C. albicans

Metal NPs are the most widely used type of nanoparticle, due to their unique properties and wide range of antifungal applications. Metal NPs are submicron-sized particles made of pure metals (e.g., gold, platinum, silver, titanium, zinc, iron etc.). Antifungal applications of metal-based NPs have been extensively investigated [11].

Silver NPs are very famous antimicrobial agent in the development of therapeutics. Silver NPs can release silver ions continuously, which may contribute to the inhibiting of *C. albicans.* Silver ions adhere to cell walls of fungus due to electrostatic attraction and affinity for sulphur proteins. Lara et al. [29] demonstrated the effect of silver NPs on *C. albicans* biofilms by ultrastructural analysis [29]. Silver NPs demonstrated a potent dose-dependent inhibitory effect on *C. albicans* biofilm formation, with 50% inhibitory concentration (IC_{50}) of 0.089 μg/mL. Furthermore, silver NPs demonstrated efficacy when tested against pre-formed *C. albicans* biofilms, displaying an IC_{50} value of 0.48 μg/mL. A change in the appearance of the yeast surface from smooth to rough was visible under scanning electron microscopy (SEM) after silver NPs treatment. This indicated outer cell wall damage. As shown in Fig. 10.3, Lara et al. concluded that silver NPs inhibited filamentation and also inhibited biofilm formation at very low concentrations [29]. Ahamad et al. [30] examined the antibiofilm activities of biogenic silver NPs against *C. albicans* [30].

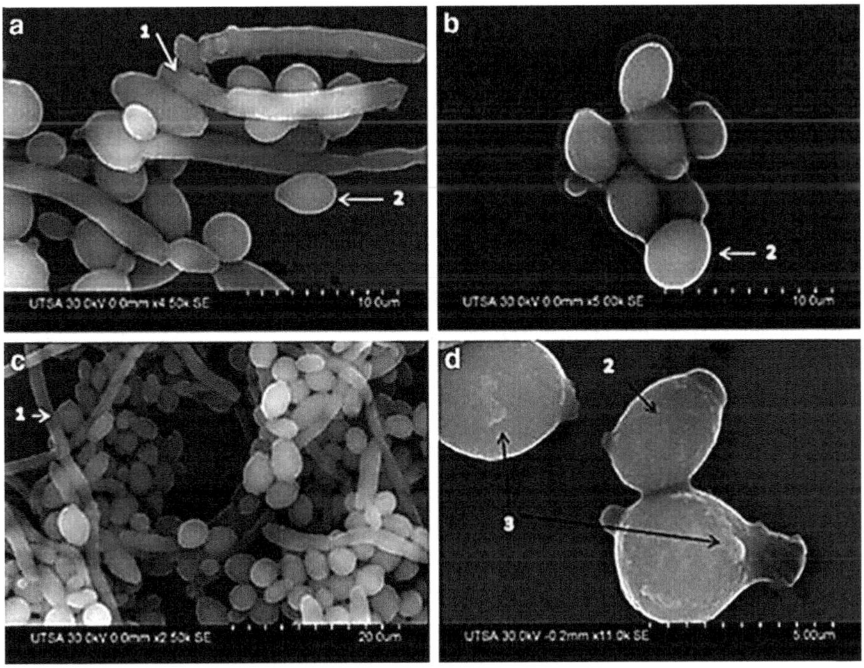

Fig. 10.3 A SEM image demonstrating inhibition of *C. albicans* biofilm formation by silver NPs. (**a**) and (**c**) are biofilm without silver NPs treatment (1. True hyphae and 2. Yeast), (**b**) and (**d**) are inhibited biofilm and hyphae formation after treatment with silver NPs. (black arrow indicated rough cell wall and disruption of *C. albicans*). (SEM images a, b, c and d are at the scale of 100, 100, 200 and 500 μm) (This is a reprinted image of Ref. [29] with permission)

A minimum inhibitory concentration (MIC) of silver NPs was determined at 12.50 μg/mL against *C. albicans*. Silver NPs at 25 μg/mL concentration displayed 79% permeability of *C. albicans* cell membranes and 22.2% ROS generation. Similarly, silver NPs (25 μg/mL) inhibited biofilm growth and degradation by 62.5%. It is therefore possible to consider silver NPs as an antifungal AUent that could be effective in controlling *C. albicans* biofilms [30]. Takamiya et al. [31] synthesized biocompatible silver NPs and incorporated them in acrylic resins for dental applications to inhibit the biofilm of *C. albicans* [31]. Silver NPs incorporated into acrylic resin at concentrations of 0.05 and 0.5% reduced the growth of *C. albicans* biofilms. Silver NPs incorporated into acrylic resin at concentrations of 0.05 and 0.5% reduced the growth of *C. albicans* biofilms. SEM study revealed that the silver NPs inhibited *C. albicans* biofilm in comparison with control. Based on their results, silver NPs addition into acrylic resin at 0.05 and 0.5% exhibited antimicrobial effects against *C. albicans* biofilm, did not interfere with flexural strength, and was considered biocompatible [31].

It has been demonstrated in a study by Li et al. [32] that silver NPs contained in an acrylic resin for denture bases inhibit *C. albicans* adhesion and biofilm formation. In the presence of increasing concentration of silver NPs, the bioactivity and biomass of the *C. albicans* biofilm gradually decreased. Low concentrations of denture base resin containing silver NPs had no effect on adhesion, but high concentrations (5%) exhibited anti-adhesion activity [32]. Increasing nano-silver concentrations slowly decreased the thickness and live/dead cell ratios of biofilms formed on resin specimens over a 72-h period [32].

There has also been considerable research on gold NPs because of their physicochemical properties, chemical resistance, ease of synthesis, and small size. Since the Food and Drug Administration (FDA) approved the use of gold NPs in medicine applications, research and development have been intensified around these gold NPs. Khan et al. [33] demonstrated that gold NPs can enhance methylene blue-induced photodynamic therapy to inhibit *C. albicans* biofilms. Microscopic studies showed a significant reduction in biofilm in the presence of conjugate and an adverse effect on *C. albicans* cells. A fluorescent spectroscopic study confirmed that type I phototoxicity attacked biofilm. The use of gold nanoparticle conjugate-mediated photodynamic therapy may be used to treat nosocomially acquired refractory *C. albicans* biofilms [33]. The effects of gold NPs on pathogenic biofilm formation and invasion of host cells were demonstrated by Yu et al. [34]. Accordingly, this study demonstrated that gold NPs formed by as-synthesized processes are strongly inhibitory of pathogenic biofilm formation and invasion of host cells. There is anti-pathogenic effect of NPs due to strong electrostatic interactions between NPs and pathogenic cells. A novel application of gold NPs for fighting clinical pathogens has been revealed by these results [34].

10.3.2 Metal-Oxide NPs for Combating Biofilm Formation of **C. albicans**

Zinc oxide NPs are one of the most common metal-oxide NPs in medicine due to their unique medicinal properties, due to this makes zinc oxide NPs ideal for biological labelling, drug delivery, gene delivery, and nanomedicine. According to Hosseini et al. [35], zinc oxide NPs exhibit antifungal activity on *C. albicans* biofilms formed on urinary catheters [35]. A higher reduction of fluconazole-resistant *C. albicans* biofilm biomass was observed with zinc oxide NPs in comparison with susceptible isolates. As a result of these findings, zinc oxide NPs may serve as a beneficial treatment for catheter-related urinary tract infections [35]. In a study published by Joshi et al. [36] demonstrated that biogenic hierarchical zinc oxide NPs inhibit *C. albicans* biofilms [36]. In this study used fragmented lignin to synthesize zinc oxide NPs to inhibit planktonic growth, biofilm formation, and morphogenesis. In addition, real-time polymerase chain reaction (RT-PCR) was used to study the molecular mechanisms of virulence inhibition of fragmented zinc NPs. Results of RT-PCR showed that down regulation of phr1, phr2, efg1, hwp1, ras1, als3 and als4 genes inhibited the virulence of *C. albicans* [36]. A study by Rosenberg et al. [37] demonstrated that zinc oxide NPs have selective antibiofilm properties and are biocompatible [37]. It was shown that nano zinc oxide coated surfaces inhibited the formation of *C. albicans* biofilms. Zinc oxide NPs coated surfaces appears to be the most effective method of inhibiting *C. albicans* biofilm formation across a broad spectrum of surfaces. In a study published by Jalal et al. [39] evaluated the anticandidal activity of bioinspired zinc oxide NPs and their effect on the growth, cell morphology, and key virulence characteristics of *Candida* species [38]. Zinc oxide NPs penetrated inside the cell and caused extensive damage to the cell wall and membrane, as shown by SEM and TEM images [38]. Zinc oxide NPs inhibit *C. albicans* growth and various virulent factors, providing insight into their therapeutic application for *Candida*-associated infections.

There are many advantages to iron oxide NPs, including their high chemical and thermal stability, low cost, and extremely high area-to-volume, as well as their superparamagnetic behaviour. The use of iron oxide NPs as drug and drug delivery systems has been explored as a way to enhance the effectiveness of antifungal drugs and to reduce the concentrations of antifungal drugs. A study conducted by Esfahani et al. [39] demonstrated that magnetic iron oxide NPs affect the expression of biofilm-associated genes in *C. albicans* [39]. It inhibited the growth of *C. albicans* and the formation of biofilms with nano-iron oxide. It was found that iron oxide NPs inhibited the growth of *C. albicans* at MICs ranging from 50 to 200 μg/mL. Iron oxide NPs may be effective in treating biofilm-associated infections by targeting biofilm-associated genes in *C. albicans* [39]. The study by Salari et al. [40] evaluated biofilm formation ability in *C. albicans* as well as the anti-biofilm effects of iron oxide NPs compared to standard drug fluconazole in vitro [40]. A reduction in biofilm formation was found in *C. albicans* after exposure to various concentrations of iron oxide NPs was greater than that of fluconazole. According to Pugazhendhi et al. [41], iron-doped copper oxide NPs possess photocatalytic and antimicrobial properties against the pathogenic yeast *C. albicans* [41]. Iron-doped copper oxide NPs were tested in vitro against the pathogenic fungus *C. albicans* for their antibiofilm potential. Hence, iron-doped copper oxide NPs are potential candidates to target the biofilm infections caused by *C. albicans*.

Additionally, metal-oxide NPs such as copper oxide have shown their potential in pharmaceutical and biomedical applications, such as antibacterial, antifungal, anticancer, and drug delivery. Copper oxide NPs inhibit growth and biofilm-forming virulence factor, yeast-to-hyphae transition in *C. albicans*, as demonstrated by Padmavathi et al. [42]. A real-time PCR analysis revealed that copper oxide NPs suppressed yeast-to-hyphae morphological switching by down-regulating cph1, hst7, and ras1 and by up-regulating the negative regulator tup1. In a study by Raj et al. [43], L-cysteine-capped copper oxide nanocarriers inhibited the growth of fluconazole-resistant *C. albicans* biofilms [43]. Antibiofilm potency of copper oxide NPs against *C. albicans* was investigated. Nanosized copper oxide has high penetration power and retention in biofilms due to its small size. A concentration of 100 μg/mL of copper oxide NPs showed a significant antibiofilm effect against *C. albicans*. A study by Garcia-Marin et al. [44] shown that biosynthesized copper oxide NPs are highly antifungal against *C. albicans* [44]. MIC of copper oxide NPs against *C. albicans* was 35.5 μg/mL. As a result of exposing *C. albicans* to different concentrations of copper oxide NPs, ROS production was identified. The Martins et al. [26] showed that copper metallic nanostructures and hierarchical copper oxide marigold microstructures grown in situ had dual antifungal properties against *C. albicans* [45]. It was found that copper oxide NPs inhibited the growth of three strains of the biofilm-forming *C. albicans*. Moreover, copper oxide NPs disrupt the cell wall, causing cytoplasmic leakage.

10.3.3 Polymeric NPs for Combating Biofilm Formation of **C. albicans**

In one study, Zhang et al. [46] investigated the antifungal effects and mechanical properties of silver bromide/cationic polymer nanocomposites supported by Poly-methyl methacrylate-based dental resins against *C. albicans* biofilm [46]. Silver bromide/cationic polymer nano-composite-modified Poly-methyl methacrylate-based dental resin significantly inhibits *C. albicans* biofilm growth. Saleh et al. [61] demonstrated that biodegradable polymer-based NPs can enhance the efficacy of fluconazole against *C. albicans*. In the present study encapsulation efficiency of fluconazole drug is about 40% for nanosized particles. Alginate/chitosan-based formulations released fluconazole more slowly over 24 h but at a greater rate than free fluconazole suspensions. In this study, fluconazole's MIC value was reduced by 20 times using alginate/chitosan-based formulations [61]. According to Iadnut et al. [47], biosynthesized ethanolic extracts of propolis-loaded poly(lactic-co-glycolic acid) NPs possess antifungal and anti-virulence properties against *C. albicans* [47]. It was found in this study that NPs inhibit the growth of *C. albicans* and their virulence by reducing the genes that encode virulence-associated hyphal-adhesion proteins, disrupting their morphology and reducing their virulence. Costa et al. [48] reported the development, characterization, and in vitro/in vivo evaluation of polymeric NPs containing miconazole and farnesol for treating candidiasis of the vulvovaginal area [48]. Based on the results, it can be concluded that chitosan NPs containing miconazole and farnesol are effective at inhibiting fungal proliferation. Moreover, chitosan NPs containing farnesol decreased the pathogenicity of infection, as demonstrated by the absence of inflammation [48]. Gürsu et al. [49] reported potential antibiofilm activity of farnesol-loaded (poly(D,L-lactide-*co*-glycolide)) (PLGA) NPs against *C. albicans* [49]. The results demonstrate that PLGA NPs can inhibit growth and biofilm formation of *C. albicans* at lower concentrations than farnesol alone by 57%, but further research is required to determine the effectiveness of PLGA NPs.

Above all, the case studies demonstrated the potential of NPs to combat biofilm infections caused by *C. albicans*. Despite this, the exact molecular mechanism of antifungal activity of NPs has yet to be fully elucidated. In the next section, we will discuss some studies that reveal possible mechanism for inhibiting biofilm formation.

10.4 Anti-biofilm Mechanism of Nanomaterials in *C. albicans* Biofilm Infection

As far we learn that, NPs are excellent candidates for treating *C. albicans* biofilm infections because of their strong antimicrobial properties. Biofilm formation is inhibited by NPs, which penetrate into the extracellular matrix of biofilms, increasing their penetration into the matrix and inhibiting their formation [19]. There's an ongoing research effort to understand the molecular mechanism of different NPs. As shown in Fig. 10.4, the antifungal effects of metal NPs are mediated by ions released, oxidative and nitrosative stress, membrane and cell wall damage, inhibition of enzymes, gene expression regulation, decreasing ATP levels, and DNA, protein, and mitochondrial dysfunction. NPs with different structural characteristics have different biological function, specifically generating ROS (ROS) at different rates. ROS production by NPs is correlated with particle size, shape, surface area, and chemistry. Free radicals and ROS can be produced by the released ions of NPs [36, 50]. Basically NPs create oxidative stress by releasing ions, oxidative stress occurs when ROS and free radicals are imbalanced. The ROS generated by metallic NPs play many roles in cellular biology, including contributing to the toxicity of metallic NPs and inhibiting cell proliferation and differentiation by modulating cellular signalling [51]. It is also possible for the ions to interact with glutathione's thiol group and convert it to glutathione disulphide. A disruption of the cell

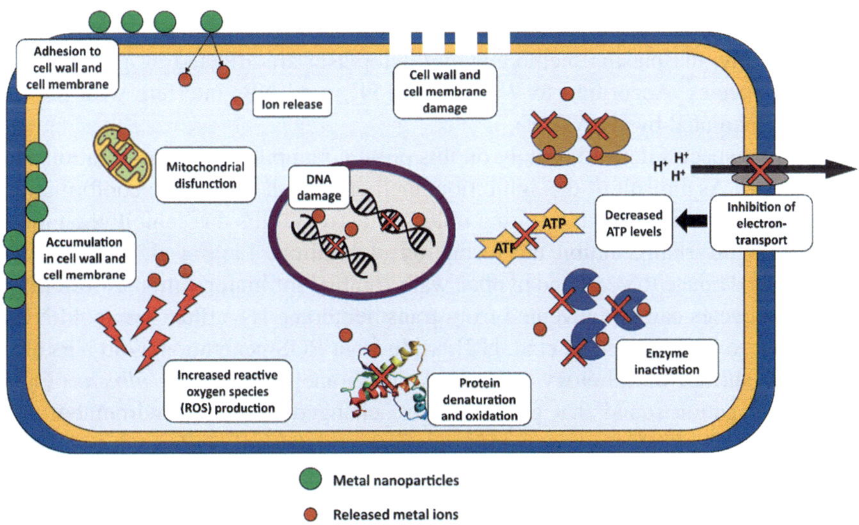

Fig. 10.4 Mechanisms of action of metallic NPs against *C. albicans*. *C albicans* are inhibited by metal NPs in the following ways: metal ion release, adhesion, accumulation, and damage to the cell wall and cell membrane, inhibition of electron transport with a reduction in ATP levels, enzyme inactivation, protein oxidation and denaturation, increased ROS, mitochondrial dysfunction, and DNA damage are among the effects. (This is a reprinted image of Ref [11] with permission)

scavenging mechanism is also reported by Zhang and Tang [52] in addition to interactions with other enzymes. In addition to disrupting electron transport chain (ETC), the ions also damage the cell membrane. Furthermore, ROS generation disrupts mitochondrial membranes, facilitates DNA breakdown, and ultimately leads to cell death [51]. In addition to this, the ions also interact with the signal transduction pathway, which is crucial for mediating many cellular processes. Nanoparticle activity depends on the quantity and quality of the ions released.

In general, NPs adhere to *C. albicans* membrane surfaces. After NPs adhere and permeate the membrane, the cell becomes highly susceptible to metal ions. NPs penetrate *C. albicans* cells easily due to the surfactant used during synthesis. This adherence leads to an increase in membrane permeability and changes in the lipid bilayer. As the NPs release ions inside the cell, these interact with lipids, DNA, proteins and a variety of other biomolecules. During nanoparticle interactions with DNA, DNA is damaged and mutations are induced that are detrimental to a *C. albicans* health. Moreover, the ions interfere with nucleic acid bases (purine and pyrimidine), which leads to a destabilization of the single-sheet and double-helix structures. These processes all work together to inhibit DNA replication [30]. As expected, transcription and translation are also inhibited, making protein synthesis extremely difficult. Metal ions can also interact with the proteins that have already been synthesized, disrupting their structure and binding properties.

Many researchers have described the molecular mechanism involved in inhibiting the formation of *C. albicans* biofilms as follows: Using ROS generated by silver NPs, Hwang et al. [53] found mitochondrial dysfunctional apoptosis, externalization of phosphatidylserine, and nuclear DNA fragmentation in *C. albicans* [53]. In another study, the plasma membrane and cytoplasm are affected by gold NPs in *Candida* species. According to Wani et al. [54], gold NPs interfere with proton pumping mediated by H^+-ATPase.

Candida species depend heavily on this protein pump for their proliferation and growth [54]. As a result of this inhibition, the normal cell membrane conformation and architecture are altered, resulting in loss of activity. NPs that inhibit yeast morphogenesis invariably inhibit the formation of biofilms. Lara et al. [29] demonstrated that damage to yeast and hyphae walls inhibit biofilm formation as well [29]. *Candida* species can be prevented from transmembrane H+ efflux with gold NPs. According to Radhakrishnan et al. [12], aside from ROS generation, gold NPs also alters cell surface morphology, as well as membrane fluidity in *C. albicans* [12]. This investigation found that gold NPs also changed the microenvironment and architecture of cells. Gomez et al. [55] demonstrated that gold NPs inhibits *C. albicans* by disrupting enzyme, β-glucan synthase. This fungicidal activity is due to silver NPs effect on *C. albicans* cell wall structure and loss of its mechanical resistance. A study published by Muthamil et al. [56] found that silver NPs synthesized via green synthesis had antifungal activity in vitro and in vivo against *Candida* species [56]. According to this study, silver NPs reduced yeast-to-hyphae transition, exopolysaccharide production, and SAP production in *C. albicans* significantly. A study by Jalal et al. [39] showed that silver NPs inhibited cell growth, hyphae formation, biofilm formation, proteinase production, and hemolysin production [38].

Results of this study suggested that, silver NPs act not only on biofilm cells, but can also penetrate and disrupt extracellular matrix, the investigation indicates. The study by Lee et al. [57] proved that silver NPs disrupt the membrane architecture of *C. albicans*. A number of processes were inhibited, including budding and subsequent cell death. There was a report that this was caused by the formation of pits and holes on the cell. As Halbandge et al. [58] demonstrate, biogenic silver NPs suppress Ras-mediated signalling pathway in *C. albicans* by suppressing Ece1, Tec, Tup1, Rfg1 (key genes that regulate yeast to hyphal transition) [58]. Rasool et al. [59] and Padmavathi et al. [42] demonstrated that copper oxide NPs (Copper oxide NPs) inhibited biofilm formation in *C. albicans*. The study found that copper oxide NPs inhibit yeast hypae transformation. In *C. albicans*, copper oxide NPs generated ROS, while cupric oxide NP induced membrane damage. Copper oxide NPs exhibited better antifungal activity than cupric oxide NPs in comparison to both types of copper oxide NPs. It is possible to increase the activity of NPs by combining them with antibiotics and other NPs. can also be coated with several compounds and materials to enhance their bioavailability, distribution, and antimicrobial effects. In a recent study, Cheong et al. [60] reported a synergistic antifungal effect between polyethylene glycol (PEG)ylated graphene oxides and copper NPs against *C. albicans* [60]. It has been found that GO-PEG copper NPs cause extensive shrinkage and porosity deformation in *C. albicans*.

In summary, various NPs are highly effective at inhibiting *C. albicans* through a variety of molecular mechanisms, after attaching NPs to the cell wall of *C. albicans*; they penetrate or invade the cell membrane, releasing metallic ions after entering the cell. These metallic ions change the conformation, architecture, and integrity of the cell membrane of *C. albicans* once they enter the cell. It is possible for NPs to disrupt cell walls and alter the surface of cells. They also affect transport activity, rupturing cells, and leaking intracellular materials. The nanoparticle reduces membrane fluidity and alters the sense of temperature; alters cellular ergosterol content and alters fatty acid composition (oleic acid) crucial for yeast to hyphal transition in *C. albicans*. Furthermore, NPs inhibit extracellular hydrolytic enzyme (proteinase, phospholipase, hemolysin, and lipase) production; inhibit genes which control morphogenesis and other critical cellular processes in *C. albicans*. NPs induce the production of ROS (ROS), which are known to cause damage at the molecular level. There are various effects of ROS on cells, including inhibiting formation of biofilms, disrupting ETC, disrupting signalling pathways, and altering metabolism in *C. albicans*. Additionally, ROS damage mitochondrial membranes, cause DNA damage, interrupt proton pumping by forming pits and holes and inhibit budding.

10.5 Conclusion and Future Prospective

The present chapter concludes that NPs have potential as treatments for *C. albicans* causing biofilm infections. NPs serve as potential antifungal agents for biofilm inhibition due to their size, shape, surface charges and colloidal nature, making them good candidates for antifungal therapy. The efficacy of various NPs against *C. albicans* has been demonstrated in several studies both at the planktonic and biofilm stages. It is possible to combat *C. albicans* drug resistance mechanism with NPs, which offers a new and effective approach for the development of therapeutics to treat fungal infections and combat their virulence in *C. albicans.* Some promising evidence are found, that suggesting metal, metal oxides and polymeric NPs inhibit *C. albicans* growth and its virulence. Several NPs have been found to inhibit critical virulence factors like morphogenesis, biofilm formation, surface adhesion, extracellular matrix formation, at extremely low concentrations without harming human tissues. Future advancements in nano antifungal treatment must take into account optimizing treatment parameters (NP dosage quantity, time, etc.) to achieve reproducible antibiofilm activity. To understand genotoxicity at the molecular level of nanoantifungals, the effects should also be studied over time. A better understanding of the antifungal properties and cytotoxicity of different NPs is still needed for its clinical applications. There is no clear understanding of the mechanisms of antifungal action of various NPs against *C. albicans* biofilm formation. NPs have been studied extensively at the molecular level, most studies suggest they induce oxidative stress in order to inhibit biofilm formation, other mechanisms of inhibition should be investigated at the omics level.

Acknowledgements R.P. and A.S. sincerely acknowledge Savitribai Phule Pune University (SPPU) (Formerly Pune University), Pune, for providing facilities. P.N.D. gratefully acknowledges the facilities provided by the University of Oxford, United Kingdom.

References

1. Coque, T. M., Cantón, R., Pérez-Cobas, A. E., Fernández-de-Bobadilla, M. D., & Baquero, F. (2023). Antimicrobial resistance in the global health network: Known unknowns and challenges for efficient responses in the 21st century. *Microorganisms, 11*, 1050.
2. Chinemerem Nwobodo, D., Ugwu, M. C., OliselokeAnie, C., Al-Ouqaili, M. T. S., ChineduIkem, J., Victor Chigozie, U., et al. (2022). Antibiotic resistance: The challenges and some emerging strategies for tackling a global menace. *Journal of Clinical Laboratory Analysis, 36*, e24655.
3. Khalsa, J. H., Treisman, G., McCance-Katz, E., & Tedaldi, E. (2008). Medical consequences of drug abuse and co-occurring infections: Research at the national institute on drug abuse. *Substance Abuse, 29*, 5–16.
4. Dhasarathan, P., AlSalhi, M. S., Devanesan, S., Subbiah, J., Ranjitsingh, A. J. A., Binsalah, M., et al. (2021). Drug resistance in Candida albicans isolates and related changes in the structural domain of Mdr1 protein. *Journal of Infection and Public Health, 14*, 1848–1853.
5. Thakre, A., Zore, G., Kodgire, S., Kazi, R., Mulange, S., Patil, R., et al. (2018). Limonene inhibits Candida albicans growth by inducing apoptosis. *Medical Mycology, 56*, 565–578.
6. Ramage, G., Saville, S. P., Wickes, B. L., & López-Ribot, J. L. (2002). Inhibition of Candida albicans biofilm formation by farnesol, a quorum-sensing molecule. *Applied and Environmental Microbiology, 68*, 5459–5463.
7. Kasabe, U. I., Kale, K. B., Rode, N. R., Shelar, A. V., Patil, R. H., Mhaske, P. C., et al. (2022). Synthesis and antifungal screening of tetramethyl hexahydro-1: H -xanthene-1,8(2 H)-dione

derivatives as potential inhibitors of morphogenesis and biofilm formation in Candida albicans. *New Journal of Chemistry, 46*, 2128–2139.
8. Gulati, M., & Nobile, C. J. (2016). Candida albicans biofilms: Development, regulation, and molecular mechanisms. *Microbes and Infection, 18*, 310–321.
9. Cavalheiro, M., & Teixeira, M. C. (2018). Candida biofilms: Threats, challenges, and promising strategies. *Front Med (Lausanne), 5*, 28.
10. Haleem, A., Javaid, M., Singh, R. P., Rab, S., & Suman, R. (2023). Applications of nanotechnology in medical field: A brief review. *Global Health Journal, 7*, 70.
11. do Carmo, P. H. F., Garcia, M. T., Figueiredo-Godoi, L. M. A., Lage, A. C. P., da Silva, N. S., & Junqueira, J. C. (2023). Metal nanoparticles to combat candida albicans infections: An update. *Microorganisms, 11*, 138.
12. Radhakrishnan, V. S., Mudiam, M. K. R., Kumar, M., Dwivedi, S. P., Singh, S. P., & Prasad, T. (2018). Silver nanoparticles induced alterations in multiple cellular targets, which are critical for drug susceptibilities and pathogenicity in fungal pathogen (Candida albicans). *International Journal of Nanomedicine, 13*, 2647–2663.
13. Wrońska, N., Anouar, A., El Achaby, M., Zawadzka, K., Kędzierska, M., Milowska, K., et al. (2020). Chitosan-functionalized graphene nanocomposite films: Interfacial interplay and biological activity. *Materials, 13*, 998.
14. Hong, Q., Huo, S., Tang, H., Qu, X., & Yue, B. (2021). Smart nanomaterials for treatment of biofilm in orthopedic implants. *Front BioengBiotechnol, 9*, 694635.
15. Ramasamy, M., & Lee, J. (2016). Recent nanotechnology approaches for prevention and treatment of biofilm-associated infections on medical devices. *BioMed Research International, 2016*, 1.
16. Kumar, L., Bisen, M., Harjai, K., Chhibber, S., Azizov, S., Lalhlenmawia, H., et al. (2023). Advances in nanotechnology for biofilm inhibition. *ACS Omega, 8*, 21391–21409.
17. Vera-González, N., & Shukla, A. (2020). Advances in biomaterials for the prevention and disruption of Candida biofilms. *Frontiers in Microbiology, 11*, 538602.
18. Berkow, E. L., & Lockhart, S. R. (2017). Fluconazole resistance in Candida species: A current perspective. *Infection and Drug Resistance, 10*, 237–245.
19. Atriwal, T., Azeem, K., Husain, F. M., Hussain, A., Khan, M. N., Alajmi, M. F., et al. (2021). Mechanistic understanding of Candida albicans biofilm formation and approaches for its inhibition. *Frontiers in Microbiology, 12*, 638609.
20. Nobile, C. J., Fox, E. P., Nett, J. E., Sorrells, T. R., Mitrovich, Q. M., Hernday, A. D., et al. (2012). A recently evolved transcriptional network controls biofilm development in Candida albicans. *Cell, 148*, 126–138.
21. Chong, P. P., Chin, V. K., Wong, W. F., Madhavan, P., Yong, V. C., & Looi, C. Y. (2018). Transcriptomic and genomic approaches for unravelling Candida albicans biofilm formation and drug resistance—An update. *Genes (Basel), 9*, 540.
22. Ali, A., Jadhav, A., Jangid, P., Patil, R., Shelar, A., & Karuppayil, S. M. (2018). The human muscarinic acetylcholine receptor antagonist, Dicyclomine targets signal transduction genes and inhibits the virulence factors in the human pathogen, Candida albicans. *Journal of Antibiotics, 71*, 456–466.
23. Jadhav, A., Bansode, B., Phule, D., Shelar, A., Patil, R., Gade, W., et al. (2017). The antibacterial agent, moxifloxacin inhibits virulence factors of Candida albicans through multitargeting. *World Journal of Microbiology and Biotechnology, 33*, 96.
24. Kruppa, M. (2009). Quorum sensing and Candida albicans. *Mycoses, 52*, 1–10.
25. Pierce, C. G., Vila, T., Romo, J. A., Montelongo-Jauregui, D., Wall, G., Ramasubramanian, A., et al. (2017). The Candida albicans biofilm matrix: Composition, structure and function. *Journal of Fungi, 3*, 14.
26. Martins, M., Uppuluri, P., Thomas, D. P., Cleary, I. A., Henriques, M., Lopez-Ribot, J. L., et al. (2010). Presence of extracellular DNA in the Candida albicans biofilm matrix and its contribution to biofilms. *Mycopathologia, 169*, 323–331.

27. Mukherjee, P. K., Chandra, J., Kuhn, D. M., & Ghannoum, M. A. (2003). Mechanism of fluconazole resistance in Candida albicans biofilms: Phase-specific role of efflux pumps and membrane sterols. *Infection and Immunity, 71*, 4333–4340.
28. Aboualigalehdari, E., Tahmasebi Birgani, M., Fatahinia, M., & Hosseinzadeh, M. (2021). Transcription factors of CAT1, EFG1, and BCR1 are effective in persister cells of Candida albicans-associated HIV-positive and chemotherapy patients. *Frontiers in Microbiology, 12*, 651221.
29. Lara, H. H., Romero-Urbina, D. G., Pierce, C., Lopez-Ribot, J. L., Arellano-Jiménez, M. J., & Jose-Yacaman, M. (2015). Effect of silver nanoparticles on Candida albicans biofilms: An ultrastructural study. *Journal of Nanobiotechnology, 13*, 91.
30. Ahamad, I., Bano, F., Anwer, R., Srivastava, P., Kumar, R., & Fatma, T. (2022). Antibiofilm activities of biogenic silver nanoparticles against Candida albicans. *Frontiers in Microbiology, 12*, 741493.
31. Takamiya, A. S., Monteiro, D. R., Gorup, L. F., Silva, E. A., de Camargo, E. R., Gomes-Filho, J. E., et al. (2021). Biocompatible silver nanoparticles incorporated in acrylic resin for dental application inhibit Candida albicans biofilm. *Materials Science and Engineering C, 118*, 111341.
32. Li, Z., Sun, J., Lan, J., & Qi, Q. (2016). Effect of a denture base acrylic resin containing silver nanoparticles on Candida albicans adhesion and biofilm formation. *Gerodontology, 33*, 209–216.
33. Khan, S., Alam, F., Azam, A., & Khan, A. U. (2012). Gold nanoparticles enhance methylene blue-induced photodynamic therapy: A novel therapeutic approach to inhibit Candida albicans biofilm. *International Journal of Nanomedicine, 7*, 3245–3257.
34. Yu, Q., Li, J., Zhang, Y., Wang, Y., Liu, L., & Li, M. (2016). Inhibition of gold nanoparticles (AuNPs) on pathogenic biofilm formation and invasion to host cells. *Scientific Reports, 6*, 26667.
35. Hosseini, S. S., Ghaemi, E., & Koohsar, F. (2018). Influence of ZnO nanoparticles on Candida albicans isolates biofilm formed on the urinary catheter. *Iranian Journal of Microbiology, 10*, 424–432.
36. Joshi, K. M., Shelar, A., Kasabe, U., Nikam, L. K., Pawar, R. A., Sangshetti, J., et al. (2021). Biofilm inhibition in Candida albicans with biogenic hierarchical zinc-oxide nanoparticles. *Materials Science and Engineering C., 134*, 112592.
37. Rosenberg, M., Visnapuu, M., Vija, H., Kisand, V., Kasemets, K., Kahru, A., et al. (2020). Selective antibiofilm properties and biocompatibility of nano-ZnO and nano-ZnO/Ag coated surfaces. *Scientific Reports, 10*, 13478.
38. Jalal, M., Ansari, M. A., Alzohairy, M. A., Ali, S. G., Khan, H. M., Almatroudi, A., et al. (2019). Anticandidal activity of biosynthesized silver nanoparticles: Effect on growth, cell morphology, and key virulence attributes of Candida species. *International Journal of Nanomedicine, 14*, 4667–4679.
39. Baghiat Esfahani, M., Khodavandi, A., Alizadeh, F., & Bahador, N. (2023). Biofilm-associated genes as potential molecular targets of nano-Fe3O4 in Candida albicans. *Pharmacological Reports, 75*, 682–694.
40. Salari, S., Sadat Seddighi, N., & Ghasemi Nejad Almani, P. (2018). Evaluation of biofilm formation ability in different Candida strains and anti-biofilm effects of Fe3O4-NPs compared with Fluconazole: An in vitro study. *Journal of Medical Mycology, 28*, 23–28.
41. Pugazhendhi, A., Kumar, S. S., Manikandan, M., & Saravanan, M. (2018). Photocatalytic properties and antimicrobial efficacy of Fe doped CuO nanoparticles against the pathogenic bacteria and fungi. *Microbial Pathogenesis, 122*, 84–89.
42. Padmavathi, A. R., Murthy, S. S., Das, A., Priya, A., Sushmitha, T. J., Pandian, S. K., et al. (2020). Impediment to growth and yeast-to-hyphae transition in Candida albicans by copper oxide nanoparticles. *Biofouling, 36*, 56–72.

43. Raj, V., Raorane, C. J., Lee, J. H., & Lee, J. (2023). Gum Arabic polysaccharide embedded L-cysteine capped copper oxide nanocarriers selectively inhibit fluconazole-resistant C. albicans biofilm and remove the toxic dye from wastewater. *International Journal of Biological Macromolecules, 244*, 244.
44. Garcia-Marin, L. E., Juarez-Moreno, K., Vilchis-Nestor, A. R., & Castro-Longoria, E. (2022). Highly antifungal activity of biosynthesized copper oxide nanoparticles against Candida albicans. *Nanomaterials, 12*, 3856.
45. Martínez, A., Apip, C., Meléndrez, M. F., Domínguez, M., Sánchez-Sanhueza, G., Marzialetti, T., et al. (2021). Dual antifungal activity against Candida albicans of copper metallic nanostructures and hierarchical copper oxide marigold-like nanostructures grown in situ in the culture medium. *Journal of Applied Microbiology, 130*, 1883–1892.
46. Zhang, Y., Chen, Y. Y., Huang, L., Chai, Z. G., Shen, L. J., & Xiao, Y. H. (2017). The antifungal effects and mechanical properties of silver bromide/cationic polymer nano-composite-modified Poly-methyl methacrylate-based dental resin. *Scientific Reports, 7*, 1547.
47. Iadnut, A., Mamoon, K., Thammasit, P., Pawichai, S., Tima, S., Preechasuth, K., et al. (2019). In vitro antifungal and antivirulence activities of biologically synthesized ethanolic extract of propolis-loaded PLGA Nanoparticles against Candida albicans. *Evidence-based Complementary and Alternative Medicine, 2019*, 1.
48. Costa, A. F., Araujo, D. E., Cabral, M. S., Brito, I. T., De Menezes Leite, L. B., Pereira, M., et al. (2019). Development, characterization, and in vitro-in vivo evaluation of polymeric nanoparticles containing miconazole and farnesol for treatment of vulvovaginal candidiasis. *Medical Mycology, 57*, 52–62.
49. Yenice, G. B. (2020). Potential antibiofilm activity of farnesol-loaded poly(DL-lactide-co-glycolide) (PLGA) nanoparticles against Candida albicans. *Journal of Analytical Science and Technology, 11*, 43.
50. Thakre, A., Jadhav, V., Kazi, R., Shelar, A., Patil, R., Kharat, K., et al. (2021). Oxidative stress induced by piperine leads to apoptosis in Candida albicans. *Medical Mycology, 59*, 366–378.
51. Christina, S., ta-Christina, H., Axel, W., Waldemar, K., & Uwe, W. (2000). Generation of reactive oxygen species by Candida albicans in relation to morphogenesis. *Archives of Dermatological Research, 292*, 260–264.
52. Zhang, Z., & Tang, W. (2018). Drug metabolism in drug discovery and development. *Acta Pharmaceutica Sinica B, 8*, 721–732.
53. Hwang, I. S., Lee, J., Hwang, J. H., Kim, K. J., & Lee, D. G. (2012). Silver nanoparticles induce apoptotic cell death in Candida albicans through the increase of hydroxyl radicals. *FEBS Journal, 279*, 1327–1338.
54. Wani, I. A., & Ahmad, T. (2013). Size and shape dependant antifungal activity of gold nanoparticles: A case study of Candida. *Colloids and Surfaces. B, Biointerfaces, 101*, 162–170.
55. Gómez-Garzón, M., Gutiérrez-Castañeda, L. D., Gil, C., Escobar, C. H., Rozo, A. P., González, M. E., et al. (2021). Inhibition of the filamentation of Candida albicans by Borojoapatinoi silver nanoparticles. *SN Applied Sciences, 3*. https://doi.org/10.1007/s42452-020-04103-0
56. Muthamil, S., Devi, V. A., Balasubramaniam, B., Balamurugan, K., & Pandian, S. K. (2018). Green synthesized silver nanoparticles demonstrating enhanced in vitro and in vivo antibiofilm activity against Candida spp. *Journal of Basic Microbiology, 58*, 343–357.
57. Lee, B., Lee, M. J., Yun, S. J., Kim, K., Choi, I. H., & Park, S. (2019). Silver nanoparticles induce reactive oxygen species-mediated cell cycle delay and synergistic cytotoxicity with 3-bromopyruvate in Candida albicans, but not in Saccharomyces cerevisiae. *International Journal of Nanomedicine, 14*, 4801–4816.
58. Halbandge, S. D., Jadhav, A. K., Jangid, P. M., Shelar, A. V., Patil, R. H., & Karuppayil, S. M. (2019). Molecular targets of biofabricated silver nanoparticles in Candida albicans. *Journal of Antibiotics, 72*, 640–644.

59. Rasool, U., Sah, S. K., & Hemalatha, S. (2019). Effect of biosynthesized copper nanoparticles (cunps) on the growth and biofilm formation of fluconazole-resistant Candida albicans. *Journal of Microbiology, Biotechnology and Food Sciences, 9*, 21–24.
60. Cheong, Y. K., Arce, M. P., Benito, A., Chen, D., Crisóstomo, N. L., Kerai, L. V., et al. (2020). Synergistic antifungal study of pegylated graphene oxides and copper nanoparticles against Candida albicans. *Nanomaterials, 10*, 819.
61. Saleh, N., Elshaer, S., & Grigis, G. (2022). Biodegradable polymers-based nanoparticles to enhance the antifungal efficacy of fluconazole against Candida albicans. *Current Pharmaceutical Biotechnology, 23*, 5.

Chapter 11
The Effect of Nano Silver Diamine Fluoride in Arresting Dental Caries

Merve Aksoy, Cenkhan Bal, Mohamed Brahmi, and Nurgul K. Bakirhan

11.1 Introduction

Dental caries is a chronic, infectious, and behavioral disease. The etiology of caries development shows that acidogenic bacteria invasion and the presence of fermentable carbohydrate decreases the bacterial biofilm's pH, which initiates the demineralization process. Once demineralization occurs, this process may lead to caries development. Demineralization should be followed by the process of remineralization to prevent the formation of dental caries [1].

Despite the recent advances in preventive dentistry, dental caries remains a worldwide chronic health problem [2]. World Health Organization (WHO) has announced in the latest oral health status report that oral diseases affect approximately 3.5 billion people worldwide. In middle-income countries, three out of four people suffer from dental diseases. Globally, an estimated two billion people have caries in permanent teeth, and 514 million children suffer from caries of primary teeth [3]. Although organizations and oral health programs continuously attempt to

M. Aksoy · C. Bal
Faculty of Dental Medicine, Department of Pediatric Dentistry, University of Health Sciences Gulhane, Ankara, Turkiye

M. Brahmi
Faculty of Sciences, Department of Chemistry, Physical Chemistry of Natural Substances and Process Team, Laboratory of Applied Chemistry and Environment (LCAE-CPSUNAP), University Mohamed Premier, Oujda, Morocco

N. K. Bakirhan (✉)
Gulhane Faculty of Pharmacy, Department of Analytical Chemistry, University of Health Sciences, Ankara, Turkiye

R. S. Chaughule, A. S. Lokur (eds.), *Applications of Nanotechnology in Microbiology*, https://doi.org/10.1007/978-3-031-49933-3_11

increase awareness of oral care worldwide, dental caries are still among the most prevalent chronic diseases [4]. The treatment of carious teeth and remaining infected dental tissues is a time-consuming and expensive procedure. Furthermore, this attempt may not be possible in some cases, especially when little children or adults with dental phobia cannot tolerate dental treatment [5]. Preventive approaches that should be applied in the early periods of life come to the fore to avoid complicated, time-consuming, and high-cost dental treatments. Oral hygiene information, diet arrangements, local and systemic fluoride applications, and fissure sealant applications are among the most preferred and recommended preventive approaches [6, 7]. Among the listed contents, fluoride has led the caries preventive programs for many years. Fluoride varnishes of 22.600 ppm $F^{-}$are the most common fluoride preparation, and it is recommended to be applied twice a year to prevent caries progression. Fluoride can be applied via two methods: Systemic and local applications. The purpose of systemic applications is to enhance the strength of tooth structure towards acid attacks that end in caries progression. In the early periods of tooth development, before the eruption period, the fluoride ions in circulation can be a part of the apatite formation to form fluorapatite, which is more stable than hydroxyapatite main component of hard dental tissues. In topical applications, fluoride ions can deposit in the dental plaque of erupted teeth and act as a barrier for preventing the displacing of Ca^{+2} ions, usually dissolved from the apatite structure of the tooth due to possible acid attacks. Fluoride ions react with Ca^{+2} ions to form CaF_2 in critical pH (<5.5). When the pH of dental biofilm reaches the physiological pH, the dissolved Ca^{+2}ions precipitate in the dental tissue, so remineralization occurs. Topical fluoride can also play a part in the sclerosis of dentinal tubules. Fluoride ions can penetrate through the gaps between the apatite crystals and enhance the strength of enamel and dentin toward carious progression [8, 9]. Despite the listed efficiencies of fluoride applications, dental caries are still a global problem, which enlightens the importance of taking new stages in preventive dentistry [10].

11.2 Nano-Sized Materials

Nanomaterials are materials characterized by their nanometric size, typically ranging between 1 and 100 nm. The unique size of nanomaterials provides a wide range of applications, particularly due to their high active site number and surface area. Nanocarriers are also a category of nanomaterials that have been extensively used in medicine for their ability to transport active substances such as drugs, genes, and plant extracts.

Nanomaterials can be classified into three categories based on their source:

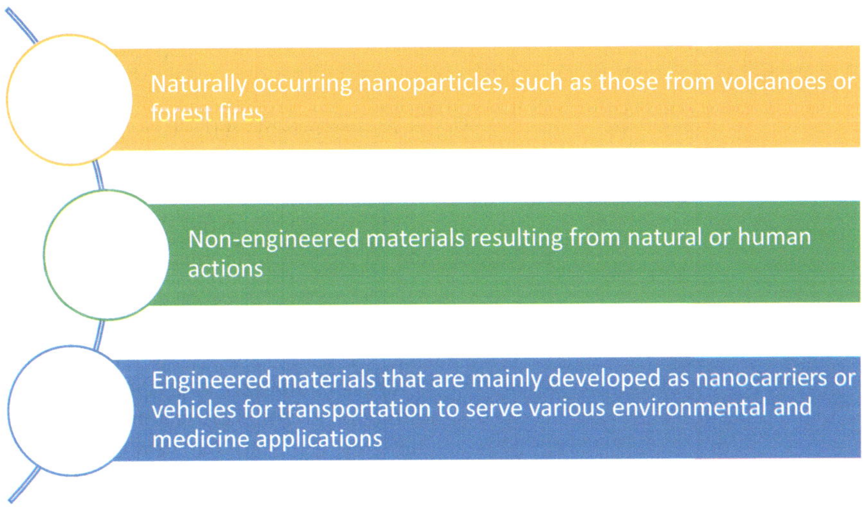

In dentistry, both organic and inorganic nanocarriers such as polymer, micelles, liposomes, metal and metal oxide nanoparticles, graphene-based nanomaterials, polymer-based nanomaterials, and silica-based nanomaterials have been used for the treatment of dental caries either as restorative or preventative materials. Table 11.1 summarizes the applications of these materials in dental caries over the past decade.

Metal and metal oxide nanoparticles have shown interesting results in their biological applications, which are mainly due to their small size and high surface area. Additionally, their ability to release metal ions into the targeted area has encouraged their use in the field of dental care. It is also worth mentioning that the increasing use of these materials is related to their high biocompatibility, which has been demonstrated by several metals such as Gold, Silver, Copper, and Zinc. However, the properties of these nanoparticles are still being studied by nanotechnology researchers. It has been reported that reducing the size of nanoparticles can make them more effective and safer to use. Pérez-Dias et al., have examined the effectiveness of silver nanoparticles against biofilms formation, and confirms that both the size and concentration of the nanoparticles are crucial for the inhibition of *Streptococcus mutans,* which was considered as the primary etiological agent in caries [11]. The incorporation of such metal oxide into resin composites in a controlled manner to avoid any lose on the mechanical properties of the resin was also reported to enhance the inhibition of oral infection [12].

The microhardness test reported in the study of Scarpelli et al., revealed that besides metal nanoparticles microbial activity, they also improved the remineralization of deciduous tooth enamel [13]. Another study reported that with an optimum serum of magnesium, a continuous release of Mg ions shows significant results as dental caries like by preventing demineralization [14].

Table 11.1 Nanomaterials type and properties applied in dental caries application over the past decade (2013_2023)

Nanomaterials class	Dental material	Size (nm)	Outcome	Refs.
Metal and metal oxide nanoparticles	Silver nanoparticles	9.5 ± 1.1	Prevent biofilm formation, antimicrobial activity against Streptococcus mutans	[11]
	Zinc Oxide nanoparticles	1–100	Antimicrobial activity against Streptococcus mutans with a MIC concentration of ZnO nanoparticles of 0. 312 mg/ml	[17]
	Silver nanoparticles mixed sealant	NS	The use of silver nanoparticles gives a notable decrease on tooth demineralization and increase in their remineralization	[18]
	Silver nanoparticles	NS	Microhardness and microbial test showed that the Ag-Nano treatment was able to promote remineralization of primary dental enamel with initial caries lesion	[13]
	Silver nanoparticle incorporated in calcium phosphate composite	116	The obtained composite greatly reduced the oral microcosm biofilm viability, metabolic activity, CFUs, and lactic acid production	[19]
	Zinc oxide nanoparticles incorporated in resin composite	20	Antibacterial activity against broad-spectrum of pathogenic bacteria in oral cavity associated with caries	[20]
	Zinc oxide nanoparticles incorporated in resin composite	NS	The incorporation of ZnO nanoparticles into resin composite mainly inhibit the growth of Streptococcus mutans	[22]
	Copper oxide nanoparticles	40	CuO nanoparticle could be used as agent to prevent dental caries	[31]
	Magnesium oxide nanoparticles incorporated in glass-ionomer cement	20.8	The cement modified with MgO shows interesting results regarding the antibacterial and antibiofilm activity which encourage their application on dental care field	[32]
	Titanium oxide nanoparticules	70–100	TiO_2 nanoparticles could be used as agent for preventing biofilms formation	[23]
Polymers-based nanoparticles	Poly(methyl methacrylate) with TiO_2 nanoparticles	65.94–170.2	The TiO_2 incorporated in polymer matrix has been found to have antibacterial activity against candida species	[24]
	Fluoride loaded chitosan nanoparticles	101	Fluoride ion released from the nanoparticles was promising for the protection against caries development	[25]
	Polyethylene-glycol coated maghemite nanoparticles	25	PEG coated maghemite nanoparticles were able to reduce the permeability of dental tubules	[26]
	Poly(ε-caprolactone) nanocapsules loaded with chlorhexidine	244 ± 3.8	PCL nanoparticles loaded Chlorhexidine have demonstrated a novel approach to demineralized dentin substrates and the resin-dentin interface	[27]
	Poly(lactic-co-glycolic acid) nanoparticles loaded photoactive drug	100–250	The polymer-based nanoparticles have could be used as promising antimicrobial endodontic treatment	[28]
	Nanoparticles-based diblock copolymers 2-(dimethylamino)ethyl methacrylate (DMAEMA), butyl methacrylate (BMA), and 2-propylacrylic acid (PAA) (p(DMAEMA)-b-p(DMAEMA-co-BMA-co-PAA)) loaded with Farnesol	21	The cationic nanoparticles were able to adsorb into the hydroxyapatite surface and release the loaded drug for biofilms treatment	[29]

(continued)

Table 11.1 (continued)

Nanomaterials class	Dental material	Size (nm)	Outcome	Refs.
Graphene based nanoparticles	Graphene oxide	200–400	Graphene oxide nanosheets have shown antimicrobial activity against dental pathogens through the distortion of cell wall and membrane	[30]
	Graphene Nanoplatelets	NS	Graphene nanoplatelets could be highly effective against Streptococcus mutans in a size dependent manner	[15]
	Silanized graphene oxide modifedTransbond XT adhesive	100–700	The modified adhesive exhibits a promising antimicrobial activity for bonding orthodontic brackets	[31]
	Polyethylene glycol/ graphene oxide/ hydroxyapatite nanocomposite	170–200	The bioactivity of the formed nanocomposites has been found to be higher that pure hydroxyapatite due to the higher release of Ca and P from the nanocomposite.	[32]
	Reduced graphene-silver nanoparticle modified glass ionomer cements	NS	Modification of glass ionomer cements with graphene silver nanoparticles provide a promising approach for dental restoration materials	[33]
	Silver-nanoparticle-decorated reduced graphene	60	The incorporation of silver nanoparticles into reduced graphene oxide enhances the antimicrobial activity of the nanocomposite against oral pathogens	[34]
	Graphene/zinc oxide nanocomposite	20–40	An effective ani-biofilm and anti-oral bacterial has been developed by the use of graphene/zinc oxide nanocomposite	[35]
	Graphene oxide-copper nanocomposites	NS	Graphene oxide/copper nanocomposite is a novel biomaterial anti-caries as it exhibit an antibacterial activity against oral pathogens specifically Streptococcus mutans	[36]
	Hydroxyapatite / graphene silver nanocomposite	NS	The mechanical properties of the nanocomposite enhance their application as restorative composite	[37]
Silica-based nanocomposite	Mesoporous silica encapsulated chlorhexidine	NS	Mesoporous silica nanocomposites provide an efficient carrier material for oral biofilm inhibition	[38]
	Poly(Bis-GMA) grafted hydroxyapatite whiskers and silica nanoparticles	40	The nanocomposite introduced with silica nanoparticles shows and improved physical and mechanical properties which enhance the in vitro bioactivity of the formed dental resin	[39]
	Silica Nanoparticles incorporated with mouthwash	NS	The evaluation of antimicrobial activity of the nanoparticles shows a good potential activity against gram positive bacteria	[40]
	Drug-silica co-assembled nanoparticles	424 ± 75	Controlled drug release has been achieved through the use of silica nanoparticle as carriers for the restoration of tooth interface	[41]
	Diazeniumdiolate-modified silica nanoparticles	NS	The release of nitric oxide from the silica nanocomposite mainly inhibits the biofilm formation	[42]
	Mesoporous silica nanoparticles incorporated in poly(methyl methacrylate)	85.2 ± 7.7	The nanocomposite could be used as denture resin as it exhibits an antimicrobial activity against *C. albicans* and *S. oralis*	[43]
	Zn doped mesoporous silica nanoparticles	138	Silica nanoparticles doped with Zn form a nanocomposite resin with improved mechanical and antibacterial activity	[44]

The dentition field has also benefited from the recent development on graphene-based nanomaterials. The use of pure graphene nanoplatelets as antibacterial agents has shown a promising result [15]. However, due to the carbonaceous structure of graphene material, which basically formed by one single carbon component, their application has faced some serious limitations [16]. For this research, the functionalization of graphene surface became of major interest.

Graphene or reduced graphene oxide with its surface rich in oxygen atoms gives more stability and dispersity to graphene-based materials and mainly enhances their microbial activity [30]. Recently, authors have combined the graphene-based with metals particles to enhance their biological activities; Mengying Mao, and co-authors have successfully investigated the activity of Graphene oxide-copper nanocomposite against *Streptococcus mutans* and found that the maintained release of copper ions from the composite was able to disturb the exopolysaccharides and consequently inhibit the growth of biofilm [36]. A similar finding was noticed by incorporating other types of metals such as zinc [35] or silver [34]. The mechanism of action of graphene-based nanomaterials as dental material was reported to be affected by both intrinsic and extrinsic factors, and further studies are intended to better optimize the biocompatibility and functionality of such nanocomposite [16].

On the other hand, silica-based materials have been used as substrates due to their high biocompatibility, providing the necessary mechanical and controlled release properties for oral bacterial treatment. The presence of singlet oxygen on the silica nanoparticles surface was suggested to be behind the antibacterial activity of these materials as this oxygen atoms could promote an oxidative damage to the bacterial membranes. Many papers have reported the use of silica nanoparticles on dentistry field [45, 46]. Properties such as stability, low toxicity, and good dispersity of these nanoparticles within organic and inorganic matrix encouraged the elaboration of nanocomposite based on silica nanoparticles as carriers [40]. The mesoporous structure of silica nanoparticle is commonly profited to encapsulate and deliver some active agent such as octenidinedihydrochloride [41] or chlorhexidine [38] which provide a bio nanocarrier with interesting properties regarding the release of the encapsulated drug which mainly fix in on hand the poor mechanical properties of pristine drugs and provide in the other hand a controlled release of drugs into the targeted area. Other authors were also reported the synthesis of metal-doped silica nanoparticles for improved antibacterial activity [44]. Overall, the use of silica-based composite in dentistry led to the development of new filling materials with improved mechanical and biological properties [47].

11.3 Synthesis of SDF

After using only silver fluoride materials for quite a time, the first report of silver diamine fluoride (SDF) with the chemical formula $Ag(NH_3)_2F$ came out in 1969 when Nishino et al. synthesized SDF and found its use in inhibiting caries progression. Following its initial use in Japan, Australia and Mexico were among the first countries to report using SDF for preventing dental caries [48]. The SDF solution is both colorless and odorless, making it highly convenient for use. The presence of ammonia ions plays a crucial role in stabilizing the solution. These ions combine with silver and fluoride ions to form a much more stable product than what can be obtained with silver and fluoride alone [49]. To initiate the process of SDF synthesis, the first step involves the complexation of silver ions with two ammonia molecules. This reaction results in the formation of $Ag(NH_3)_2$, which is a metal ammine complex. This complex is highly desirable due to its superior chemical stability and low oxidation rate. Following this, the metal ammine complex is mixed with fluoride ions, ultimately resulting in the production of the desired SDF solution [50]. Depending on the concentration of fluoride and silver ions, different commercial SDF products are available on the market. These options range from SDF with lower fluoride concentrations, containing only 12%, to the highest concentration SDF with 38% fluoride, which has been proven to be the most effective in arresting dental caries among all SDF products [51].

11.4 Silver Diamine Fluoride

Silver diamine fluoride (SDF) is an alkaline, colorless, and topically applicable solution containing silver and fluoride ions. This solution is a newly developed caries-arresting material that acts as an antimicrobial and remineralization-enhancing agent thanks to its high fluoride and silver ion concentration. Fluoride acts as a remineralization-promoting factor, and silver ions decrease the number of microorganisms that play a role in the progression of dental caries. It has been suggested that these synergistic effects of silver and fluoride may inhibit the caries process and prevent new caries development [52].

Silver compounds have been used in different fields of medicine for centuries. Silver ions were first used in the disinfection of potable water around 1000 BC and the early 1900s in disinfection procedures [1]. These ions were popular agents for treating tetanus and rheumatoid arthritis before antibiotics were discovered [52]. With the introduction of penicillin and other antibiotics in the 1930s, clinical interest and research in silver have decreased significantly. However, in the 1970s, the use of silver compounds was revived due to the worldwide awareness of antibacterial resistance. Recently, silver has gained popularity due to its broad antimicrobial spectrum, low toxicity, and lack of cross-spectrum bacterial resistance [52].

In the early applications of dental treatments, it was detected that silver-containing amalgam restorations had caused dark staining in the dental tissue. In the 1840s, silver compounds were used in preventive approaches in primary dentition [1]. Silver was used for cavity disinfection and in the treatment of dentin sensitivity. In 1914, Howe applied silver nitrate, which caused black staining and sclerotic dentin, an indicator of caries arresting [1]. Silver-fluoride-containing restorations have been preferred for preventive approaches in clinical and in-vitro studies in the following 40 years [48]. In the 1960s, the synergistic effect of silver with fluoride was detected and SDF was started to be used as an anti-caries agent. The co-administration of silver with fluoride ions has been advocated considering this benefit. Nishino has claimed that SDF has reduced the progression of carious lesions [53].

SDF can be preferred in various dental applications such as desensitizing treatments and arresting root surface caries in adult patients, enhancing the root canal disinfection process in endodontic treatments, preventing pit-fissure caries and secondary carious lesions in pediatric patients [52]. Although the beneficial effects of SDF in caries arrest have been known for a long time, the official permissions for using SDF in the United States did not come to the fore until the last decade. In 2015, the US Food and Drug Administration (FDA) approved the product [54], and the Advantage Arrest (Elevate Oral Care) was offered as the first approved SDF product to the dental market [1]. This solution was first used for hypersensitivity treatments. The fluoride ion in the solution is responsible for the agent's remineralization effect, while silver acts as an antimicrobial component, and ammonium acts as a stabilization agent [48].

Following this stage, clinical and laboratory studies have proven the effectiveness of SDF in comparison to other fluoride varnishes, which are still recognized as the most effective and preferred caries-preventing agents. Rosenblatt et al. declared that SDF is a reliable, effective, and efficient caries control agent and more effective than fluoride varnishes in caries arresting [52]. Silver diamine fluoride was directly applied to the carious lesions, and successful results were detected [55]. SDF was suggested to be applied twice a year. The fluoride amount of the solution can be varied, and different fluoride ratios can be found in the dental market (10%, 30%, 38%). However, the form with 38% fluoride is the most effective fluoride solution [1].

11.5 Mechanism of Action of SDF

Fluoride acts as a remineralization-promoting factor, and silver ions decrease the number of microorganisms that play a role in the progression of dental caries. Following the agent's application to the dental tissue, SDF reacts with hydroxyapatite, which ends with the formation of silver phosphate and calcium fluoride. These compounds can be a fluoride and phosphate ions reservoir for remineralization [56]. It is known that silver ions can easily penetrate porous tissues. Silver compounds such as silver oxide and silver phosphate are responsible for the staining effect of SDF products [57]. A recent review has successfully explained SDF's mechanism of action by indicating the agent's three significant actions [58]. The mechanism of action of the agent can be seen in Fig. 11.1. These can also be listed as follows: (1) The antibacterial effect of SDF on carious-causing bacteria Streptococcus mutans [59, 60]. (2) inducing the remineralization and decreasing the demineralization of enamel and dentine [59, 61], and (3) Due to the collagenase inhibition, SDF reduces the collagen matrix inhibition of the dentine [62].

Furthermore, SDF also has the advantage of fluoride release. The anti-carious effect of fluoride, mainly established through topical applications, has been well studied. In acid attacks, intra-oral pH decreases, and a demineralization process occurs. Following the pH decreases, Ca^+ ions leave the hydroxyapatite structure, the main component of the hard dental tissues, and the possible fluoride ions in the biofilm react with the dissolved Ca^+ ions. When the oral pH turns to physiological pH, the dissolved Ca^+ ions precipitate on the tooth structure, leading to the remineralization process. Fluoride ions also have the advantage of penetrating through the porous enamel and dentine, which ends with the hardening of the tooth structure. The new form of apatite crystals with fluoride addition is named fluor hydroxyapatite, which is known to be more stable to acid attacks. Moreover, fluoride ions also perform antibacterial activity on dental plaque microorganisms, enhancing silver ions' antimicrobial effect [56]. However, the clinical studies showed that the effectiveness of SDF was superior to the topical fluoride applications [59, 63], and the silver ions can be accepted as the main factor for the successful effect of SDF on decreasing the carious formation and progression [64]. Silver ions penetrate through the cell membrane, causing a decompensation in the cell structure, leading to the agent's bactericidal effect [65]. Moreover, ammonium components also affect bacterial inhibition with the features of quaternary ammonium compounds as an effective disinfection agent [66].

11.6 Adverse Effects and Safety of SDF

The most considered side-effect of SDF is the blackening due to the silver ions in carious dental tissue. Following the chemical reaction of silver diamine (AgF^-) with hydroxyapatite $Ca_{10}(PO_4)_6(OH)_2$, Silver phosphate Ag_3PO_4 forms as a side-product which is responsible for the blackening effect of SDF. This effect may cause esthetic concerns among children and patients. The parental acceptance of SDF's tooth staining effect varies among countries. American parents were questioned, and the ratio of participants acceptable with the application of SDF on the anterior region was 30%. In the same study, 68% declared that the posterior region staining was acceptable. However, SDF applications still show a higher parental acceptable compared to the procedure of dental treatments under general anesthesia.

This disadvantage of the agent may be why clinicians do not choose this method in treating the dental caries of young children. However, considering the non-cooperative patients with dental treatment procedures that include local anesthesia and drilling, general anesthesia, and sedation would be indicated, SDF seems to be an appropriate and preferred method. This technique is pain-free and acceptable for non-cooperative children and adult patients with dental phobia [67]. No need for local anesthesia and patient cooperation with low treatment expenses are among the technique's advantages [6]. Parental acceptance of the blackening effect of SDF differs in various cultures. Hong Kong parents showed an overall acceptance of 61–71% for their child's appearance following the SDF application; moreover, they also declared accepting posterior staining compared to anterior [63]. Parents and patients were both agreeable to the acceptance of SDF treatment. However, obtaining informed consent regarding the information on SDF's staining effect is essential, mainly when it is used in the anterior region.

SDF has a high fluoride concentration (44,800 ppm), which may cause toxicity. Silver ions can also be absorbed and deposited in the organism, activating allergic reactions [64].

In a previous study, fluoride safety limits were not exceeded when SDF was applied to three anterior teeth. However, silver limits were detected as higher than the oral daily reference dose. The amount was considered negligible compared to the lifetime limit, and this exceeding was not expected to cause an adverse effect [68]. Previously, 888 children aged 3–4 years were treated with SDF, and only minor adverse effects such as tooth and gum pain (6.6%), soft tissue swelling (2.8%), and bleaching (4.7%) were reported. No other major drawbacks were mentioned in this study [63]. Although the usage of SDF in young patients seems to be a priority in some cases, limiting the amount of SDF used in a single visit can be a reasonable attempt to avoid the reactions that can occur in a possible fluoride toxicity scenario.

11.7 Indications and Contraindications

Caries management in patients with high caries risk and non-cooperative children with behavioral problems needing dental treatments under general anesthesia are among the primer indications of SDF applications. There is no limitation; the solution can be applied on any dentinal surface if the area can be reachable by micro brush application. Carious lesions with spontaneous pain, which indicate the irreversible form of pulpal inflammation, and carious lesions in close relation with dental pulp in accordance with the radiographic images and clinical outcomes are among the contraindications of this method. Silver allergy and objection to the blackening effect of SDF are also essential topics that should be considered before choosing SDF applications as a treatment approach [54].

11.8 Clinical Protocol

SDF can be applied rapidly, and special equipment and technical precision in treatment sessions are unnecessary. However, there are still some issues that should be considered before SDF applications. One of them is to be sure that patients and parents agree with the chosen procedure in case of the staining effects. The other concern is that SDF can also cause a chemical reaction and a staining effect in possible contact with soft tissue. Considering this warning, isolation with cotton pellets

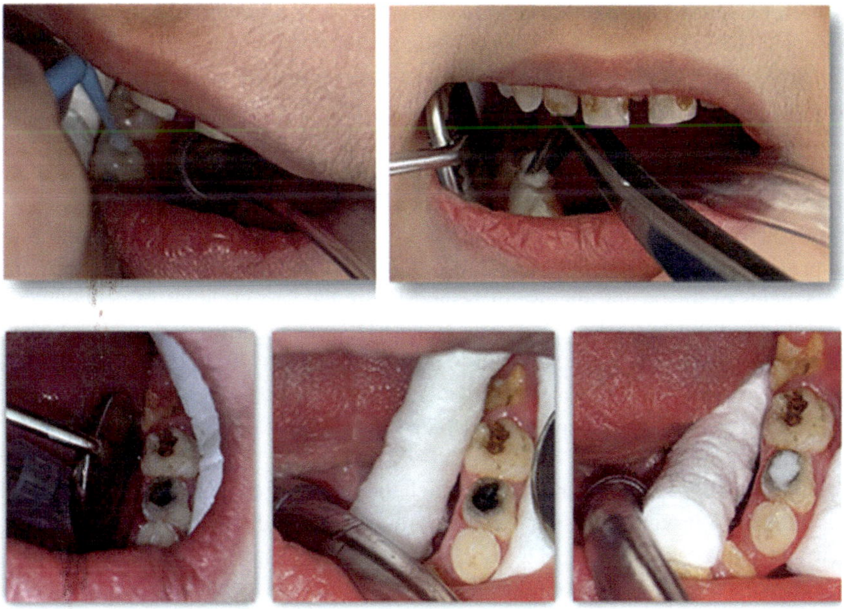

Fig. 11.1 Clinical application stages of SDF with permission of Makbule Buse Dündar Sarı

should be obtained even with uncooperative patients (Fig. 11.1). The dosage of used SDF is also essential to end the treatment in safety limits regarding chemical toxicity [54].

11.9 Nano Silver Diamine Fluoride (Nano SDF)

Nano Silver Fluoride (NSF) is a new experimental formulation containing silver nanoparticles. This new formulation is a promising caries-arresting agent with no side effects like dental tissue staining. This new substance is known to have excellent antimicrobial properties against *Mutans streptococci* and *Lactobacilli,* the primary pathogens responsible for developing dental caries. Long-term clinical studies were also conducted to assess the caries prevention effect and the blackening activity of nano SDF. The results showed that nano SDF did not cause staining in porous dental tissue. The antimicrobial and caries-arresting activity was not affected by particle size changes, and nano SDF showed a successful antimicrobial activity in the arresting of dental caries [69].

As mentioned above, caries arresting ability of SDF has been proven. However, this agent has the disadvantage of staining the porous, carious hard tissues black. This staining effect is relevant to the silver ions' oxidation process. Painful soft tissue lesions were also reported in a possible contact of the solution with oral mucosa. However, these lesions were reported to disappear within 48 h [48].

Various approaches have been offered to overcome these drawbacks, and developing a novel silver diamine solution with nanoparticles is one of them. Nano SDF was produced, claiming to be an effective carious arresting agent without staining the porous hard tissues black, not likely the other SDF solutions [70].

This new agent was used in-vivo and reported to perform excellent antimicrobial properties on caries-causing microorganisms, *Mutans streptococci,* and *Lactobacilli.* In a previous study, the antibacterial effect of Nano SDF was detected as higher than chlorhexidine. Furthermore, Nano SDF solutions founded to have an additional suppressing effect on viruses and fungi. Considering these proven antimicrobial efficiencies of this agent, the use of Nano SDF in the various areas of dentistry has come to the fore. In a previous study, Nano SDF was incorporated into varnishes, and successful results were detected [71]. The outcomes of another clinical study held by Tirupathi et al. revealed that the Nano SDF solution had a better efficiency in arresting carious lesions in comparison to the SDF solution in 12 months follow-up [70].

Santos et al. held a similar study in which the caries-arresting effect of Nano SDF was examined in-vivo. One hundred thirty decayed deciduous teeth were randomized into two groups: Nano SDF as examination and water as control. Teeth were treated with one blinded examiner. In the 12 months of control examination, the success in caries arresting of the Nano SDF group was 66.7%, while a 34.7% success rate was detected in the control group. These results revealed the successful efficiency of Nano SDF in comparison to the water control group in deciduous teeth (Figs. 11.2 and 11.3) [69].

Targino et al. have investigated the antimicrobial efficiency of Nano SDF versus SDF solution, in-vitro. The results revealed that the antimicrobial effect of Nano SDF added fluoride varnish was superior to SDF [72]. In addition to that, the cytotoxicity levels of Nano SDF were found to be less than SDF solutions. Previously it has also been reported that Nano-Silver has lower cytotoxicity compared to other dental materials [73].

In previous studies, Nano Silver was added to the composite restorative materials, and significant results were detected in reducing the number of *S.mutans*and *L.bacillus* [74–76]. Nano-silver was also added into the acrylic resins of the denture base, and an enhanced anti-candidal effect was detected. Nano SDF has better contact with the microorganisms due to the particle size. Nano-meter scale silver ensures a larger surface area for contact with microorganisms.

The antimicrobial effect of Nano SDF is inversely proportional to the particle sizes. When the particle size gets smaller, antimicrobial activity tends to enhance. The nanoparticles easily attach to the bacteria and penetrate the bacterial cell membrane. This membrane includes sulfur and phosphorus-containing proteins. Nano Ag^+ can react with them and inhibit their function. Silver nanoparticles can also interact with the respiratory chain in mitochondria and cause cell death. Once the death of the microorganism occurs, the release of Ag^+ is known to be continued, and free radicals induce oxidative stress. Therefore, the detrimental effect of Nano SDF is sustained with the enhanced bactericidal activity. To sum up, these activities in the cell membrane prevent DNA replication, leading to the death of bacteria.

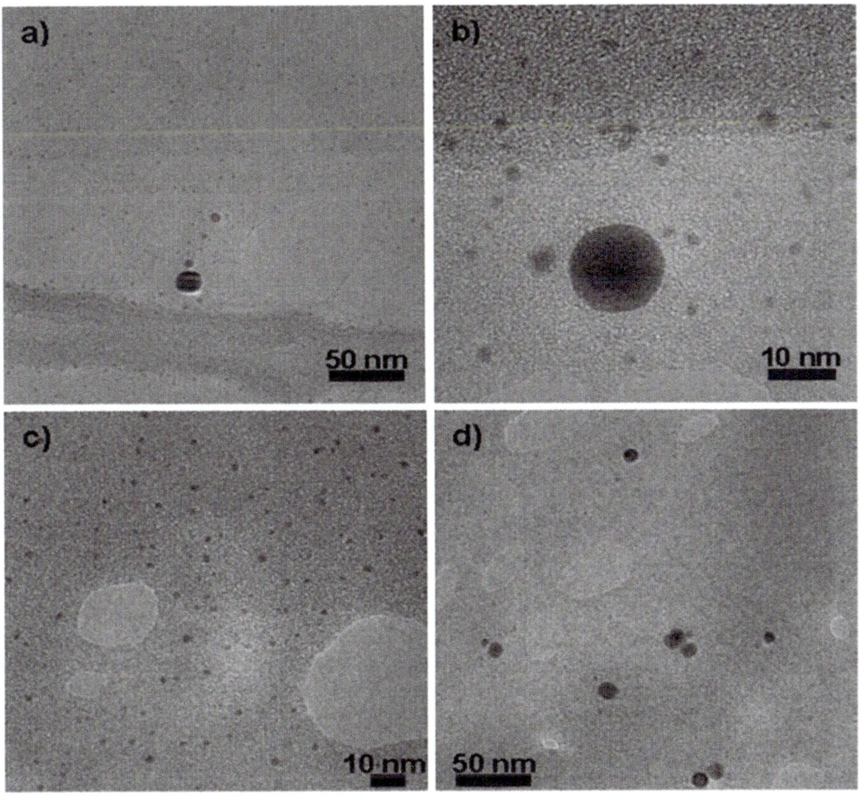

Fig. 11.2 Transmission electron microscopy of nano SDF. (Reprinted with permission from Elsevier Ref. [69])

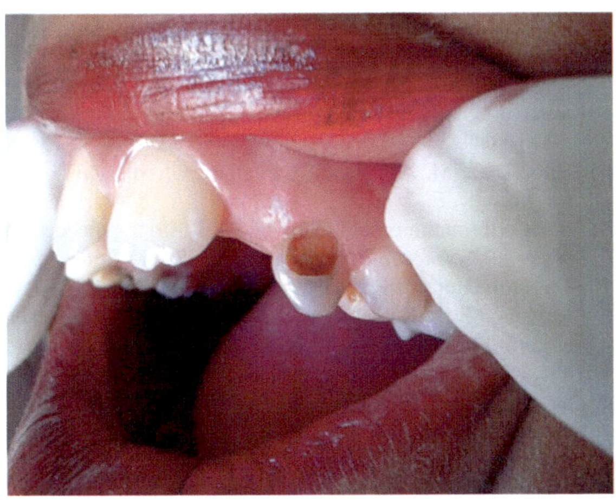

Fig. 11.3 12 months follow-up of the tooth treated with nano SDF. (Reprinted with permission from Elsevier Ref. [69])

11.10 Conclusion

Nano SDF should be preferred mostly among pediatric dentists who are dealing with young patients who cannot tolerate advanced dental treatments with drilling, and they should give a chance to this agent, especially versus general anesthesia and dental sedation treatments in young children.

References

1. Burgess, J. O., & Vaghela, P. M. (2018). Silver diamine fluoride: A successful Anticarious solution with limits. *Advances in Dental Research, 29*, 131–134.
2. Marinho, V. C. C., Worthington, H. V., Walsh, T., & Clarkson, J. E. (2013). Fluoride varnishes for preventing dental caries in children and adolescents. *Cochrane Database of Systematic Reviews, 7*, CD002279.
3. The global status report on oral health (2022). Towards universal health coverage for oral health by 2030, World Health Organization (WHO).
4. Selwitz, R. H., Ismail, A. I., & Pitts, N. B. (2007). Dental caries. *Lancet (London, England), 369*, 51–59.
5. Vollú, A. L., Rodrigues, G. F., Rougemount Teixeira, R. V., Cruz, L. R., dos Santos Massa, G., de Lima Moreira, J. P., Luiz, R. R., Barja-Fidalgo, F., & Fonseca-Gonçalves, A. (2019). Efficacy of 30% silver diamine fluoride compared to atraumatic restorative treatment on dentine caries arrestment in primary molars of preschool children: A 12-months parallel randomized controlled clinical trial. *Journal of Dentistry, 88*, 103165.
6. American Academy of Pediatric Dentistry (2023). Pediatric restorative dentistry. The reference manual of pediatric dentistry. Chicago, Ill.: American Academy of Pediatric Dentistry, 443–456.
7. Aksoy, M., & Kübra Topsakal, G. (2022). YouTube ™ for information on paediatric oral health instructions. *International Journal of Dental Hygiene, 20*, 496–503.
8. Chu, C., Mei, M. L., & Lo, E. C. M. (2010). Use of fluorides in dental caries management. *General Dentistry, 58*, 37–43; quiz 44.
9. Ogard, B., Seppä, L., & Rølla, G. (1994). Professional topical fluoride applications–Clinical efficacy and mechanism of action. *Advances in Dental Research, 8*, 190–201.
10. Tirupathi, S., Nirmala, S. V. S. G., Rajasekhar, S., & Nuvvula, S. (2019). Comparative cariostatic efficacy of a novel Nano-silver fluoride varnish with 38% silver diamine fluoride varnish a double-blind randomized clinical trial. *Journal of Clinical and Experimental Dentistry, 11*, e105.
11. Pérez-Díaz, M. A., Boegli, L., James, G., Velasquillo, C., Sánchez-Sánchez, R., Martínez-Martínez, R. E., Martínez-Castañón, G. A., & Martinez-Gutierrez, F. (2015). Silver nanoparticles with antimicrobial activities against streptococcus Mutans and their cytotoxic effect. *Materials Science and Engineering C, 55*, 360–366.
12. Tavassoli Hojati, S., Alaghemand, H., Hamze, F., Ahmadian Babaki, F., Rajab-Nia, R., Rezvani, M. B., Kaviani, M., & Atai, M. (2013). Antibacterial, physical and mechanical properties of Flowable resin composites containing zinc oxide nanoparticles. *Dental Materials, 29*, 495–505.
13. Scarpelli, B. B., Punhagui, M. F., Hoeppner, M. G., de Almeida, R. S. C., Juliani, F. A., Guiraldo, R. D., & Berger, S. B. (2017). In vitro evaluation of the remineralizing potential and antimicrobial activity of a Cariostatic agent with silver nanoparticles. *Brazilian Dental Journal, 28*, 738–743.

14. Jawed, M., Al Abdulmonem, W., Alkhamiss, A., Alghsham, R., Alsaeed, T., Alhumaydhi, F. A., Hershan, A. A., & Shahid, S. M. (2021). Role of serum magnesium in dental caries. *Bahrain Medical Bulletin, 43*, 327–330.
15. Rago, I., Bregnocchi, A., Zanni, E., D'Aloia, A. G., De Angelis, F., Bossu, M., De Bellis, G., Polimeni, A., Uccelletti, D., & Sarto, M. S. (2015). Antimicrobial activity of graphene nanoplatelets against streptococcus mutans. In *IEEE-NANO 2015 – 15th international conference on nanotechnology* (pp. 9–12). IEEE.
16. Radhi, A., Mohamad, D., Abdul Rahman, F. S., Abdullah, A. M., & Hasan, H. (2021). Mechanism and factors influence of graphene-based nanomaterials antimicrobial activities and application in dentistry. *Journal of Materials Research and Technology, 11*, 1290–1307.
17. Hamad, A. M., & Mahmood Atiyea, Q. (2021). In vitro study of the effect of zinc oxide nanoparticles on streptococcus Mutans isolated from human dental caries. *Journal of Physics: Conference Series, 1879*. https://doi.org/10.1088/1742
18. Salas-López, E. K., Pierdant-Pérez, M., Hernández-Sierra, J. F., Ruíz, F., Mandeville, P., & Pozos-Guillén, A. J. (2017). Effect of silver nanoparticle-added pit and fissure sealant in the prevention of dental caries in children. *Journal of Clinical Pediatric Dentistry, 41*, 48–52.
19. Cheng, L., Zhang, K., Zhou, C. C., Weir, M. D., Zhou, X. D., & Xu, H. H. K. (2016). One-year water-ageing of calcium phosphate composite containing nano-silver and quaternary ammonium to inhibit biofilms. *International Journal of Oral Science, 8*, 172–181.
20. Al-Mosawi, R. M., & Al-Badr, R. M. (2017). The study effects of dental composite resin as antibacterial agent which contain nanoparticles of zinc oxide on the bacteria associated with Oral infection. *IOSR Journal of Dental and Medical Sciences, 16*, 49–55.
21. Amiri, M., Etemadifar, Z., Daneshkazemi, A., & Nateghi, M. (2017). Antimicrobial effect of copper oxide nanoparticles on some Oral bacteria and Candida species. *Journal of Dental Biomaterials, 4*, 347–352.
22. Noori, A. J., & Kareem, F. A. (2019). The effect of magnesium oxide nanoparticles on the antibacterial and Antibiofilm properties of glass-ionomer cement. *Heliyon, 5*, e02568.
23. Haghighi, F., Roudbar Mohammadi, S., Mohammadi, P., Hosseinkhani, S., & Shipour, R. (2013). Antifungal activity of TiO_2 nanoparticles and EDTA on Candida Albicans biofilms related papers. *Infection, Epidemiology and Microbiology, 1*, 33–38.
24. Totu, E. E., Nechifor, A. C., Nechifor, G., Aboul-Enein, H. Y., & Cristache, C. M. (2017). Poly(methyl methacrylate) with TiO_2 nanoparticles inclusion for Stereolitographic complete denture manufacturing – The Fututre in Dental Care for Elderly Edentulous Patients? *Journal of Dentistry, 59*, 68–77.
25. Nguyen, S., Escudero, C., Sediqi, N., Smistad, G., & Hiorth, M. (2017). Fluoride loaded polymeric nanoparticles for dental delivery. *European Journal of Pharmaceutical Sciences, 104*, 326–334.
26. Dabbagh, A., Abu Kasim, N. H., Bakri, M. M., Wakily, H., Ramasindarum, C., & Abdullah, B. J. J. (2014). Polyethylene-glycol coated Maghemite nanoparticles for treatment of dental hypersensitivity. *Materials Letters, 121*, 89–92.
27. Priyadarshini, B. M., Selvan, S. T., Lu, T. B., Xie, H., Neo, J., & Fawzy, A. S. (2016). Chlorhexidine Nanocapsule drug delivery approach to the resin-dentin Interface. *Journal of Dental Research, 95*, 1065–1072.
28. Pagonis, T. C., Chen, J., Fontana, C. R., Devalapally, H., Ruggiero, K., Song, X., Foschi, F., Dunham, J., Skobe, Z., & Hajime Yamazaki, R. K. (2010). Nanoparticle-based endodontic antimicrobial photodynamic therapy – Clinical key. *Journal of Endodontia, 36*, 322–328.
29. Horev, B., Klein, M. I., Hwang, G., Li, Y., Kim, D., Koo, H., & Benoit, D. S. W. (2015). PH-activated nanoparticles for controlled topical delivery of Farnesol to disrupt Oral biofilm virulence. *ACS Nano, 9*, 2390–2404.
30. He, J., Zhu, X., Qi, Z., Wang, C., Mao, X., Zhu, C., He, Z., Li, M., & Tang, Z. (2015). Killing dental pathogens using antibacterial graphene oxide. *ACS Applied Materials and Interfaces, 7*, 5605–5611.

31. Alnatheer, M., Alqerban, A., & Alhazmi, H. (2021). Graphene oxide-modified dental adhesive for bonding orthodontic brackets. *International Journal of Adhesion and Adhesives, 110*, 102928.
32. Mohandes, F., & Salavati-Niasari, M. (2014). In vitro comparative study of pure hydroxyapatite Nanorods and novel polyethylene glycol/graphene oxide/hydroxyapatite nanocomposite. *Journal of Nanoparticle Research, 16*, 2604.
33. Chen, J., Zhao, Q., Peng, J., Yang, X., Yu, D., & Zhao, W. (2020). Antibacterial and mechanical properties of reduced graphene-silver nanoparticle nanocomposite modified glass ionomer cements. *Journal of Dentistry, 96*, 103332.
34. Peng, J. M., Lin, J. C., Chen, Z. Y., Wei, M. C., Fu, Y. X., Lu, S. S., Yu, D. S., & Zhao, W. (2017). Enhanced antimicrobial activities of silver-nanoparticle-decorated reduced graphene nanocomposites against Oral pathogens. *Materials Science and Engineering C, 71*, 10–16.
35. Kulshrestha, S., Khan, S., Meena, R., Singh, B. R., & Khan, A. U. (2014). A graphene/zinc oxide nanocomposite film protects dental implant surfaces against cariogenic streptococcus Mutans. *Biofouling, 30*, 1281–1294.
36. Mao, M., Zhang, W., Huang, Z., Huang, J., Wang, J., Li, W., & Gu, S. (2021). Graphene oxide-copper nanocomposites suppress cariogenic streptococcus Mutans biofilm formation. *International Journal of Nanomedicine, 16*, 7727–7739.
37. Sava, S., Moldovan, M., Sarosi, C., Mesaros, A., Dudea, D., & Alb, C. (2015). Effects of graphene addition on the mechanical properties of composites for dental restoration. *Materiale Plastice, 52*, 90–92.
38. Zhang, J. F., Wu, R., Fan, Y., Liao, S., Wang, Y., Wen, Z. T., & Xu, X. (2014). Antibacterial dental composites with chlorhexidine and mesoporous silica. *Journal of Dental Research, 93*, 1283–1289.
39. Liu, F., Jiang, X., Zhang, Q., & Zhu, M. (2014). Strong and bioactive dental resin composite containing poly(Bis-GMA) grafted hydroxyapatite whiskers and silica nanoparticles. *Composites Science and Technology, 101*, 86–93.
40. Barma, M. D., Kannan, S. D., Indiran, M. A., Rajeshkumar, S., & Kumar, R. P. (2020). Antibacterial activity of Mouthwash Incorporated with Silica Nanoparticles against S. Aureus, S. Mutans, E. Faecalis: An in-Vitro Study. *Journal of Pharmaceutical Research International, 32*, 25–33.
41. Stewart, C., Hong, J., Hatton, B., & Finer, Y. (2018). Responsive antimicrobial dental adhesive based on drug-Templated mesoporous nanoparticles. *Acta Biomaterialia, 76*, 283–294.
42. Carpenter, A. W., Reighard, K. P., Saavedra, J. E., & Schoenfisch, M. H. (2013). O2-Protected Diazeniumdiolate-modified silica nanoparticles for extended nitric oxide release from dental composites. *Biomaterials Science, 1*, 456–459.
43. Lee, J. H., El-Fiqi, A., Jo, J. K., Kim, D. A., Kim, S. C., Jun, S. K., Kim, H. W., & Lee, H. H. (2016). Development of long-term antimicrobial poly(methyl methacrylate) by incorporating mesoporous silica Nanocarriers. *Dental Materials, 32*, 1564–1574.
44. Bai, X., Lin, C., Wang, Y., Ma, J., Wang, X., Yao, X., & Tang, B. (2020). Preparation of Zn doped mesoporous silica nanoparticles (Zn-MSNs) for the improvement of mechanical and antibacterial properties of dental resin composites. *Dental Materials, 36*, 794–807.
45. Chen, M.-H. (2010). Update on dental. *Journal of Dental Research, 89*, 549–560.
46. Mitra, S. B., Wu, D., & Holmes, B. N. (2003). An application of nanotechnology in advanced dental materials. *Journal of the American Dental Association, 134*, 1382–1390.
47. Lührs, A. K., & Geurtsen, W. (2009). The application of silicon and silicates in dentistry: A review. *Progress in Molecular and Subcellular Biology, 47*, 359–380.
48. Rosenblatt, A., Stamford, T. C. M., & Niederman, R. (2009). Silver diamine fluoride: A caries "silver-fluoride bullet". *Journal of Dental Research, 88*, 116–125.
49. Greenwall-Cohen, J., Greenwall, L., & Barry, S. (2020). Silver diamine fluoride - an overview of the literature and current clinical techniques. *British Dental Journal, 228*, 831–838.
50. Contractor, I. A., Girish, M. S., & Indira, M. D. (2021). Silver diamine fluoride: Extending the Spectrum of preventive dentistry, a literature review. *Pediatric Dental Journal, 31*, 17–24.

51. Hiraishi, N., Sayed, M., Takahashi, M., Nikaido, T., & Tagami, J. (2022). Clinical and primary evidence of silver diamine fluoride on root caries management. *Japanese Dental Science Review, 58*, 1–8.
52. Ballikaya, E., & Çehreli, Z. C. (2020). Gümüş Diamin Florürün Çürük Yönetimindeki Yeri. *Turkiye Klinikleri Journal of Dental Sciences, 26*, 276–284.
53. Nishino, M. (1969). Studies on the topical application of Ammoniacal silver fluoride for the arrest of dental caries. *The Journal of Osaka University Dental School, 14*, 1–14.
54. Hu, S., Meyer, B., & Duggal, M. (2018). A silver renaissance in dentistry. *European Archives of Paediatric Dentistry, 19*, 221–227.
55. Cheng, L. L. (2017). Limited evidence suggesting silver diamine fluoride may arrest dental caries in children. *Journal of the American Dental Association (1939), 148*, 120–122.
56. Buzalaf, A. R., Pessan, J. P., Honório, H. M., & Jacob Martien ten Cate, M. (2011). Impact of fluoride in the prevention of caries and erosion. In M. A. R. Buzalaf (Ed.), *Fluoride and the oral mechanisms of action of fluoride for caries control* (Environment: 22 (Monographs in oral science)) (pp. 97–114). Karger.
57. Lou, Y. L., Botelho, M. G., & Darvell, B. W. (2011). Reaction of silver diamine [corrected] fluoride with hydroxyapatite and protein. *Journal of Dentistry, 39*, 612–618.
58. Zhao, I. S., Gao, S. S., Hiraishi, N., Burrow, M. F., Duangthip, D., Mei, M. L., Lo, E. C. M., & Chu, C. H. (2018). Mechanisms of silver diamine fluoride on arresting caries: A literature review. *International Dental Journal, 68*, 67–76.
59. Chu, C. H., Mei, L., Seneviratne, C. J., & Lo, E. C. M. (2012). Effects of silver diamine fluoride on dentine carious lesions induced by streptococcus Mutans and Actinomyces Naeslundii biofilms. *International Journal of Paediatric Dentistry, 22*, 2–10.
60. Mei, M. L., Li, Q. L., Chu, C. H., Lo, E. C. M., & Samaranayake, L. P. (2013). Antibacterial effects of silver diamine fluoride on multi-species cariogenic biofilm on caries. *Annals of Clinical Microbiology and Antimicrobials, 12*, 4.
61. Mei, M. L., Ito, L., Cao, Y., Lo, E. C. M., Li, Q. L., & Chu, C. H. (2014). An ex vivo study of arrested primary teeth caries with silver diamine fluoride therapy. *Journal of Dentistry, 42*, 395–402.
62. Mei, M. L., Ito, L., Cao, Y., Li, Q. L., Lo, E. C. M., & Chu, C. H. (2013). Inhibitory effect of silver diamine fluoride on dentine demineralisation and collagen degradation. *Journal of Dentistry, 41*, 809–817.
63. Duangthip, D., Chu, C. H., & Lo, E. C. M. (2016). A randomized clinical trial on arresting dentine caries in preschool children by topical fluorides—18 month results. *Journal of Dentistry, 44*, 57–63.
64. Lansdown, A. B. G. (2006). Silver in health care: antimicrobial effects and safety in use. *Current Problems in Dermatology, 33*, 17–34.
65. Woo, K. J., Hye, C. K., Ki, W. K., Shin, S., So, H. K., & Yong, H. P. (2008). Antibacterial activity and mechanism of action of the silver ion in staphylococcus aureus and Escherichia Coli. *Applied and Environmental Microbiology, 74*, 2171–2178.
66. Jennings, M. C., Minbiole, K. P. C., & Wuest, W. M. (2016). Quaternary ammonium compounds: An antimicrobial mainstay and platform for innovation to address bacterial resistance. *ACS Infectious Diseases, 1*, 288–303.
67. Clemens, J., & Gold, J. (2018). Effect and acceptance of silver diamine fluoride treatment on dental caries in primary teeth. *Journal of Public Health Dentistry, 78*, 63–68. Wiley Online Library.
68. Vasquez, E., Zegarra, G., Chirinos, E., Castillo, J. L., Taves, D. R., Watson, G. E., Dills, R., Mancl, L. L., & Milgrom, P. (2012). Short term serum pharmacokinetics of Diammine silver fluoride after Oral application. *BMC Oral Health, 12*, 60.
69. Dos Santos, V. E., Filho, A. V., Ribeiro Targino, A. G., Pelagio Flores, M. A., Galembeck, A., Caldas, A. F., & Rosenblatt, A. (2014). A new "Silver-Bullet" to treat caries in children – Nano silver fluoride: A randomised clinical trial. *Journal of Dentistry, 42*, 945–951.

70. Tirupathi, S., Nirmala, S. V. S. G., Rajasekhar, S., & Nuvvula, S. (2019). Comparative Cariostatic efficacy of a novel Nano-silver fluoride varnish with 38% silver diamine fluoride varnish a double-blind randomized clinical trial. *Journal of Clinical and Experimental Dentistry, 11*, e105–e112.
71. Haghgoo, R., Saderi, H., Eskandari, M., Haghshenas, H., & Rezvani, M. (2014). Evaluation of the antimicrobial effect of conventional and Nanosilver-containing varnishes on Oral streptococci. *Journal of Dentistry, 15*, 57.
72. Targino, A. G. R., Flores, M. A. P., Dos Santos, V. E., De Godoy Bené Bezerra, F., De Luna Freire, H., Galembeck, A., & Rosenblatt, A. (2014). An innovative approach to treating dental decay in children. A new anti-caries agent. *Journal of Materials Science. Materials in Medicine, 25*, 2041–2047.
73. Chan, E. L. K., Zhang, C., & Cheung, G. S. P. (2015). Cytotoxicity of a novel Nano-silver particle endodontic Irrigant. *Clinical, Cosmetic and Investigational Dentistry, 7*, 65.
74. Kasraei, S., Sami, L., Hendi, S., AliKhani, M.-Y., Rezaei-Soufi, L., & Khamverdi, Z. (2014). Antibacterial properties of composite resins incorporating silver and zinc oxide nanoparticles on streptococcus Mutans and lactobacillus. *Restorative Dentistry and Endodontics, 39*, 109.
75. Li, F., Weir, M. D., Chen, J., & Xu, H. H. K. (2013). Comparison of quaternary ammonium-containing with Nano-silver-containing adhesive in antibacterial properties and cytotoxicity. *Dental Materials, 29*, 450–461.
76. Azarsina, M., Kasraei, S., Yousefi-Mashouf, R., Dehghani, N., & Shirinzad, M. (2013). The antibacterial properties of composite resin containing Nanosilver against streptococcus Mutans and lactobacillus. *The Journal of Contemporary Dental Practice, 14*, 1014–1018.

Chapter 12
Emerging Microfluidics Devices for Microbial Studies

Saurabh Khachane, Vaibhav Lagad, Rutuja Vikhe, and Saurabh Kumar

12.1 Introduction

Microfluidics has seen rapid development in microbiology, resulting in the miniaturization of techniques and devices. It is a rapidly evolving field that deals with manipulation and control over small volumes of fluids at the microscale level. The scientific and technological field known as microfluidics processes or manipulates small volumes of fluids between 10^{-6} and 10^{-12} liters in channels or structures with at least one dimension in the micrometer range or between 1 μm and 1 mm [1]. Applications of microfluidics centered on fluid characteristics, rules, and alterations specified by fluid dimension ranging from sub-millimeter to micrometer and connected with microstructure and nanostructures of the device. Different types of scientific instruments, including flow reactors and gas chromatography, already have microchannels built in [2]. It provides extensively new possibilities for manipulating molecule concentrations in both time as well as space [3]. We can do multiple studies simultaneously and reduce the errors caused by manual handling by using microfluidic equipment to manage tiny quantities of liquids precisely, automatically, and accurately. The experiment volume can be reduced through miniaturization, which lowers the possible reagent cost and reaction time. This exceptional level of precision permits designs that can hold individual cells in place or direct the cells to a specific area while the entire experiment is continuously monitored. It has been proven that microfluidics is useful for biotechnology and microbiology [1].

Considering the above-mentioned advantages of microfluidics, the complicated dynamic between bacterial cells and their surroundings can be seen and understood using it as a magnifying glass. It provides a controlled and reproducible

S. Khachane · V. Lagad · R. Vikhe · S. Kumar (✉)
Department of Medical Devices, National Institute of Pharmaceutical Education and Research (NIPER), Guwahati, Assam, India
e-mail: saurabh@niperguwahati.in

R. S. Chaughule, A. S. Lokur (eds.), *Applications of Nanotechnology in Microbiology*, https://doi.org/10.1007/978-3-031-49933-3_12

environment for microbial cultures, allowing researchers to study the growth, metabolism, and interactions of microorganisms at the microscale level. By using microfluidics, researchers can also create a range of different microenvironments, each with different nutrient and electron transfer rates, to mimic the natural conditions of microbial communities [4]. In terms of biocompatibility, microfluidics also offers a better environment for cell cultivation. It gives a higher surface-to-volume ratio when compared to a cell culture flask. The media, which is stagnant in a cell culture flask, is continuously refreshed due to laminar circulation in the microfluidic channel. The temperature and pressure of the channel can also be adjusted, allowing the device to be integrated with other instruments [5].

Overall, microfluidics could completely transform the field of microbiology by providing new tools for studying the molecular mechanisms of microbial behavior and interactions. As research progresses, microfluidic devices are likely to play a significant role in understanding the complex processes that underlie microbial life. In this chapter, we will discuss the fluid flow at a micro-scale, various aspects of micro-fabrication, such as 3D printing, soft lithography, and in the last sections, we will cover several applications of microfluidic in microbiology such as microbiome host interaction, microbial fuel cell study, antibiotic-resistant mutant study, pathogen detection, screening and diagnosis, microbial culture, PCR-based study, biosensors integration, and various application of microbiology.

12.1.1 Fluid Flow at Micro-scale

Before studying the use of microfluidics in microbiological research, we need to understand the basic fluid mechanics in microfluidics. When fluid flows through the channel, depending upon the channel dimensions flow behavior will change. The flow of fluid refers to the movement of a fluid (liquid or gas) in response to external or internal forces. Fluids are characterized by their ability to flow and take the shape of their container. The fluid flow of fluids is described by the fluid dynamics principles, which govern the behavior of fluids in motion.

The flow of fluids is classified as laminar and turbulent flow. Smooth, parallel layers of fluid that move predictably and in an organized way are known as laminar flow (Fig. 12.1a). This type of flow occurs at low velocities and is typically observed

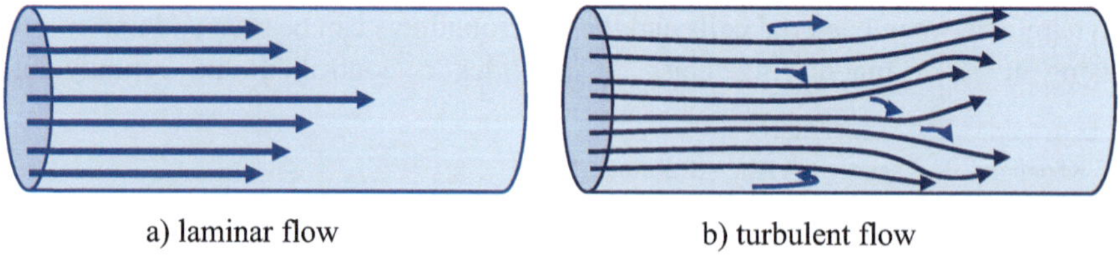

Fig. 12.1 Schematic representation of laminar and turbulent flow (**a**) laminar flow (**b**) turbulent flow

in small channels or tubes. Turbulent flow, on the other hand, is characterized by irregular and chaotic movement of fluid (Fig. 12.1b), which results in mixing and turbulence. This type of flow occurs at high velocities and is typically observed in large channels or open spaces.

In microfluidics, channel dimensions usually range from 100 nm to 1 mm. At the micro level viscous forces become more dominant than the inertial force, this will reduce the ratio of inertial force/viscous force. Reynolds number (Re) is the measure of the inertial to viscous force ratio.

$$\mathrm{Re} = {\rho VL}/{\eta}$$

Where, **Re** = Reynolds No., $\boldsymbol{\rho}$ = density of fluid, $\boldsymbol{V}$ = Velocity of fluid, $\boldsymbol{\eta}$ = Viscosity of fluid

The Reynolds no. value up to 2000 states the laminar behavior and values more than 2000 usually consider the non-laminar (turbulent) behavior of fluid in the channel. In the microfluidic channel, viscous forces are dominate [6]. Therefore, fluid will flow in layers resulting in a laminar flow pattern. This results in mass transfer by diffusion mode rather than a convection mode and will also increase the surface area for mass transfer. These are the characteristic features taken into consideration for the microfluidic device.

12.2 Device Fabrication

The success of the microfluidic device is greatly dependent on design and fabrication. Soft lithography and 3D printing are popular and common techniques to fabricate microfluidic chips. To provide a thorough picture, we shall talk about each of these techniques. The user can choose the fabrication technique based on the required conditions for microbiological application.

12.2.1 Soft Lithography

Soft lithography is an important technique for the fabrication of microchannels and micro/nanostructures inside a microchannel (Fig. 12.2). This technique was developed by Whiteside's group [7]. Basically, it is a patterning technique that applies ink to a substrate using an elastomeric stamp. Soft lithography is advantageous over traditional patterning techniques, such as significantly large cost savings, a simpler setup, excellent efficiency, and a pattern resolution that can be as accurate as nanometers or micrometers. The stamp master can be created using either photolithography or e-beam lithography; however, once the master is created, it can be used repeatedly to create stamps, so this process only needs to be done once. PDMS

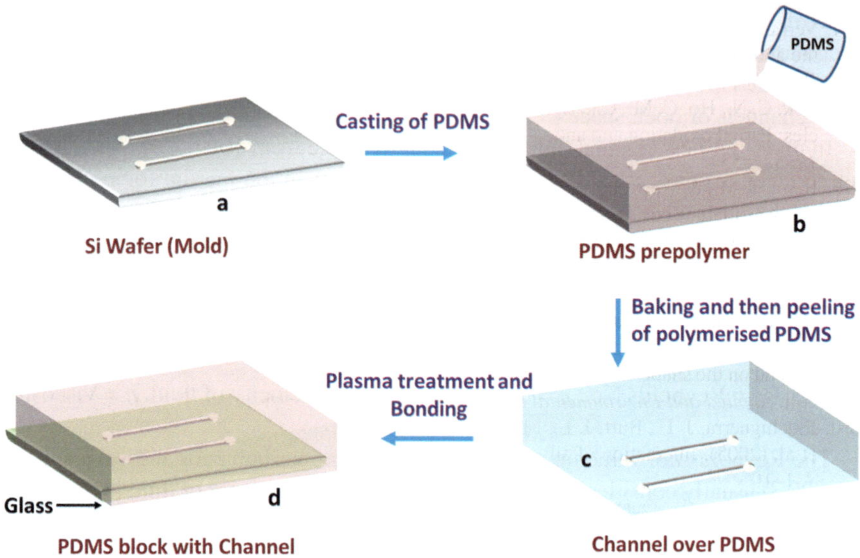

Fig. 12.2 Schematic representation of soft lithography technique

(most popular), photocurable perfluoro polyether, photosensitive polymers, and polymerized hydrogels are different polymers that are being utilized to fabricate microfluidic channels. The properties of PDMS such as elasticity, optical transparency, hydrophobicity, biocompatibility, and gaseous permeability, make it a popular choice to fabricate mold. These properties can be tuned according to the experiment.

Figure 12.2 represents the schematic diagram of microfluidics device fabrication steps. An integrated microfluidic device contains a channel layer bonded to a plane surface like glass (step d) or a micro/nanostructured layer. Prepolymer PDMS is cured on the mold (steps a and b). This mold is often constructed of SU8 or Si using photolithography. The ratio of prepolymer agents such as silicone elastomer and curing agent (10:1, depending on the stiffness of the channel required) is determined by the required stiffness. The polymer is fully cured before being taken from the master mold (step c) and plasma-bonded to a completely cleaned glass substrate (step d). This simple bonding without high temperature or pressure makes it possible for the fabrication of multiple-layer stacking channels.

12.2.2 3D Printing

3D printing is a relatively new technique being used. As the structure complexity increases, the fabrication time and cost also increase. It also hinders the automated system for fabrication. Thus, generating 3D structures in a single step is beneficial for rapid and high-throughput manufacturing. In this approach, 3D geometrical

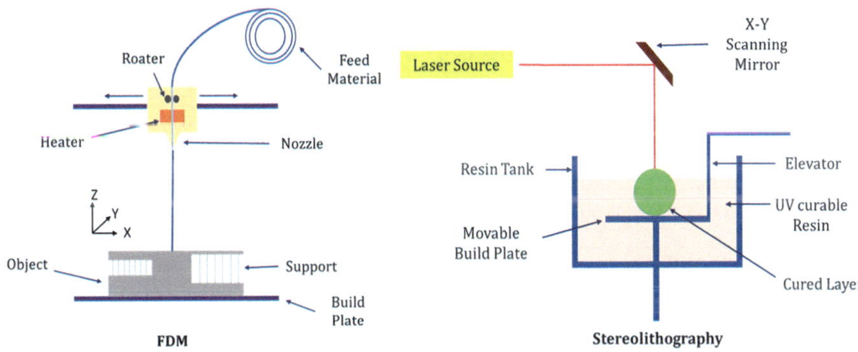

Fig. 12.3 Schematic representation of fuse deposition modeling and stereolithography technique

information is stored in a computer as a CAD file that runs the attached printer. Unlike Soft lithography and Micromachining, 3D printing is an additive manufacturing method that interfaces with an industry-grade user interface and embedded control system for manufacturing. The 3D printing method known as Fused Deposition Modeling (FDM) employs a heated nozzle to liquefy and extrude thermoplastic materials, like Polylactic acid (PLA), Polycarbonate, Polyethylene terephthalate (PET) and, acrylonitrile butadiene styrene (ABS) which gets hardened by spontaneous cooling to construct a three-dimensional (3D) structure in a layer-by-layer approach. Scott Crump invented FDM (patent granted in 1992), and was commercialized by a corporation named Stratasys [8]. A schematic illustration of the FDM 3D printer is shown in Fig. 12.3. To create microfluidic devices using FDM, a series of steps are necessary. The initial step involves utilizing computer-aided design (CAD) software to design the microfluidic device, including the channels, reservoirs, and other elements required to manipulate and analyze fluids at the microscale. The design must also consider the constraints of FDM, such as the layer height and nozzle diameter, to guarantee the final device's precision and functionality. Once the design is completed, the FDM printer must be prepared. This process involves loading the thermoplastic material into the printer and adjusting the printing parameters, like layer height and printing speed. The printer must also be calibrated to ensure that the extruded material is deposited with accuracy and consistency. The next stage is to print the microfluidic device layer by layer until the entire structure is completed. During the printing process, it is critical to ensure that the channels are properly aligned and connected to the reservoirs and other elements. Following printing, the device may require post-processing to enhance its performance, including cleaning the device to remove any residual material or support structures, smoothing the surfaces to minimize friction, or adding coatings to render the surfaces hydrophobic or hydrophilic. The final step in constructing a microfluidic device using FDM is to examine and validate its performance. This may involve conducting fluidic tests to confirm the proper connection of channels and device functionality. Testing the device under various flow rates or fluid properties may also be necessary to evaluate its performance under diverse conditions [9].

The nozzle size and materials used in FDM play an important role in the resolution of printing parts. Another issue is the lack of more advanced printing materials like glass and ceramics, which restricts its universal use in industrial settings. Despite using engineering-grade thermoplastics, FDM cannot manufacture parts that are as strong and long-lasting as those made using other 3D printing processes. Due to the asymmetry of FDM-printed parts, they are easily broken by compressive forces that are parallel to their layering.

Stereolithography (SL), also known as Vat Photopolymerization, is a photochemical process utilized in 3D printing technology to fabricate prototypes, structures, and manufactured parts in a layer-by-layer approach. In this process, chemical monomers and oligomers are linked together by light to create polymers, which subsequently make up the 3D structure of a solid. Exploration in this field started in the 1970s, and in 1984 Chuck Hull coined the phrase while submitting a patent application for the method [10]. Architecture of stereolithography can be seen in Fig. 12.3. The most popular way that SL operates is by projecting a UV laser onto a container of photopolymer resin. A designed pattern or shape is inked with the UV laser onto the surface of the container containing a photopolymer using computer-aided design (CAM/CAD) software. The resin is photochemically solidified in response to ultraviolet light, forming a unit layer of the desired 3D object. A layer of the build platform goes down, and then a blade coats the top of the tank with resin once more. Up to the finished 3D product, this process is repeated. To remove wet resin from finished parts' surfaces, a solvent wash is required. The SL process offers high precision, a smooth surface finish, the capability for customization, integration of multiple components, and rapid prototyping, making it a suitable technology for microfluidics fabrication [11].

Here are a few ways in which SL helps in the fabrication of microfluidics:

- High Resolution: SL has a high resolution and can produce features down to tens of micrometers in size. This allows for the creation of intricate microfluidic channels, valves, and other components.
- Smooth Surface Finish: SL produces parts with a smooth surface finish, which is important in microfluidics to minimize frictional forces and prevent the formation of bubbles.
- Customization: SL allows for the customization of microfluidic devices, enabling the creation of unique geometries that are optimized for specific applications.
- Integration of Multiple Components: SL can be used to fabricate microfluidic devices that integrate multiple components, such as valves, pumps, and sensors, into a single device. This reduces the number of external components required for a microfluidic system, making it more compact and easier to use.
- Rapid Prototyping: SL allows for the rapid prototyping of microfluidic devices, enabling researchers to quickly iterate and optimize their designs.

The fabricated microchannel using above mentioned technique can be used in various applications of microbiology such as making PCR, study microbiome host interaction, producing fuel cell energy in the microfluidic channel from microbes, diagnosis of microbes in the human gut by producing microbial environment on chip (Lab on Chip and Organ on chip) and many more applications.

12.3 Applications

This section will discuss about some of the most important areas of microbiology research where microfluidics offers significant experimental advantages over traditional methods. There are various applications of microfluidics devices for microbial studies. These include studies of microbiome-host interaction, antibiotic resistance mutants, molecular analysis, fuel cells, and screening and diagnostics applications. Each application demands a specific design of microfluidics devices. Therefore, the design of microfluidics devices depends on what they are used for. This section includes several examples from current research that give the readers a broad view and help them to understand the real benefits of using microfluidic devices in microbiology.

12.3.1 Microbiome Host Interaction

A community of microbes known as the "microbiota" that coexist harmoniously with humans, together with their genetic material and their immediate surroundings, collectively make up the human microbiome [12]. The role of the human microbiome in health and disease is widely acknowledged. Dysregulation of the gut microbiota has been associated with a number of disease states, such as neurodegenerative, cardiovascular, and metabolic illnesses [13]. Additionally, the gut microbiome's makeup can influence how the body reacts to medicines, most notably those used to treat cancer. Unintended adverse pharmacological responses can occur because of the gut microbiota's impact on drug metabolism [14].

Here, we will highlight a few possibilities of using microfluidic technology in organ-specific microbiome research and discuss particular issues that microfluidics must deal with when working with biological components that are important to the microbiome, such as microbes, host tissues, and fluids [15]. Moreover, organs-on-chips offer researchers a way to better comprehend the significance of microbiomes and how they affect host health. It is essential to remember that research on host-microbe interactions in organs-on-chips devices is still in the preliminary stages. Therefore, we must be aware of their existing limits and challenges. The gut microbiome is the most exciting area for studies because the Human gut comprises a large community of microbiomes, and this microbiome helps in the development of many diseases like cancer and other metabolic disturbances. There is induction of inflammation and progression of disease due to an imbalance of microbiota. However, establishing a connection between the gut microbiota and cancer cells using standard methods is challenging. Microfluidics is developing as a promising method to simulate this environment. Mittal et al. has developed a microfluidics device to co-culture colorectal carcinoma cell line and bacterial cells to learn how inflammatory stress and microbiota in the gut affect the cancer cell [16]. For this study, they developed a PDMS-based microfluidics device that consists of one middle chamber (0.8 cm wide × 2 cm length × height 250 μm) and two side channels (0.1 cm wide × 1.5 cm long × 250 μm high). Both side channels were connected with the middle chamber via six side channels (W × L × H, 0.5 cm wide × 1.5 cm length × 250 μm high) at a certain angle to create a gradient of oxygen and/or bacteria in the middle chamber. Middle chambers were used to culture colorectal cancer cells (HCT-116) line, side channels were used to diffuse media and bacterial cells in the middle chamber, and cancer cells and bacteria were cultured in a similar environment. This experimental setup based on a microfluidics device helped to mimic gut physiological conditions to recognize that abundant bacteria promote cancer progression (Fig. 12.4). It is clearly observed in colorectal cancer that Gram-positive (*Streptococcus, Gemella*) microflora is abundant. Next, to recreate a physiologically realistic intestinal oxygen gradient profile in Intestine Jalili-Firoozinezhad et al. fabricated a microfluidics intestine-on-a-chip that establishesand regulates oxygen gradient in the chip [17]. Thus, it is used to co-culture living human intestinal epithelium with the complex microbiome of gut (aerobic and anaerobic). They successfully established a higher level of a microbial diverse population (~200 unique operational taxonomic units) closely mimicking the human gut microbiome. Similar to this, Mahajan et al. developed the "vagina chip," a microchip culture that replicates the human vaginal mucosal linings using *Gardnerella vaginalis* and *Lactobacillus crispatus* [18]. In the vaginal microbiome, it is well known that a dominating lactobacillus species is linked to good gynecological and reproductive health, while a low or missing population is regarded to be abnormal. For this investigation, a two-channel microfluidics chip was built with a top channel that is 1 mm wide and 1 mm high and a bottom channel that is 1 mm wide and 0.2 mm high. These two channels are separated along the length of the chip, 16.7 mm, by a porous membrane with pores that are 7 μm in diameter. To replicate the vaginal epithelial-stromal interface, which is essential for the development of the vaginal epithelium,

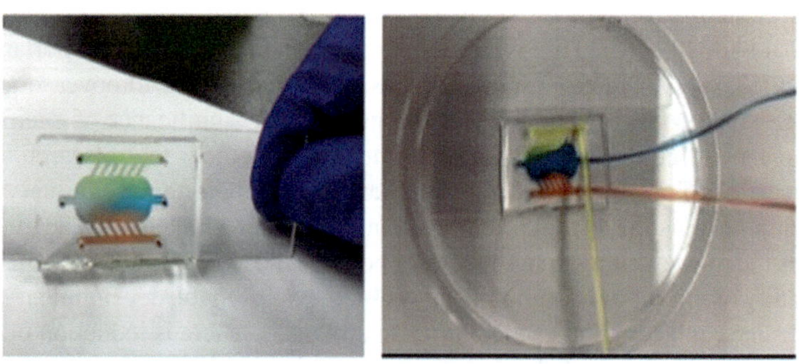

Fig. 12.4 The test for device function under static and dynamic situations using food color [16]. (Adapted under the terms of the CC-BY creative commons license)

the top channel was cultured with primary human vaginal epithelium and primary human uterine fibroblasts were cultivated on the base of the porous membrane (basal channel). The proper media and bacteria were passed through the top and bottom tubes. It has been observed from the study that co-culture with *Lactobacillus crispatus* directly influence the epithelium to suppress inflammation by reducing production of inflammatory cytokines while *Gardnerella vaginalis* induced inflammation. Such kind of human-microbiome model is helpful for the pre-clinical validation of drugs, testing, and basic science research to understand the physiology associated with the disease.

12.3.2 Antibiotic-Resistant Study

Healthcare professionals are becoming increasingly concerned about antibiotic resistance as more microorganisms develop resistance to the antibiotics we have used for years. Due to ability to precisely regulate and manipulate small amounts of fluids at the microscale level, microfluidic presents a promising route for exploring antibiotic resistance. This technology makes it possible to analyze the interactions between bacteria and antibiotics in controlled microenvironments, helping researchers to understand the mechanisms underlying antibiotic resistance and create more potent defenses against it. In this context, the study of antibiotic resistance in microfluidics has emerged as a fast-expanding area of study, with several studies being carried out to investigate the dynamics of bacterial growth and antibiotic efficacy in microfluidic systems [19]. Usually, in conventional studies of antibiotic resistance, bacteria are often cultured in vitro in bulk liquid cultures or on agar plates. Subjecting the cultures to various antibiotic doses determines their antibiotic susceptibility or resistance. Disk diffusion or minimum inhibitory concentration (MIC) tests are standard assays to measure the outcomes. However, investigations on antibiotic resistance based on microfluidics devices (microscopic channels or chambers) enable accurate control and handling of tiny amounts of fluid. These microenvironments can be selectively populated with bacteria and antibiotics, and real-time imaging like microscopy can be used to observe how the two of them interact. As a result, scientists can examine the dynamics of bacterial growth and antibiotic effectiveness in considerably more depth, as well as the influence of variables like spatial confinement and nutritional gradients.

In one study, *E. coli* biofilm is grown inside a microfluidic device, and its growth and death rates are determined [20]. Each device contained two biofilm growth chambers at the junction of the inlet and outlet channels (Fig. 12.5). The biofilm is exposed to different antibiotic concentrations (ciprofloxacin) to study its resistance. Additionally, excellent quality direct observation of fluorescence-labeled microorganisms in the biofilm was carried out using confocal microscopy to produce time-lapse images in order to gain a deeper understanding how ciprofloxacin functions.

Zoheir et al. developed a concentration gradient across the different zone of microfluidics devices [21]. In the interconnected compartment, microbial cells can

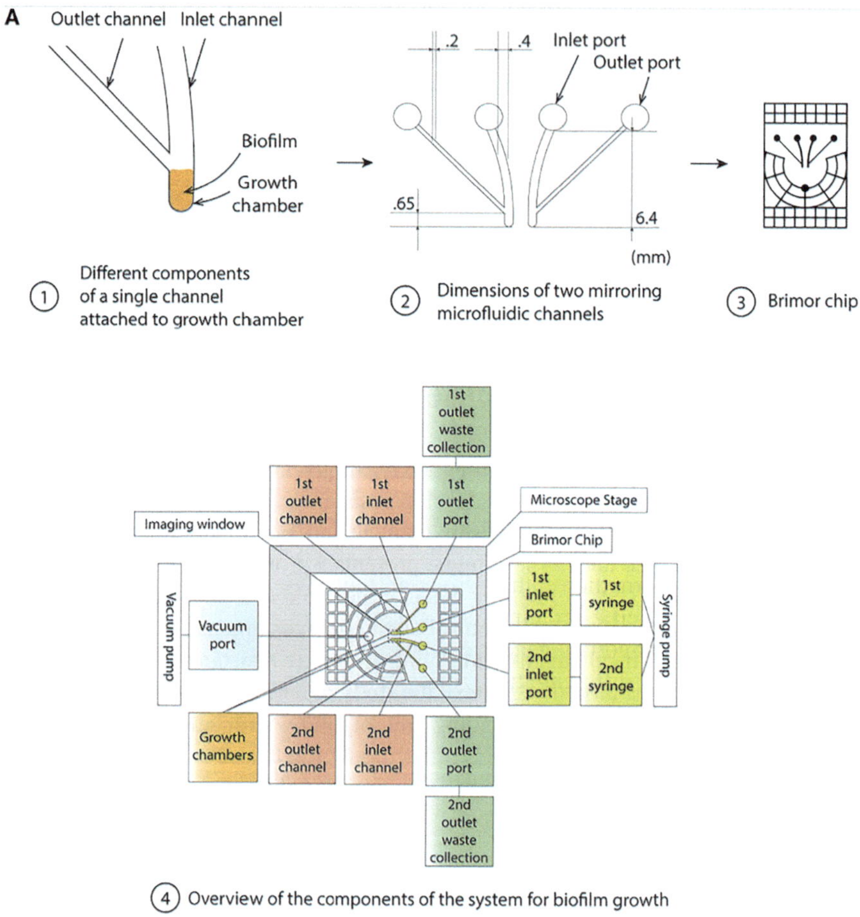

Fig. 12.5 Schematic of the microfluidic chip and its components for the biofilm studies: (A) Illustrations of the microfluidic chip's essential components are shown. (1) The microfluidic chips with inlet and outlet channel and growth chamber are 45° angle to one another. Cells are removed efficiently from a peripheral layer of biofilm due to the high shear stress of fluid flow in a channel, cell aggregates formed are shown in orange. (2) Dimensions and relative placement of the two growing chambers and flow channels on a chip. (3) An illustration of the entire Brimor chip, demonstrating the growth chambers and flow channels that are positioned in the centre, (4) A schematic representation of the entire system [20]. (Adapted under the terms of the CC-BY creative commons license)

be subjected to the specified profile of stressors, like antibiotics, with regulated and stable concentration gradients (Fig. 12.6). It is possible to precisely regulate the antibiotic concentration and adaptation to stress in the chip resistance. Thus, a concentration gradient was formed across the device. This kind of setup helps to disclose the previously unknown mutation in *microbes*. Further, it may help modifications to occur by creating a specific microenvironment. This allows scientists to learn more about the factors that affect how bacteria become resistant to antibiotics.

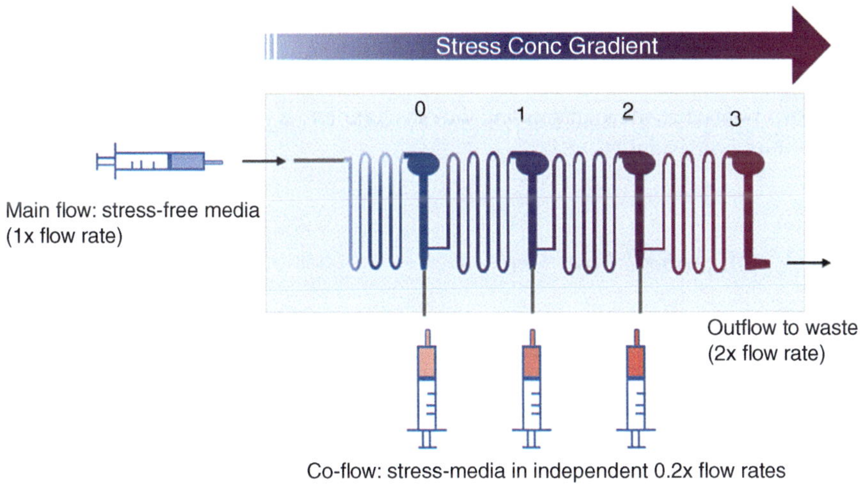

Fig. 12.6 The schematic representation of stress concentration gradient developed in a microfluidic device. [21]

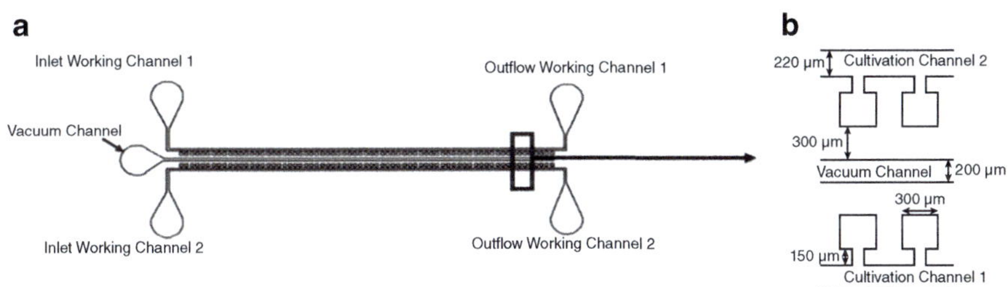

Fig. 12.7 (**a**) Design of the microfluidic system used; (**b**) magnified view of the cell chambers showing the dimensions [22]. (Adapted under the terms of the CC-BY creative commons license)

Antibiotic resistance is typically monitored using a variety of techniques. Most resistant gene detection methods, such as broth dilution, disc diffusion, and polymerase chain reaction (PCR), are based on the growth of bacteria on agar media plate and molecular identification of resistant genes. Long incubation times and limited quantification are the primary limitations of these methods. As a result, accurate and timely analysis of infectious strains is essential to decreasing antibiotic use worldwide and slowing the spread of antibiotic-resistant strains. Emil et al. created a microfluidic platform incorporating antibiotic susceptibility testing (AST) [22]. They used a vacuum in an inventive way to aid in growth inoculation, where all the channels are interconnected together (Fig. 12.7). To create a microenvironment, cells are loaded into these channels under a vacuum. To observe the effect of the antibiotic on growth rate of cell culture, an antibiotic is introduced in channel one and nutrient-rich media in channel.

Overall, the key difference between traditional antibiotic resistance studies and microfluidics-based studies is the level of control and precision afforded by microfluidic technology, which can enable more detailed and accurate investigation of antibiotic resistance mechanisms and the development of new strategies for combating antibiotic-resistant bacteria.

12.3.3 *Microbial Fuel Cell*

Microbial fuel cells (MFCs) are current-producing fuel cell systems, that generate electricity by the microbial conversion of organic matter, such as wastewater or food waste. The cell system is based on self-organized microorganisms. They are normally inoculated into an anodic chamber to bio-catalyze the chemicals and release electrons, protons, and different metabolic products. Electrons move across the external circuit as protons move to the cathode chamber and take part in reduction processes [23]. Basically, these anodic and cathodic compartments were divided by a membrane called a proton exchange membrane. The conventional setup of microbial fuel cells is shown in Fig. 12.8.

MFCs use microorganisms as catalysts, instead of using extracted enzymes or metallic catalysts. The use of microorganisms in MFCs ensures the system's long-term viability [24]. In recent years, microfluidics has emerged as a promising tool in

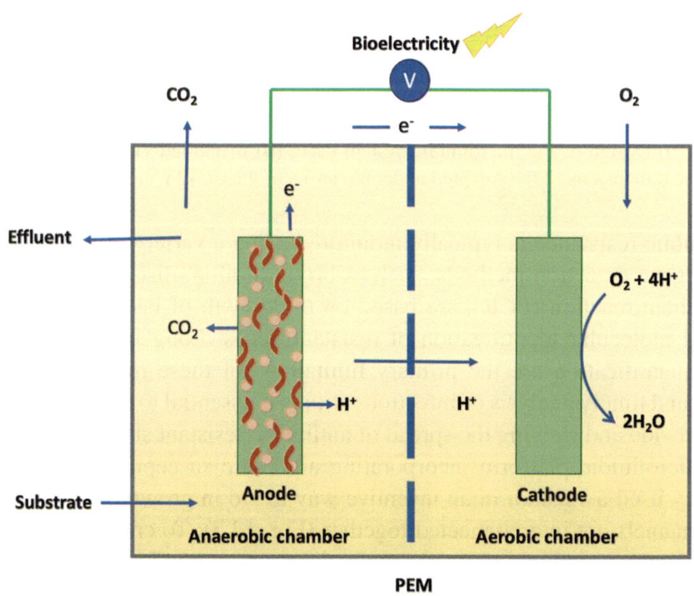

Fig. 12.8 Working principles of a traditional two-chamber micro MFC [23]

the development of MFCs.Microfluidic channels can be integrated into MFCs to create a 3D network of microchannels that act as a bioreactor. The high surface-to-volume ratio is obtained due to miniature size of channel, which enhances the mass transfer of substrates and products. Additionally, microfluidics enables precise regulation of the substrate flow rate and concentration, which may enhance MFC performance. One of the major challenges in designing MFCs is optimizing the transport of electrons and nutrients between the microbial community and the electrode. Microfluidics can address this challenge by providing a well-controlled microenvironment that enables precise manipulation of the MFC's mass transport and reaction rates. Also, fluid flow at the micro-scale is characterized by low Reynolds no. resulting in laminar flow. When anolyte and catholyte are injected in a microfluidic channel, co-laminar flow forms a functional interface between them. It creates a narrow mixing region between two layers eliminating the use membrane as in Fig. 12.9.

Additionally, micro MFCs are far more oxygen susceptible than large-scale MFCs. Poor mass loading of bacteria in micro space results in poor oxygen tolerance. The usage of PDMS, a polymer frequently used to create MFC, is constrained by its high oxygen permeability. To address the aforementioned issue and improve cell performance, Mehran Abbaszadeh Amirdehi et al. developed high-performance membrane-free microfluidics MFCs for steady, over time bench top operation under strong flow [25]. In order to prevent cross-contamination at various flow rates, they offer a membrane-free MFC that comprises (i) O_2 elimination via a gas diffusion barrier, (ii) linked graphite electrodes, and (iii) optimized electrode location. In comparison to earlier reports, the MFC containing a pure-culture anaerobic

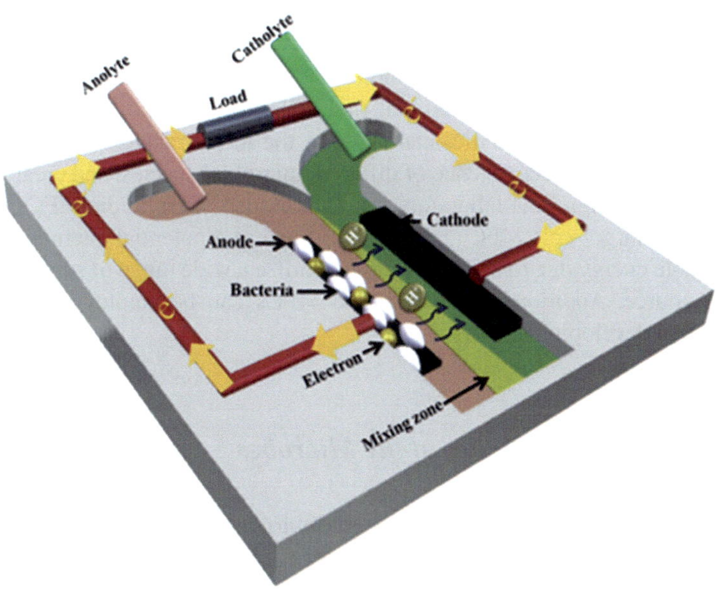

Fig. 12.9 Working principle of Co-laminar MFC [23]. (Reproduced with permission © Elsevier)

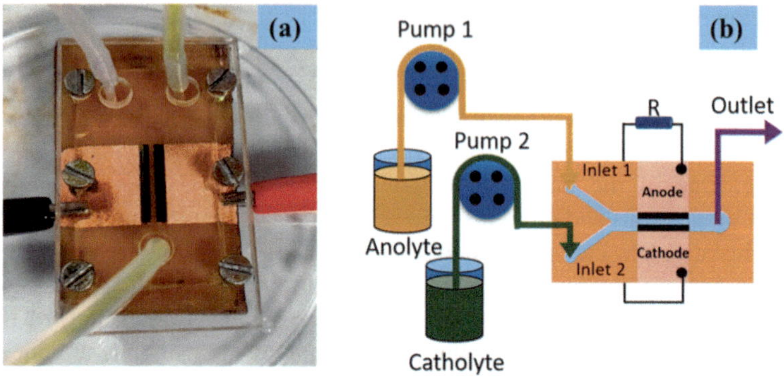

Fig. 12.10 (**a**) Optical image of M-MFC architecture with a (**b**) schematic that is fully integrated on a PCB [26]. (Reproduced with permission © Elsevier)

Geobacter sulfurreducens biofilm operated six times longer without the need of an oxygen scavenger. The fabricated MFC has greater stability at high flow rates and produces four times more power density.

In another work, P. Rewatkar et al. described how soft-lithographic microchannels and etched PCB electrodes were used to manufacture micro-MFC. *Shewanella putrefaciens*, one of several electrochemically active bacterial species, was utilized in this study as a power source [26]. As illustrated in Fig. 12.10, the printed circuit board fuel cell electrodes were tested in a co-laminar microfluidic flow environment after being patterned utilizing multiwalled carbon nanotubes (MWCNT) on a polymer-based Y-shaped microchannel. It produced a maximum power density of 239.2 W/cm^2 under ideal circumstances.

Microfluidics is an emerging technology that has the potential to significantly improve the performance of microbial fuel cells. By providing a well-controlled microenvironment, microfluidics can enhance the mass transfer of substrates and products, promote the coexistence of different microbial species, and create a more efficient MFC design. The laboratory-scale developed microfluidic MFCs serve as a tool to lead large-scale MFC and were identified as a potential alternative source of renewable energy due to their capacity to oxidize a wide range of substrates as an electron source. Additionally, microfluidic MFCs can be employed as a power source when developing sensors.

12.3.4 Molecular Studies of the Microbes

Due to precise and controlled manipulation of biological samples at the microscale level microfluidics emerge as a useful tool for studying microbes at the molecular level. Further, microfluidic systems may be used to isolate and examine individual cells or microbial communities, enabling researchers to study the molecular mechanisms of microbial interactions and the dynamics of microbial populations. The amplification of brief DNA fragments using the polymerase chain reaction (PCR) is one way that microfluidics is used in molecular studies of microbiology. The PCR has become an essential tool in molecular biology because of its ability to exponentially multiply. The molecular photocopying technique, often known as the PCR method, is an effective tool which is frequently employed in related molecular and genetic research. For amplifying the DNA samples, the PCR proceeds through three steps: denaturation (94 °C), also known as melting step, where the double-stranded DNA (dsDNA) is converted into a single strand. Further temperature lowers to the (60 °C) for annealing which includes adding small fragments of template DNA called primers to increase primer activity and produce single-strand DNA (ssDNA) at a lower temperature. A whole strand is extended during the extension (72 °C) phase, creating two separate DNA strands. This cycle is repeated as many times as necessary to accomplish nucleic acid amplification. To amplify DNA sequences effectively different approaches have been reported. We can differentiate microfluidic PCR devices into space domain, flow-through type, and isothermal nucleic acid amplification, as described in Fig. 12.11.

In the space domain, the sample is passed via a long, slender microchannel necessary for DNA amplification. This microchannel has distinct temperature zones. While in the case of the time domain, the sample remains stationary in the channel, and the temperature of the sample changes over time. Adjustments to the PCR thermal cycling can be made by modifying the heating and cooling protocols [27, 28]. In one study, Nam Ho Bae et al. reported the development of a disposable multichamber film-based PCR chip for the identification of foodborne pathogens [29]. They showed amplification of genomic DNA isolated from *Bacillus cereus* with a concentration of 10 CFU/ml using a PCR chip made of polyethylene terephthalate (PET) and polyvinyl chloride (PVC), allowing faster thermal transfer from heat blocks made up of copper (Fig. 12.12). Further, gel electrophoresis was used to quantify the amplified gene. The actual detection of a foodborne pathogen may benefit from this chip. We can precisely regulate heat and mass movement in

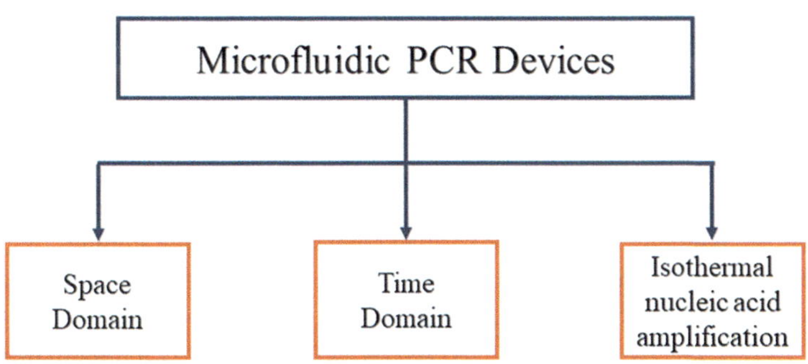

Fig. 12.11 Classification of Microfluidic based PCR devices

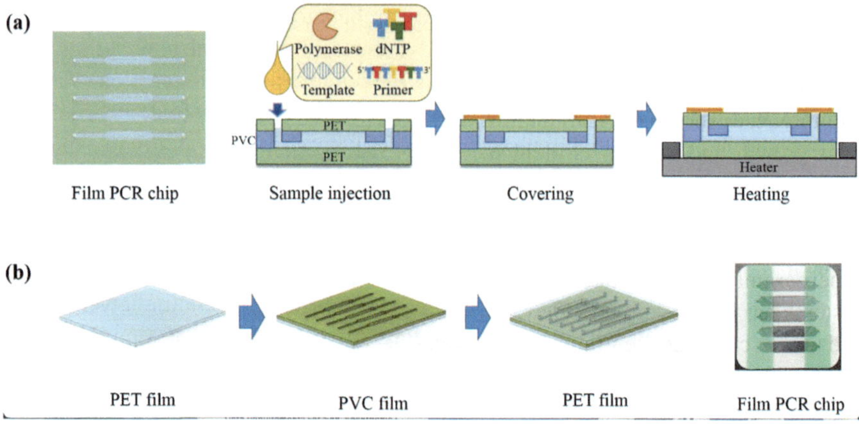

Fig. 12.12 (**a**) Gene amplification utilizing a film polymerase chain reaction (PCR) chip is shown graphically. (**b**) By stacking each manufactured film layer by layer, a basic designed multiplex film PCR chip made of polyethylene terephthalate (PET) and PVC was created [29]. (Adapted under the terms of the CC-BY creative commons license)

microfluidic devices. Short diffusion length in the channel also improves reactions between enzymes and nucleic acid resulting in less assay time.

Isothermal nucleic acid amplification is a potential alternative to conventional PCR because of its simplicity and low energy needs. As the name suggests reaction happens at a constant temperature therefore it eliminates the temperature stages require during conventional PCR [30]. Microfluidic devices have been used in conjunction with isothermal amplification techniques. In this method elimination of heating and cooling systems, during PCR significantly simplifies the process. Amplification carried out at constant temperature eliminates the need of thermo cycles. Use of the LAMP method was described in the study carried out by Li Liu et al. *Salmonella* species are typical pathogenic bacteria found in food [31]. They used the CRISPR/Cas12a technique with recombinant polymerase replication (RPA) to create a unique method for the on-site detection of Salmonella in food. One tube was used for the detection, and a portable blue-light trans-illuminator was used to examine the fluorescence that resulted from the detection. The RPA-CRISPR/Cas12a technique had a detection limit of 1×10^4 ng/L for genomic DNA (gDNA) and 10^2 CFU/mL for bacterial liquid (Fig. 12.13). Thus, this class of portable devices has the potential to be used as a microbe detection test.

Another advantage of microfluidics in gene expression studies is that has the ability to perform multiplexed analysis of multiple genes or samples simultaneously. Microfluidic devices can be designed to accommodate multiple reaction chambers or channels, allowing for parallel analysis of multiple samples or genes in a single experiment. This can greatly increase the throughput and efficiency of gene expression analysis while reducing the amount of samples and reagents required.

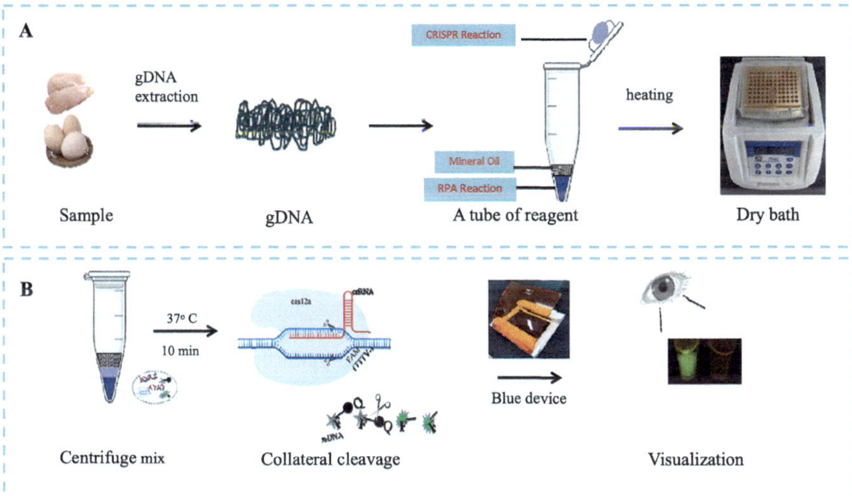

Fig. 12.13 Graphical representation of LAMP experiment carried on salmonella species [31]. (Adapted under the terms of the CC-BY creative commons license)

12.3.5 *Screening and Diagnostics*

12.3.5.1 Microbial Culture and Screening

Microfluidic-based culture is a multidisciplinary concept from biology, engineering, and physics to develop tools and methods for cultivating, analyzing, maintaining, and conducting experiments with cells at the microscopic scale. Cell culture entails the growth and upkeep of cells in a carefully controlled laboratory setting. Because of this, microfluidic cell culture has become a viable method for studying the physiology and behavior of individual cells at the microscale. A vital element of microfluidic cell culture is the ability to replicate the microenvironment of cells. This includes reproducing the soluble factors that regulate the structure, function, behavior, and growth of cells. It is crucial to create stable gradients in the devices that mimic those found in vivo.

If we talk about the comparison of conventional cell culture and microfluidic cell culture there are lot of merits we can see in microfluidics over conventional. In conventional cell culture large fluid volumes hinders the temperature and gasses regulation while in microfluidics, due to miniaturized chamber/channel small volumes of fluids can allow dynamic control. Nutrient supplementation and waste management can be precisely done in microfluidic environment. Also, Drug/protein -induced imaging is feasible in microfluidics cell culture. In microfluidics, single cell manipulation and analysis can be done accurately with high-throughput. In conventional cell culture automation of cell culture tasks is bulky process and expensive while the microfluidic has capability to handle the automation in compact and inexpensive format [32].

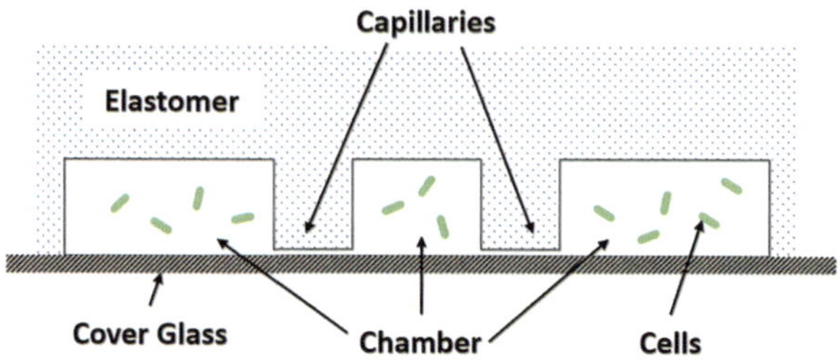

Fig. 12.14 Schematic drawing of the two-layer microfluidic device created by Groisman et al. [33]

The abundance of bacterial and yeast colonies in nature, such as the ***Vibrio fischeri*** cells (gram-negative bacteria) found in the light organs of *Euprymna scolopes,* achieve densities of 10^{10}–10^{11} cells/ml. It is challenging to reach this density level in a lab setting using conventional methods because the nutrients required for development are quickly depleted, and waste products accumulate, which can restrict the growth of the colonies. In other words, even though very high densities of bacteria and yeast can exist in the natural world, laboratory conditions may not be able to reproduce those densities due to nutrient and waste management restrictions. As a result, A. Groisman et al. created a microfluidic device made of the silicone elastomer polydimethylsiloxane (PDMS) using soft lithography, which enables the culture of cells in a variety of shallow microscopic compartments under thermostatic and chemostatic conditions (Fig. 12.14) [33]. To briefly expose the cells to an exogenous signal, the medium in the chambers can be rapidly changed with help of a microfluidics device. Thus, starting with just one cell, the apparatus was used to grow cell colonies to high densities.

Fungi are eukaryotic creatures, including yeast and mold-like microorganisms, and their proper identification is essential for classification and taxonomy. There are two categories of fungi detection techniques: direct inspection and staining. Conventional methods for identifying fungi are complex, costly, time-consuming, and unable to perform single-cell analysis. With the advancement in microfluidic-based devices and micro-imaging technology, it is now possible to observe the germination and growth of fungi in real time, overcoming the limitations of conventional methods. Due to their capacity to release enormous amounts of proteins, filamentous fungus is a key source of industrial enzymes. The poor throughput and elevated prices associated with the functional screening of fungi significantly limit the discovery of novel enzymatic activity and enhanced production strains. Beneyton et al. used droplet sorting by fluorescence and droplet generation technique to implement a nanoliter-range microfluidic device [34]. This chip generated extremely monodisperse droplets that each contained a single yeast cell. They were followed by single spores (contained in 10 nl droplets) incubation and classification based on fluorescence (Fig. 12.15). This tool allowed it to incubate filamentous fungus for 24 h while time-lapse microscopy was being monitored. For the high-throughput identification and screening of filamentous fungus, it was a convincing new approach.

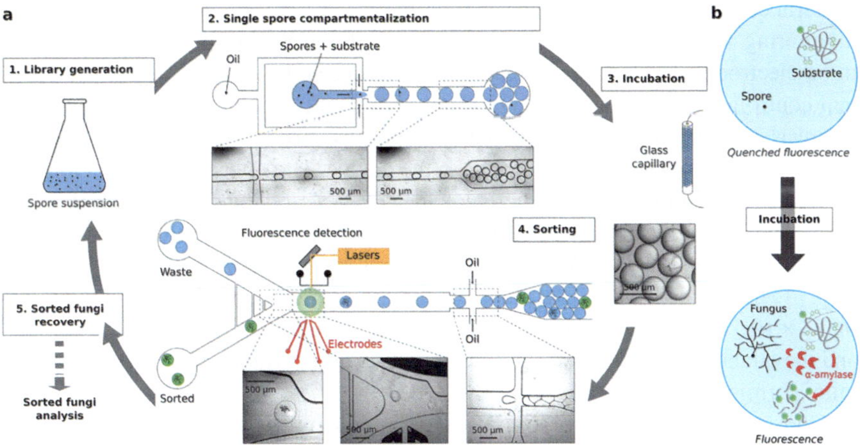

Fig. 12.15 Microfluidics screening platform based on droplets. (**a**) The system's schematic. 1. Creation of a library of whole-genome mutant fungus. 2. The spores are compartmentalized in droplets of fewer than 10 nl employing a fluorogenic protein substrate and an appropriate spore-to-droplet ratio, allowing the compartmentalization of individual spores. 3. To aid in the growth of fungi, the synthesis of enzymes, and substrate digestion, the emulsion is incubated for 24 h at 30 °C outside of the chip in a glass capillary. 4. Reloaded droplets are sorted in a sorting device by fluorescence intensity and enzyme activity. 5. After being extracted from the sorted droplets, the fungus has been characterized or subjected to another cycle of mutagenesis or screening. (**b**) Fluorogenic test for α-amylase. Many quenched BODIPY®FL fluorophores are bonded to a starch backbone to form the substrate. The hydrolysis of the starch backbone by α-amylase releases the fluorophores and causes fluorescence [34]. (Adapted under the terms of the CC-BY creative commons license)

12.3.5.2 Point-of-Care Biosensor

Point-of-care (POC) devices are portable, low-cost, and easy to use, making them particularly suitable for use in resource-limited settings. These devices can detect bacterial, viral, and fungal infections using small volumes of biological samples, such as blood, urine, or saliva. Microfluidics-based POC devices have numerous advantages over traditional microbiological methods, such as culture-based assays, which can take days or weeks to provide results. These devices have the potential to provide results within minutes, allowing healthcare providers to diagnose and treat infectious diseases quickly, improving patient outcomes, and reducing the spread of disease. In addition to rapid diagnosis, microfluidics-based POC devices have the potential to enable personalized medicine by providing clinicians with detailed information about the specific strain of microorganisms causing the infection, allowing for targeted treatment strategies. Further, it can potentially reduce antibiotic use by providing clinicians with rapid diagnostic tools that can differentiate between viral and bacterial infections, thereby reducing unnecessary antibiotic prescriptions and mitigating antibiotic resistance [35, 36]. In the study conducted by Sonal Fande et al., a microfluidic-based electrochemical device was developed for the rapid and vulnerable identification of *E. coli* [37]. The device not only enabled bacteria culturing without the need for an incubator but also allowed simultaneous monitoring and detection of bacterial growth through electrochemical means. A three-electrode system was assembled with storage and a handheld thermostat thermal controller to create the gadget. The working electrode of the microfluidic system was strengthened by graphitized mesoporous carbon. Screen-printing carbon paste was used to add three electrodes to the device. The microfluidic reservoir was created by combining polydimethylsiloxane (PDMS) and the curing agent in a 10:1 ratio. The response of the device was investigated in relation to the quantity and development of *Escherichia coli* in the reservoir using cyclic voltammetry (Fig. 12.16). The linear bacterial concentration range of the instrument was 0.333×10^{12} to 40×10^{12} CFU mL^{-1}, the level of detection limit was 0.35 CFU mL^{-1}, and the quantification limit was 1.05 CFU mL^{-1}, all of which were below the maximum permissible limit.

Ghosh Dastider S. et al. reported a work in which they demonstrated a low-cost and simply produced biosensor capable of promptly and correctly detecting *Salmonella typhimurium* [38]. The biosensor used a microfluidic chip with a dense interdigitated electrode array for identifying Salmonella cells. On a glass substrate, surface micromachining and photolithography techniques were used in the fabrication process. In order to create the interdigitated electrode arrays, chromium (Cr) followed by gold (Au) were sputtered and patterned. Polydimethylsiloxane (PDMS) covers were then used to create the microchannel and fluidic inlet/outlet (Fig. 12.17). To enable targeted detection, monoclonal anti-Salmonella antibodies were immobilized on the electrode array surface. The biosensor's response signal was measured and recorded using an impedance analyzer. The microfluidic biosensor provided within 3 h, both qualitative and quantitative outcomes, eliminating the need for enrichment steps. Compared to the non-microfluidic biosensor, the sensing capability of a microfluidic biosensor was increased by ten times, with a lower detection limit of 3×10^3 CFU mL^{-1}.

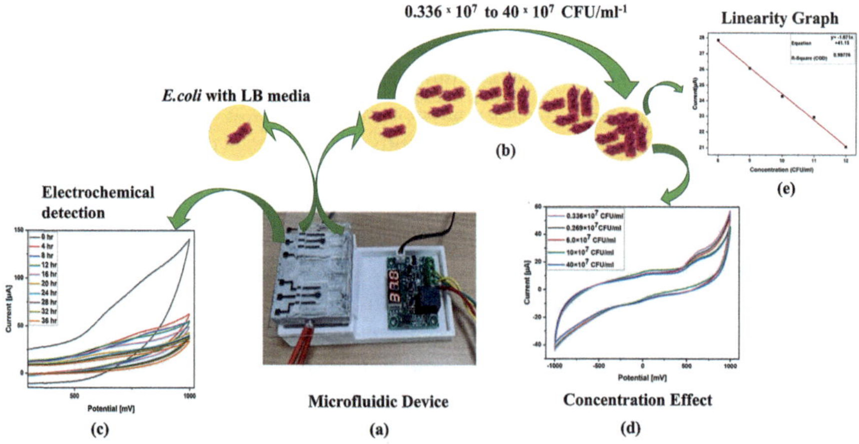

Fig. 12.16 Schematic of the microfluidic device used by Sonal Fande et al. to detect E. coli that includes screen-printed electrodes, an aluminum block, and a thermal management system (heater and temperature sensor) [37]. (Reproduced with permission © Elsevier)

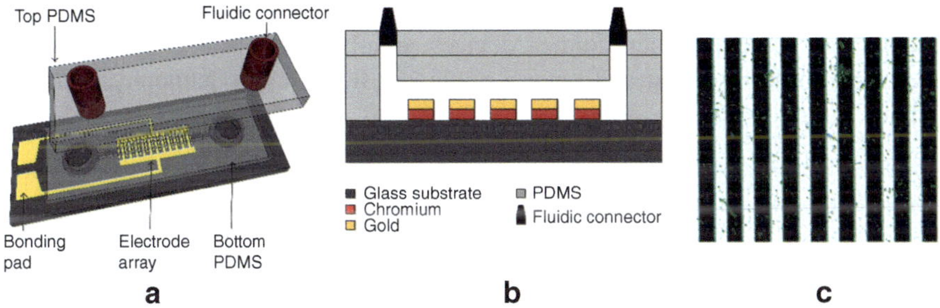

Fig. 12.17 (**a**) A 3D schematic of an impedance biosensor with an embedded electrode array and a microchannel with an inlet and an outlet. (**b**) A cross-sectional profile showing the impedance biosensor's multiple layers. Bacteria are shown in an optical picture in (**c**) [38]. (Adapted under the terms of the CC-BY creative commons license)

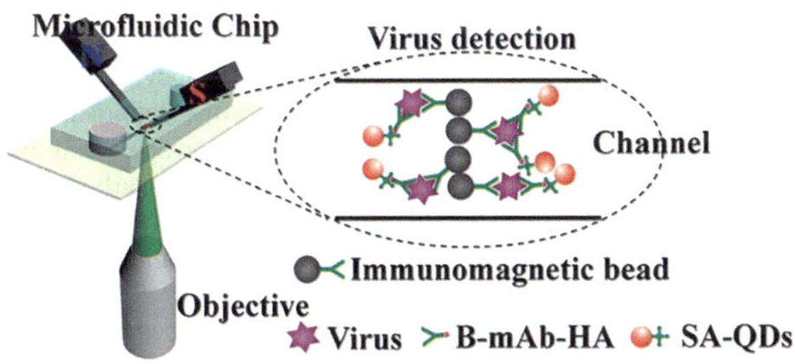

Fig. 12.18 Zhang et al.'s microfluidic device with optical fiber spectrometer [39]. (Reproduced with permission © ACS Publications)

Zhang et al. developed a simple, rapid, and point-of-care magnetic immunofluorescence technique for confirming the presence of the avian influenza virus (AIV) [39]. A portable experimental set-up including a microfluidic device and an optical fiber spectrometer was created. Soft lithography along with rapid prototyping with PDMS technology was used to develop the microfluidic device. The microfluidic chip successfully integrated immunomagnetic target capture, concentration, and fluorescence detection (Fig. 12.18). The researchers were able to reach a low limit of detection of 3.7×10^4 copy/μL with a sample consumption of 2 μL and a total assay duration of under 55 min by optimizing flow rate and incubation time. The created assay was created especially for AIV, a respiratory virus that poses serious risks to both human health and the economy.

12.4 Conclusions

In this chapter, we have discussed the importance of microfluidics devices in microbial studies. Next, we have discussed the fabrication methods for microfluidics devices and their various applications such as microbiome host interaction, antibiotic-resistance study, fuel cell, molecular studies, and POC device. The various technical aspects were discussed while designing the device for specific applications. Overall, microfluidics devices greatly reduce the consumption of chemicals, time, and cost and open the window for complex pathophysiological studies, which are usually difficult in real scenarios.

References

1. Scheler, O., Postek, W., & Garstecki, P. (2019). Recent developments of microfluidics as a tool for biotechnology and microbiology. *Current Opinion in Biotechnology, 55*, 60–67.
2. Chen, L., Yang, C., Xiao, Y., Yan, X., Hu, L., Eggersdorfer, M., Chen, D., Weitz, D. A., & Ye, F. (2021). Millifluidics, microfluidics, and nanofluidics: Manipulating fluids at varying length scales. *Materials Today Nano, 16*, 100136.
3. Whitesides, G. M. (2006). The origins and the future of microfluidics. *Nature, 442*(7101), 368–373.
4. Siddiqui, S., Tufenkji, N., & Moraes, C. (2016). Microfluidics in microbiology: Putting a magnifying glass on microbes. *Integrative Biology, 8*(9), 914–917.
5. Preetam, S., Nahak, B. K., Patra, S., Toncu, D. C., Park, S., Syväjärvi, M., Orive, G., & Tiwari, A. (2022). Emergence of microfluidics for next generation biomedical devices. *Biosensors and Bioelectronics: X, 10*, 100106.
6. Song, Y., Zhao, X., Tian, Q., & Liang, H. (2018). Fundamental concepts and physics in microfluidics. Microfluidics: Fundamentals, devices, and applications. In *Microfluidics: Fundamental, devices and applications*. https://doi.org/10.1002/9783527800643.ch2
7. Zhao, X. M., Xia, Y., & Whitesides, G. M. (1997). Soft lithographic methods for nano-fabrication. *Journal of Materials Chemistry, 7*(7), 1069–1074.
8. Crump SS, inventor; Stratasys Inc, assignee. (1992, June 9). Apparatus and method for creating three-dimensional objects. United States patent US 5,121,329.
9. Garcia-Rey, S., Nielsen, J. B., Nordin, G. P., Woolley, A. T., Basabe-Desmonts, L., & Benito-Lopez, F. (2022). High-resolution 3D printing fabrication of a microfluidic platform for blood plasma separation. *Polymers, 14*(13), 2537.
10. Hull, C. W. (1984). Apparatus for production of three-dimensional objects by stereolithography. United States Patent, Appl., No. 638905, Filed.
11. Prabhakar, P., Sen, R. K., Dwivedi, N., Khan, R., Solanki, P. R., Srivastava, A. K., & Dhand, C. (2021). 3D-Printed microfluidics and potential biomedical applications. *Frontiers in Nanotechnology, 3*, 609355.
12. Marchesi, J. R., & Ravel, J. (2015). The vocabulary of microbiome research: A proposal. *Microbiome, 3*, 1–3.
13. Flint, H. J., Scott, K. P., Louis, P., & Duncan, S. H. (2012). The role of the gut microbiota in nutrition and health. *Nature Reviews Gastroenterology & Hepatology, 9*(10), 577–589.
14. Guthrie, L., Gupta, S., Daily, J., & Kelly, L. (2017). Human microbiome signatures of differential colorectal cancer drug metabolism. *NPJ Biofilms and Microbiomes, 3*(1), 27.
15. Tan, H. Y., & Toh, Y. C. (2020). What can microfluidics do for human microbiome research? *Biomicrofluidics, 14*(5), 051303.
16. Mittal, E., Cupp, G., & Kang, Y. (2023). Simulating the effect of gut microbiome on cancer cell growth using a microfluidic device. *Sensors, 23*(3), 1265.
17. Jalili-Firoozinezhad, S., Gazzaniga, F. S., Calamari, E. L., Camacho, D. M., Fadel, C. W., Bein, A., Swenor, B., Nestor, B., Cronce, M. J., Tovaglieri, A., & Levy, O. (2019). A complex human gut microbiome cultured in an anaerobic intestine-on-a-chip. *Nature Biomedical Engineering, 3*(7), 520–531.
18. Mahajan, G., Doherty, E., To T, Sutherland, A., Grant, J., Junaid, A., Gulati, A., LoGrande, N., Izadifar, Z., Timilsina, S. S., & Horváth, V. (2022). Vaginal microbiome-host interactions modeled in a human vagina-on-a-chip. *Microbiome, 10*(1), 201.

19. Postek, W., Pacocha, N., & Garstecki, P. (2022). Microfluidics for antibiotic susceptibility testing. *Lab on a Chip, 22*(19), 3637–3662.
20. Tang, P. C., Eriksson, O., Sjögren, J., Fatsis-Kavalopoulos, N., Kreuger, J., & Andersson, D. I. (2022). A microfluidic chip for studies of the dynamics of antibiotic resistance selection in bacterial biofilms. *Frontiers in Cellular and Infection Microbiology, 12*, 896149.
21. Zoheir, A. E., Späth, G. P., Niemeyer, C. M., & Rabe, K. S. (2021). Microfluidic evolution-on-A-Chip reveals new mutations that cause antibiotic resistance. *Small, 17*(10), 2007166.
22. Grigorov, E., Peykov, S., & Kirov, B. (2022). Novel microfluidics device for rapid antibiotics susceptibility screening. *Applied Sciences, 12*(4), 2198.
23. Yang, Y., Ye, D., Li, J., Zhu, X., Liao, Q., & Zhang, B. (2016). Microfluidic microbial fuel cells: From membrane to membrane free. *Journal of Power Sources, 324*, 113–125.
24. Cheng, S. K., Li, T., Meena, S. S., Cao, Q., Li, B., Kosgei, B. K., Cheng, T., Luo, P., Liu, Q., Zhu, G., & Liu, Q. (2022). Advances in microfluidic technologies for energy storage and release systems. *Advanced Energy and Sustainability Research, 3*(10), 2200060.
25. Amirdehi, M. A., Khodaparastasgarabad, N., Landari, H., Zarabadi, M. P., Miled, A., & Greener, J. (2020). A high-performance membraneless microfluidic microbial fuel cell for stable, long-term benchtop operation under strong flow. *ChemElectroChem, 7*(10), 2227–2235.
26. Rewatkar, P., & Goel, S. (2022). Shewanella putrefaciens powered microfluidic microbial fuel cell with printed circuit board electrodes and soft-lithographic microchannel. *Chemosphere, 286*, 131855.
27. Ahrberg, C. D., Manz, A., & Chung, B. G. (2016). Polymerase chain reaction in microfluidic devices. *Lab on a Chip, 16*(20), 3866–3884.
28. Kulkarni, M. B., & Goel, S. (2020). Advances in continuous-flow based microfluidic PCR devices—A review. *Engineering Research Express, 2*(4), 042001.
29. Bae, N. H., Lim, S. Y., Song, Y., Jeong, S. W., Shin, S. Y., Kim, Y. T., Lee, T. J., Lee, K. G., Lee, S. J., Oh, Y. J., & Park, Y. M. (2018). A disposable and multi-chamber film-based PCR chip for detection of foodborne pathogen. *Sensors, 18*(9), 3158.
30. Trinh, T. N., & Lee, N. Y. (2022). Advances in nucleic acid amplification-based microfluidic devices for clinical microbial detection. *Chemosensors, 10*(4), 123.
31. Liu, L., Zhao, G., Li, X., Xu, Z., Lei, H., & Shen, X. (2022). Development of rapid and easy detection of salmonella in food matrics using RPA-CRISPR/Cas12a method. *LWT, 162*, 113443.
32. Mehling, M., & Tay, S. (2014). Microfluidic cell culture. *Current Opinion in Biotechnology, 25*, 95–102.
33. Groisman, A., Lobo, C., Cho, H., Campbell, J. K., Dufour, Y. S., Stevens, A. M., & Levchenko, A. (2005). A microfluidic chemostat for experiments with bacterial and yeast cells. *Nature Methods, 2*(9), 685–689.
34. Beneyton, T., Wijaya, I. P., Postros, P., Najah, M., Leblond, P., Couvent, A., Mayot, E., Griffiths, A. D., & Drevelle, A. (2016). High-throughput screening of filamentous fungi using nanoliter-range droplet-based microfluidics. *Scientific Reports, 6*(1), 27223.
35. Park, J., Han, D. H., & Park, J. K. (2020). Towards practical sample preparation in point-of-care testing: User-friendly microfluidic devices. *Lab on a Chip, 20*(7), 1191–1203.
36. Yang, S. M., Lv, S., Zhang, W., & Cui, Y. (2022). Microfluidic point-of-care (POC) devices in early diagnosis: A review of opportunities and challenges. *Sensors, 22*(4), 1620.
37. Fande, S., Amreen, K., Sriram, D., & Goel, S. (2023). Microfluidic electrochemical device for real-time culturing and interference-free detection of Escherichia coli. *Analytica Chimica Acta, 1237*, 340591.
38. Ghosh Dastider, S., Barizuddin, S., Yuksek, N. S., Dweik, M., & Almasri, M. F. (2015). Efficient and rapid detection of Salmonella using microfluidic impedance based sensing. *Journal of Sensors, 2015*. https://doi.org/10.1155/2015/293461
39. Zhang, R. Q., Liu, S. L., Zhao, W., Zhang, W. P., Yu, X., Li, Y., Li, A. J., Pang, D. W., & Zhang, Z. L. (2013). A simple point-of-care microfluidic immunomagnetic fluorescence assay for pathogens. *Analytical Chemistry, 85*(5), 2645–2651.

Chapter 13
Nanotechnology Approaches for Microbe-Based Formulations and Drug Delivery

Vaishali A. Shirsat, Achyut Chalodiya, Rutuja Kadam, and Divya Jaiswal

13.1 Introduction

Nanotechnology offers a potent drug delivery system for effective therapy (Fig. 13.1). Lipid-based nanoparticles, polymer-based nanomaterials, and metal-based nanomaterials are investigated as potential nanocarriers for the delivery of drugs. Utilizing two major techniques, namely, intracellular synthesis and extracellular synthesis, certain microbes are used to create nanoparticles. In conveying viral genetic material like mRNA and triggering an immune response against COVID-19, nanoparticles in vaccines have demonstrated a significant role.

13.1.1 Microbes as Drug Products

Microbes can be a boon or a curse in this healthcare industry. On one hand, microbes act as pathogens, infecting human beings, and on the other hand, we exploit their explicit properties to combat a wide range of diseases. In the pharmaceutical industry, microorganisms act as a prolific source of structurally diverse bioactive metabolites which are used for antibacterial, antifungal, and antiviral infections. Multiple antibiotics are created using bacteria like Streptomycin, Neomycin, Tetracycline, Vancomycin, and Rifamycin from the bacteria streptomyces. Microorganisms are also used in the production domain for the synthesis of chemical drugs, chemical compounds, and other compounds. The discovery of cell mechanisms is being carried out that allows pharmacists to discover antimicrobial drugs with the capability to prevent escalating communicable diseases. Different hormones are manufactured

V. A. Shirsat (✉) · A. Chalodiya · R. Kadam · D. Jaiswal
Bombay College of Pharmacy (Autonomous), Mumbai, India

R. S. Chaughule, A. S. Lokur (eds.), *Applications of Nanotechnology in Microbiology*, https://doi.org/10.1007/978-3-031-49933-3_13

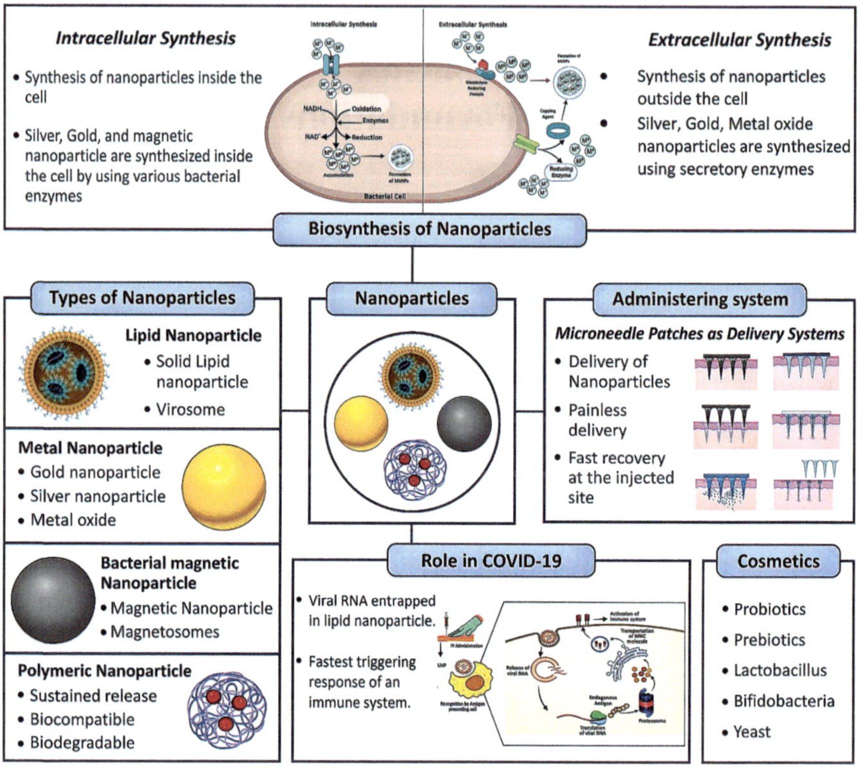

Fig. 13.1 Graphical abstract. Schematics illustrating the comparison of types of Nanoparticles, Biosynthesis of Nanoparticles, Administration Systems, Role of nanoparticles in COVID-19, and its Evaluation for Therapeutic Activity

utilizing bacteria which in turn execute the control of lethal diseases. Commercial products that substantially demand the use of microorganisms are numerous in today's world, especially in the healthcare industry. Examples of such economic products in practice include human insulin for the treatment of diabetes and somatotropin (human growth hormone) for the treatment of pituitary dwarfism [1]. Some of the important products prepared using microorganisms include antifungal, anticancer, anti-inflammatory, immunosuppressive antibiotics, probiotics, enzyme, and vitamin products, bacteriocins, chelates, phage therapy agents, antimicrobial activity, and disinfection products [2]. Pharmacological microbiology has advanced and stretched expressively to comprehend numerous other sides pertaining to the use of microbes, e.g., examination and expansion of novel anti-infective representatives, the use of microbes to perceive mutagenic and oncogenic perspective in medications, and the usage of microbes in the production of medicinal formulations [3]. Microorganisms can be exploited in two ways, either directly using them or a part of their biological structure or, using the secondary metabolites made by them and devising them into useful formulations. Recent advancements in recombinant DNA technology have sparked the development of a wide array of biopharmaceutical products like recombinant proteins that significantly treat a broad spectrum of medical illnesses and conditions.

Natural products made from the secondary metabolites by microbes are a prolific source for the preparation of numerous medical agents which have widely divergent chemical structures and biological activities. These include antimicrobial, immunosuppressive, anticancer, and anti-inflammatory activities, many of which have been developed as treatments and have potential therapeutic applications for human diseases [4]. Another recent and notable approach to drug discovery from microorganisms is co-cultivation. In co-cultivation, two or more organisms from different species are cultivated through which their physiology can be changed to produce cryptic compounds that cannot otherwise be produced in routine cultivation media [5].

13.1.2 Conventional Formulations and Nanotechnology Approaches

The conventional available delivery systems have given researchers a chance to develop a good and effective drug treatment, demonstrating the promising potential of microorganisms as a therapeutic agent. These microbes incorporated formulations can be delivered by topical, oral, parenteral, implants, etc. Microbial APIs range from well-characterized, engineered strains to undefined, fecal-derived microbiota Different types of single specified microbial formulations, which employ just a single microbe are available. e.g., *Lactobacillus reuteri* incorporated in suspension formulation is used for necrotizing enterocolitis. Various microbial formulations contain multiple microorganisms as therapeutic agents like the combined formulations of *Lactobacillus acidophilus* and *Bifidobacterium lactis* used against necrotizing enterocolitis disorder.

Engineered microorganisms are genetically modified microorganisms using various in vitro techniques to increase the desired therapeutic effect. Genetically modified *Escherichia coli* loaded in pharmaceutical suspension acts as an NH_3 scavenger for urea cycle disorders. Other examples of genetically modified microbes as therapeutics are *Lactococcus lactis* incorporated in Capsules used for Diabetes mellitus Type 1 and suspension formulation used for oral mucositis [6]. Current studies show the prominent antiviral effect of microorganisms as therapeutics in the Upper respiratory tract (URT) [7].

Despite all the above-mentioned advantages of microbial formulations, conventional microbe-based therapeutics still face major challenges in terms of stability, targeting, and efficacy. It is established that the immigrant microbes, antimicrobials, disease, and diet can alter the composition of a microbial community. There is still no clear explanation of a set of prediction criteria that would explain how these disturbances might have an effect.

It can be difficult to determine which microbe is most suited for a certain application, making it difficult to select the proper microbial substrate for a microbiota-based treatment. As a result, a therapeutic microbe may not thrive in the target environment and engraft in the endogenous microbiota. To avoid the loss or malfunction of recombinant genetic material, engineered bacteria must be both evolutionary and phenotypically resilient.

Microbial APIs are susceptible to heat and chemical denaturation, just as modern biologics. Microbial APIs, however, also have fragile lipid walls. This fragility creates a difficulty for stability throughout production, storage, and administration. Thus, an improved drug delivery system of microbial APIs is critical in developing effective microbial therapeutics. Nanotechnology refers to the development or application of particles with dimension(s) that fall into the nanometre range (10^{-9} or one billionths of a meter). The interaction between nanoscience and biological systems is known as 'Nanobiotechnology,' while the associated area known as 'nanomedicine' deals with the application of nanostructured materials to diagnose, treat and prevent diseases [8].

The medical industry has studied the longevity, efficiency, durability, flexibility, and inimitable physicochemical characteristics of nanoparticles. They are being utilized in numerous therapeutic approaches, such as the targeted delivery of medications, prognostic visual monitoring of therapy, and even tumor identification. Several conventional approaches have been used to synthesize nanoparticles, for example, effective techniques such as physical vapor deposition, laser ablation, sputtering, melt mixing, and chemical methods such as photo-reduction, sol-gel, thermolysis, and microemulsion. As a result of these techniques, nanoparticles can become unstable, harmful compounds can attach to nanoparticles' surfaces, and hazardous by-products can develop [9]. Nanomaterials produced by microbes like polymers, magnetosomes, and engineered systems like proteins, peptides, and customized metallic nanoparticles are greatly recognized in the field of nanotechnology [10].

13.2 Types of Nanoparticles Incorporating Microbial Formulations

Microbial fabricated nanoparticles show a wide range of applications in drug delivery because of their biocompatibility and high specificity. Microbial incorporated Lipid-based Nanoparticles (LNPs), Metal Nanoparticles (MtNPs), and Polymer-based Nanoparticles along their its application are discussed below.

13.2.1 Lipid-Based Nanoparticles (LNPs)

Lipid Nanoparticles (LNPs) have a major role in effectively protecting and transporting mRNA to cells. Amongst LNPs, liposomes which are an early version of LNPs are an extremely versatile nanocarrier platform. Liposomes are potential nanocarriers for the delivery of microorganism metabolites owing to their ability to carry hydrophobic or hydrophilic molecules, including small molecules, proteins, and nucleic acids. Liposomes have a fair clinical application in various formulations in medical practice. Next generations of LNPs, including solid lipid nanoparticles, nanostructured lipid carriers, and cationic lipid-nucleic acid complexes, possess more complex internal frameworks and enhanced physical stabilities. LNPs can be used to deliver treatments due to their ability to control the location and timing of drug delivery in the body. Currently, scientists are moving beyond traditional biopharmaceuticals to more complex and specialized approaches that reach out to genetic levels [11].

Liposomes have a cell membrane-like bilayer structure possessing the unique capability of fusing directly with bacterial membranes. Efflux pumps that otherwise cause drug resistance by preferentially pumping antibiotics out of the cells are used to burst and release a high dose of antibiotics through such a fusion process. The approach of combinatorial antibiotic release against multiple drug targets is a common strategy in the picture for broadening the antimicrobial spectrum and generating synergy to counteract antibiotic resistance with maximal efficacy [12]. LNPs offer several benefits over other vectors, such as protection of non-stabilized mRNA, delivery of large payload, attachment of adjuvants like targeting ligands that can be co-delivered to attain more steered delivery, and the lack of sophistication in synthesis [13].

Zhang LF et al. engineered LNPs like liposomes with small charged nanoparticles onto surfaces in order to release the active metabolite. In this design was achieved by adsorbing charged nanoparticles non-specifically onto phospholipid bilayer surface stabilization and subsequently providing steric repulsion and reduced surface tension. Substantial fraction of their surfaces which are accessible to membrane-active bacterial toxins can still be exposed even when the changes of pH are unavailable to trigger nanoparticle detachment. This feature made the nanoparticle-stabilized liposomes responsive to pathogens those secret pore-forming toxins (PFTs) to damage lipid membranes and trigger drug release such as group A hemolytic *Streptococcus* and *Staphylococcus aureus* [14].

13.2.1.1 Virosomes

Virosomes are reconstituted envelopes that contain unilamellar phospholipid membrane vesicles that are prepared by incorporation of virus-derived proteins which allows the virosomes to fuse with target cells. They can fill in as vaccines and as vehicles for cell conveyance of different macromolecules. The prospect of drug delivery and targeting systems utilizing virosomes is an intriguing innovative work field. Endeavors have been made to use them as antibodies or adjuvants and carriage frameworks for drugs and organic agents for remedial purposes due to their biocompatible, biodegradable, non-poisonous, and non-auto immunogenic nature. Achievement of virosomal medicate conveyance relies upon strategy used to set up the typified bioactive materials and fuse them into the virosomes. Virosome innovation could conceivably be utilized to convey peptides, nucleic acids or, then again qualities and medications like anti-toxins, anticancer agents, and steroids.

Virosomes can be used as a carrier and serve as safe and effective vaccine and adjuvant models justifying their biomimetic property making them FDA approved nanocarriers for pharmaceutical applications. Virosomes can easily be combined with other adjuvants comfortably. RNA viruses are accompanied by high mutation rates in many instances. Virosomes can develop multi-antigens epitopes using hydrophobic domains or lipid linkers on their surface and inside, against all the strains of RNA viruses, thus versatile in behavior offering a significant advantage. Virosomes cannot replicate and incorporate into the host genome. Thus, this nanovaccine has an excellent safety profile and provides a new horizon in vaccine development that eliminates medical complications with their concerns about next generation products on similar lines [15].

For the formation of virosomes, viral glycoprotein epitopes are incorporated in the virosome surface (PeviPROTM) or positioned in the hollow membrane vesicles (PeviTERTM). Other agents are also incorporated while formulating for the surface modification of virosomes by hydrophilic polymers such as polyethylene glycol (PEG) and polyvinyl pyrrolidone (PVP). This in turn increased their circulation time and eliminated its significant disadvantages. Subcellular or secondary level of targeting was achieved in tumor cells, respiratory and immune system cell components [16].

One such application of virosomes is Inflexal®V, a trivalent influenza virus vaccine. The formulation of Inflexal®V consists of a mixture of trivalent (three monovalent) virosome pools, each formed with one influenza strain's specific hemagglutinin and neuraminidase glycoproteins [17]. Figure 13.2 recreated and

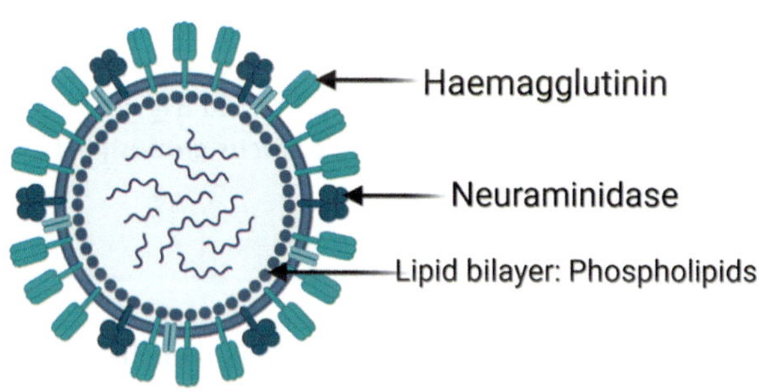

Fig. 13.2 Influenza virosome. Illustration of the basic structure of Influenza Virosome showing the presence of glycoproteins hemagglutinin and neuraminidase with internal phospholipid presence

referred from Singh N et al. shows assembly of virosome components against the flu virus.

Influenza virosome preparation includes optimizing the number of detergents that have solubilized influenza viruses followed by removing their nucleocapsid; consequently, Immunostimulating Reconstituted Influenza Virosomes (IRIV) is naturally formed in the presence of viral lipids and glycoproteins. Phospholipids (PL), especially phosphatidylcholines (PC), attain virosome reconstitutions. About 70% of the virosomal structure is made up of phosphatidyl choline (PC) whereas, 30% of the membrane's components are made up of phospholipids that form the influenza virus's envelope and supply neuraminidase (NA) and hemagglutinin (HA) glycol proteins. Limitations of conventional influenza vaccines were eliminated due to development of Inflexal®V which contained no thiomersal or formaldehyde in formulations, and reduced allergic reactions. Varying storage conditions were provided while producing the vaccine formulation which significantly improved vaccination safety and decreased preservation cost. Inflexal®V has been shown to maintain high levels of HA content for 24 months, and is another advantage over conventional vaccines [18].

The protein substances in the vaccine which are the actives called hemagglutinins are isolated from the surface of the flu virus. To make the vaccine, the flu viruses are grown in chicken eggs, then they are inactivated (with ß-propiolactone) followed by purification of the hemagglutinins. They are then combined with natural lipids to form particles called virosomes. The virosomes act both as carrier and adjuvant in the vaccine. The other excipients in the formulation include sodium chloride, disodium phosphate dihydrate, potassium dihydrogen phosphate, lecithin, and water for injections [19].

Reconstituted influenza virus envelopes (virosomes) were used in several preclinical studies and clinical trials for cancer treatment, such as ovarian carcinoma (OVCAR-3). Viral HA membrane fusion activity was inhibited by incorporating PEG-derivatized lipids into the virosome membrane, and then FabP fragments of mAb 323/A3 (anti-epithelial glycoprotein-2) were conjugated to the distal end of PEG on the virosomes. This study suggested that influenza virosomes had desirable properties in cytosolic delivery. Virosomes have wider applications in the field of pharmaceuticals due to their biomimetic nature [20].

TRANSVAC2 is a new virosomal vaccine designed against SARS-CoV-2 virus in phase II clinical trials. TRANSVAC2 is a follow-up project after its successful predecessor project TRANSVAC, by the European Network of Vaccine Research and Development [21].TRANSVAC2 proposes to establish a fully operational and sustainable European vaccine R&D infrastructure. TRANSVAC2 supports innovation for both prophylactic and therapeutic vaccine development thereby advancing the knowledge and expertise gained during the development of both human and animal vaccines. The influence of TRANSVAC2 will be maximized by two external advisory bodies. To surround scientific-technical and ethical issues, an independent Scientific & Ethics Advisory Committee will provide recommendations and the Board of Stakeholders comprising representatives of policy and decision makers, industry associations and European infrastructures coordinate activities of

TRANSVAC2 with other related initiatives for the further promotion of the long-term stability of the European vaccine.

For the project, Biomedical Metabolomics Facility Leiden (BMFL) has offered a highly structured environment for advanced metabolomics studies acting as a service provider involving vaccine candidates. Profiling was based on liquid chromatography – mass spectrometry. Suitable sample matrices for metabolomics analyses included; in vitro: cell lysate and medium, and in vivo: plasma, serum, urine, CSF, micro-dialysate, and tissues.

13.2.1.2 Solid Lipid Nanoparticles (SLNs)

Solid lipid nanoparticles (SLNs) can also be used as lipid nanocarriers, more specifically solid core nanocarriers. They can be made up of biocompatible ingredients and therefore are one of the preferred choices for drug delivery. Unique features like mucoadhesion or targeting capability can be inculcated as a part of surface modifications in SLNs [11]. Figure 13.3 referred from Rumian T. et al. and redrawn shows how SLNs are made using excipients like surfactants and phospholipids to achieve the goal of targeted and protected delivery of the Active Pharmaceutical Ingredient (API).

Safeena Inama et al. retrieved the Hepatitis B Surface Antigen gene's, also known as the HBsAg gene's complete coding sequence (CDS) from nucleotide database and analyzed for restriction sites through Bio Edit software. Unique BamH1 and Sac1 enzymes were selected to clone the HBsAg gene under 35S promoter in pBI121. Using the above unique restriction sites, full-length primers were engineered. The forward primer was added with a BamH1 site beside 5′-CG-3′ protective bases, while the reverse primer was added with a Sac1 restriction site beside 5′-C-3′ protective bases [22].

The HBsAg is synthesized in bacteria and hence is not able to properly assemble as particles like the ones that assemble in natural infection in humans. To tackle this, in another study, the expression of the S gene was tried in eukaryotic systems. The HBsAg synthesized in the yeast *Saccharomyces cerevisiae* was able to assemble into particles like the 22-nm particles produced in humans. The plasma-derived hepatitis B vaccine was prepared by several steps that can exclude and destruct all

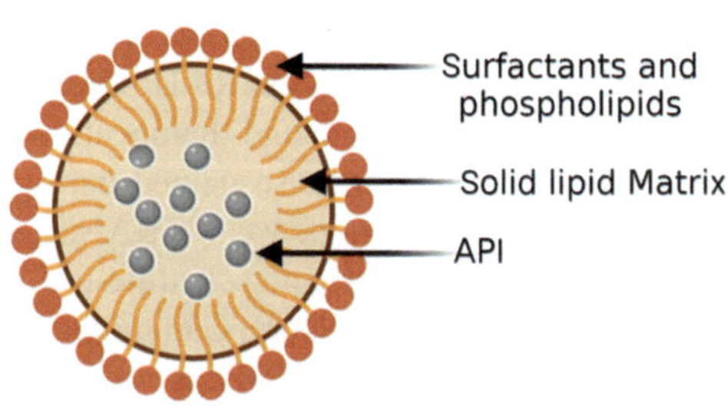

Fig. 13.3 Solid Lipid Nanoparticle (SLN). Structure of API* incorporated in Solid Lipid Matrix with surfactants and phospholipids to make an effective SLN system, *API (Active Pharmaceutical Ingredient)

known animal viruses then tested. These steps involved ultracentrifugation, digesting the partially purified HBsAg with pepsin having pH 2, unfolding the HBsAg in 8 M urea solution. Next, renaturation was carried out, gel filtration and then treating the purified HBsAg in formalin finally [23].

For the intravenous administration, dispersion can be prepared by dissolving tristearin (50 mg) in 3.0 mL of acetone at room temperature, followed by separate addition of, 2.5% w/v lactose monohydrate to 4 mL of saline solution of antigen, and then adding this solution to aqueous phase containing Tween-80 (0.5% v/v). An injection needle can be used to rapidly add the organic phase into the continuously stirred aqueous phase followed by subsequent filtration and collection [24].

The solvent injection technique utilized by SLNs for the formulation was found to depend on rapid diffusion of solvent across the solvent lipid interface with the aqueous phase. Hence, use of an emulsifier was necessitated to reduce the surface tension between aqueous and organic phase. A smaller size of particles was obtained as it also led to the formation of smaller solvent droplets. Moreover, with increasing lipid concentration, viscosity of the organic phase increases, which causes slower diffusion of the organic solvent in the outer phase leading to a larger particle size and lower entrapment. Increase in surfactant concentration led to decrease in particle size, meaning at a lower Tween-80 concentration, higher particle size will be obtained [25].

Jain N et al, evaluated this formulation of HBsAg fabricated with SLNs for their functional characteristics, in vitro cellular uptake, and internalization studies by human dendritic cells (DCs), macrophages and fibroblasts, T-cell proliferation, and T-helper (TH)1/TH2 response as compared to the conventional formulations pertaining Hepatitis B. Nevertheless, it was found that for vaccination against Hepatitis B, subcutaneous immunization was found to be an effective alternative approach in which SLNs could be used as an efficient carrier system for immunization. SLNs act as a signal for phagocytic cells and possess less diffusivity and restricted movement accounting for targeted delivery and release at the site of action. Furthermore, when the antigen is prepared by surface modification using yeast, a greater amount of antigen can be entrapped within the SLN. Better immunological potential of the system was inferred due to the following results; in vitro T cell proliferation, induction of TH1 type of immune response and more sustained antibody titer [26].

13.2.2 Metal-Based Nanoparticles

Metal Nanoparticles (MtNP) have been synthesized using a variety of techniques, including physical, chemical, and biological ones. The utilization of expensive machinery, high heat generation, high energy consumption, and low manufacturing yield are the drawbacks of the physical and chemical processes for making MtNP. The major drawback is the use of a toxic chemical that is environmentally hazardous. The green nanoparticle synthesis by using microorganisms, microbial enzymes, and polysaccharide makes the nanoparticle eco-friendlier. Additionally, they are toxic-free, need less equipment, and are highly biocompatible with in vivo conditions. The eco-friendly and low-cost synthesis of diverse MtNPs including metals like silver, gold, copper, zinc, titanium, palladium, and nickel may be accomplished using a variety of microorganisms as possible biofactories.

Metallic nanoformulations of microbial origin show wide application in drug delivery due to their high pharmacological and pharmacokinetic profile, biocompatibility, anti-inflammatory, and antimicrobial action, bioactivity, bioavailability, tumor targeting, and biological absorption. Various mechanisms like solubility changes, biosorption, bioaccumulation, intracellular & extracellular precipitation, metal complexation, and chelation are used by the microbes to synthesize metal nanoparticles. Various microbial metabolites and enzymes like NAD(P)H- dependent enzymes viz. nitrate reductase, sulfite reductase, cysteine desulfhydrase, and glutathione are been utilized for the fabrication of MtNPs. As metal nanoparticles utilize cellular and secretary enzymes by microbes, not all microorganisms are able to produce Metal nanoparticles as they lack these enzymes.

13.2.2.1 Extracellular and Intracellular Synthesis

Bacteria utilizes mainly two pathways for the synthesis of nanoparticle i.e., extracellular synthesis and intracellular synthesis pathway. Figure 13.4 modified and referred from Hagos et al shows intracellular and extracellular synthesis of nanoparticle within the bacterial cell using enzymes.

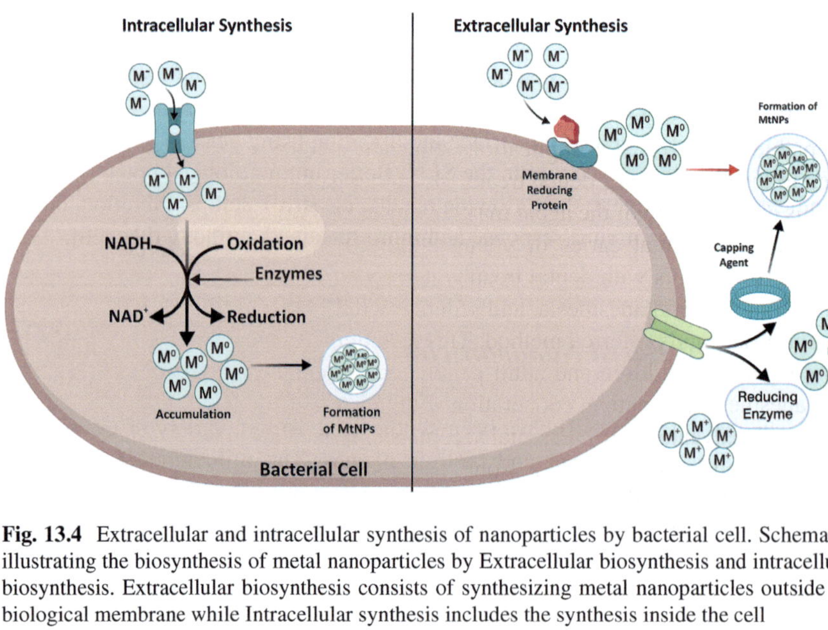

Fig. 13.4 Extracellular and intracellular synthesis of nanoparticles by bacterial cell. Schematics illustrating the biosynthesis of metal nanoparticles by Extracellular biosynthesis and intracellular biosynthesis. Extracellular biosynthesis consists of synthesizing metal nanoparticles outside the biological membrane while Intracellular synthesis includes the synthesis inside the cell

Extracellular synthesis of Metal nanoparticles utilizes secretary enzymes and the cellular enzymes by microbes and reduces them after trapping on the cell wall of microbes. NADH-dependent reductases, acting as the electron carrier, appear to transfer an electron from NADH at the onset of the reduction processes. Example is synthesis of gold nanoparticles. Nanoparticles produced extracellularly have more applications in optoelectronics, electronics, bioimaging, and sensor technologies than those produced within cells. Prokaryotic *Rhodopseudomonas capsulata* was discovered to convert Au^{+3} to Au^0 at room temperature. Magnetite particles have also been produced extracellularly by non-magnetotactic bacteria *Geobacter metallireducens* GS-15.

In Intracellular Synthesis, Metal nanoparticles are synthesized after being transferred into the cell cytoplasm as a result of metabolic interactions with enzymes like nitrate reductase. *Bacillussubtilis* 168 reduced water-soluble Au^{+3} ions to Au^0 producing octahedral morphology inside the cell walls in the dimensions of 5–25 nm [27].

13.2.2.2 Gold Nanoparticles (AuNPs)

Gold Nanoparticles are highly complex and have wide applications. Microbes-mediated synthesis of gold nanoparticles recently gained attention due to its eco-friendly nature, are non-toxic, and biocompatible than physiological and chemical methods. The microbial source for the synthesis of AuNPs includes species of bacteria, fungi and actinomycetes. The synthesis of nanoparticles has been reported both inside and outside the cells. Both the actinomycete *Thermomonospora* sp. and the fungus *Fusarium oxysporum* produce gold nanoparticles outside of cells. It is also revealed that the fungus *Verticillium* sp. produced gold nanoparticles inside its cells. The cytotoxic effects of AuNPs associated with *Penicillium brevicompactum* have been linked to their ability to inhibit the development of C2C12 cells in mice [28].

13.2.2.3 Silver Nanoparticles (AgNPs)

Microbe mediated synthesis of silver nanoparticle (AgNPs) comprises both processes-Intracellular, and extracellular process. *Bacillus subtilis* culture supernatant combined with microwave irradiation in water is a unique combinational synthesis method for the green biosynthesis of silver nanoparticles. The emission of Ag^+, a substance poisonous to bacteria, may be responsible for AgNPs bactericidal effects. Use of AgNPs can potentially reduce the antibiotic resistance problem [29]. Silver nanoparticles effectively combat both Gram-positive and Gram-negative bacteria, including highly multiresistant species like methicillin-resistant *Staphylococcus aureus*. *Streptomyces hygroscopicus* might be used to produce silver nanoparticles as a potential nanomedicine to eradicate bacteria that cause human disease. It has been proven that extracellularly produced silver or gold nanoparticles (NPs) made by *Fusarium oxysporum* may be integrated into a variety of materials, including cloth. These sterile, silver nanoparticle-infused clothes have antibacterial qualities and can be helpful in hospitals to prevent or limit infection with dangerous microorganisms like *Staphylococcus aureus* [30].

The release of silver cation from Ag NPs has been implicated in the highly antibacterial action. Ag^+ enters the bacterium through the cell wall; as a result, the cell wall ruptures, causing protein denaturation and death. Positively charged Ag ions may easily interact with electron-rich biomolecules in the bacterial cell wall that include S or P and N because they are smaller and positively charged than neutral AgNPs. According to some studies, the key to stopping the development of bacteria is an interaction between the negative charge on the cell membrane of the microorganisms and the positive charge on AgNPs [31].

13.2.2.4 Bacterial Magnetic Nanoparticles

Fe_3O_4 (magnetite) and Fe_3O_4 (maghemite), two types of magnetic nanoparticles, are known to be biocompatible. They have been actively investigated for guided drug delivery, targeted cancer treatment (magnetic hyperthermia), gene therapy, DNA analysis, and magnetic resonance imaging (MRI). Magnetotactic bacteria are a heterogeneous group of prokaryotes that orients and migrates along geomagnetic field lines. This migration is based on an intracellular magnetic structure, called Magnetosomes. Iron oxide, iron sulfides, or both are used to make intracellular magnetic nanoparticles by magnetotactic bacteria.

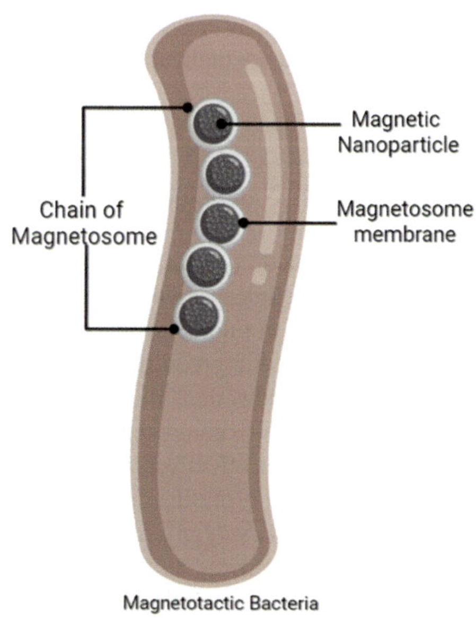

Fig. 13.5 Magnetotactic Bacteria containing chain of magnetosomes. Assembly of chain of magnetosomes comprising of magnetic nanoparticle in magnetotactic bacteria

Magnetosomes, a rare type of lipid-bound intracellular organelle is found only in magnetotactic bacteria. Magnetosomes consist of both crystalline and noncrystalline magnetic crystals [32]. Figure 13.5 modified and referred from Lin et al. illustrate typical magnetotactic bacteria having an intracellular chain of the magnetic nanoparticle, wherein each magnetic nanoparticle is encoated with magnetosome membrane which assists the chain in maintaining its alignment. They have several unique qualities and are highly valued by researchers studying drug delivery. Regular shape, a restricted size distribution, a low toxicity profile, and resistance to agglomeration make them ideal carriers for applications involving the delivery of drugs or genes. These particles, which may be extracted from bacterial cells, have shown helpful in medical procedures including gene therapy, anticancer medication administration, and peptide screening in drug development. These microorganisms are now widely used as efficient drug delivery systems for cancer patients.

Magnetospirillum gryphiswaldense MSR-1 a bacterial strain having magnetosome's chain alignment is ideal for enhancing the hyperthermia impact in cancer therapies. In another study, Sun et al. loaded doxorubicin (DOX) onto bacterial magnetosomes (BMs) through covalent attachment and evaluated the ability of these particles to inhibit tumor growth. The particles were administered subcutaneously into the solid tumor in this work on H22tumour-bearing mice. It is achievable to magnetically control these drug-loaded BMs to cause them to collect and only have therapeutic effects at the disease locations. The tumor suppression rate in DOX-loaded BMs was like that of DOX alone (86.8% vs. 78.6%), but there was much less cardiac damage [33].

Up to 70 drug-loaded nanoliposomes were delivered to difficult-to-reach oxygen-depleted areas using the *Magnetococcus marinus* strain MC-1 [33]. Other examples of Magnetic bacteria are Magnetotactic *magnetotacticum*, *M. Magnetotacticum* (MS-1), and *M. Gryphiswaldense* [27].

13.2.2.5 Metal Oxide Nanoparticles (MtNPs)

ZnO nanoparticles were developed utilizing *Serratia ureilytica* to impart antibacterial properties to cotton textiles specifically targeting *E. coli* and *S. aureus*. Biosynthesis of ZnO nanoparticle has also been associated with *Lactobacillus plantarum*. *Aeromonas hydrophila*, a gram-negative bacterial strain, has been investigated to produce ZnO nanoparticles with additional antibacterial uses. *Halomonas elongata* was used to create triangular CuO nanoparticles, which demonstrated antibacterial efficacy against *S. aureus* and *E. coli* [34].

13.2.3 Polymer-Based Nanoparticles

Good solubility, stability, safety, and prolonged release characteristics of polymer-based nanoparticles boost the absorption of loaded medications, shield them from deterioration, and extend their circulation duration and targeted administration.

Table 13.1 Merits and demerits of polymer-based formulation methods [34]

Formulation methods	Merits	Demerits
Solvent evaporation	Simple and multifaceted	Liposoluble excipients cannot be incorporated due to possibility of coalescence
Nanoprecipitation	Simple, fast, and reproducible	Poor encapsulation efficacy, and limited to certain actives and excipients
Supercritical fluid	Easy to scale up and homogeneity is maintained	Solubility issues with many polymers
Salting method	Reduces stress to protein based encapsulants	Multiple steps of washing nanoparticles
Solvent diffusion-emulsification	Simple, easy scale up, reproducible batch wise, particle size distribution is narrow, encapsulates efficiently	Removal of solvent after the formulation process and hydrophilic excipients may pose a risk of leakage
Dialysis	Most used and simple	Formulating hydrophilic components using this method is not recommended

These vaccines, however, including live attenuated vaccines, inactivated vaccines, protein subunit vaccines, recombinant subunit vaccines, synthetic peptide vaccines, and DNA vaccines, have a few flaws, including immune tolerance, poor immunogenicity, low expression level, and induction of respiration pathological changes [35]. Since, they may pass the biological barrier when supplied parenterally and thanks to their tiny particle size, polymer-based nanoparticles prove useful in the treatment and prevention of infectious disorders [36].

Additionally, polymeric nanoparticles offer a method for prolonged release and increase the durability of labile medications against in vivo enzymatic degradation [37]. Table 13.1 describes the comparative merits and demerits of various method for the formulation of Polymer-based Nanoparticle. Currently, a range of biodegradable polymer-based nanoparticles, both natural and synthetic, can be employed to boost immunogenicity for vaccine administration. Also, vaccination antigens encapsulated in polymer-based nanoparticles given via the mucosal channel can shield antigens from deterioration and guarantee that the encapsulated antigen is released at the action site to trigger effective immune responses [37]. The creation of nanovaccines and increasing the immunogenicity of vaccinations has both shown significant promise for polymer-based nanoparticles [38].

13.2.3.1 Formulations of Polymer-Based Nanoparticles

The chemical makeup and structural characteristics of coronas and membranes, such as the permeability and homogeneity of the membrane, the symmetricity of the corona, and other factors, are primarily considered when designing polymer vesicles to satisfy the needs of diverse applications [36–38]. Figure 13.6 recreated and

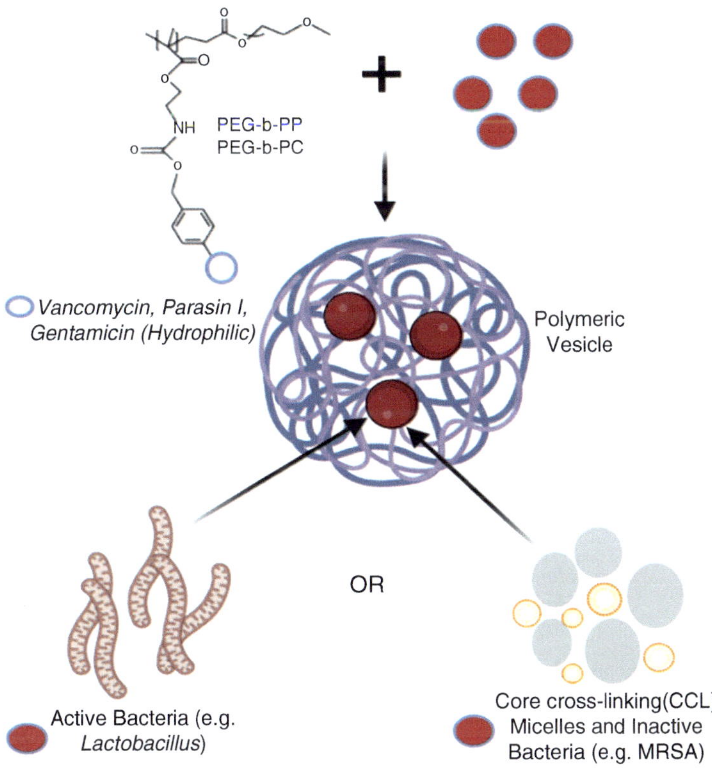

Fig. 13.6 Formation of polymeric micelles containing bacteria-A general technique for building a bacterial strain-selective antibiotic delivery system based on responsive polymeric vesicles [40]

referred from Yamin Li. et al shows formation of polymeric micelles comprising of active bacteria like Lactobacillus or an inactive bacteria like MRSA. Polymer micelles and vesicles have several uses in cancer therapy, gene transfer, and cell imaging, but they also have significant potential for the effective administration of antimicrobial medicines [39–41].

Botulism is a lethal neuroparalytic disease produced by *Clostridium botulinum* toxins. The ability of cationic PLGA NPs to transport plasmid DNA expressing the BoNT heavy-chain (Hc) fragment and the nontoxicity of the resulting material were both shown by Ruwona et al. After 5 to 9 weeks, highly titrated antibodies were generated in immunized mice [42].

Table 13.2 describes the types of Nanoparticles based on the nature of the particle i.e., metal-based nanoparticles, lipid-based nanoparticles, and polymeric-based nanoparticles. Metal-based nanoparticles like gold nanoparticle, silver nanoparticle, metal oxide nanoparticle and magnetic nanoparticle obtained from various species of *Lactobacillus*, *bacillus*, *Magnetococcus*, etc. shows a wide range of antimicrobial and anticancer activity. Lipid-based nanoparticles like SLNs and NLCs are mostly used as a delivery system in Vaccinology. Polymer-based nanoparticles can entrap

Table 13.2 Summary of types of biosynthetic nanoparticle

Category	Microorganism	Application
Metal-Based Nanoparticles		
Gold nanoparticle 	*Lactobacillus plantarum, Lactobacillus plantarum, Gluconacetobacter liquefaciens, Oscillatoria limnetica, Ustilago maydis, Sargassum wightii, Shewanella oneidensis, Candida utilis, Rhodop seudomonas, Capsulate, Ureibacillus thermosphaericus* [43]	MDR(Antibiotic-AuNP-EPS), Antibiotic (Bacterial EPS stabilized NP), Biomedical (Mannosylerythritol lipid)
Silver nanoparticle 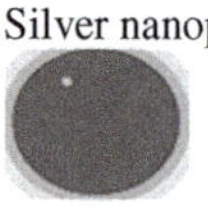	*Bacillus cereus, Aspergillus flavus, Aspergillus fumigatus, Verticillium* sp., *Fusarium oxysporum, Oscillatoria limnetica, Pseudomonas Stutzeri* [44]	Anticancer drug delivery, Active targeting, Gene delivery
Metal oxide 	*Diatom, Shewanella oneidensis, Yeast cells,* *Lactobacillus* sp., *Fusarium Oxysporum, Shewanellaoneidensis MR-1* [44]	Antimicrobial activity, cytotoxicity
Magnetic 	*Magnetococcus maneticum, Magnetospirillum magnetotacticum, Magnetospirillum gryphiswaldense, Magnetospirillum magneticum* [45]	Anticancer drug delivery, gene delivery
Lipid-Based Nanoparticles		
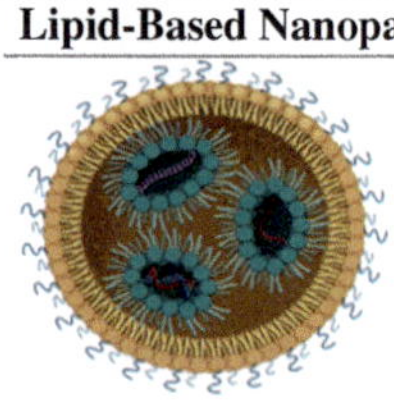	**Liposome**: SARS-CoV-2 [46]	Pfizer-BioNTech vaccine and Moderna-Spikevax vaccine for COVID-19
	Virosome: Trivalent influenza virus [46]	Inflexal®V vaccine for Hepatitis B, Epaxal® and Epaxal junior® for Crucell's Hepatitis A
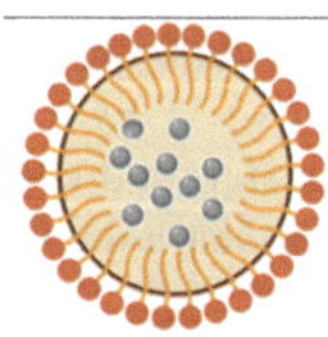	**Solid-Lipid Nanoparticle (SLN)**: *Saccharomyces cerevisiae* [47]	HBsAg vaccine for Hepatitis B
	Nanostructured Lipid Carriers (NLC): Zika virus (alphavirus genus) [48]	Replicating viral RNA (rvRNA) Zika vaccine for Zika virus infection
Polymer-Based Nanoparticles		
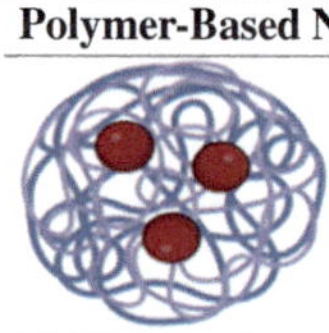	*Eimeria falciformis, Plasmodium berghei,* Mycobacterium lipids [38]	Rodent malarial parasitic infection, Hepatitis B, Anthrax, Tuberculosis

various microbial metabolites which can be used therapeutically in diseases like Tuberculosis, Anthrax, Hepatitis, etc.

13.3 Role of Nanotechnology in COVID-19

The world's healthcare system has been severely taxed by the extremely contagious and ubiquitous illness, COVID-19. This makes the urgent need of a COVID-19 vaccine to pull the world out of the current crises. The convenient method for discovery of a new vaccine is a long process and takes many years. This convenient method comprises use of live attenuated virus vaccine by reducing its virulence and making it harmless. This process requires many approval stages and is slow. Scientists and healthcare researchers tend to focus on creating mRNA vaccines, considering the current situation where the globe needs a rapid and effective vaccination. The goal of the mRNA vaccination is to trigger an immunological response and cause the body to produce its own defense mechanism. Almost approximately 1 year ago, UK authorities approved the COVID-19 mRNA vaccine created by Pfizer and BioNTech for emergency use. This was followed by the approval of the mRNA vaccine created by Moderna.

In case of SARS CoV-2, the virus that causes COVID-19, the RNA was isolated and the viral genome was decoded early in the epidemic to deduce the RNA sequence codes for the surface protein i.e., Spike protein. Then many copies of these RNA sequences were formed by using RNA polymerase. These RNA strands need to travel the host cell and promote the cellular machinery to produce the viral protein which will elicit the immune response. The major challenge faced by these mRNA techniques for inducing an immune response is the transportation. RNA is a large and polar molecule and does not readily pass through the cell membrane. Another major challenge is the action of Ribonucleases. Ribonucleases play a major role in catalyzing the degradation of RNA into smaller components. For an mRNA vaccine to work, the incorporated viral RNA must be protected from ribonucleases. This protection of viral RNA can be done by incorporating it into nanoparticles. Lipid nanoparticles and silver nanoparticles are widely used in mRNA vaccines. The COVID-19 mRNA vaccines approval unquestionably represented a significant step forward for nanotechnology. The development of COVID-19 mRNA vaccines owes its success to the utilization of lipid nanoparticles [49].

Lipid nanoparticles are tiny particles of about 80–200 nm which are formed with walls made of lipid bilayer with a fluid center. When compared to liposomes, LNPs demonstrated comparably reduced cytotoxicity and immunogenicity. These are created to encapsulate viral RNA and shield it from ribonucleases. This Lipid Nanoparticle containing viral RNA payload also allows passage through the cell membrane, and the delivery is mostly intravenous, cutaneous, or subcutaneous. LNPs are used to deliver nucleic acids to cells by adhering to the plasma membrane of the cells, being taken up via endocytosis, and then releasing the nucleic acids once within the cell. These NPs are absorbed by the body and are recognized by the

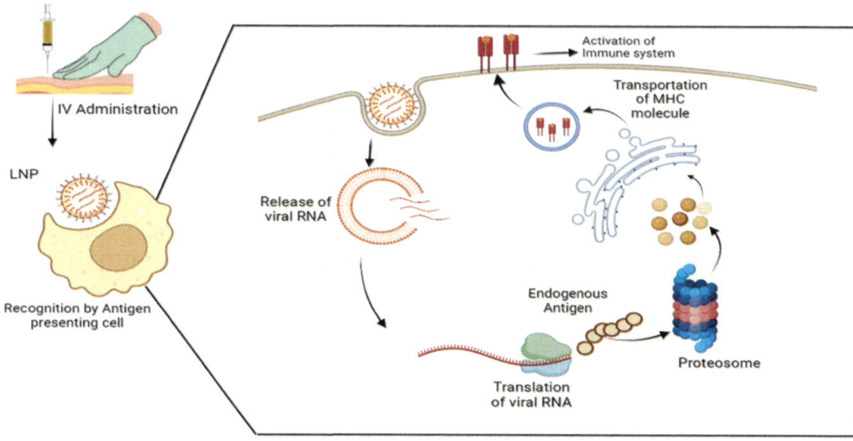

Fig. 13.7 Delivery of viral mRNA incorporated in lipid nanoparticle and its processing

Antigen presenting cells (i.e., Dendritic cell or B cell) by the process of endocytosis. The membrane of the lipid bilayer and the endosome merges causes the release of viral RNA strands in the host cell. These RNA strands undergo a series of processing steps, leading to viral protein synthesis. The synthesized viral proteins are then cleaved into fragments by the proteasome and are further processed to major histocompatibility complex (MHC) molecules which are transported to the cell membrane surface. The CD8+ T- cell binds to these MHC molecules creating memory cells of the viral antigen. This activation cascade leads to the attack of the infected cells, resulting in the production of antibodies to the virus and thereby activates the immune system for defense in case of future viral infection. Additionally, these lipid nanoparticles can be modified by inducing the ligands on its surface for targeting specific cells. Additionally, there is less contact between the neutral lipids and the blood cell membranes, which increases the biocompatibility of the LNPs.

Silver nanoparticles (AgNPs) also show a great contribution to COVID-19 management because of their potential antimicrobial, antiviral, immunomodulatory, and biosensing properties. The antiviral activity of AgNPs was found to be due to the generation of free radicals and reactive oxygen species (ROS) which leads to apoptosis-mediated cell death, and inhibits the viral infection [50].

The delivery of viral mRNA incorporated in lipid nanoparticles and its mechanism involves the following steps for a more targeted approach and has been illustrated in Fig. 13.7:

I. *Synthesis of viral mRNA:* The viral mRNA is synthesized in the laboratory using a variety of techniques, including in vitro transcription or chemical synthesis.
II. *Encapsulation of mRNA into lipid nanoparticles:* Methods like thin-film hydration method or lipid film hydration method can be used to incorporate the synthesized mRNA into lipid nanoparticles.

III. *Delivery of lipid nanoparticles to target cells:* This system of lipid nanoparticles containing viral mRNA is delivered to target cells. Depending on the application, the delivery can be performed via different routes, including intravenous injection, intramuscular injection, or topical application.
IV. *Internalization of lipid nanoparticles:* The lipid nanoparticles interact with the cell membrane of the target cells and are recognized by Antigen-Presenting Cells (APC). They are engulfed through a process called endocytosis or phagocytosis, depending on the size and composition of the particles. The nanoparticles are then internalized into vesicles. This recognition of lipid nanoparticles by Antigen-Presenting Cells (APC) facilitates their internalization.
V. *Release of mRNA into the cytoplasm:* The mRNA is released from the lipid nanoparticles into the cytoplasm. The mRNA is then processed and translated into viral proteins.
VI. *Processing of mRNA by ribosomes:* The mRNA is processed by ribosomes to produce the desired protein product. Transportation of MHC molecules takes place which stimulates the immune system by imitating an infection and creating memory in the body for future infections.

13.4 Role of Nanotechnology in Cosmetics and Microbe-Based Cosmetic Formulations

An increasingly popular field of study in recent years has been the use of microbe-based compositions in cosmetics. Cosmetic goods with additional therapeutic advantages are known as cosmeceuticals. They contain physiologically active chemicals. They are applied as cosmetics because they promise to enhance beauty. The gap between personal care products and medications is filled by cosmetics. Cosmeceuticals products have measurable medicinal effects on the skin because medications and formulations have been developed for the skin, the body, and the hair and are used to treat a variety of conditions, including hair damage, wrinkles, photoaging, dryness of the skin, dark spots, an uneven complexion, hyper pigmentation, and others. The personal care sector is expanding substantially, with cosmetics being the area of the business that is thought to be rising the quickest. By leveraging the beneficial properties of microorganisms, such as bacteria and fungi, these microbe-based cosmetic formulations aim to provide unique skincare benefits and improve overall skin health. The potential for microbe-based cosmetic formulations to provide the skin a variety of advantages has increased in recent years.

Microbe-based cosmetic formulations are provided below:

Probiotics: Live bacteria and yeasts called probiotics are beneficial to your health, particularly your digestive system. Probiotics usually contain bacteria belonging to the groups called *Lactobacillus* and *Bifidobacterium*. It can regulate the skin's microbiota and strengthen the skin's natural defenses in cosmetics. Probiotics are used in various areas of medicine, specifically in Atopic dermatitis (AD) and wound healing.

Prebiotics: Prebiotics feed probiotics and aid in the development of healthy microorganisms on the skin. In cosmetic compositions, they are frequently used with probiotics. Probiotics has a positive effect on several skin conditions. While convincing data show that probiotics and prebiotics decrease the incidence of AD in infants. Prebiotics, like probiotics, have recently demonstrated to reduce the production of harmful fermentation products and modify immunological parameters, including the TH1/TH2 balance, which may be helpful for AD patients [51]. Prebiotic-based cream help individuals with dry, sensitive skin maintain a healthy skin barrier and decreased inflammation. There are currently a few cosmetic products in the market, including moisturizers and cleansers with prebiotic bases.

Lactobacillus topical formulation: Yogurt and other foods that have been fermented frequently include a kind of bacterium called *lactobacillus*. In cosmetics, it can aid in enhancing the skin's built-in defenses and in lowering inflammation. It has been investigated that probiotic *Lactobacillus* strains may be administered vaginally using suppositories or vaginal ovules to restore the vaginal health [52]. There are currently several cosmetic products in the market, including lotions and serums based on Lactobacillus.

***Bifidobacterialysate products*:** Lysate preparations have been in medical practice as immunomodulators for more than 50 years. In the stomach, *Bifidobacteria* are a particular kind of bacteria that can strengthen the skin's defensive mechanisms and delay skin aging. In fact, scientists from a renowned cosmetics company have demonstrated that a lysate from the probiotic *Bifidobacterium longum* reuter strain could reduce vasodilation, edema, mast cell degranulation, and TNF-alpha release upon the application of this lysate containing product. They also used trans-epidermal water loss to assess barrier function, which improved with application of the lysate-containing product [52].

Yeast: Yeast is a kind of fungus that can aid in enhancing the texture and look of the skin. Inflammation can be decreased and the skin's natural defenses can be strengthened. Yeast, specifically certain strains of *Saccharomyces cerevisiae*, has been studies extensively and incorporated into cosmetic products [52].

The effectiveness of microbe-based formulations in cosmetics can be improved using nanotechnology methods. Following are the few advantages of nanotechnology uses in microbe-based formulations:

- *Stability & viability:* Microorganisms, such as probiotics and prebiotics, can be protected from environmental stresses by being encapsulated in nanoparticles, which also increases their stability and effectiveness in cosmetic formulations. Also, it ensures that the microbes reach the skin in an active state.

- A study published by Kim H, Lee J, et al. demonstrated that encapsulating *Lactobacillus curvatus*, which is used as a preservative in cosmetic products within liposomes is a prospectively useful strategy for retaining its multifunctional natural preservative effect in O/W cosmetic emulsions. Encapsulation improved its stability and viability in a cosmetic formulation. Several cosmetic products, such as encapsulated probiotic-based serums and masks, are now available in the market [53].

- *Targeted delivery:* Nanoparticles may be made to target certain skin structures, such hair follicles, or sebaceous glands, and distribute microbes there for the best results.
- *Enhancing penetration:* Nanoparticles may also be used to increase the penetration of microorganisms into the skin. This enables them to get to deeper layers of the skin, where they can have a bigger influence on skin health.
- *Controlled release:* Nanoparticles are capable of being created to release microorganisms gradually over time, resulting in a sustained and long-lasting impact on the skin.

In general, the stability, potency, and targeted administration of formulations based on microbes in cosmetic goods can be enhanced with the application of nanotechnology. But it is crucial to make sure that these nanoparticles are carefully vetted for safety before being used in cosmetic formulas.

Although nanoparticles have many advantages, little is known about how they will affect the environment and other animals' and humans' short- and long-term health. Security issues were brought up due to the nanomaterials' known toxicity and possible dangers. Many nanocarriers utilized to deliver nano cosmeceuticals, including liposomes, niosomes, solid lipid nanoparticles, nanostructured lipid carriers, and nanoemulsion, are discussed, along with the products offered and any positive or undesirable traits. Many advantages are offered by nano cosmeceuticals, alternatively put, by controlling the drug's release from carriers, occlusion boosts penetration and hydrates the skin more. Cosmeceuticals are more durable than conventional cosmetics and have strong sensory qualities as well as a high entrapment quality [54].

A few marketed microbe-based compositions in their cosmetic goods includes the following:

- *A live culture of ammonia-oxidizing bacteria*-aids in re-establishing the skin microbiome's normal equilibrium.
- *Probiotic and prebiotic skincare products*-Probiotic cleanser, probiotic moisturizer, and prebiotic exfoliating powder, Prebiotic face cream, probiotic foamy facial cleanser, and prebiotic scalp and hair serum, Probiotic gel cleansers, hydrating serums, and moisturizers.
- *Probiotic and botanical-based skincare products*-A probiotic serum, a probiotic day and night cream, and a probiotic eye cream.

The purpose of these formulations is to take advantages of the probiotic and prebiotic bacteria positive effects on the skin's microbiome, which can then aid to enhance the skin's general health and look. Studies have shown that microbe-based formulations in cosmetics may offer advantages such as better skin hydration, decreased inflammation, and improved barrier function. For instance, studies have indicated that using probiotic-based skincare products can assist to balance the skin's microbiota, which helps reduce skin irritation and redness.

However, additional investigation is required to completely comprehend the skin-related effects of formulations based on microbes as well as to determine the most efficient formulations and delivery methods. While microbe-based products have shown promise in enhancing skin health and appearance, it is vital to keep in mind that not all skin types and issues are compatible with them. Ultimately, the utilization of microbe-based formulations in cosmetics is a fascinating field of study that has the potential to completely change how we think about skincare. Like with any skincare product, it is crucial to conduct research and speak with a dermatologist or other skincare expert to ascertain whether microbe-based formulas are suitable for you.

13.5 Routes for Delivery of Microbe-Based Nanosystems

The route of administration is an important factor as it likely governs both the immune response and side effects. Route of delivery affects various pharmacokinetics parameters like Clearance, duration of action, immune response. Investigational studies by Anderluzzi G et al demonstrated that the delivery method, such as intra-muscular, intra-dermal, or intra-nasal, as well as the nanoparticle format, has an impact on the immunogenicity of self-amplifying RNA vaccines. Despite receiving a ten-fold larger dose, the immune responses produced following intranasal delivery were minimal. Hence, the route of administration and delivery system of self-amplifying mRNA–LNPs vaccine dictates its efficacy and potency, as it affects the total amount of protein produced and duration of expression. These are the two important considerations that must be taken while establishing a route of administration for a specific vaccine [55].

13.5.1 Administration Routes in Practice

In order to achieve a more targeted delivery, it is essential that nanoparticles infused with microbes or their metabolites reach the lymph nodes or the lymphatic parts. Hence, factors like particle size, particle size distribution, charge, and colloidal stability with respect to the formulation affect the straightforward delivery of nanoparticles at the site of action [56]. Intramuscular injection (IM) and subcutaneous injection (SC) are the most practiced routes of administration. The former vaccination method is simple to execute with limited significant training. The subcutaneous tissue has as few blood vessels, and hence the delivery occurs at a slow rate with a sustained rate of absorption [13].

Directly injecting immunogenic material like mRNA or protein formulations in the bloodstream can cause tissue damage and inflammation due to production of cytokine storms. Intravenous injections (IV) invite potential of systemic side effects if mRNA vaccines incorporated into lipid nanoparticles are administered via this route [57]. Mucosal route of administration for delivery is better alternative to injection vaccine, being non-invasive and avoiding spread of infectious agents via contaminated syringes as in the parenteral route. Potential mucosal options include ocular, nasal, oral, rectal, and vaginal routes. Oral and nasal vaccine delivery routes are most accessible and elementary to implement [58].

Besides being a convenient route for vaccine administration, this method also provides additional benefits of mucosal immunity, including the secretion of IgA antibodies. Cytotoxic T-cell response has also been witnessed in the case of intranasal route of administration [59]. Mucosal penetration requires lipophilic character and so, LNPs like liposomes, Solid-lipid nanoparticles (SLNs) and Nanostructured-lipid carriers (NLCs) would be suitable candidates for delivering microbe-based formulations via these methods.

Intra-dermal injection (ID) delivers the formulation into the outer layers of the dermis underneath the epidermis (upper skin layer) directly [60]. ID route of administration significantly induces a Th1 type immune response and cytotoxic T-cell induction for mRNA–LNP vaccines. Additionally, studies with conventional vaccines have revealed that ID administration might require as slightly as one-fifth of a standard IM dose to produce an equivalent immune response [61]. ID applications may have great potential for delivery, collectively with recently developed transdermal drug delivery technologies like *Microneedle patches.*

13.5.2 Microneedle Patches as Delivery Systems

Even though hypodermic needles have been the gold standard for drug delivery and vaccination, they suffer from several drawbacks such as pain, needle phobia, hazardous sharp waste, and injury [62]. Hence, MNs have been developed to overcome the limitations of hypodermic needles [63]. Microneedle devices are composed of arrays of micron-size needles. The dimensions of these microprojections generally range from lengths of only a few micrometers to those as long as 2000 μm [64]. Several microneedles are designed based on its mechanism of drug delivery like solid microneedles, hollow microneedles, and coated microneedle. Figure 13.8 has been redrawn, and referred from Alimardani, V. et al. Figure 13.8a describes the drug delivery through Poke and flow mechanism using hollow microneedle wherein drug loaded patch is applied for the flow of drug. Figure 13.8b describes the delivery of drug using solid microneedle, wherein these microneedles are first applied for the formation of microchannels by puncturing stratum corneum followed by the application of drug loaded transdermal patch.

Different types of MNs have been used to enhance the permeability of NPs and microparticles (MPs). A wide variety of research studies can be found in the

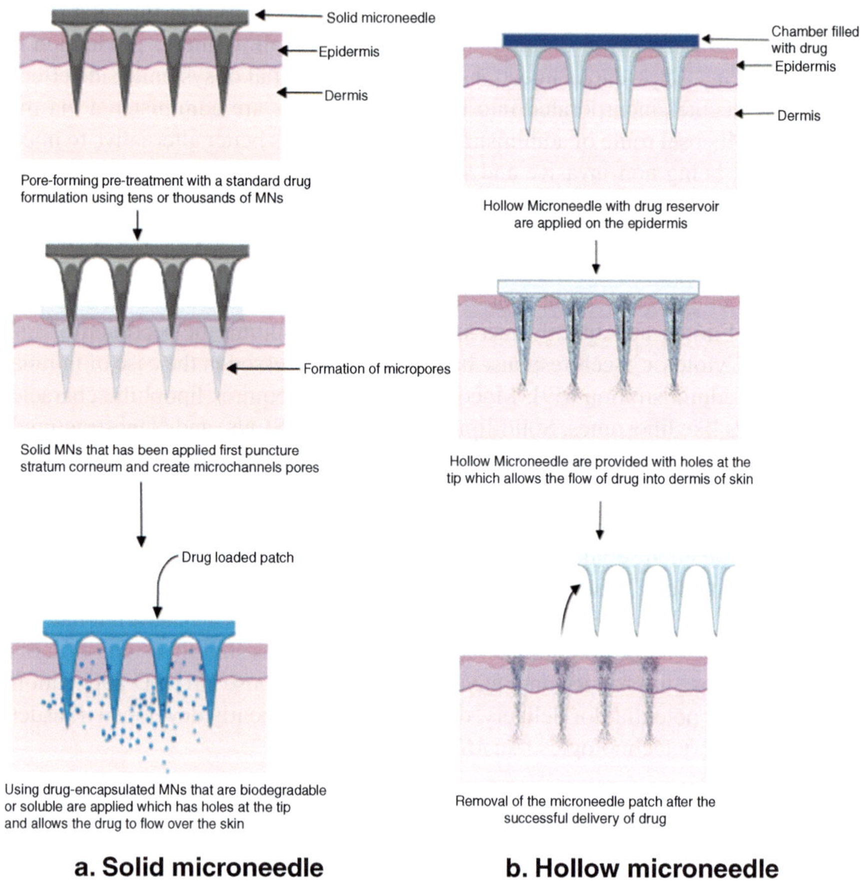

Fig. 13.8 Microneedle delivery mechanism using solid microneedle. (**a**) Solid microneedle. (**b**) Hollow microneedle

literature dealing with MN-assisted permeation of nanometer and micrometer particles [65, 66]. The first report of NP permeation through the skin using MNs was published in 2003 by McAllister et al. [67].

13.5.2.1 Approved Products

The first microneedle product was derma roller. Many microneedle products are coming in the market and are approved for medical and cosmetic use. Some of them are mentioned in Table 13.3. Many companies in Germany, US, Europe, Japan are marketing microneedle products.

Table 13.3 Approved microneedle products

Product name	Company name	Description	Use	References
Soluvia	Sanofi Pasteur Europe	Hollow microneedle	Influenza vaccination	[68]
Dermaroller	Dermaroller Germany, White Lotus	A cylindrical roller with solid or metal microneedles, 0.2–2.5 mm in length	Improve skin texture	[68]
CIT-8 (Collagen Induction Therapy)	The Dermaroller Series by Anastassakis	A needle length of 0.5 mm (500 microns)	Collagen induction and skin remodeling	[69]
h-patch	Valeritas	Small adhesive machine-like patch is used	Deliver drugs in subcutaneous tissue (insulin)	[69]
Microstructured transdermal system	3 M	Hollow microneedle	Deliver biologics and other small molecules	[70]

13.6 Future Applications

Apart from using nanoparticles to encapsulate drugs or other therapeutic agents for targeted delivery to specific tissues or cells, nanotechnology has also added more advanced formulations. Using current technology, *nanopatterned surfaces* can be explored by creating nanoscale patterns on surfaces to control the growth and behavior of microorganisms [71]. This concept is used to enhance their adhesion to surfaces or altering their release properties to efficiently deliver the formulation. It also considers the fact that different cell types respond differently to various types of nanopatterns while designing the formulation. Another approach, *nanoscale imaging* can also be used which involves using nanoscale imaging techniques to visualize microorganisms at high resolution and manipulate their behavior, such as guiding their movement or controlling their interactions with other cells [72]. This technology is estimated to reach nanoscale spatial resolutions for the visualization of the molecular components of specific cellular structures. *Biosensors* have a physicochemical detector which is combined with a biological component for the detection and analysis of a chemical agent. It is basically an analytical device with applications in disease monitoring, drug discovery and detection of disease-causing microorganisms [73]. Now, nanomaterial-based biosensors using nanoscale materials are generated that can detect and respond to changes in the environment, such as changes in pH or the presence of specific chemicals. Nanotechnology has led to another breakthrough called *bacterial nanobots* which are engineered bacteria for performing tasks such as targeted delivery of drugs or sensing and responding to environmental changes owing to release of the active at the site of action. The inherent nature of bacteria is exploited here which the ability to steer due to flagellar is bundling, especially in multi-flagellated nanoswimmers [74]. These approaches are still in the early stages of research and development and their practical applications are still being explored. There are several challenges associated with their development and use, including the need to ensure their safety and stability, and the need to control their behavior and function in vivo.

13.7 Summary

Microorganisms have a high potential to meet both prevention and cure purposes in medicine. Although they are utilized for the cure goal via antibiotics, much of their ability is exploited for the prevention of diseases using vaccines. They are used in numerous forms in vaccines, such as whole organism-live attenuated, dead, or inactive forms like heat-killed microbes, particular parts like DNA, mRNA, protein subunits, spike proteins, recombinant microbes, conjugated, using inactivated toxins produced by microbes called toxoids, viral vectors, etc. Nonetheless, more research is needed to develop new vaccine types and improve current approaches. Traditional immunization methods can be replaced with nanotechnology-based formulations that offer advantages like improved bioavailability, faster release, reduced toxicity, greater dose response, effective targeting, enhanced solubility, and MDR combating. Nanosystems based on lipidic materials are promising delivery systems due to their biocompatible and biodegradable excipients, minimalism in manufacturing processes, enhanced tissue penetration, and the capacity to scale up. Polymeric NP based formulations could be investigated more rationally as they exhibit an enhanced therapeutic index, offer potential for controlled release, and safeguard the active microorganism or its metabolite, along with other excipients with biologic activity, against the environment. Metal nanoparticles provide stability, and large-scale production can be done while avoiding organic solvents. They can be further exploited for microbe-based formulation delivery in both, pure form and as oxides. Although specific routes of administration offer a variety of advantages for vaccine delivery, microneedle-patches, being transdermal delivery systems, stand out. Despite the barrier presented by the stratum corneum layer in the skin; it can provide less pain and invasion as compared to parenteral systems, ease of self-administration, and higher drug bioavailability than oral or mucosal pathways for efficacious delivery of microbial-based nanoformulations. Limited nanosystem-based formulations incorporating microorganisms are commercially available in the pharmaceutical industry, and many are still undergoing clinical trials, especially vaccines. The use of microbe-based cosmetics is based on the idea that a healthy skin microbiome can contribute to overall skin health. While they hold promise for some individuals, it is essential to make informed choices based on personal skin needs, consult professionals, and prioritize product safety and quality. Nanotechnology approaches, such as nanoparticle-based delivery systems, bacterial nanobots, nanopattern surfaces, nanoscale imaging and manipulation, and nanomaterial-based biosensors, are being developed to enhance the capability of microbe-based formulations and to improve their efficacy and specificity in drug delivery. However, there are also challenges associated with microbe-based formulations, including ensuring of their safety and stability, and controlling their behavior and function in vivo. Despite these challenges, the potential benefits of microbe-based formulations make them an important area of research for the development of new and innovative microbe-based formulations. With the evolution of resistance development against conventional dosage forms, advancing day by day, these types of formulations should be practically implemented soon in the market to alleviate the sufferings of humanity.

References

1. Rasab, J., Arslan, A., Muhammad, I., Khan, S., Hassan, A., Khan, S., & Sharafat, A. (2021). Microbes: Role in industries, medical field and impact on health. *Saudi Journal of Medical and Pharmaceutical Sciences, 7*(6), 278–282.
2. Pham, J. V., Yilma, M. A., Feliz, A., Majid, M. T., Maffetone, N., Walker, J. R., Kim, E., Cho, H. J., Reynolds, J. M., Song, M. C., Park, S. R., & Yoon, Y. J. (2019). A review of the microbial production of bioactive natural products and biologics. *Frontiers in Microbiology, 10*, 1404.
3. Kalsoom, M., Ur Rehman, F., Shafique, T., Junaid, S., Khalid, N., Adnan, M., Zafar, I., Abdullah Tariq, M., Raza, M. A., Zahra, A., & Ali, H. (2020). Biological importance of microbes in agriculture, food and pharmaceutical industry: A review. *Innovare Journal of Life Sciences, 8*(6), 1–4.
4. Lacey, L. A., Grzywacz, D., Shapiro-Ilan, D. I., Frutos, R., Brownbridge, M., & Goettel, M. S. (2015). Insect pathogens as biological control agents: Back to the future. *Journal of Invertebrate Pathology, 132*, 1–41.
5. Abdel-Razek, A. S., El-Naggar, M. E., Allam, A., Morsy, O. M., & Othman, S. I. (2020). Microbial natural products in drug discovery. *PRO, 8*(4), 470.
6. Jimenez, M., Langer, R., & Traverso, G. (2019). Microbial therapeutics: New opportunities for drug delivery. *Journal of Experimental Medicine, 216*(5), 1005–1009.
7. Spacova, I., De Boeck, I., Bron, P. A., Delputte, P., & Lebeer, S. (2021). Topical microbial therapeutics against respiratory viral infections. *Trends in Molecular Medicine, 27*(6), 538–553.
8. Singh, L., Kruger, H. G., Maguire, G. E. M., Govender, T., & Parboosing, R. (2017). The role of nanotechnology in the treatment of viral infections. *Therapeutic Advances in Infectious Disease, 4*(4), 105–131.
9. Mubeen, B., Ansar, A. N., Rasool, R., Ullah, I., Imam, S. S., Alshehri, S., Ghoneim, M. M., Alzarea, S. I., Nadeem, M. S., & Kazmi, I. (2021). Nanotechnology as a novel approach in combating microbes providing an alternative to antibiotics. *Antibiotics, 10*(12), 1473.
10. Sarikaya, M., Tamerler, C., Jen, A. K., Schulten, K., & Baneyx, F. (2003). Molecular biomimetics: Nanotechnology through biology. *Nature Materials, 2*(9), 577–585.
11. Tenchov, R., Bird, R., Curtze, A. E., & Zhou, Q. (2021). Lipid nanoparticles—from liposomes to mRNA vaccine delivery, a landscape of research diversity and advancement. *ACS Nano, 15*(11), 16982–17015.
12. Gao, W., Chen, Y., Zhang, Y., Zhang, Q., & Zhang, L. (2018). Nanoparticle-based local antimicrobial drug delivery. *Advanced Drug Delivery Reviews, 127*, 46–57.
13. Reichmuth, A. M., Oberli, M. A., Jaklenec, A., Langer, R., & Blankschtein, D. (2016). mRNA vaccine delivery using lipid nanoparticles. *Therapeutic Delivery, 7*(5), 319–334.
14. Thamphiwatana, S., Gao, W. W., Pornpattananangkul, D., Zhang, Q. Z., Fu, V., Li, J. Y., Li, J. M., Obonyo, M., & Zhang, L. F. (2014). Phospholipase A2-responsive antibiotic delivery via nanoparticle-stabilized liposomes for the treatment of bacterial infection. *Journal of Materials Chemistry. B, 2*(46), 8201–8207.
15. Singh, N., Gautam, S. P., Kumari, N., Kaur, R., & Kaur, M. (2017). Virosomes as novel drug delivery system: An overview. *Pharma Tutor, 5*(9), 47–55.

16. Wang, Z. B., & Xu, J. (2020). Better adjuvants for better vaccines: Progress in adjuvant delivery systems, modifications, and adjuvant-antigen codelivery. *Vaccines (Basel), 8*(1), 128.
17. Otten, G., Matassa, V., Ciarlet, M., & Leav, B. (2020). A phase 1, randomized, observer blind, antigen and adjuvant dosage finding clinical trial to evaluate the safety and immunogenicity of an adjuvanted, trivalent subunit influenza vaccine in adults ≥ 65 years of age. *Vaccine, 38*(3), 578–587.
18. Herzog, C., Hartmann, K., Künzi, V., Kürsteiner, O., Mischler, R., Lazar, H., & Glück, R. (2009). Eleven years of Inflexal V-a virosomal adjuvanted influenza vaccine. *Vaccine, 27*(33), 4381–4387.
19. Wong, S. S., & Webby, R. J. (2013). Traditional and new influenza vaccines. *Clinical Microbiology Reviews, 26*(3), 476–492.
20. Asadi, K., & Gholami, A. (2021). Virosome-based nanovaccines; a promising bioinspiration and biomimetic approach for preventing viral diseases: A review. *International Journal of Biological Macromolecules, 182*, 648–658.
21. European Vaccine Research and Development Infrastructure. TRANSVAC2. [https://cordis.europa.eu/project/id/730964/results].
22. Inam, S., Abbas, Z., Noor, S., Rehman, N., Adeel Zafar, S., Ramzan Khan, M., Ali Kaimkhani, Z., Al-Misned, F., Shah, M., Mahboob, S., & Muhammad, A. G. (2022). Isolation, cloning and transgenic expression of hepatitis B surface antigen (HBsAg) in Solanum lycopersicum L. *Saudi Journal of Biological Sciences, 29*(3), 1559–1564.
23. Zhao, H., Zhou, X., & Zhou, Y. H. (2020). Hepatitis B vaccine development and implementation. *Human Vaccines & Immunotherapeutics, 16*(7), 1533–1544.
24. Maurya, L., Vijayakumar, M. R., Dewangan, H. K., & Singh, S. (2018). Lipid based aqueous core nanocapsules (ACNs) for encapsulating hydrophilic Vinorelbine bitartrate: Preparation, optimization, characterization, and in vitro safety assessment for intravenous administration. *Current Drug Delivery, 15*(9), 1284–1293.
25. Dewangan, H. (2021). Different approaches for nanovaccine formulation and characterization. *IOP Conference Series: Materials Science and Engineering, 1116*, 012042. https://doi.org/10.1088/1757-899X/1116/1/012042
26. Mishra, H., Mishra, D., Mishra, P. K., Nahar, M., Dubey, V., & Jain, N. K. (2010). Evaluation of solid lipid nanoparticles as carriers for delivery of hepatitis B surface antigen for vaccination using subcutaneous route. *Journal of Pharmacy & Pharmaceutical Sciences (www.csps-Canada.org), 13*(4), 495–509.
27. Narayanan, K. B., & Sakthivel, N. (2010). Biological synthesis of metal nanoparticles by microbes. *Advances in Colloid and Interface Science, 156*(1–2), 1–13.
28. Chowdhury, N. K., Choudhury, R., Gogoi, B., Chang, C. M., & Pandey, R. P. (2022). Microbial synthesis of gold nanoparticles and their application. *Current Drug Targets, 23*(7), 752–760.
29. Iravani, S. (2014). Bacteria in nanoparticle synthesis: Current status and future prospects. *International Scholarly Research Notices, 2014*, 359316, 18 pages.
30. Medhi, R., Srinoi, P., Ngo, N., Tran, H.-V., & Randall Lee, T. (2020). Nanoparticle-based strategies to combat COVID-19. *ACS Applied Nano Materials, 3*, 8557–8580.
31. Siddiqi, K. S., Husen, A., & Rao, R. A. K. (2018). A review on biosynthesis of silver nanoparticles and their biocidal properties. *Journal of Nanobiotechnology, 16*(14):28 pages.
32. Bruna, T., Maldonado-Bravo, F., Jara, P., & Caro, N. (2021). Silver nanoparticles and their antibacterial applications. *International Journal of Molecular Sciences, 22*(13), 7202.
33. Sun, J. B., Duan, J. H., Dai, S. L., et al. (2007). In vitro and in vivo antitumor effects of doxorubicin loaded with bacterial magnetosomes (DBMs) on H22 cells: The magnetic bio-nanoparticles as drug carriers. *Cancer Letters, 258*(1), 109–117.
34. Ghosh, S., Ahmad, R., Zeyaullah, M., & Khare, S. K. (2021). Microbial nano-factories: Synthesis and biomedical applications. *Frontiers in Chemistry, 9*, 626834.
35. Guo, S., Fu, D., Utupova, A., Sun, D., Zhou, M., Jin, Z., & Zhao, K. (2019). Applications of polymer-based nanoparticles in the vaccine field. *Nanotechnology Reviews, 8*, 143–155.

36. Torres, M. P., Wilson-Welder, J. H., Lopac, S. K., et al. (2011). Polyanhydride microparticles enhance dendritic cell antigen presentation and activation. *Acta Biomaterialia, 7*(7), 2857–2864.
37. Puri, V., Chaudhary, K. R., Singh, A., & Singh, C. (2022). Inhalation potential of N-Acetylcysteine loaded PLGA nanoparticles for the management of tuberculosis: In vitro lung deposition and efficacy studies. *Current Research in Pharmacology and Drug Discovery, 3*, 100084.
38. Zhao, K., Li, S., Li, W., et al. (2017). Quaternized chitosan nanoparticles loaded with the combined attenuated live vaccine against Newcastle disease and infectious bronchitis elicit immune response in chicken after intranasal administration. *Drug Delivery, 24*(1), 1574–1586.
39. Bhardwaj, B., Singh, P., Kumar, A., Kumar, S., & Budhwar, V. (2020). Eco-friendly greener synthesis of nanoparticles. *Advanced Pharmaceutical Bulletin, 10*(4), 566–576.
40. Li, Y., Liu, G., Wang, X., Hu, J., & Liu, S. (2016). Enzyme-responsive polymeric vesicles for bacterial-strain-selective delivery of antimicrobial agents. *Angewandte Chemie, 55*(5), 1760–1764.
41. Wang, Y., & Sun, H. (2021). Polymeric nanomaterials for efficient delivery of antimicrobial agents. *Pharmaceutics, 13*(12), 2108.
42. Ruwona, T. B., Johnson, V. J., Schmechel, D., Simoyi, R. H., Beezhold, D., & Siegel, P. D. (2010). Monoclonal antibodies against toluene diisocyanate haptenated proteins from vapor-exposed mice. *Hybridoma (Larchmt), 29*(3), 221–229.
43. Sachin, K., & Karn, S. K. (2021). Microbial fabricated nanosystems: Applications in drug delivery and targeting. *Frontiers in Chemistry, 9*, 617353.
44. Jacinto, M. J., Silva, V. C., Valladão, D. M. S., & Souto, R. S. (2021). Biosynthesis of magnetic iron oxide nanoparticles: A review. *Biotechnology Letters, 43*(1), 1–12.
45. Attia, M. A., Essa, E. A., Elebyary, T. T., Faheem, A. M., & Elkordy, A. A. (2021). Brief on recent application of liposomal vaccines for lower respiratory tract viral infections: From influenza to COVID-19 vaccines. *Pharmaceuticals, 14*(11), 1173.
46. Gasparini, R., Amicizia, D., Lai, P. L., Rossi, S., & Panatto, D. (2013). Effectiveness of adjuvanted seasonal influenza vaccines (Inflexal V ® and Fluad ®) in preventing hospitalization for influenza and pneumonia in the elderly: A matched case-control study. *Human Vaccines & Immunotherapeutics, 9*(1), 144–152.
47. Cargill, T., & Barnes, E. (2021). Therapeutic vaccination for treatment of chronic hepatitis B. *Clinical and Experimental Immunology, 205*(2), 106–118.
48. Erasmus, J. H., Khandhar, A. P., Guderian, J., et al. (2018). A nanostructured lipid carrier for delivery of a replicating viral RNA provides single, low-dose protection against Zika. *Molecular Therapy: The Journal of the American Society of Gene Therapy, 26*(10), 2507–2522.
49. Fehr, A. R., & Perlman, S. (2015). Coronaviruses: An overview of their replication and pathogenesis. *Methods in Molecular Biology, 1282*, 1–23.
50. Allawadhi, P., Singh, V., Khurana, A., et al. (2021). Silver nanoparticle based multifunctional approach for combating COVID-19. *Sensors International, 2*, 100101.
51. Baquerizo Nole, K. L., Yim, E., & Keri, J. E. (2014). Probiotics and prebiotics in dermatology. *Journal of the American Academy of Dermatology, 71*(4), 814–821.
52. Puebla-Barragan, S., & Reid, G. (2021). Probiotics in cosmetic and personal care products: Trends and challenges. *Molecules (Basel, Switzerland), 26*(5), 1249.
53. Kim, H. T., Lee, J., Jo, Y. J., & Choi, M. J. (2021). Application of liposome encapsulating *Lactobacillus curvatus* extract in cosmetic emulsion lotion. *Materials (Basel), 14*(24), 7571.
54. Bhat, E. A., Sajjad, N., & Rather, I. (2021). *Cosmetic and medical applications of microbial nanotechnology*. Springer. https://doi.org/10.1007/978-981-16-1923-6_16
55. Anderluzzi, G., Lou, G., Woods, S., et al. (2022). The role of nanoparticle format and route of administration on self-amplifying mRNA vaccine potency. *Journal of Controlled Release: Official Journal of the Controlled Release Society, 342*, 388–399.

56. Pardi, N., Tuyishime, S., Muramatsu, H., et al. (2015). Expression kinetics of nucleoside-modified mRNA delivered in lipid nanoparticles to mice by various routes. *Journal of Controlled Release: Official Journal of the Controlled Release Society, 217*, 345–351.
57. Manolova, V., Flace, A., Bauer, M., Schwarz, K., Saudan, P., & Bachmann, M. F. (2008). Nanoparticles target distinct dendritic cell populations according to their size. *European Journal of Immunology, 38*(5), 1404–1413.
58. Mauro, V. P., & Chappell, S. A. (2014). A critical analysis of codon optimization in human therapeutics. *Trends in Molecular Medicine, 20*(11), 604–613.
59. Ramvikas, M., Arumugam, M., Chakrabarti, S. R., & Jaganathan, K. S. (2017). Nasal vaccine delivery. In *Micro and nanotechnology in vaccine development* (pp. 279–301).
60. Phua, K. K., Staats, H. F., Leong, K. W., & Nair, S. K. (2014). Intranasal mRNA nanoparticle vaccination induces prophylactic and therapeutic anti-tumor immunity. *Scientific Reports, 4*, 5128.
61. Stewart, K. L. (2017). Chapter 8: Common technical procedures in rodents. In M. A. Suckow & K. L. Stewart (Eds.), *Principles of animal research for graduate and undergraduate students* (pp. 177–198). Imprint: Academic.
62. Midoux, P., & Pichon, C. (2015). Lipid-based mRNA vaccine delivery systems. *Expert Review of Vaccines, 14*(2), 221–234.
63. Indermun, S., Luttge, R., Choonara, Y. E., et al. (2014). Current advances in the fabrication of microneedles for transdermal delivery. *Journal of Controlled Release: Official Journal of the Controlled Release Society, 185*, 130–138.
64. Alimardani, V., Abolmaali, S. S., Yousefi, G., et al. (2021). Microneedle arrays combined with nanomedicine approaches for transdermal delivery of therapeutics. *Journal of Clinical Medicine, 10*(2), 181.
65. Tuan-Mahmood, T. M., McCrudden, M. T., Torrisi, B. M., et al. (2013). Microneedles for intradermal and transdermal drug delivery. *European Journal of Pharmaceutical Sciences: Official Journal of the European Federation for Pharmaceutical Sciences, 50*(5), 623–637.
66. Larrañeta, E., McCrudden, M. T., Courtenay, A. J., & Donnelly, R. F. (2016). Microneedles: A new frontier in nanomedicine delivery. *Pharmaceutical Research, 33*(5), 1055–1073.
67. McAllister, D. V., Wang, P. M., Davis, S. P., et al. (2003). Microfabricated needles for transdermal delivery of macromolecules and nanoparticles: Fabrication methods and transport studies. *Proceedings of the National Academy of Sciences of the United States of America, 100*(24), 13755–13760.
68. Jung, J. H., & Jin, S. G. (2021). Microneedle for transdermal drug delivery: Current trends and fabrication. *Journal of Pharmaceutical Investigation, 51*(5), 503–517.
69. Li, J., Zeng, M., Shan, H., & Tong, C. (2017). Microneedle patches as drug and vaccine delivery platform. *Current Medicinal Chemistry, 24*(22), 2413–2422.
70. Singh, A., & Yadav, S. (2016). Microneedling: Advances and widening horizons. *Indian Dermatology Online Journal, 7*(4), 244–254.
71. Shen, M. Y., Yuran, S., Aviv, Y., et al. (2019). Electrically responsive, nanopatterned surfaces for triggered delivery of biologically active molecules into cells. *ACS Applied Materials and Interfaces, 11*(1), 1201–1208.
72. Zhao, Y., Bucur, O., Irshad, H., et al. (2017). Nanoscale imaging of clinical specimens using pathology-optimized expansion microscopy. *Nature Biotechnology, 35*, 757–764.
73. Vigneshvar, S., Sudhakumari, C. C., Senthilkumaran, B., & Prakash, H. (2016). Recent advances in biosensor technology for potential applications - an overview. *Frontiers in Bioengineering and Biotechnology, 4*(11), 1–9.
74. Ali, J., Cheang, U. K., Martindale, J. D., Jabbarzadeh, M., Fu, H. C., & Jun, K. M. (2017). Bacteria-inspired nanorobots with flagellar polymorphic transformations and bundling. *Scientific Reports, 7*(1), 14098.

Chapter 14
Nanotechnology-Based Electrochemical Diagnostic Tools for the Detection of Viral Diseases: Advantages and Disadvantages

Ahmet Cetinkaya, S. Irem Kaya, Banu Kaskatepe, Nurgul K. Bakirhan, and Sibel A. Ozkan

14.1 Introduction

Viruses have simple structures with small sizes. These structures cause many risky diseases like SARS, AIDS, EBOLA, COVID-19, and MERS. Considerable risk of transmission, replication capacities, and proliferation rates are the most notable features that make viruses pathogenic. As seen in the pandemic process, the COVID-19 virus caused a lot of damage. One of the most important steps to control the disease in the pandemic was its diagnosis. Diagnosis is one of the most important steps in treatment. One of the most important steps in the diagnosis of diseases, especially viral diseases, is "POCT" (point of care testing). POCT can be given as an example of blood glucose measurement, a home pregnancy test, or PT/INR tests. Viruses contain different diagnostic methods thanks to the characteristics they carry. Electron microscopy, molecular methods, and morphological methods can be given as examples of the previously used methods. The higher selectivity of the biosensors used today and their greater openness to method development depending on the

A. Cetinkaya
Ankara University, Faculty of Pharmacy, Department of Analytical Chemistry, Ankara, Türkiye

Ankara University, Graduate School of Health Sciences, Ankara, Türkiye

S. I. Kaya · N. K. Bakirhan (✉)
University of Health Sciences, Gülhane Faculty of Pharmacy, Department of Analytical Chemistry, Ankara, Türkiye

B. Kaskatepe
Ankara University, Faculty of Pharmacy, Department of Pharmaceutical Microbiology, Ankara, Türkiye

S. A. Ozkan
Ankara University, Faculty of Pharmacy, Department of Analytical Chemistry, Ankara, Türkiye

R. S. Chaughule, A. S. Lokur (eds.), *Applications of Nanotechnology in Microbiology*, https://doi.org/10.1007/978-3-031-49933-3_14

virus type have been turning points for POCT. It has been one of the most important steps for treatment. Electrochemical biosensors are among the most preferred types of POCT. The applied electrochemical techniques include potentiometry, amperometry, voltammetry, and electrochemical impedance spectroscopy. Among the most important advantages of electrochemical method, it has a vital role in the rapid detection of infectious diseases due to its rapid reaction, high sensitivity, selectivity, low cost and ease of miniaturization. In this study, different detection methods related to viral agents were used, and the most effective method was selected depending on the virus type. The electrochemical method used for the diagnosis of adenoviruses that cause epidemic diseases is 3DNRE-treated tungsten rods. Among the advantages that make this method special, the absence of additional washing and the need for additional purification are the most important reasons [1]. CTV (Citrus tristeza virus) is one of the other viruses with an excessive economic cost and has caused the death of numerous citrus trees. One of the methods used in the diagnosis of CTV was the use of unlabeled impedimetric biosensors. It has been observed that conductivity increases when using gold nanoparticles (AuNP). This method has been used to immobilize single-stranded deoxyribonucleic acid (ssDNA) probes [2]. Among the methods for foot and mouth disease virus (FMDV) detection, we can draw attention to the virally suppressed polymer (VIP). Among the most important advantages of this biosensor are its high capture efficiency, binding stability, and selectivity [3]. In another study, a biosensor was developed against the M1 (Influenza A matrix protein) influenza virus protein. This biosensor has high biocompatibility due to the NCD-substrate approach [4]. Depending on the H9N2 serotype diagnosis of influenza virus, two different receptors were used. The use of another receptor to isolate the first receptor at a lower concentration caused it to be detected by causing a Fetuin-Hemagglutinin interaction [5].

Nanosensors are fundamental for the development and progress of research in electrochemical analysis of various compounds, molecules, and biological samples. These mini devices have excellent features because of their fast detection, sensitive responses, and ability to measure at the molecular level the chemical and physical features of analytes and targets. They can be classified according to their structure and functions. Optical and electrochemical nanosensors can be mentioned as structural nanosensors. On the other side, chemical nanosensors, biosensors, deployable sensors, and electrometers can be used for a variety of applications [6].

The aim of this chapter is to present the methods developed based on innovative approaches for accurate and rapid diagnosis based on pathogen freedom, depending on the disease agent.

14.2 Viral Infections

Viruses are the primary cause of morbidity and mortality worldwide and cause diseases ranging from self-resolving conditions to fatal infections such as AIDS. Laboratory diagnosis is crucial in patients with especially severe diseases, as signs and symptoms of viral infections overlap and are not always specific to any viral agent, or the same virus can have different clinical symptoms. Due to the different structures of viruses from other microorganisms, the diagnosis of viral infections, which are common around the world, varies. Reasons such as the fact that most viral diseases are self-limited, viruses causing latent and/or chronic infections, and limited specific anti-viral treatment options also affect the diagnosis of these infections [7]. Rapid and accurate identification of viral infections is of valuable importance for using appropriate anti-viral drugs, preventing inappropriate antimicrobial therapy, taking infection control measures, epidemiological monitoring of the disease, and optimal patient/disease management. A detailed discussion of the laboratory diagnosis of viral infections is beyond the scope of this review. So, the advantages and limitations of these methods will be discussed by mentioning them in general terms.

14.2.1 Clinical Detection Techniques for Viruses in Common Infection

Laboratory diagnosis of viral infections can be divided into direct methods for detecting the agent in the material and indirect methods for determining the immune response, and these methods include (1) the isolation of viruses with cell culture systems; (2) The detection of viruses in clinical samples (a): the detection of viral antigen (direct serological tests), (b): the detection of virus particles (electron microscopy), (c): histological and cytological techniques for the detection of morphological changes in viruses, and (d): the detection of the nucleic acid of viruses (molecular methods); (3) Indirect serological methods that evaluate the patient's immune response to the virus and the detection of antibodies [8–10].

The importance of these methods has changed over the years due to advances in the scientific world. Since the use and performance of these tests have considerable differences, which tests are to be performed and the selection of samples to be taken for testing should be made with care based on laboratory capacity and conditions, patient population, and clinical situation.

14.2.1.1 The Isolation of Viruses with Cell Culture Systems

Cell culture methods involving the isolation and identification of viruses used to be the "gold standard" of the diagnostic laboratory, but in recent years, they have been used more in research studies. In culture methods, at least theoretically, a single live virion in a material can be grown in cultured cells so that it is possible to produce enough material to allow for more detailed characterization, but the presence of a live virus in the sample is required. The virus may not be isolated from a sample if the sample is improperly handled, contains a neutralizing antibody, or is acquired before or after viral shedding. Thus, the sample must be collected during active viral replication. Another limitation is that some viruses do not grow in cell culture [8, 11].

14.2.1.2 Serological Methods

Numerous immunological assays that detect viral antigens in clinical materials using well-defined reagents containing known antibody specificities are now commercially available, and some of them are routinely used in most clinical laboratories for many viruses. The important advantages of these tests are that they are fast, inexpensive, simple to perform, and do not necessarily detect live viruses. The sensitivity and specificity of assays vary depending on the virus, the test format, and the quality of the sample to be detected, and the methods are not as accurate as molecular methods and cell culture [9, 12]. Serological techniques are methods that can target the antigen during the acute phase of the infection or the virus-specific antibody later in the infection. It is often used to identify viruses that are difficult to grow and isolate in cell culture. With serology, it is possible to identify the virus and its serotype and to understand whether the infection is acute or chronic, primary, or secondary (reinfection). However, false positive and negative results in serological tests may confuse the diagnosis. In addition, the patient's antibodies may have formed immune complexes with viral antigens, in which case the detection of antibodies is prevented. Serological cross-reactions between different viruses may also pose a problem in identifying the agent [8, 10, 12–14].

14.2.1.3 Detection of Morphological Changes in Viruses

Direct microscopy of stained histology or cytology specimens is one of the oldest methods of detecting viruses, in some cases revealing the first signs of viral involvement including cellular changes. The cytologic examination of the sample provides a rapid initial diagnosis for viral infections that produce characteristic cytopathologic effects including changes in cell morphology. Although it is simple and cost-effective, it has low sensitivity and specificity compared to direct antigen or nucleic acid determination methods [8].

14.2.1.4 Electron Microscopy

Electron microscopy is not a standard laboratory technique but is used to detect and identify some viruses according to their characteristic morphology. The combination of electron microscopy with culture-based methods and serology testing for the detection of targeted antibodies to the virus makes a major contribution to the diagnosis of viral infections. However, the electron microscopy method has some limitations, such as detecting samples containing sufficient viral particles; the prerequisite sample preparation step can reduce the virus concentration; and it requires significant technical skill and expertise [13, 15].

14.2.1.5 Molecular Methods

Molecular biological methods used in the diagnosis and research of viral infections are nucleic acid-based methods such as amplification, hybridization, polymorphism analysis, DNA sequence, and phylogenetic analysis. These methods can be used alone or in combination, depending on the purpose. Over the past few decades, the development of molecular technologies and making them accessible to diagnostic laboratories has revolutionized the operation of virus diagnostic laboratories, and molecular methods have replaced traditional culture and antigen-based procedures to become the new "gold standard" for the identification of most viruses of medical importance. These methods successfully detect new viruses for which current tests are much less accurate or for which no tests are available. Viruses that cannot be cultured are fastidious, slow-growing, or too dangerous to grow. It is also particularly suitable for identifying viruses present in small sample volumes, low in numbers, or nonviable in clinical samples. Nowadays, with technological improvements, it is possible to detect more than one virus from a single sample with the multiplex procedure developed and commercialized. Quantitative molecular methods have become priceless for assessing disease progression and prognosis, monitoring treatment, and the emergence of drug resistance. The use of molecular genotyping analyses comprising direct sequencing of the amplified products and nucleic acid amplification of specific viral genes can also provide precious information about the evolution and phylogenetic relationships among closely related viruses. New-generation rapid molecular amplification methods for the identification of viruses have been advanced by reason of new technological developments as well. Therefore, currently used methods may be faster at diagnosing known viruses. Next-generation sequencing is the most important method to diagnose a previously unknown virus or to identify unexpected viruses in samples. Although nucleic acid-based analyses are highly sensitive and specific, they often have limitations such as the need for trained personnel, the associated costs, and the carry-on contamination that can cause problems in the analysis of the results, and calibration problems [10, 14–16].

Currently, no single method meets all demands for viral diagnosis. Therefore, the decision as to which method to use or not will be determined by the requirements of the analysis. At this point, conditions such as laboratory conditions, test cost, trained personnel, virus type, sensitivity and specificity of the test, and the detection of a new epidemic agent will be effective in determining the test.

14.3 Nanotechnology and Electrochemical Progress on the Assay of Pathogens

Nanotechnology is related to materials at the nanoscale to create materials with unique properties and purposes [17]. The term "nanomaterial" means "nanoproducts in the form of materials containing structural nanoelements that greatly improve physical, chemical, biological, mechanical, and other properties or cause qualitatively new properties" [18]. The last few decades have seen remarkable advances in nanotechnology. Particles, fibers, or particles smaller than 100 nm are called nanomaterials, which have increased researchers' interest in the field of nanotechnology, particularly in the development of methods for the early detection, prevention, and treatment of disease [19]. The use of nano-biotechnology for nano-scale molecular diagnostics is called "nano-diagnostics." The use of nanotechnologies in diagnostics provides strong assurance that the clinical laboratory's requirements for precision, multiplexing, and cost-effectiveness are met. Nanotechnology provides several technological developments for pathogen determination and therapeutics by immobilizing certain ligands on surfaces. Due to their superior properties, such as high surface area/volume ratio, high thermal conductivity, and fast signal transmission, nanomaterials are now receiving great interest in therapeutic applications [20]. Also, the miniaturization of sensing devices, which simply calls for reduced sampling volumes and speeds up reaction time, would benefit from using nanomaterials [21].

The rapid development of nanotechnology enables the creation of more effective chemo/biodetection platforms for the identification of target analytes. Numerous nanomaterials are often preferred due to their unique magnetic, optical, electrical, and other related properties in efficient and effective electrochemical processes [22, 23]. Nanomaterials are well suited for attaching different recognition moieties via covalent or non-covalent bonding modes and allow for the modification and control of their properties due to their high surface-to-volume ratio [24, 25]. Compared with traditional pathogen detection methods, nanomaterial-based assays have many advantages, especially high throughput screening, low cost, rapidity, and high accuracy [26]. Moreover, while it is undeniable that nanomaterials have enormous potential to improve the precision of pathogen detection and reaction kinetics, their practical application in biomedical applications involving complex biofluids is a major challenge. Due to the high surface-to-volume ratio of nanoparticles, direct contact with all untreated biofluids may cause false positive findings due to non-specific binding attempts in assays based on nanomaterials [27]. The development of high-performance pathogenic biosensors, the detection of pathogens when biocompatible polymers [28, 29], liposomes [30, 31], hydrogels [32, 33], Si-based capsules [34, 35], or DNA structures with nanomaterials [36, 37] are used in combination with nanomaterials have excellent potential for surface plasmon resonance (SPR), surface-enhanced Raman scattering (SERS), colorimetry, fluorimetry, electrochemistry, etc. It is also suitable for the simultaneous determination of multiple pathogen-based nanomaterials using different methods [38, 39]. This part focuses on electrochemical assays and point-of-care testing technologies (POCTs) developed using different nanoparticles for the simultaneous monitoring of various pathogens and their current challenges. Recent innovative studies for the simultaneous determination of multiple pathogens were summarized, and trending analyses were presented in detail.

14.3.1 Nanosensors/Biosensors

In recent years, noteworthy progress has been seen in the field of biosensing technology, especially in biomedical applications. Instruments used for biosensing are typically sensitive, fast, selective, accessible, and do not require laborious sample pretreatment, unlike space-consuming laboratory equipment. In addition, the versatile physical and chemical properties of nanomaterials make them suitable for the creation of biosensors. Nanomaterials like carbon-based nanomaterials, metal nanomaterials, silica nanoparticles (NPs), quantum dots (QDs), and other functionalized NPs are frequently used to build both biosensors and immunosensors [40]. Nanosensors, nanoparticle-based devices that can detect signals at the nanoscale, have three main components. These are a signal converter, a receiver, and a detector with a monitoring output. The biological molecule to be analyzed interacts with the receptor, and the detector chemically reacts with the target molecule to identify it. Immediately after detection, the converter collects the signal from the detector and digitizes it for the digital monitor. Extremely sensitive nanosensors can increase detection efficiency when used with other analytical tools. These nanotechnological methods are compatible with commonly used POCT sensing equipment [41].

Biosensors are instruments that can measure and/or determine the quantity of specific biomarkers for pathogens. These sensors, which are formed by the combination of the constituent elements, first perform ligand-selective recognition. Nucleic acids, antibodies, and enzymes are used as recognition ligands. Second, it is a sensitive transducer that converts the biochemical signals produced by the targeted analytes and the bioreceptor into measurable electrical signals for the identification and quantification of the analyte. In addition, biosensors provide fast, efficient, and cost-effective detection of a particular analyte. Real-time analyte analysis without the need for time-consuming and expensive sample preparation is one of the key benefits of biosensors. They have the potential to make portable in-situ analysis possible, which is a crucial characteristic for POC diagnostics. In addition to their employment in medical and POC applications, they are also used in drug discovery, forensics, monitoring prognosis, illness treatment, quality control for food and environmental samples, and biological research. Biosensors have several types according to their transducers; for instance, the most popular types are electrochemical and optical [42, 43]. Table 14.1 summarizes studies on different nanosensor/biosensor applications for pathogen detection. The advantages of the developed methods, such as fast, sensitive, and appropriate selectivity, made them frequently preferred in biomedical applications.

Table 14.1 Electrochemical detection of pathogens with different methods

Target pathogen	Recognition element	Electrode materials and method	Linear range	LOD	References
H5N1	MAb	BSA/ME EIS	2^{-1}–2^{4} HAU/50 μL	0.5 HAU/50 μL	[44]
H5N1	Specific Anti-H5N1 Ab	Fe_3O_4 MNPs/ AuE CV	0.0025–0.16 HAU	0.0022 HAU in 6 μL	[45]
H5N1	DNA probe	pAuNPs/AuE CV	1 pM–100 nM	1 pM	[46]
H5N1	Anti-AIV NP aptamer	Aptcon-MB@3DNRE/ CV	2–12 nM	1.13 nM	[1]
CTV	Polyclonal antibody	MUA/MPA/ AuE DPV	1 nM–5 μM	0.27 nM	[47]
CTV	Thiolated ssDNA probe	AuNPs/SPCE EIS	0.1–10 μM	100 nM	[2]
Chikungunya Virus	ssDNA probe	MoS_2NSs/ SPGEs EIS, CV	0.1 nM–100 μM	3.4 nM	[48]
Dengue Virus	Anti-nonstructural antibody MAbs	BSA/SPCE EIS	1–200 ng/mL	0.3 ng/mL	[49]
FMDV	Viral imprinted polymer	O-AP/SPGE LSV	4–75 ng/mL	1.98 ng/mL	[3]
Hepatitis A Virus	ssDNA capture probe	BSA/AuE CV	10 fg/μL–10 pg/ μL	6.94 fg/mL	[50]
Hepatitis B Virus	Peptide nucleic acid	acpcPNA/SPE ELFA	10 pM–2 μM	7.23 pM	[51]
HBsAg	Primary antibody of HBs (Ab1)	Fe_3O_4 MNPs/ GCE SWV	0.3–1000 pg/mL	0.19 pg/mL	[52]
Hepatitis C Virus	ssDNA probe	CNT/ferrocene / AuE CV	0.1 fM–1 pM	0.01 fM	[53]
Hepatitis C Virus	DNA aptamer	MWCNTs-Chit/ GCE DPV	5 fg/mL–1 pg/ mL	1.67 fg/mL	[54]
H1N1 Influenza Virus	MAb	PDMS/SiNPs/ StPCE CA	10–10^4 PFU/mL	113 PFU/ mL	[55]
H1N1 Influenza Virus	Specific MAb	DNA aptamer/ AuE DPV	12.8–0.00128 HAU	0.0128 HAU	[56]

(continued)

Table 14.1 (continued)

Target pathogen	Recognition element	Electrode materials and method	Linear range	LOD	References
H1N1 mini-HA protein	ssDNA aptamer	ITO/glass electrode DPV	10–10^4 PFU/mL	3.7 PFU/mL	[57]
EV71	MAbs of EV71	TMB/AuNPs/ITOE CA	0.01–1 ng/mL	0.01 ng/mL	[58]
Human T-lymphotropic Virus	Hairpin capture DNA probe	P(DEB-DSDA)/AuE EIS	1 pM–1 nM	17.1 pM	[59]
Inactivated H1N1 Virus	DNA aptamers against the inactivated H1N1 virus	SELEX/DNA-aptasensor/AuE EIS	NA	0.9 pg/mL	[60]
Inactivated H1N1 Virus	Anti-Influenza A Ab	Shellac-derived TrGO/ITO EIS	NA	PBS: 26.04 PFU/mL Saliva: 33.11 PFU/mL	[61]
H1N1 antigen	Anti-Influenza A HA Ab	GPE EIS	NA	<5 PFU/mL	[62]
H1N1	HA gene-specific ssDNA probe	Cysteine/SPGE EIS, CV	0.1–400 ng/6 μL	0.004 ng in 6 μL	[63]
Influenza Virus M1 protein	Polyclonal aM1Ab	Nanocrystalline BDDE EIS	NA	5 × 10^{-14} g/mL	[4]
Influenza Virus AM1 protein	Anti-M1Ab	Nanoscale BDDE EIS	1–100 fg/mL	1 fg/mL	[64]
Influenza A H9N2	Fetuin A	SPGE/Gr/Au hybrid EIS	10^{-8}–10^{-1} U/mL	10^{-8} U/mL	[65]
Influenza A H9N2	Anti-matrix protein 2 Ab and Feutin A	AuNPs/SPCE CA	8–128 HAU	8 HAU	[5]
MERS-CoV	MERS-CoV antigen	AuNPs/CE SWV	0.001–100 ng/mL 0.01–10,000 ng/mL	0.4 pg/mL 1.0 pg/mL	[66]
SARS-CoV-2	Immobilized with nCoV19 Ab	FTO/AuNPs CV, DPV	1 fM–1 μM	10 fM at nCoV19 Ag	[67]
SARS-CoV-2	Anti-His MAb	Co-TNTs Amperometry	14–1400 nM	0.7 nM	[68]

(continued)

Table 14.1 (continued)

Target pathogen	Recognition element	Electrode materials and method	Linear range	LOD	References
RSV	Antibody specific to the F protein	4-ATP/PLL/GCE CV, EIS	1.0×10^5–1.0×10^7 PFU/mL	1.1×10^3 PFU/mL	[69]
Zika Virus	SH-probe ssDNA	Gold-PET/DE EIS, CV, DPV	54–340 nM	25 nM	[70]
Zika Virus	SIP	SIPs/GO/SPGE EIS, CV	0–100 PFU/mL	2×10^{-4} PFU/mL	[71]

NA not available, *H5N1* avian influenza virus, *CTV* Citrus tristeza virus, *FMDV* foot and mouth disease virus, *HBsAg* hepatitis B virus surface antigen, *EV71* human enterovirus 71, *MERS-CoV* middle east respiratory syndrome corona virus, *RSV* respiratory syncytial virus, *MAb* monoclonal antibody, *DPV* differential pulse voltammetry, *SWV* square wave voltammetry, *EIS* electrochemical impedance spectroscopy, *CV* cyclic voltammetry, *LSV* linear sweep voltammetry, *CA* chronoamperometry, *ELFA* electrochemical lateral flow assay, *GCE* glassy carbon electrode, *AuE* gold electrode, *SPGE* screen printed gold electrode, *ME* microelectrode, *SPCE* screen-printed carbon electrode, *SPE* screen-printed electrode, *StPCE* stencil-printed carbon electrodes, *ITOE* indium tin oxide electrode, *CE* carbon electrode, *DE* disposable electrode, *GPE* gold paper electrode, *BDDE* boron-doped diamond electrode, *CdS QDs* cadmium sulfide quantum dots, *ZIF-8* zeolitic imidazolate framework-8, *p-AP* p-aminophenol, *AuNPs* gold nanoparticles, *T4-MES* T4 bacteriophage-based micro electrochemical sensor, *BSA* bovine serum albumin, *NiNWs* nickel nanowires, *rGO* reduced graphene oxide, *AuNUs* gold nano-urchins, *GNPs* gold nanoparticles, *RPA* recombinase polymerase amplification, *MUAA* 11-mercaptoundecanoic acid, *MPA* 3-mercapto propionic acid, *MoS_2NSs* molybdenum disulfide nanosheets, *O-AP* O-aminophenol, *acpcPNA* a pyrrolidinyl peptide nucleic acid, *CNT* carbon nanotube, *MWCNTs-Chit* multi-walled carbon nanotubes-chitosan nanocomposite, *PDMS* polydimethylsiloxane, *SiNPs* silica nanoparticles, *TMB* 3,3′,5,5′-tetramethylbenzidine, *DEB* 1,4-diacetylenebenzene, *DSDA* 4,4′-Diazido-2,2′-stilbenedisulfonic acid disodium salt tetrahydrate, *SELEX* systematic evolution of ligands by exponential enrichment, *4-ATP* 4-aminothiophenol, *PLL* poly-L-lysine, *PET* polyethylene terephthalate, *SIPs* surface imprinted polymers, *GO* graphene oxide, *TrGO* thermally decomposed rGO, *Gr* graphene, *pAuNPs* porous Au nanoparticles, *3DNRE* 3D nanostructured porous silica film, *MB* methylene blue, *FTO* fluorine-doped tin oxide electrode, *Co-TNTs* cobalt-functionalized TiO_2 nanotubes, *SIP* surface-imprinted polymer

Reliable and independent detection of AIVs, the main cause of highly contagious respiratory infections, using a self-calibrating and dual-electrode platform-based electrochemical aptasensor was performed by Lee et al. [1]. 3DNRE-treated tungsten rods were used to construct both electrodes. Then, after loading the pores with a redox-active substance, the appropriate aptamer was placed on the MB. An anti-AIV nucleoprotein (NP) aptamer (AptAIV-MB@3DNRE) was used to seal an electrode, enabling specific binding to the target that alters the electrochemical signal upon the diffusional release of charged redox molecules (Fig. 14.1). The other electrode was coated with a control aptamer (Aptcon-MB@3DNRE). It provided a reference task to correct non-specific aptamer separation and false responses caused by MB release in the presence of non-target molecules from cell lysis residues. In addition, the Aptcon-MB@3DNRE on the dual-electrode platform offers a rectified basis for the erratic original output signals from the AptAIV-MB@3DNRE. In

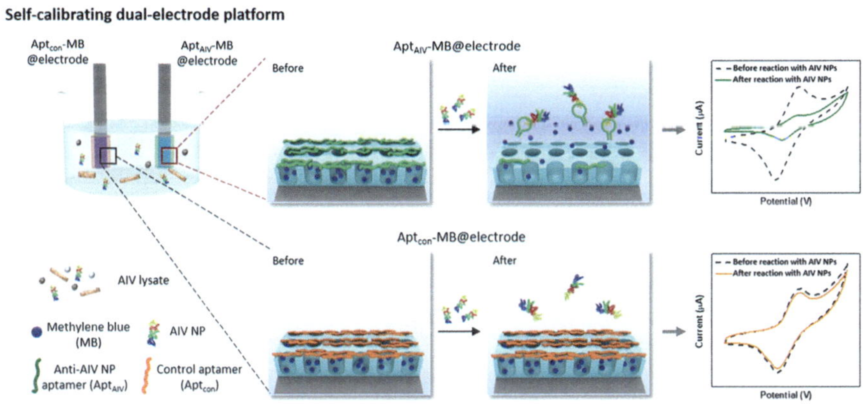

Fig. 14.1 Schematic presentation of a self-calibrating electrochemical aptasensor platform for the detection of AIVs. (Reprinted from Ref. [1] with permission from Elsevier)

conclusion, at equivalent concentrations of AIV NP samples under various reaction buffer conditions, the dual-electrode platform (RSD: 5.86%) exhibited excellent results compared to a conventional single-electrode platform (RSD: 30.13%). In addition, the lack of additional purification and washing processes in the technique suggests that the method can be used as a dependable platform for the electrochemical determination of different biomolecules.

A label-free impedimetric biosensor was developed for the first time by Khater et al. [2] for CTV nucleic acid detection. To effectively immobilize thiolated ssDNA probes and improve electrode conductivity, electrodeposited gold AuNPs were added to the SPCE-based detection platform (Fig. 14.2). Scanning electron microscopy (SEM), CV, and EIS were used to optimize and characterize the formation of AuNPs. EIS measurements in a $[Fe(CN)_6]^{3/-4}$ redox system, the behavior of the thiolated ssDNA probe layer, and its hybridization with target DNA on AuNP surfaces were examined in detail. AuNPs size, probe DNA concentration, immobilization, and DNA hybridization times were optimized to achieve the highest performance. When synthetic DNA associated with CTV was present in higher concentrations, DNA hybridization impedance values also increased, exhibiting a logarithmic relationship between 0.1 and 10 μM. Furthermore, the hybridization of the sensor on MCH/poly(AT) thiolate DNA probes was confirmed due to the excellent repeatability and reproducibility of the sensor.

Developed by Hussein et al. [3], this sensor has demonstrated a number of limitations in sensitivity, specificity, and cross-reactivity compared to other FMDV diagnostic techniques. In the current study, a new biosensor based on viral imprinted polymer (VIP) was developed for rapid and accurate detection of FMDV. The oxidized O-AP film stamped with the FMDV O serotype on a SPGE was electrochemically polymerized to form the biorecognition components (Fig. 14.3). Morphological characterizations were performed to examine the overall changes in the design pattern. For this, CV, atomic force microscope (AFM), field emission scanning electron

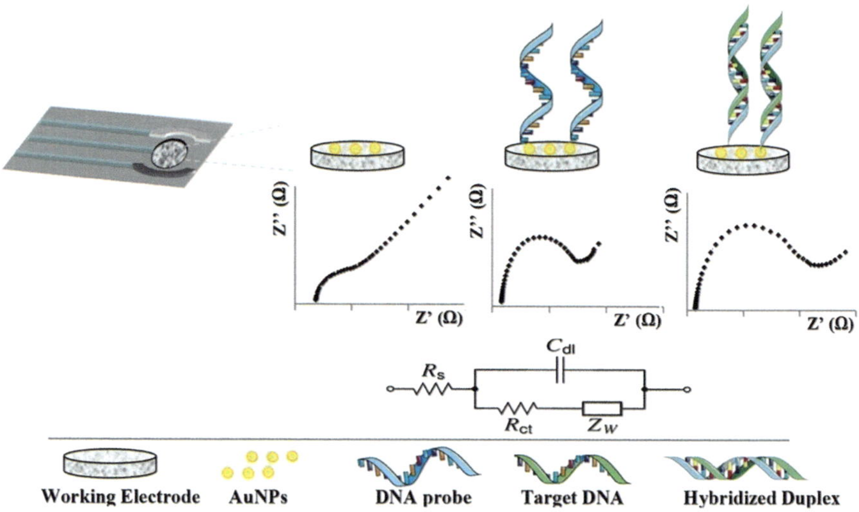

Fig. 14.2 Schematic representation of an improved DNA hybridization sensor based on AuNPs-modified SPCE using the EIS method for the detection of CTV. (Reprinted from Ref. [2] with permission from Elsevier)

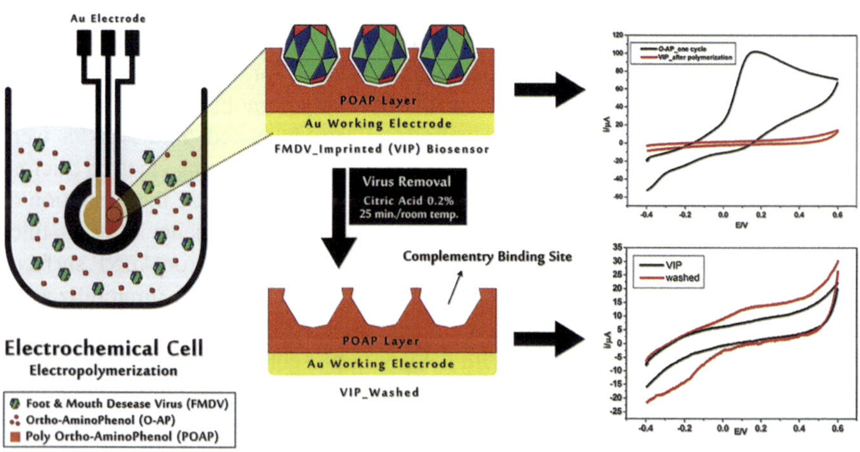

Fig. 14.3 Schematic representation of the sensor developed using the LSV method for the detection of FMDV. (Reprinted from Ref. [3] with permission from Elsevier)

microscope (FE-SEM), and Fourier transform infrared spectroscopy (FT-IR) were used. Parameters such as the capture efficiency, binding stability, selectivity, and lifetime of the developed biosensor under optimum conditions were optimized. The findings showed that the biosensor has a high selectivity for serotype O over other serotypes A, SAT2, Lumpy skin disease virus (LSDV), and inactivated serotype O. In addition to the repeatability and reproducibility of the measurement with a coefficient of variance of 1.0% and 3.6%, respectively, the limits of detection (LOD) and quantization (LOQ) were found to be 2 ng/mL and 6 ng/mL, respectively. This inexpensive biosensor analysis of original saliva samples yielded a LOD 50 times lower than reference methods (ELISA and PCR), providing the potential for online monitoring in the field without preprocessing the sample.

In another study, Siuzdak et al. [4] presented a new, sensitive, label-free approach for the detection of M1 influenza virus protein. The electrode used in this method is based on the B: NCD-substrate approach, and its surface was chemically modified to facilitate covalent bonds binding the anti-M1 antibody (Fig. 14.4). The developed biosensor primarily exhibited outstanding properties, including low background current, intrinsic biocompatibility, a wide potential window, and high stability. The detection methodology was designed to account for the varied protein concentrations, incubation times, non-specific interactions, and temperature conditions in the electrode response. A significant shift in charge transfer resistance was observed after the electrode was exposed to a sample containing real viruses. However, no effect was seen after incubation with PBS solution, Triton X-100/PBS solution, or a negative biological sample taken from a healthy person. These findings support the modified B:NCD substrate's potential as a biosensing electrode. The experiments conducted herein verified the universality and specificity of the chosen antibodies for a particular strategy. This demonstrates that a particular detection technology may be used to accurately identify influenza virus particles at a concentration of 50 fg/mL. The electrode reported here can be considered a useful biosensor created for the detection of the influenza virus, given the short incubation time, low detection limit, and suggested ambient temperature settings.

Traditional approaches, which are typically laborious and time-consuming, are insufficient for field identification of the virus. Therefore, there is an urgent need for studies that try to create effective alternatives to existing practices. In this study, a sensor for isolating and identifying Influenza A virus subtype H9N2 was developed

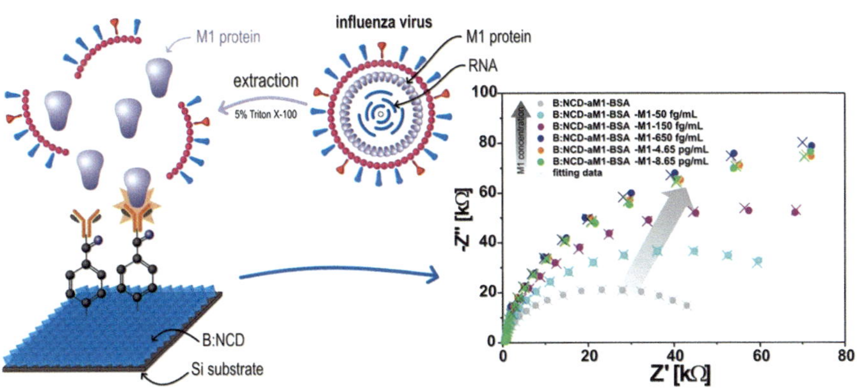

Fig. 14.4 Schematic representation of the biosensor developed using EIS with a nanocrystalline boron-doped diamond electrode for the detection of influenza virus M1 protein. (Reprinted from Ref. [4] with permission from Elsevier)

by Sayhi et al. [5]. Two different influenza receptors were used to achieve this goal. First, an anti-matrix protein 2 (M2) antibody was used to isolate the virus from the allantoid fluid and bind it to MNPs. The second biomolecule, Fetuin A was used to detect the virus splicing advantage caused by the fetuin-hemagglutinin interaction. The MNP-Influenza virus-AuNP complex was purified and treated with an acid solution, and the recovered gold nanoparticles were then applied to a carbon electrode via screen printing (Fig. 14.5). This method enables the quick identification of influenza virus A/H9N2 at a titer of less than 16 HAU. In addition, among influenza, there are differences in sub-types or within pathogenic and nonpathogenic strains that is currently restricted and can be produced utilizing other biomolecules like aptamers or antibodies.

14.3.2 Point of Care Testing Technologies (POCTs): Current Challenges

Pathogenic microorganisms, such as bacteria, fungi, viruses, and parasites, bring on most infectious diseases. Comparatively quickly compared to other diseases, infectious diseases can spread rapidly across communities, endangering both the general public's health and the economy. Infectious diseases are among the most critical risks to humanity because it is estimated that more than 50% of the world's population is at risk [72]. Adequate and rapid treatment of diseases cannot be provided

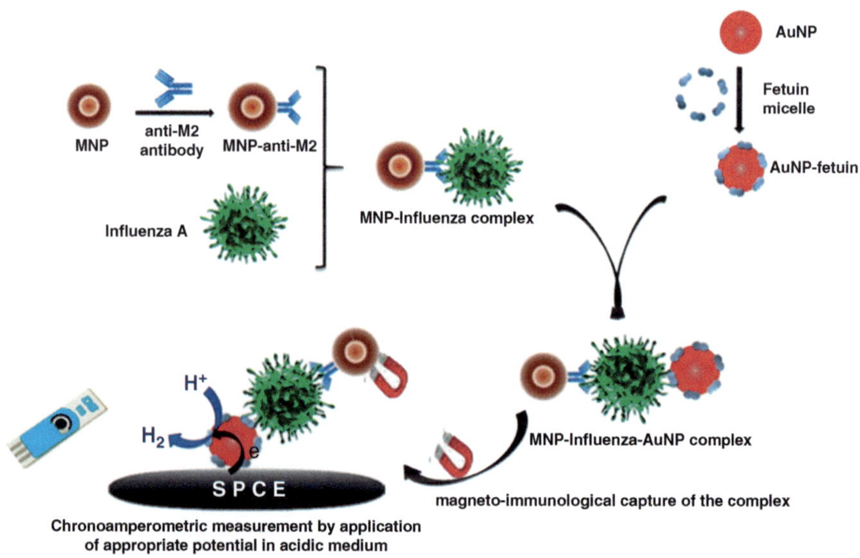

Fig. 14.5 Schematic illustration of the gold nanoparticle-based chronoamperometric immunosensor strategy developed for the detection of influenza virus. (Reprinted from Ref. [5] with permission from Elsevier)

without a correct diagnosis. Rapid, accurate, and sensitive diagnostic testing is essential to both facilitating effective treatment and reducing the spread of infectious diseases. Although blood cultures, high-throughput immunoassays, polymerase chain reactions (PCR), and mass spectrometry (MS) tests are among the sensitive and specific tests offered by central clinical laboratories, they often have disadvantages such as being expensive, needing sophisticated instruments, and requiring specialized equipment. POC tests, on the other hand, offer quick 'on-site' results at the location where care is delivered and in settings with limited resources, facilitating prompt and appropriate treatment [73]. The World Health Organization (WHO) states that POC tests must meet the criteria of "RELIABLE." These criteria include features such as cheap, sensitive, specific, user-friendly, fast, and robust [74]. Also, among the key innovations in the development of POC tests for infectious diseases over the past decade have included developments in two technologies, namely, microfluidics and plasmonic. These technologies, along with others in the "POCT Toolbox" (Fig. 14.6), serve as personal radar in the fight against infectious diseases towards the goal of patient-centered diagnosis and treatment, as illustrated in Fig. 14.6 [75].

In conclusion, POC tests make it possible to diagnose infectious diseases quickly, especially in settings with limited resources. This, in turn, allows for early and efficient patient-centered therapy and care. Biomarkers with greater sensitivity and specificity are still required, despite several biomarkers being effectively used as targets in POC tests for infectious illnesses. Future biomarker screening may find it advantageous to systematically characterize a group of biomarker signatures for a

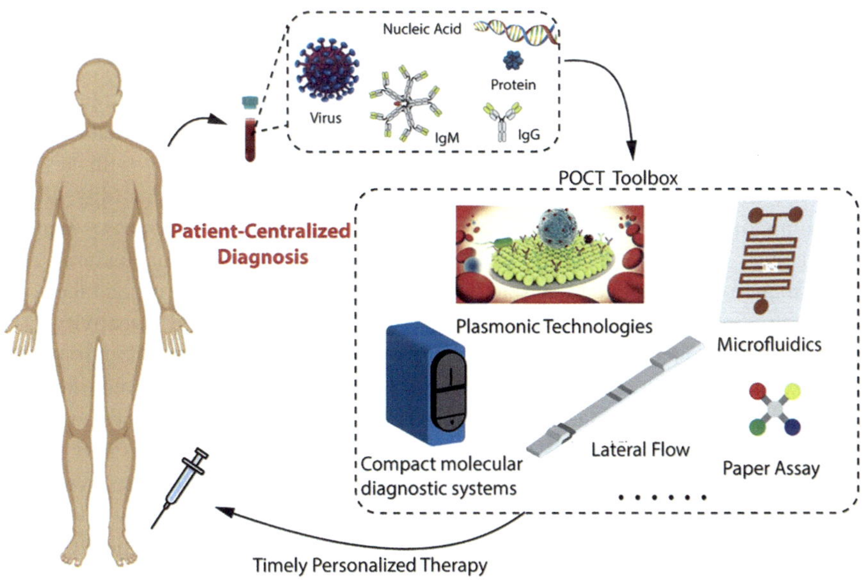

Fig. 14.6 General schematic representation of Point of Care Tests (POCT). (Reprinted from Ref. [75] with permission from Elsevier)

single infectious disease. Diagnosis of POC with multiple abilities is also commonly requested, as several infectious diseases may present with the same clinical manifestations. In the last decade, significant advances have been made in innovative technologies such as microfluidics and plasmonics for POC infectious disease testing. Tight clinical validation is still required to transform these technologies from research into clinical practice. Many practical concerns, such as infection control, small-area testing, and clinical route optimization, are also crucial factors in successful adoption to address clinical barriers [76].

14.3.3 Advantages and Disadvantages of Electrochemical Sensors

Pathogens that cause infectious diseases, especially viruses and bacteria, pose a great threat to global health [77, 78]. It is stated in some sources that pathogen-related deaths can reach up to 15 million worldwide [77]. In particular, the high rate of transmission of viruses and the presence of asymptomatic patients make studies on the control of diseases with early diagnosis important [79]. Therefore, sensor development studies for the detection of viral diseases and other pathogens have been the focus of researchers. Currently available conventional methods require tedious sample preparation procedures, long analysis times, expensive equipment, and lab-based analysis [79–81]. Considering all of these, the aim is to enable highly selective and sensitive, reproducible, cheap, and easy-to-use sensors [82]. In terms of clinical applications and determination, these targets can be expressed as accurate detection of the disease/pathogen and accurate detection of the negative results of non-patients [78].

Compared to other available methods, electrochemical sensors have been quite popular thanks to their significant advantages. The most notable advantages are providing user-friendly, straightforward application, rapid analysis with a short response time, miniaturized and portable equipment, flexibility, versatility, ultra-low limit of detection (LOD) values, high reproducibility, and accuracy performance [79, 83, 84]. There are several factors that can affect the features and performance of electrochemical sensors. For example, double-layer capacitance can influence the LOD value; the electron transfer rate affects the sensitivity and response time, and the type of electrode material influences the immobilization [78].

Another advantage of electrochemical sensors is the feasibility of modification with various materials. Nanomaterials are used to improve sensor performance and to increase and highlight the advantages of electrochemistry. Gold, carbon, or magnetic-based nanomaterials are utilized for the modification of the electrode surface, and they enable faster electron transfer, enhanced electrical conductivity and active surface area, and signal amplification [79]. In the analysis of pathogens such as viruses, the analyzed samples are biological fluids such as saliva and blood; hence, the determination is difficult due to the interference effect. At this stage, one of the significant advantages is that the electrochemical sensors are not affected by the color or turbidity of the sample media. Also, they provide high selectivity with low LOD values [85].

Various techniques can be selected for electrochemical sensors, and each of them has its own advantages. For example, amperometry-based sensors stand out with a noticeably short response time and good reproducibility. Cyclic voltammetry (CV) offers valuable information on the redox behavior of target compounds and biorecognition elements. Pulse techniques such as differential pulse voltammetry (DPV) and square wave voltammetry (SWV) come to the forefront with superior sensitivity, allowing detection up to pico- and femtomolar levels. Electrochemical impedance spectroscopy (EIS) techniques give valuable information on mechanistic and kinetic characteristics and are used in immunosensor applications [84].

When searching the Web of Science database with the keywords "electrochemical virus detection" to evaluate the distribution of studies on this subject in the literature in recent years, it is seen that there has been a significant increase in the last 2 years, and there are approximately 250 studies conducted each year.

Despite all these advantageous characteristics and the increasing number of studies on electrochemical sensors, some challenges and disadvantages should be overcome [86]. In some cases, stability can be a major issue, especially when using the sensor repeatedly and storing it for a long time. This can also cause problems in the case of accuracy. Particularly in biosensors, due to the irreversible interactions between the target analyte and biorecognition element, electrochemical biosensors cannot be used more than once. This increases both workload and cost [84].

One of the greatest issues with electrochemical biosensors is their extremely limited commercialization. Some problems prevent conversion to commercial devices used in routine analysis: As mentioned above, stability issues related to the biorecognition element layer constitute a major obstacle to commercialization. Another major aspect is the requirement of expensive materials and complicated fabrication procedures for commercial devices. Lastly, it is a critical drawback that, in some cases, the real samples cannot be analyzed without preprocessing [87]. As a solution to the issue of working with raw biological samples, the latest applications of microfluidics technologies can offer innovative ways. To overcome the issues associated with the biorecognition layer, some antifouling agents, such as zwitterionic polymers, biomimetic materials, polyethylene glycol, etc., can be utilized to improve accuracy. Finally, sensitivity and selectivity issues with commercial electrochemical devices can be enhanced by nanomaterials [84, 88].

14.4 Future Perspectives

Electrochemical sensors have a great advantage in terms of portability and miniaturization. By this means, it can be possible to adapt electrochemical sensors for point-of-care (POC) analysis. Nowadays, the pandemic process that we have experienced has clearly shown that POC applications have gained significant importance in the detection of pathogens such as viruses. Therefore, the greatest aim for future applications is to develop electrochemical sensors that can be easily used by inexperienced users, can give fast responses, allowing early detection of diseases, and can be produced at a low cost. Thus, patients can acquire results/diagnoses easily and rapidly without hospital administration. These goals are also critical steps toward commercialization. For this purpose, the researchers aim to overcome several issues soon: eliminating the complexity and difficulty of sample preparation and improving the sensitivity and selectivity of the results; enabling miniaturization and portability of the electrochemical sensor system; and enhancing the stability and reproducibility of the sensor.

In the most general sense, it is expected that there will be great advantages in terms of the determination of viruses and other pathogens, early diagnosis of diseases, and improvement of public health thanks to future studies on electrochemical biosensors.

14.5 Conclusion

The comfortable use and innovative nature-friendly approaches of biosensors developed based on electrochemical reactions are seen as a potential open target for development. The effects of the importance of diagnosis on disease prevention and treatment, especially in the pandemic period, were clearly observed. In diseases that develop due to pathogens, hospital stuff or professional teams are needed for the detection of infections, especially viral-related infections, but biosensors developed with POC targets can save time; thanks to their ability to be used at home or anywhere. In this study, we observed the adaptability of electrochemical reactions to today's problems thanks to biosensors, both in terms of direct targeting of leaves and pathogen detection. Electrochemical reactions, one of the ways of solving the need for biosensors, will develop in the future and solve the diverse needs of science.

References

1. Lee, I., Kim, S. E., Lee, J., Woo, D. H., Lee, S., Pyo, H., et al. (2020). A self-calibrating electrochemical aptasensing platform: Correcting external interference errors for the reliable and stable detection of avian influenza viruses. *Biosensors and Bioelectronics, 152*, 112010.
2. Khater, M., de la Escosura-Muñiz, A., Quesada-González, D., & Merkoçi, A. (2019). Electrochemical detection of plant virus using gold nanoparticle-modified electrodes. *Analytica Chimica Acta, 1046*, 123–131.
3. Hussein, H. A., Hassan, R. Y. A., El Nashar, R. M., Khalil, S. A., Salem, S. A., & El-Sherbiny, I. M. (2019). Designing and fabrication of new VIP biosensor for the rapid and selective detection of foot-and-mouth disease virus (FMDV). *Biosensors and Bioelectronics, 141*, 111467.
4. Siuzdak, K., Niedziałkowski, P., Sobaszek, M., Łęga, T., Sawczak, M., Czaczyk, E., et al. (2019). Biomolecular influenza virus detection based on the electrochemical impedance spectroscopy using the nanocrystalline boron-doped diamond electrodes with covalently bound antibodies. *Sensors and Actuators B: Chemical, 280*, 263–271.

5. Sayhi, M., Ouerghi, O., Belgacem, K., Arbi, M., Tepeli, Y., Ghram, A., et al. (2018). Electrochemical detection of influenza virus H9N2 based on both immunomagnetic extraction and gold catalysis using an immobilization-free screen-printed carbon microelectrode. *Biosensors and Bioelectronics, 107*, 170–177.
6. Yang, T., & Duncan, T. V. (2021). Challenges and potential solutions for nanosensors intended for use with foods. *Nature Nanotechnology, 16*, 251–265.
7. Kudesia, G., & Wreghitt, T. (2009). *Clinical and diagnostic virology*. Cambridge University Press.
8. Murray, P. R., Rosenthal, K. S., & Pfaller, M. A. (2021). *Medical microbiology* (9th ed., pp. 278–285). Elsevier.
9. Burrell, C. J., Howard, C. R., & Murphy, F. A. (2016). *Fenner and White's medical virology* (5th ed., pp. 1–583). Academic Press.
10. Hodinka, R. L. (2010). *Serologic tests in clinical virology* (4th ed., pp. 133–150). Lennette's Laboratory Diagnosis of Viral Infections.
11. Kennedy, M. (2005). Methodology in diagnostic virology. *Veterinary Clinics of North America: Exotic Animal Practice, 8*, 7–26.
12. Pretorius, M., & Venter, M. (2017). Diagnosis of viral infections. In *Viral infections in children* (Vol. 1, p. 151). Springer.
13. Al-Hajjar, S. (2012). Laboratory diagnosis of viral disease. In *Textbook of clinical pediatrics* (pp. 923–928). Springer.
14. Cassedy, A., Parle-McDermott, A., & O'Kennedy, R. (2021). Virus detection: A review of the current and emerging molecular and immunological methods. *Frontiers in Molecular Biosciences, 8*, 637559.
15. Dronina, J., Samukaite-Bubniene, U., & Ramanavicius, A. (2021). Advances and insights in the diagnosis of viral infections. *Journal of Nanobiotechnology, 19*, 348.
16. Souf, S. (2016). Recent advances in diagnostic testing for viral infections. *Bioscience Horizons: The International Journal of Student Research, 9*, hzw010.
17. Sanchez, F., & Sobolev, K. (2010). Nanotechnology in concrete – A review. *Construction and Building Materials, 24*, 2060–2071.
18. Ramakrishnan, S. G., Robert, B., Salim, A., Ananthan, P., Sivaramakrishnan, M., Subramaniam, S., et al. (2021). Nanotechnology based solutions to combat zoonotic viruses with special attention to SARS, MERS, and COVID 19: Detection, protection and medication. *Microbial Pathogenesis, 159*, 105133.
19. Zhang, L., & Webster, T. J. (2009). Nanotechnology and nanomaterials: Promises for improved tissue regeneration. *Nano Today, 4*, 66–80.
20. Sheikhzadeh, E., Beni, V., & Zourob, M. (2021). Nanomaterial application in bio/sensors for the detection of infectious diseases. *Talanta, 230*, 122026.
21. Liu, X., Li, Y., Xu, X., Li, P., Nie, Z., Huang, Y., et al. (2014). Nanomaterial-based tools for protein kinase bioanalysis. *TrAC – Trends in Analytical Chemistry, 58*, 40–53.
22. Chandra Ray, P., Afrin Khan, S., Kumar Singh, A., Senapati, D., & Fan, Z. (2012). Nanomaterials for targeted detection and photothermal killing of bacteria. *Chemical Society Reviews, 41*, 3193–3209.
23. Sai-Anand, G., Sivanesan, A., Benzigar, M. R., Singh, G., Gopalan, A. I., Baskar, A. V., et al. (2019). Recent progress on the sensing of pathogenic bacteria using advanced nanostructures. *Bulletin of the Chemical Society of Japan, 92*, 216–244.
24. Muniandy, S., Teh, S. J., Thong, K. L., Thiha, A., Dinshaw, I. J., Lai, C. W., et al. (2019). Carbon nanomaterial-based electrochemical biosensors for foodborne bacterial detection. *Critical Reviews in Analytical Chemistry, 49*, 510–533.
25. Chen, J., Andler, S. M., Goddard, J. M., Nugen, S. R., & Rotello, V. M. (2017). Integrating recognition elements with nanomaterials for bacteria sensing. *Chemical Society Reviews, 46*, 1272–1283.
26. Zhang, R., Belwal, T., Li, L., Lin, X., Xu, Y., & Luo, Z. (2020). Nanomaterial-based biosensors for sensing key foodborne pathogens: Advances from recent decades. *Comprehensive Reviews in Food Science and Food Safety, 19*, 1465–1487.

27. Fenzl, C., Genslein, C., Domonkos, C., Edwards, K. A., Hirsch, T., & Baeumner, A. J. (2016). Investigating non-specific binding to chemically engineered sensor surfaces using liposomes as models. *The Analyst, 141*, 5265–5273.
28. IslamyMazrad, Z. A., In, I., Lee, K. D., & Park, S. Y. (2017). Rapid fluorometric bacteria detection assay and photothermal effect by fluorescent polymer of coated surfaces and aqueous state. *Biosensors and Bioelectronics, 89*, 1026–1033.
29. Bezdekova, J., Zemankova, K., Hutarova, J., Kociova, S., Smerkova, K., Adam, V., et al. (2020). Magnetic molecularly imprinted polymers used for selective isolation and detection of Staphylococcus aureus. *Food Chemistry, 321*, 126673.
30. Petaccia, M., Bombelli, C., ParoniSterbini, F., Papi, M., Giansanti, L., Bugli, F., et al. (2017). Liposome-based sensor for the detection of bacteria. *Sensors and Actuators, B: Chemical, 248*, 247–256.
31. Edwards, K. A., & Baeumner, A. J. (2014). Enhancement of heterogeneous assays using fluorescent magnetic liposomes. *Analytical Chemistry, 86*, 6610–6616.
32. Vemula, P. K., Wiradharma, N., Ankrum, J. A., Miranda, O. R., John, G., & Karp, J. M. (2013). Prodrugs as self-assembled hydrogels: A new paradigm for biomaterials. *Current Opinion in Biotechnology, 24*, 1174–1182.
33. Yang, L., Deng, W., Cheng, C., Tan, Y., Xie, Q., & Yao, S. (2018). Fluorescent immunoassay for the detection of pathogenic bacteria at the single-cell level using carbon dots-encapsulated breakable organosilicananocapsule as labels. *ACS Applied Materials and Interfaces, 10*, 3441–3448.
34. Gu, W., Zhang, Q., Zhang, T., Li, Y., Xiang, J., Peng, R., et al. (2016). Hybrid polymeric nanocapsules loaded with gold nanoclusters and indocyanine green for dual-modal imaging and photothermal therapy. *Journal of Materials Chemistry B, 4*, 910–919.
35. Amstad, E. (2017). Capsules: Their past and opportunities for their future. *ACS Macro Letters, 6*, 841–847.
36. Norouzi, A., Ravan, H., Mohammadi, A., Hosseinzadeh, E., Norouzi, M., & Fozooni, T. (2018). Aptamer–integrated DNA nanoassembly: A simple and sensitive DNA framework to detect cancer cells. *Analytica Chimica Acta, 1017*, 26–33.
37. Setyawati, M. I., Kutty, R. V., Tay, C. Y., Yuan, X., Xie, J., & Leong, D. T. (2014). Novel theranostic DNA nanoscaffolds for the simultaneous detection and killing of escherichia coli and staphylococcus aureus. *ACS Applied Materials and Interfaces, 6*, 21822–21831.
38. Zhang, H., Ma, X., Liu, Y., Duan, N., Wu, S., Wang, Z., et al. (2015). Gold nanoparticles enhanced SERS aptasensor for the simultaneous detection of Salmonella typhimurium and Staphylococcus aureus. *Biosensors and Bioelectronics, 74*, 872–877.
39. Yin, B., Wang, Y., Dong, M., Wu, J., Ran, B., Xie, M., et al. (2016). One-step multiplexed detection of foodborne pathogens: Combining a quantum dot-mediated reverse assaying strategy and magnetic separation. *Biosensors and Bioelectronics, 86*, 996–1002.
40. Ge, X., Asiri, A. M., Du, D., Wen, W., Wang, S., & Lin, Y. (2014). Nanomaterial-enhanced paper-based biosensors. *TrAC – Trends in Analytical Chemistry, 58*, 31–39.
41. Munawar, A., Ong, Y., Schirhagl, R., Tahir, M. A., Khan, W. S., & Bajwa, S. Z. (2019). Nanosensors for diagnosis with optical, electric and mechanical transducers. *RSC Advances, 9*, 6793–6803.
42. Xu, L., Shoaie, N., Jahanpeyma, F., Zhao, J., Azimzadeh, M., & Al–Jamal, K. T. (2020). Optical, electrochemical and electrical (nano)biosensors for detection of exosomes: A comprehensive overview. *Biosensors and Bioelectronics, 161*, 112222.
43. Kaya, H. O., Cetin, A. E., Azimzadeh, M., & Topkaya, S. N. (2021). Pathogen detection with electrochemical biosensors: Advantages, challenges and future perspectives. *Journal of Electroanalytical Chemistry, 882*, 114989.
44. Lin, J., Wang, R., Jiao, P., Li, Y., Li, Y., Liao, M., et al. (2015). An impedance immunosensor based on low-cost microelectrodes and specific monoclonal antibodies for rapid detection of avian influenza virus H5N1 in chicken swabs. *Biosensors and Bioelectronics, 67*, 546–552.
45. Zhang, Q., Li, L., Qiao, Z., Lei, C., Fu, Y., Xie, Q., et al. (2017). Electrochemical conversion of Fe3O4 magnetic nanoparticles to electroactive prussian blue analogues for self-sacrificial label biosensing of avian influenza virus H5N1. *Analytical Chemistry, 89*, 12145–12151.

46. Lee, T., Park, S. Y., Jang, H., Kim, G. H., Lee, Y., Park, C., et al. (2019). Fabrication of electrochemical biosensor consisted of multi-functional DNA structure/porous au nanoparticle for avian influenza virus (H5N1) in chicken serum. *Materials Science and Engineering C, 99*, 511–519.
47. Haji-Hashemi, H., Norouzi, P., Safarnejad, M. R., & Ganjali, M. R. (2017). Label-free electrochemical immunosensor for direct detection of Citrus tristeza virus using modified gold electrode. *Sensors and Actuators, B: Chemical, 244*, 211–216.
48. Singhal, C., Khanuja, M., Chaudhary, N., Pundir, C. S., & Narang, J. (2018). Detection of chikungunya virus DNA using two-dimensional MoS2 nanosheets based disposable biosensor. *Scientific Reports, 8*, 1–11.
49. Nawaz, M. H., Hayat, A., Catanante, G., Latif, U., & Marty, J. L. (2018). Development of a portable and disposable NS1 based electrochemical immunosensor for early diagnosis of dengue virus. *Analytica Chimica Acta, 1026*, 1–7.
50. Manzano, M., Viezzi, S., Mazerat, S., Marks, R. S., & Vidic, J. (2018). Rapid and label-free electrochemical DNA biosensor for detecting hepatitis A virus. *Biosensors and Bioelectronics, 100*, 89–95.
51. Srisomwat, C., Yakoh, A., Chuaypen, N., Tangkijvanich, P., Vilaivan, T., & Chailapakul, O. (2021). Amplification-free DNA sensor for the one-step detection of the hepatitis B virus using an automated paper-based lateral flow electrochemical device. *Analytical Chemistry, 93*, 2879–2887.
52. Alizadeh, N., Hallaj, R., & Salimi, A. (2017). A highly sensitive electrochemical immunosensor for hepatitis B virus surface antigen detection based on Hemin/G-quadruplex horseradish peroxidase-mimicking DNAzyme-signal amplification. *Biosensors and Bioelectronics, 94*, 184–192.
53. Zribi, B., Roy, E., Pallandre, A., Chebil, S., Koubaa, M., Mejri, N., et al. (2016). A microfluidic electrochemical biosensor based on multiwall carbon nanotube/ferrocene for genomic DNA detection of Mycobacterium tuberculosis in clinical isolates. *Biomicrofluidics, 10*, 10.
54. Ghanbari, K., & Roushani, M. (2018). A nanohybrid probe based on double recognition of an aptamer MIP grafted onto a MWCNTs-Chit nanocomposite for sensing hepatitis C virus core antigen. *Sensors and Actuators, B: Chemical, 258*, 1066–1071.
55. Devarakonda, S., Singh, R., Bhardwaj, J., & Jang, J. (2017). Cost-effective and handmade paper-based immunosensing device for electrochemical detection of influenza virus. *Sensors (Switzerland), 17*, 2597.
56. Lum, J., Wang, R., Hargis, B., Tung, S., Bottje, W., Lu, H., et al. (2015). An impedance aptasensor with microfluidic chips for specific detection of H5N1 avian influenza virus. *Sensors (Switzerland), 15*, 18565–18578.
57. Bhardwaj, J., Chaudhary, N., Kim, H., & Jang, J. (2019). Subtyping of influenza A H1N1 virus using a label-free electrochemical biosensor based on the DNA aptamer targeting the stem region of HA protein. *Analytica Chimica Acta, 1064*, 94–103.
58. Hou, Y. H., Wang, J. J., Jiang, Y. Z., Lv, C., Xia, L., Hong, S. L., et al. (2018). A colorimetric and electrochemical immunosensor for point-of-care detection of enterovirus 71. *Biosensors and Bioelectronics, 99*, 186–192. https://doi.org/10.1016/j.bios.2017.07.035
59. Cheng, D., Zhang, Y., Wen, D., Guo, Z., Yang, H., Liu, Y., et al. (2019). Hairpin probes based click polymerization for label-free electrochemical detection of human T-lymphotropic virus types II. *Analytica Chimica Acta, 1059*, 86–93.
60. Bai, C., Lu, Z., Jiang, H., Yang, Z., Liu, X., Ding, H., et al. (2018). Aptamer selection and application in multivalent binding-based electrical impedance detection of inactivated H1N1 virus. *Biosensors and Bioelectronics, 110*, 162–167.
61. Joshi, S. R., Sharma, A., Kim, G. H., & Jang, J. (2020). Low cost synthesis of reduced graphene oxide using biopolymer for influenza virus sensor. *Materials Science and Engineering C, 108*, 110465.
62. Bhardwaj, J., Sharma, A., & Jang, J. (2019). Vertical flow-based paper immunosensor for rapid electrochemical and colorimetric detection of influenza virus using a different pore size sample pad. *Biosensors and Bioelectronics, 126*, 36–43.

63. Ravina, M. H., Gill, P. S., & Kumar, A. (2019). Hemagglutinin gene based biosensor for early detection of swine flu (H1N1) infection in human. *International Journal of Biological Macromolecules, 130*, 720–726.
64. Nidzworski, D., Siuzdak, K., Niedziałkowski, P., Bogdanowicz, R., Sobaszek, M., Ryl, J., et al. (2017). A rapid-response ultrasensitive biosensor for influenza virus detection using antibody modified boron-doped diamond. *Scientific Reports, 7*, 1–10.
65. Anik, Ü., Tepeli, Y., Sayhi, M., Nsiri, J., & Diouani, M. F. (2018). Towards the electrochemical diagnostic of influenza virus: Development of a graphene-Au hybrid nanocomposite modified influenza virus biosensor based on neuraminidase activity. *The Analyst, 143*, 150–156.
66. Layqah, L. A., & Eissa, S. (2019). An electrochemical immunosensor for the corona virus associated with the Middle East respiratory syndrome using an array of gold nanoparticle-modified carbon electrodes. *Microchimica Acta, 186*, 2–11.
67. Mahari, S., Roberts, A., Shahdeo, D., & Gandhi, S. (2020). eCovSens-ultrasensitive novel in-house built printed circuit board based electrochemical device for rapid detection of nCovid-19 antigen, a spike protein domain 1 of SARS-CoV-2. *BioRxiv*, 2020.04.24.059204.
68. Vadlamani, B. S., Uppal, T., Verma, S. C., & Misra, M. (2020). Functionalized tio2 nanotube-based electrochemical biosensor for rapid detection of SARS-CoV-2. *Sensors (Switzerland), 20*, 1–10.
69. Białobrzeska, W., Firganek, D., Czerkies, M., Lipniacki, T., Skwarecka, M., Dziąbowska, K., et al. (2020). Electrochemical immunosensors based on screen-printed gold and glassy carbon electrodes: Comparison of performance for respiratory syncytial virus detection. *Biosensors (Basel), 10*, 1–13.
70. Faria, H. A. M., & Zucolotto, V. (2019). Label-free electrochemical DNA biosensor for Zika virus identification. *Biosensors and Bioelectronics, 131*, 149–155.
71. Tancharoen, C., Sukjee, W., Thepparit, C., Jaimipuk, T., Auewarakul, P., Thitithanyanont, A., et al. (2019). Electrochemical biosensor based on surface imprinting for Zika virus detection in serum. *ACS Sensors, 4*, 69–75.
72. Hwang, H., Hwang, B. Y., & Bueno, J. (2018). Biomarkers in infectious diseases. *Disease Markers, 2018*, 2–4.
73. Chen, H., Hagström, A. E. V., Kim, J., Garvey, G., Paterson, A., Ruiz-Ruiz, F., et al. (2016). Flotation immunoassay: Masking the signal from free reporters in sandwich immunoassays. *Scientific Reports, 6*, 1–8.
74. Chen, H., Liu, K., Li, Z., & Wang, P. (2019). Point of care testing for infectious diseases. *Clinica Chimica Acta, 493*, 138–147.
75. Tokel, O., Inci, F., & Demirci, U. (2014). Advances in plasmonic technologies for point of care applications. *Chemical Reviews, 114*, 5728–5752.
76. Wang, P., & Kricka, L. J. (2018). Current and emerging trends in point-of-care technology and strategies for clinical validation and implementation. *Clinical Chemistry, 64*, 1439–1452.
77. Cesewski, E., & Johnson, B. N. (2020). Electrochemical biosensors for pathogen detection. *Biosensors and Bioelectronics, 159*, 112214.
78. de Eguilaz, M. R., Cumba, L. R., & Forster, R. J. (2020). Electrochemical detection of viruses and antibodies: A mini review. *Electrochemistry Communications, 116*, 106762.
79. Zhao, Z., Huang, C., Huang, Z., Lin, F., He, Q., Tao, D., et al. (2021). Advancements in electrochemical biosensing for respiratory virus detection: A review. *TrAC – Trends in Analytical Chemistry, 139*, 116253.
80. Ozer, T., & Henry, C. S. (2021). Paper-based analytical devices for virus detection: Recent strategies for current and future pandemics. *TrAC – Trends in Analytical Chemistry, 144*, 116424.
81. Nelson, P. P., Rath, B. A., Fragkou, P. C., & Antalis, E. (2020). Current and future point-of-care tests for emerging and new respiratory viruses and future perspectives. *Frontiers in Cellular and Infection Microbiology, 10*, 181.
82. Ribeiro, B. V., Cordeiro, T. A. R., Oliveira e Freitas, G. R., Ferreira, L. F., & Franco, D. L. (2020). Biosensors for the detection of respiratory viruses: A review. *Talanta Open, 2*, 100007.

83. Tepeli, Y., & Ülkü, A. (2018). Electrochemical biosensors for influenza virus a detection: The potential of adaptation of these devices to POC systems. *Sensors and Actuators, B: Chemical, 254*, 377–384.
84. Goud, K. Y., Reddy, K. K., Khorshed, A., Kumar, V. S., Mishra, R. K., Oraby, M., et al. (2021). Electrochemical diagnostics of infectious viral diseases: Trends and challenges. *Biosensors and Bioelectronics, 180*, 113112.
85. Sher, M., Faheem, A., Asghar, W., & Cinti, S. (2021). Nano-engineered screen-printed electrodes: A dynamic tool for detection of viruses. *TrAC – Trends in Analytical Chemistry, 143*, 116374.
86. Sengupta, J., Adhikari, A., & Hussain, C. M. (2021). Graphene-based analytical lab-on-chip devices for detection of viruses: A review. *Carbon Trends, 4*, 100072.
87. Brazaca, L. C., dos Santos, P. L., de Oliveira, P. R., Rocha, D. P., Stefano, J. S., Kalinke, C., et al. (2021). Biosensing strategies for the electrochemical detection of viruses and viral diseases – A review. *Analytica Chimica Acta, 1159*, 338384.
88. Pilevar, M., Kim, K. T., & Lee, W. H. (2021). Recent advances in biosensors for detecting viruses in water and wastewater. *Journal of Hazardous Materials, 410*, 124656.

Chapter 15
Next-Generation Sequencing and Solid-State Nanopores

Snehal Mahajan-Mallakmir, Sanket Mahajan, Anushree S. Lokur, and Ramesh S. Chaughule

15.1 Introduction

Genome sequencing technology has rapidly evolved over the last few years and has found wide applications in various life sciences. The applications to human biology and its implications in human disease are vastly expanding especially after the success of the human genome project [1]. Complexities of the human genome, its interaction within itself and with the factors in the environment are fundamental to the understanding of human diseases. Equal is the importance of over-viewing the technical information about what happens at the laboratory end of the analysis.

Next-generation sequencing (NGS) refers to high throughput, in-parallel DNA sequencing technologies developed later than Sanger DNA sequencing. They differ from the Sanger method as billions of DNA nucleotides can be sequenced in parallel and offer cost and time-effective solutions to detect multiple variants across the human genome in a comparatively short period and less cost.

The correlation of knowledge between the research, laboratory end, and practical applications gives a bird's eye view of this subject which is necessary in view of this rapidly changing field and widely increasing applications in various specialties.

This is needed to minimize the errors of interpretation of the results coming from the technology end as well as to gain maximum out of such information so that the collective efforts of its application to health in mankind achieves its true potential.

S. Mahajan-Mallakmir (✉)
Clinical Geneticist, Apollo Genomics Institutes, Apollo Hospitals, Navi Mumbai, India

Clinic of Genetic Medicine, Nerul, Navi Mumbai, India

S. Mahajan
NanoAge Consultancy, Nashik, Maharashtra, India

A. S. Lokur · R. S. Chaughule
Ramnarain Ruia Autonomous College, Mumbai, India

R. S. Chaughule, A. S. Lokur (eds.), *Applications of Nanotechnology in Microbiology*, https://doi.org/10.1007/978-3-031-49933-3_15

The information from the DNA molecule is transcribed to mRNA which is further translated to proteins that are important in various cellular functions. The study of the structure and function of genes as well as the functional assembly of many genes within themselves and at the gene protein level needs accuracy in view of the vast assembly of the human genome because many structural and functional details are still unknown. DNA sequencing technology has evolved rapidly over the last few decades [2] and after the initial generations now it is exploring nanopore technology. Nanotechnology-based nanopore sequencing provides a high throughput technique with detailing due to ultra-long reads and at a level like never before and is hence enhancing the accuracy and reducing the cost [3].

15.2 Next-Generation Sequencing (NGS)

15.2.1 *Overview of NGS*

Information contained in the DNA molecule is translated to form proteins via RNA molecules. Understanding the sequence of nucleotides in the DNA molecule has helped to decode the gene structure and hence enhanced our understanding of functions. This is important in medicine where sometimes changes in structural proteins have helped us to know more about gene mutations and vis a versa. Over the last few years, sequencing technologies have evolved through various methods and next-generation sequencing has increased speed and accuracy. The interpretation of the information generated has a huge impact on biological research and implications for personalized medicine. DNA sequencing is significant for the advancement of many fields such as forensic sciences, biology, genetics, molecular biology, and archeology.

15.2.2 *Brief History of Sequencing*

After the discovery of the 3D structure of DNA by Watson and Crick in 1953 and later by Zallen in 2003, many strategies were researched to enable the sequencing of DNA. Initially, it was done using RNA and later DNA in bacteriophages. The breakthrough in DNA sequencing was achieved in 1977, with the development of Sanger 'chain-termination' or dideoxy technique [4]. The chain-termination technique uses chemical analogs of the deoxyribonucleotides (dNTPs) that are the monomers of DNA strands. These sequencing methods continued to have improvements like fluorometry-based detection, capillary-based detection, shotgun sequencing, polymerase chain reaction, and recombinant technology allowing the sequencing of hundreds of samples at the same time. Further technological advances led to the rise of DNA sequencing and the first automated DNA sequencer (ABI

PRISMAB370A) was launched in 1986, which allowed the draft of the human genome project. The human genome project was launched in 1990 and took 13 years to complete cost of $300 million worldwide, in comparison to the estimated cost of human genome sequencing in 2021 as less than $800. Apart from the advantage regarding costs, the completion of the human genome project paved the way to fasten the research into the human genome and its implications for human diseases [1]. Year-wise developments in sequencing DNA are shown in Fig. 15.1.

Single-molecule sequencing is argued by many as a mark of third-generation sequencing. As a matter of fact, the ability to sequence the human genome has outpaced our ability to interpret genetic variations. It has gained superabundant attention as the arrangement of nucleic acids in polynucleotide chains encompasses the information for the patrimonial and biochemical traits of living species. Sequencing technology has experienced three generations of evolution which will be overviewed in this review. Nucleic acid sequencing is categorized into three generations. The third generation continues to surpass technological boundaries with capabilities in sequencing single molecules without prior amplification that was previously inconceivable. The second and the third generation are often referred to as "next-generation sequencing" (NGS). Second-generation sequencing is based on massive parallel and clonal amplification of molecules and the third-generation sequencing

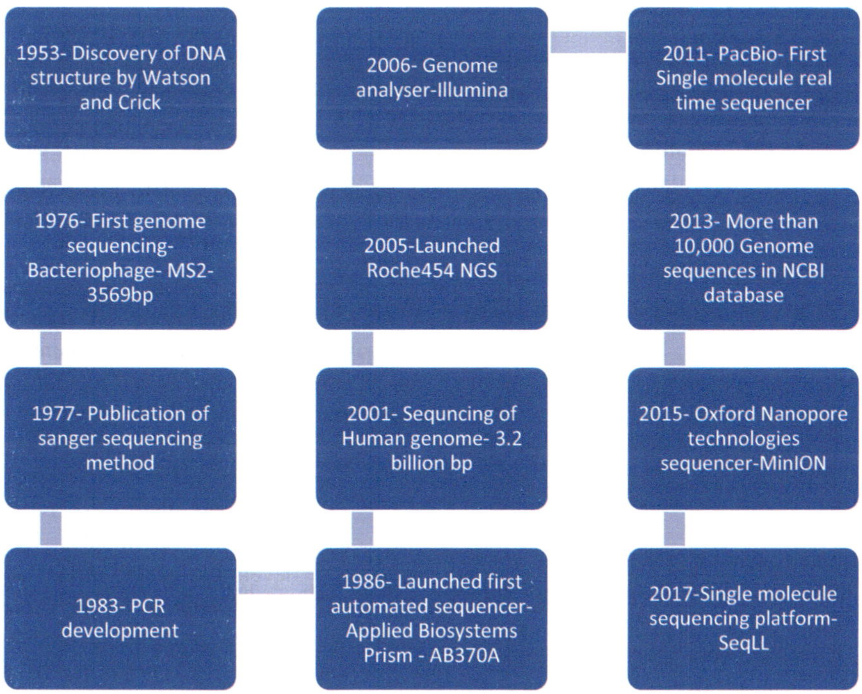

Fig. 15.1 Timeline of developments in sequencing DNA

relies on single-molecule sequencing without prior clonal amplification [2]. New technologies like nanopore sequencing, which is being addressed by many as the fourth generation in sequencing technology, have advantages over previous methods in having high throughput, improved read lengths, and decreased cost [5].

15.2.3 NGS Data Analysis

15.2.3.1 Workflow

NGS workflow has elements such as sample preparation, library preparation, sequencing, and then data analysis with bioinformatics [6]. Sample preparation includes nucleic acid extraction, fragmentation, quantification, and characterization of DNA. Robust quality control methods are essential to be laid out to prevent waste of time and resources and errors. Library preparation consists of fragmenting and/or sizing the target sequences to a desired length, converting it to double-stranded DNA, attaching oligonucleotide adapters to the ends of the fragments, and then finally quantitation of the product for sequencing. Fragments are sequenced and mapped back to a known reference sequence. A good library preparation means that all fragments of interest should be equally represented in the library and should not contain random errors, but it is challenging as some genomic regions may not be prone to sequencing in detail [7].

Four major technologies [8] for this purpose are:

1. Complementary metal-oxide semiconductor is used by Ion Torrent Personal Genome Machine
2. Single-molecular real-time (SMRT) sequencing introduced by Pacific Biosciences (PacBio)
3. Incorporation of a fluorescently labeled reversible terminator (FLRT) in the synthesis process is used by Illumina Genome Analyzer (IGA)
4. The combination of emPCR and pyro sequencing is used by Roche/454.

For whole-genome sequencing, the longer fragments are preferable, while for RNA-sequencing and exome smaller fragments are feasible because most of the human exons are under 200 base pairs in length. Table 15.1.

Table 15.1 Comparison of various novel sequencing technologies

Technology	Method	Read length	Error rate (%)
Illumina	Sequencing by synthesis	Short (100–300 bp)	0.1
PacBio	Single-molecule real-time	Long (10–100 kb)	5–15
Oxford Nanopore	Nanopore	Variable (up to 1000 kb)	5–20

Adapted from the source Bansal and Boucher [8]

15.2.3.2 Data Analysis and Interpretation

As applications of genomics in human biology are increasing and refinement of techniques happening at a rapid pace, it is leading to more complexity. There has been a paradigm shift in various test platforms from single gene analysis, gene panels, exome to genome sequencing, high-definition arrays, transcriptome, and epigenetic analysis. Hence, there is an immediate need for accurate data analysis, interpretation tools, and curation methods which are very necessary to interpret individual genome variants and their significance in a particular disease. Many mutations may be identified from samples, but differentiating clinically significant variations is a challenging task and it needs an appropriate use of validation methods [9].

The primary data analysis consists of the detection and analysis of raw data (signal analysis) targeting the generation of legible sequencing reads (base calling) and scoring base quality. Outputs from the primary analysis are FASTQ file (Illumina) or unmapped binary alignment map (uBAM) file (Ion Torrent). The base calling is the final step of primary analysis and is performed by a base-caller module. In Illumina principle of signal detection relies on fluorescence and base calling is apparently much simpler. It is made from fluorescent signal intensity measurements resulting from the incorporated nucleotides during each cycle.

The quality of the raw sequence is critical for the success of NGS analysis. Bioinformatic tools developed to evaluate the quality of raw data are the NGS QC toolkit, QC-Chain, and Fast QC. Fast QC is one of the most popular, as it gives a report containing well-structured and graphically illustrated information about different aspects of the read quality.

On the secondary analysis, the reads are aligned, or de novo assembled, and the calling is performed. Read alignment is the alignment of sequenced fragments against a reference human genome (typically hg19 or hg 38). De novo assembly is assembling a genome from scratch without the use of external data.

The variant call step has the main objective of identifying variants using the post-processed BAM file. Several tools are available for variant calling and machine learning algorithms have evolved greatly in recent years and will be critical to assist scientists and clinicians in handling large amounts of data.

Algorithms for the detection of structural variation from NGS data depend on one or more of the criteria: discordantly paired-end reads, split reads, or depth of coverage (Table 15.2).

The tertiary analysis includes variant annotation, variant filtering, prioritization, data visualization, and reporting. The variant annotation is a key initial step for the analysis of sequencing variants. Given the massive amount of NGS data, data annotation is performed automatically. Several tools are currently available, and each uses different methodologies and databases for variant annotation.

After annotation of a VCF file from exome sequencing, for example, the total number of variants may range between 30,000 and 50,000 and to know clinically significant variants, filtering strategies are required. Population databases such as

Table 15.2 Common terminologies used in sequencing experiments

Common terminologies used	Explanation
Read length	A single piece of output or number of base pairs(bp) sequenced from a DNA length
Single read sequencing	Sequencing DNA from only one end
Paired-end sequencing	Sequencing DNA from both ends of the fragment generates high-quality, alignable sequence data.
Coverage	Average number of reads that align to, or cover known reference bases
Targeted sequencing	A selected set of genes or specific genomic elements; for example, CpG islands and promoter/enhancer regions, usually applicable in exome sequencing
Amplicon sequencing	Allows the pooling of fragments from multiple samples and involves making NGS libraries from PCR products. It is applicable in microbiomes, surveying 16S rRNA sequences in complex bacterial mixtures, analysis of antibody diversity, T cell receptor gene repertoires, etc.
Variant detection	To find genetic differences between the studied sample and the reference used. The differences may be from single nucleotide variants (SNVs) to large genomic deletions, insertions, or rearrangements
Genome assembly	Genome assembly is the reconstruction of genomes from the smaller DNA segments called *reads* which are generated by a sequencing experiment

1000 Genome Project, Exome Aggregation Consortium (ExAC), and the Genome Aggregation Database (gnomAD; are some of the most widely used databases [10].

In 2015, guidelines were released by the American College of Medical Genetics and Genomics (ACMG) and the Association for Molecular Pathology (AMP) for the interpretation of germline genetic variants. Clinical genetic testing laboratories, geneticists, and researchers worldwide, use these guidelines. It is recommended that specific standard terminology should be used that is: 'pathogenic', 'likely pathogenic', 'uncertain significance', 'likely benign', and 'benign' to describe the variants in a particular Mendelian disorder being studied. The process for classification of variants into specific categories is based on various criteria using evidence like population data, computational data, functional data, and segregation data [11].

15.2.3.3 Single Molecule Real Time (SMRT) Sequencing

PacBio introduced the concept of single-molecule real-time (SMRT) sequencing and this technology enables the sequence of long reads (with average read lengths up to 30 kb). DNA polymerases are attached to the zero-mode waveguide (ZMW) wells, which are nanoholes where a single molecule of the DNA polymerase enzyme can be directly placed. A single DNA molecule is used as a template for fluorescently labeled nucleotides. Each base has a different fluorescent dye, thereby emitting a signal out of the ZMW. A second approach to single-molecule sequencing was given by Oxford Nanopore Technology named MinION which later became commercially available [9, 12].

15.3 Nanopores in Next-Generation Sequencing

Nanopore sequencing is a modern and promising technique because of the label-free sequencing and long reads help to ease sequencing requirements. Longer reads are very important to sequence repetitive elements and complex sequences, such as transposable elements, segmental duplications, and telomeric/centromeric regions that are difficult to address with short reads [12, 13].

15.3.1 Nanopore and Types of Nanopore

A nanopore is a tiny hole in an electro-resistant membrane. The size of the hole is around one nanometer equivalent to the size of DNA or protein molecules. Each nanopore works like an electrode and is connected to a channel and sensor chip. The working principle is when a voltage is applied across the membrane; the generated electric current flowing through the nanopore is measured. When the DNA or RNA molecule passes through the nanopore, changes in ion current take place leading to the determination of sequence and modification of bases [13]. Nucleotides from DNA or RNA are identified based on shifts in current as the strand enters the pore. The interruption of the open current level is termed a translocation event signal. DNA's small up or down movements within pore's constriction have a significant change in ion current. DNA goes through the pore at a speed of one million bases per second which is too fast to get a read of individual bases. Helicases and polymerases are the biological enzymes bigger in size that hold the DNA and allow it go through one base at a time.

Once the DNA enters the nanopore in a membrane, each nucleotide has a characteristic change, uncovering the sequence of the biomolecule. The membrane is immersed between two fluidic reservoirs having electrolyte solution and connected with the membrane. The charged biomolecules are driven through the pore and produce current blockages which finally the biomolecular information is obtained. The sequence accuracy is also improved because DNA is analyzed since it comes directly from the cell (Fig. 15.2).

Basically, there are two major classes of nanopores. The first consists of biological protein pores embedded in the lipid bilayer. These protein pores are derived from cell membrane proteins. Alfa-hemolysin (αHL), pores derived from *Staphylococcus aureus* and *Mycobacterium smegmatis porin* A (MspA) are used to perform DNA translocation.

The second class is solid-state pores in synthetic membranes such as Si_2N_4, Al_2O_3, TiO_2, and graphene, having tiny holes. Being robust and flexible, these solid-state nanopores can be easily transformed into solid-state electronic devices [14, 15]. While research on synthetic polymers in the field of information storage has been concluded, achieving a sufficient level of motion control is still necessary for the accurate interpretation of translocation signals of complex biopolymers [16, 17].

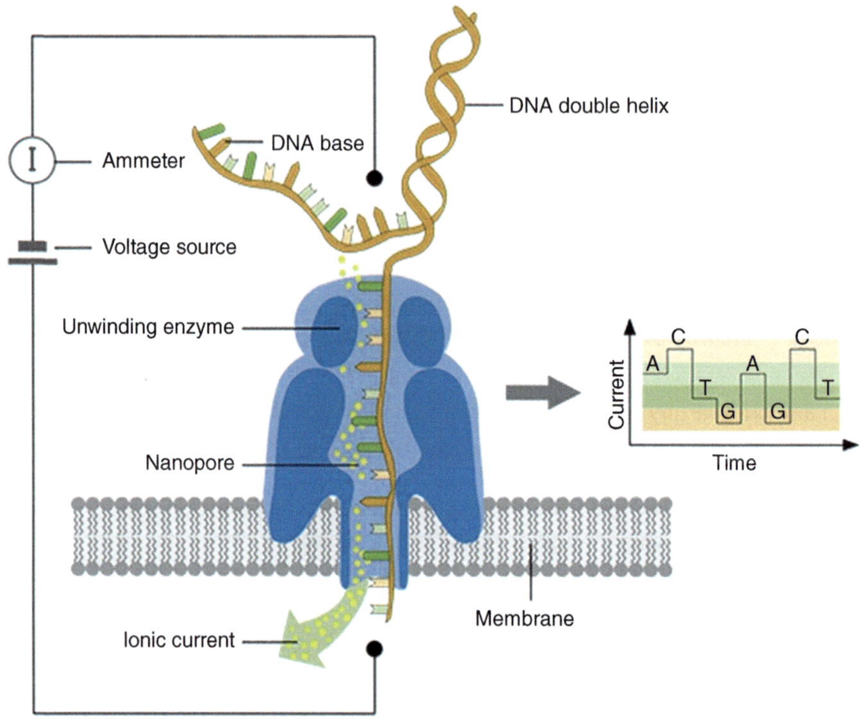

Fig. 15.2 Ref. Decoding DNA with a pocket-sized sequencer. (https://www.scienceinschool.org/article/2018/decoding-dna-pocket-sized-sequencer/)

The third class is a hybrid of biological and synthetic nanopores. A planar pore is one subclass of solid-state nanopores which is most used. Thin glass capillaries used in nanoscale opening in solid-state nanopores reduce special sensitivity and thus have an additional advantage.

15.3.2 Next-Generation Sequencing Methods by Nanopore Technology

15.3.2.1 Oxford Nanopore

The development of nanopore sequencing started in 1995. Oxford Nanopore Technologies (ONT), UK released the first data of their technology in 2012 [8]. As mentioned earlier, when a moving nucleotide passes through the nanopore immersed in conducting fluid, the amount of change in current is a characteristic value of each nucleotide that is read. ONT has sequenced the entire 5.4 kilobase genome of the virus in one continuous read. Initially, they read DNA to the tune of 100 kilobases per second [18]. The biggest advantage of the ONT is that it delivers real-time sequencing of DNA molecules at a very low cost with a fast sequencing process. A device based on an array of biological nanopores is developed by ONT [19]. This device decodes long sequences with an acceptable error rate.

15.3.2.2 2D Materials

Solid-state nanopores are also made using two-dimensional materials made of graphene and molybdenum sulfide MoS_2 by chemical vapor deposition method reaching the material thickness to 0.2 nm which is useful with single-stranded DNA [20–22]. MoS_2 is prepared in atom-thick layers consisting of nanopores. These MoS_2 nanopores have a better signal-to-noise ratio due to high current change caused by DNA translocation compared to graphene [23, 24]. For still better signal-to-noise ratio in 2D materials, quartz substrates are used [25].

15.3.3 *Molecular Dynamic Simulations of DNA Translocation Through Nanopores*

For fast, yet accurate DNA sequencing it is essential to understand the effect of nanopore material properties as well as nanopore geometry on the sequencing output. To ascertain the effect of these characteristics, it is preferred to model and simulate these nanopore sequencing systems to help better prepare the actual physical apparatus. A molecular Dynamics (MD) simulation is a fundamental atomistic simulation technique to model such atomic and molecular level systems. A simulated single-strand or double-strand DNA is made to pass through a simulated nanopore material and certain boundary conditions are applied such as temperature, salt concentration, and external bias voltage. The transport of the DNA through the nanopore causes a change in the ionic current and this blockade current is measured to quantify the status of the DNA translocation.

A variety of biological as well as solid-state nanopore systems have been tried for use in DNA sequencing [26–29]. Biological nanopores, such as α-hemolysin, though promising [26] have shown limited stability to temperature, pH, and applied bias salt concentration. Solid-state nanopores, such as those made from membranes of silicon nitride, aluminum oxide, and silicon oxide have been assessed to be used in DNA sequencing [27]. Though robust and easily fabricated, they are an order of magnitude or even greater in thickness, compared to the DNA molecules. This limits the ability to detect base-specific changes in ionic current, as multiple DNA base pairs interact with the nanopore at the same time, simply due to the bigger thickness of the nanopore.

Considering these issues, it has been proposed that graphene nanopores may prove to be useful in DNA sequencing as graphene is the thinnest known material with a thickness equal to one atomic layer. The DNA translocation through a nanopore in a graphene sheet would result in distinguishable interactions between the nanopore and the base pairs. Additionally, due to its crystalline geometry, it is relatively easier to model and observe graphene nanopores using MD simulations.

15.3.3.1 MD Simulations of DNA with Graphene Nanopores

A typical simulation system consists of a structure with dimensions in tens of angstroms (Fig. 15.3). The system consists of poly (A-T) 45 translocated through a graphene nanopore in a KCl solution [28].

Liang et al. [28] observed that the DNA moves faster when it passes through the nanopore than when it is outside the pore, as a stronger interaction exists between graphene and DNA when DNA is outside of the nanopore. However, the concentration of KCl has essentially no effect on the DNA translocation time through the nanopore. The translocation time for a DNA molecule to pass through a nanopore can be changed to a large extent and can be increased significantly with the decrease of the bias voltage. However, the rate of reduction in the pore current during the blockade is almost the same at different bias voltages, which ranged from 2.2 V to 22.0 V. This implies that the bias voltage affects the absolute value of the blockade current dramatically but has essentially no effect on the rate of reduction in the pore current during the blockade.

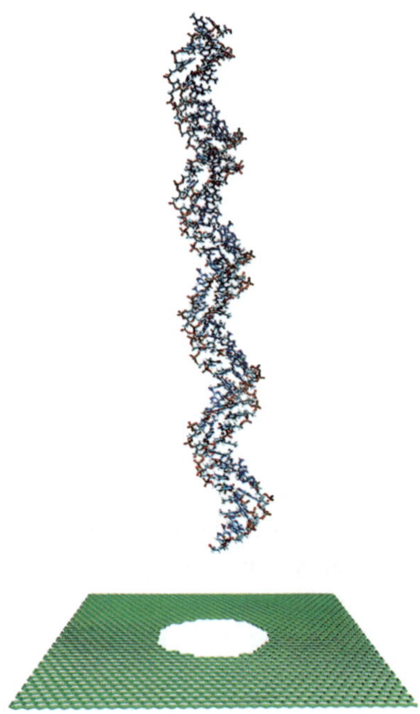

Fig. 15.3 Initial Structure of a Graphene Nanopore and Poly(A-T)45 [28]; ions and water molecules are not shown for clarity. The system dimensions are 90.0 Å x 90.0 Å x 220.0 Å

It was suggested by Sathe et al. [27] that DNA molecules can be more readily captured by a larger pore than by a smaller pore beyond the effect expected by pore area only and pore charge can be used to control the kinetics of DNA translocation through the graphene nanopore. They further concluded that the force experienced by nucleotides in the nanopore can be tailored by varying the applied electric bias voltage to discriminate between different DNA molecule types. However, there are some hurdles in the use of graphene nanopores in DNA sequencing, such as avoidance of DNA adherence to the graphene sheet to keep the DNA stretched in the nanopore. But these can be overcome with the use of optical tweezers to keep the DNA stretched.

Zhang et al. [29] characterized the force profile acting on the ssDNA molecule translocating through graphene nanopores and studied the difference in DNA sequencing results with different nanopore geometries. They noted that the four basic nucleotides (A, T, C, G) and 5-methyl-cytosine can be identified by the characteristic peaks on the force profile only with circular graphene nanopores within a certain range of diameters. The effective area of a nanopore for the ssDNA to pass through is also an important factor. It was also observed that an axisymmetric (circular) nanopore is better suited to DNA sequencing via force profiles than an asymmetric (rhombic) nanopore and that the graphene nanopore surface should be modified to become as axisymmetric as possible in order to utilize atomic force microscopy or optical tweezing for DNA sequencing (Fig. 15.4).

In general, the study of DNA translocation through graphene nanopores using molecular dynamics (MD) simulations provides valuable insights into its potential usage for DNA sequencing. It has been demonstrated through MD simulations [26–29] that it is possible to measure ionic blockade current during DNA translocation through the graphene nanopore with certain accuracy and speed. In addition, due to the specific properties of a graphene sheet, such as atomic-scale thickness, the interactions between the DNA and the nanopore can be characterized better as compared to other solid-state nanopores. Further, it has been observed that by varying the physical conditions of the sequencing set-up such as nanopore size, nanopore geometry, and applied bias voltage, requisite system parameters can be obtained to obtain better and more accurate sequencing output at increased speed.

15.4 Applications of NGS

Since its initial developments, NGS technology has come very far from its initial anticipated use to very wide applications in various fields from personal medicine to environment [30]. Recent advances like nanopore technology in NGS are projected to take them further in precision and expand their horizons.

We present a brief overview of the growing applications of NGS here.

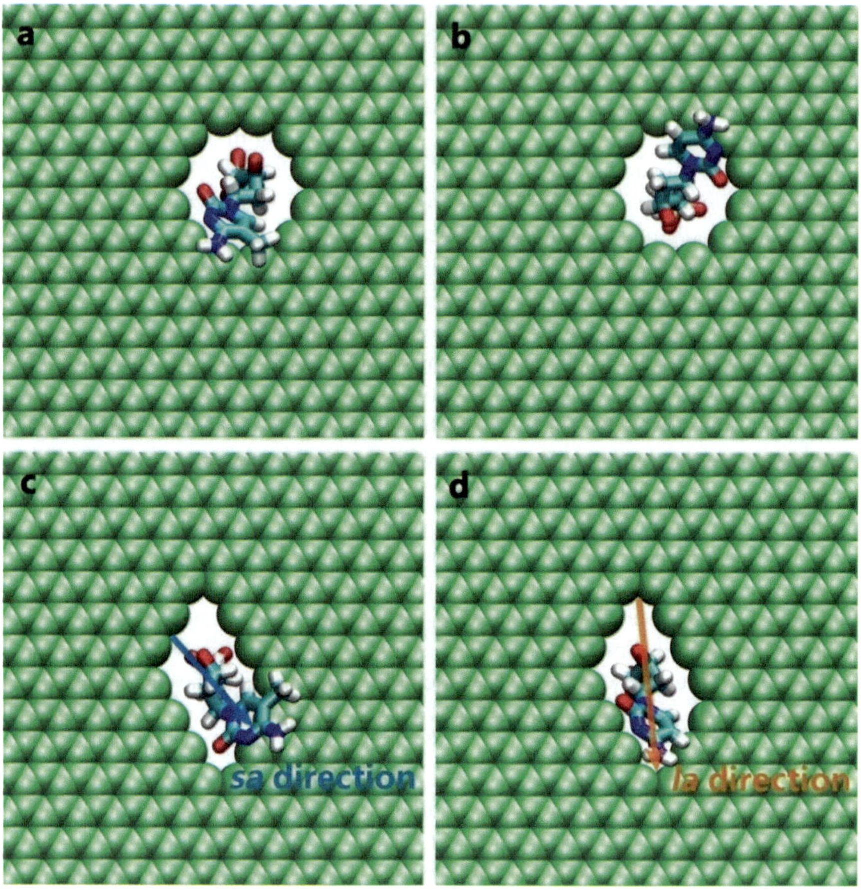

Fig. 15.4 Illustration of nucleotides passing through the circular graphene nanopore (**a** & **b**) and the rhombic graphene nanopore (**c** & **d**). In the circular nanopore, there is only a small difference in the force curves for the same nucleotides to be pulled through in different directions (**a** & **b**) since the pore is axisymmetric. In contrast, a significant difference in the force curves can occur for the same nucleotides being pulled through the rhombic nanopore in different directions; (**c**) from the short axis {sa} direction and (**d**) from the long axis {la} direction

15.4.1 Medicine and Disease Diagnosis

Changes in the DNA sequence earlier called mutation and now more precisely 'pathogenic variation' in the gene may completely alter the cellular protein function and lead to the transformation of healthy cells into cells with a particular disease as happens in rare single gene disorders or tumor cells as may happen in cancers. In immunology, we know that genetic mutations may alter the way the immune cells resist pathogenic organisms and may lead to immunodeficiency diseases. In microbiology, we study how bacterial cells can develop resistance to chemicals present in antibiotics leading to mutated species resistant to that drug. Hence sequencing the human genome helps us understand genetic variations causative of disease and sequencing bacterial, viral, parasitic, and other genomes of pathogenic organisms helps us understand microbiological factors affecting host-disease response, host-pathogen interaction, etc. Gaining insights into the disease processes is giving us better tools for improvising diagnosis, personalized medical care, and innovating precise therapies.

NGS technologies are being applied in various ways in research as well as clinical settings such as single gene sequencing if the suspected clinical phenotype of the disease correlates with a particular single gene disorder with an associated known genotype. In case of diseases having similar phenotypes associated with multiple genes, a gene panel is used. Exome and genome sequencing have provided us with an opportunity to investigate variations in the protein-coding region and protein coding with noncoding regions, respectively, rather than a few select genes and hence find very wide applications. Human exome represents less than 2% of the genome but has approximately more than 85% of known diseases-related gene variants and is commonly used. Whole genome sequencing gives more insights into deeper intronic, non-coding regions of genes as well as novel genes if associated. Many times, biochemical studies like special tissue stains or enzyme studies are not available in commercial settings and are only available in research settings. In such situations, NGS studies from exome sequencing are helpful for precise diagnosis of rare diseases, and may avert the need for invasive biopsies for diagnosis. Definitive diagnosis of any heritable disease helps in screening of family members, detecting asymptomatic members in certain diseases where early treatment can prevent complications, and has applications for prenatal diagnosis to prevent recurrence of severe disease in next progeny.

Most of the NGS applications focus on the identification of single nucleotide variations (SNVs) or small insertions/deletions (indels). Structural variations like translocations, big deletion duplications, and copy number variation (CNV) can also be recognized.

NGS technologies combined with other technologies such as RNA extraction (RNA-Seq), enrichment for exome (Exome-seq) or other genomic regions of interest, chromatin immuno-precipitation (ChIP-Seq), and bisulfate conversion (BS-seq), can provide rich information about genetic variants, transcriptome dynamics, transcription factor binding profile, epigenetic modifications, and other information.

15.4.2 Pharmacogenomics

Pharmacogenomics studies gene variation and its effect on the response of an individual to a particular drug [31]. It provides information about optimal drug response to drugs in individuals with a particular gene variation because some gene variations can have an effect on the enzymes involved in drug metabolism. This will help in optimizing dosage regimens and avoiding adverse events in various genotypes. This is an important step toward personalized medical care for precision therapeutics.

15.4.3 Metagenomics for Pathogen Identification

Metagenomics focuses on the analysis of sequencing data derived from mixtures of organisms. Characterization of microorganisms is essential to identify precise pathogens, especially in critical care to prevent delay in or inadequacy in the treatment, prolonged stays, readmissions, and increased mortality and morbidity. The rapidity of detection is also important, and the new molecular diagnostic methods have reduced the time considerably from receiving the sample to the result. In some situations, immune-compromised patients either due to cancer, hereditary syndromes, or transplantation, may be vulnerable to infections. In this setting, the causative organisms can be both common and uncommon pathogens, ranging from viruses to bacteria, fungi, and parasites [32]. Routine culture methods may have limitations due to the early administration of broad-spectrum or prophylactic antimicrobial drugs, as well as organisms that are fastidious, slow–growing, or atypical.

The study of the gut microbiota is helpful in understanding its composition, diversity, and influences because of factors such as diet, age, environment, and use of medications. This is providing new insights into its role in developing chronic diseases related to the gastrointestinal tract as well as various lifestyle disorders, metabolic diseases, etc.

As compared to other diagnostic tests like PCR, and multiplex PCR, metagenomic NGS has multiple advantages. Major advantages include broad identification of known as well as unexpected pathogens or finding new pathogenic strains. NGS can also be coupled to the use of primers from conserved 16S ribosomal RNA and internal transcribed spacer sequences which can allow for species identification. Evolutionary tracing of the species of the organism, strain identification, and prediction of drug resistance helps in developing better drugs. NGS is also useful for polymicrobial samples or in cases in which more than one pathogen has been implicated. Drawbacks include the detection of microbial contaminants present in the sample, reagents used for processing, or laboratory environment, which can complicate the analysis and interpretation of results and human host nucleic acid reads in the sequencing.

15.4.4 Metagenomics in Environmental Science

The studies of the ocean microbiome, soil microbiome, and microbiome from a particular environmental sample like water from a resource have highlighted the broad geographical distribution of phylogenetically similar organisms, raising the question of whether specific genomic variants can be identified that correlate with or contribute to the geographical location of microbes, contamination effects, effects of temperature change, etc. [33]

15.4.5 Epigenetics

Epigenetics is the study of reversible, heritable changes in gene expression that do not involve changes to the underlying DNA sequence. Epigenetic changes are mediated by various mechanisms like chemical modifications of the DNA or by modifications of proteins such as chromatin that are closely associated with DNA, DNA methylation, chromatin remodeling, histone modification, and other non-coding RNA (ncRNA)-associated mechanisms. NGS can be coupled with chromatin studies, gene expression studies, and microarray studies for better gain. These epigenetic studies along with genome-wide association studies (GWAS) are throwing light on various details for understanding the molecular pathology of multiple diseases such as cardiac diseases, diabetes, cancer, and neurodegenerative disorders [34].

15.4.6 Plant, Agriculture, and Food Industry

NGS and bioinformatics analysis on the plant genomes over the years have generated a large amount of data which is publicly available online worldwide. Each database is unique and has its focus, for example, the Cotton Gen database gives information on genomics and breeding information of any cotton species of interest.

Plant breeding is a tool for changing or improving plant traits to produce improved new crops for the benefit of mankind in view of climate change, infestations, use of fertilizers and their impact, adding nutritional value, etc. The NGS revolution has enabled rapid advances in plant breeding by applying the knowledge and integrating the biological data obtained from genomics research on various crops and hybrids. The tracing of plant origins helps in detecting food mixtures or contamination.

15.4.7 Forensic Studies

NGS applications in forensics help in solving several challenges such as mixture analysis, dealing with minute degraded samples, and multiple samples [30]. Many markers can be examined at one time and in great detail to get important biological data such as age, geographical origin, identification of tissue type, and correlation to phenotype traits, monozygotic twins' identification. As it helps in collecting data related to microbes, insects, plants, and soil there is great medico-legal importance also.

15.5 Challenges and Future Direction

Keeping up with the pace of discoveries, human beings will have to prepare for the challenges of this ever-refining technology considering various aspects as well as preparing for guidelines and advocacies for probable future applications [35].

15.5.1 Data Storage and Security

The advent of NGS is also developing a challenge as to how to manage large volumes of patient or individual data from various studies and at the same time to maintain the privacy of the data along with long-term archival capacity. Data will have to be maintained in a manner that provides access to those individuals who are authorized to have access to their own data, their caretaking medical practitioners, or researchers while also providing protection against unauthorized disclosure as well as unlawful tampering either for current use or for future research use. There are many aspects to it that connect the individuals and the investigators or medical practitioners like biobanks (i.e., storage of biological samples like DNA), genomic platforms (i.e., platforms generating genetic data from biological samples), bioinformatic platforms (clinical genotype or phenotype information or raw data related to it) and databases which store the information about genetic variations and analysis. Hence, fine structure is essential for data storage securely [36].

15.5.2 Ethical and Legal Issues

Apart from issues in enormous data storage and its handling, there is the potential for unlawful acts as well as technological difficulties posed by data corruption and hardware failures. There is a need for data integrity, security, and framing its legal perspectives. Ethical issues are involved in many situations in clinical diagnostics like prenatal, preimplantation genetic tests, predictive testing, germline tests in childhood cancers, etc. Interpretation of test results, their implications for medical decision making, and increasing burden of interpretation of variants in view of phenotypical variabilities pose ethical challenges in tests as well as counseling. Medical, forensic, and research data of all forms need to be kept for extended periods of time and hence secure handling of data is a huge challenge that needs immediate attention. To maintain and improve the trust between patients, clinicians, researchers, and private industry, legal norms must be laid down at various levels of communication [37]. In the era of digital communication, it is compelling to lay down effective system networks that include informed consent of patients and families along with electronic systems which may provide dynamic options to cope with rapid advances.

15.6 Summary

Next-generation sequencing has revolutionized molecular biology research and its applications to human health and the surrounding environment. The technique of sequencing had enormous changes in terms of fineness, detailing, platforms, cost, and applications. It is giving insights into disease mechanisms that are changing the face of medicine and therapeutics taking it toward a personalized approach. With increasing opportunities as discussed, it is also posing challenges that need to be addressed in clinical as well as research contexts.

Nanopore technology, as a very recent advanced addition to sequencing techniques, offers new hope in terms of increasing speed, accuracy, and decreasing cost. It has many other advantages including minimum sample preparation, eliminating the need for amplifications, long read lengths, and decreasing cost.

The sequencing technique still needs to be improved in its resolution to detect single bases with the rapid movement of DNA through the hole. Using parallel processing of samples, the next-generation sequencing reduces the cost and processing time reasonably well. Earlier methods were capable of sequencing short DNA lengths, but it was not possible to sequence repetitive regions of accurate genome assemblies due to gaps and there was difficulty in resolving large structural variations. There are challenges that still need to be addressed like single-base recognition, slowing down DNA velocity, etc., but overall, this technology has the potential to have a major impact on the diagnosis of diseases and hence human health.

Algorithms for long-read assembly are still in the preliminary stage even for single genomes, and further algorithm and software development needs to take place before these technologies can be used effectively in a metagenomic setting.

Data storage, handling, authorization, and individual communications are also issues with their ethical and legal aspects which need better advocacies and much remains to be regulated in the times to come.

References

1. Gibbs, R. A. (2020). The human genome project changed everything. *Nature Reviews Genetics, 21*, 575–576.
2. Heather, J. M., & Chain, B. (2016). The sequence of sequencers: The history of sequencing DNA. *Genomics, 107*, 1–8.
3. Feng, Y., Zhang, Y., Ying, C., Wang, D., & Chunlei, D. (2015). Nanopore-based fourth-generation DNA sequencing technology. *Genomics, Proteomics & Bioinformatics, 13*(1), 4–16.
4. Sanger, F., Nicklen, S., & Coulson, A. R. (1997). DNA sequencing with chain-terminating inhibitors. *Proceedings of the National Academy of Sciences of the United States of America, 12*, 5463–5467.
5. Feng, Y., Zhang, Y., Ying, C., et al. (2015). Nanopore-based fourth-generation DNASequencing technology. *Genomics, Proteomics & Bioinformatics, 13*, 4–16.
6. Li, J., Mohamed, A., Batcha, N., Grüning, B., et al. (2015). An NGS workflow blueprint for DNA sequencing data and its application in individualized molecular oncology. *Cancer Informatics, 14*, S5.

7. Hess, J. F., Kohl, T. A., Kotrova, M., et al. (2020). Library preparation for next generation sequencing: A review of automationstrategies. *Biotechnology Advances, 41*, 107537.
8. Vikas, B., & Christina, B. (2019). Sequencing technologies and analyses: Where have we been and where are we going? *iScience, 18*, 37–41.
9. Pereira, R., Oliveira, J., & Sousa, M. (2020). Bioinformatics and computational tools forNext-generation sequencing analysis inClinical genetics. *Journal of Clinical Medicine, 9*, 132.
10. Wadapurkar, R. M., & Vyas, R. (2018). Computational analysis of next generation sequencing data and itsapplications in clinical oncology. *Informatics in Medicine Unlocked, 11*, 75–82.
11. Richards, S., et al. (2015). Standards and guidelines for the interpretation of sequence variants: A joint consensus recommendation of the American College of Medical Genetics and Genomics and the Association for Molecular Pathology. *Genetics in Medicine, 17*(5), 405–424.
12. Jain, M., Olsen, H. E., Paten, B., et al. (2016). The Oxford nanopore MinION: Delivery of nanopore sequencing to the genomics community. *Genome Biology, 17*, Article no. 239, pp 1–11.
13. Deamer, D., Akeson, M., & Branton, D. (2016). Three decades of nanopore sequencing. *Nature Biotechnology, 34*(5), 518–524.
14. Kwok, H., Briggs, K., & Tabard-Cossa, V. J. (2014). Nanopore fabrication by controlled dielectric breakdown. *PLoS One, 9*(3), e92880.
15. Li, J., Stein, D., McMullan, C., Branton, D., Aziz, M. J., & Golovchenko, J. A. (2001). Ion-beam sculpting at nanometre length scales. *Nature, 412*(6843), 166–169.
16. Chen, K., Kong, J., Zhu, J., Ermann, N., Predki, P., & Keyser, U. F. (2019). Digital data storage using DNA nanostructures and solid-state nanopores. *Nano Letters, 19*, 1210–1215.
17. Lopez, R., et al. (2019). DNA assembly for nanopore data storage readout. *Nature Communications, 10*(1), 1–9.
18. Clarke, J., Wu, H. C., Jayasinghe, L., et al. (2009). Continuous base identification for single-molecule nanopore DNA sequencing. *Nature Nanotechnology, 4*, 265–270.
19. Mikheyev, A. S., & Tin, M. M. Y. A. (2014). First look at the Oxford nanopore MinION sequencer. *Molecular Ecology Resources, 14*, 1097–1102.
20. Schneider, G. F., et al. (2010). DNA translocation through graphene nanopores. *Nano Letters, 10*, 3163–3167.
21. Garaj, S., et al. (2010). Graphene as a subnanometre trans-electrode membrane. *Nature, 467*, 190–193.
22. Merchant, C. A., et al. (2010). DNA translocation through graphene nanopores. *Nano Letters, 10*, 2915–2921.
23. Liu, K., Feng, J., Kis, A., & Radenovic, A. (2014). Atomically thin molybdenum disulfide nanopores with high sensitivity for DNA translocation. *ACS Nano, 8*, 2504–2511.
24. Farimani, A. B., Min, K., & Aluru, N. R. (2014). DNA base detection using a single-layer MoS2. *ACS Nano, 8*, 7914–7922.
25. Steinbock, L. J., Bulushev, R. D., Krishnan, S., Raillon, C., & Radenovic, A. (2013). DNA translocation through low-noise glass nanopores. *ACS Nano, 7*, 11255–11262.
26. Mathé, J., Aksimentiev, A., Nelson, D. R., et al. (2005). Orientation discrimination of single-stranded DNA inside the alpha-hemolysin membrane channel. *Proceedings of the National Academy of Sciences of the United States of America, 102*(35), 12377–12382.
27. Sathe, C., Zou, X. Q., Leburton, J. P., & Schulten, K. (2011). Computational investigation of DNA detection using graphene nanopores. *ACS Nano, 5*(11), 8842–8851.
28. Liang, L., Zhang, Z., Shen, J., et al. (2013). Theoretical studies on the dynamics of DNA fragment translocation through multilayer graphene nanopores. *RSC Advances, 3*, 2445–2453.
29. Zhang, Z., Shen, J.-W., & Wang, H. (2014). Effects of graphene Nanopore geometry on DNA sequencing. *Journal of Physical Chemistry Letters, 5*(9), 1602–1607.
30. van Dijk, E. L., Auger, H., Jaszczyszyn, Y., & Thermes, C. (2014). Ten years of next-generation sequencing technology. *Trends in Genetics, 30*(9), 418–426.
31. Bahareh, R., Hirofumi, N., Shahin, A., Mustafa, T., & Nejat, M. (2016). Next generation sequencing: Implications in personalized medicine and pharmacogenomics. *Molecular BioSystems, 12*(6), 1818–1830.

32. Wei, G., Miller, S., & Chiu, C. Y. (2019). Clinical metagenomic next-generation sequencing for pathogen detection. *Annual Review of Pathology, 14*, 319–338.
33. Joly, D., & Faure, D. (2015). Next-generation sequencing propels environmental genomics to the front line of research. *Heredity, 114*, 429–430.
34. Peixoto, P., Cartron, P.-F., & Serandour, A. A. (2020). From 1957 to nowadays: A brief history of epigenetics. *International Journal of Molecular Sciences, 21*(20), 7571.
35. Yin, Y., Butler, C., & Zhang, Q. (2021). Challenges in the application of NGS in the clinical laboratory. *Human Immunology, 82*(11), 812–819.
36. Sebastian, H. (2018). Eck challenges in data storage and data management in a clinical diagnostic setting. *Journal of Laboratory Medicine, 42*(6), 219–224.
37. Nicole, M.-M., & David, M. (2019). Privacy and ethical challenges in next-generation sequencing. *Expert Review of Precision Medicine and Drug Development, 4*(2), 95–104.

Further Reading

Wang, J., Tan, A. C., & Tian, T. (2012). *Next generation microarray bioinformatics: Methods and protocols* (Vol. 802). ISBN: 978-1-61779-399-8.

Turnpenny, P. D. (2016, November 30). *Emery's elements of medical genetics*, Sian Ellard Elsevier Health Sciences

Chapter 16
An Overview of Anode in Microbial Fuel Cell: Current Challenges and Opportunities

S. Khiran, K. Vignaprasad, Manickam Matheswaran, and Samsudeen Naina Mohamed

16.1 Introduction

The significant need of today's modern era is the twin goal of achieving global energy demands, together with controlling the looming threat of global warming. As the world population is increased to 7.8 billion and expected to cross 11 billion in year 2100, the current energy reserves will certainly fall short, plunging us into an age of energy shortage [1]. Although the current fossil fuel reserves are sustaining a huge industrial revolution, it undoubtedly is not sufficient to meet increasing energy demands [2]. Microbial fuel cell (MFC) is a green energy technology and immensely promising solution, which have evolved extremely fast, since their inception in 1911 by the botanist Michael Cresse Potter [3]. MFC produces electricity and simultaneously treats the wastewater to a significant level, as typical wastewater contains almost 10 times the energy which is actually required to properly treat it [4, 5]. It comes with advantages that the substrate/raw material costs are negligible and abundant, as wastewater is generated in almost all kinds of industries (typically 10–15 L) and is completely pollution-free type of energy production [6, 7]. As it is an immensely clean and economical way of producing energy, the environmental effects by MFCs are negligible and serve to improve the ecosystem by reducing the overall carbon footprint [8]. As the MFCs begin to be produced in a large scale, the price of components will begin to reduce drastically, making the market for such type of fuel cells (FCs) more incentivized for large-scale and small-scale industries alike, maybe even for domestic purposes [9]. This paper specifically addresses the anode side of MFC, and various factors affecting its performance, such as design, material used on anode side, and surface area.

S. Khiran · K. Vignaprasad · M. Matheswaran · S. N. Mohamed (✉)
Department of Chemical Engineering, National Institute of Technology, Tiruchirappalli, India
e-mail: samsudeen@nitt.edu

R. S. Chaughule, A. S. Lokur (eds.), *Applications of Nanotechnology in Microbiology*, https://doi.org/10.1007/978-3-031-49933-3_16

16.2 Importance of MFC as an Upcoming Source of Renewable Energy

In the search of green energy, technology related to biological resources is gaining more demand. Thus, MFC has gained lot of attention, as a clean replacement for fossil fuels. MFC is a perfect choice as compared to other bioeconomic technologies because, wastewater is available in abundance and simultaneously treating the wastewater to remove impurities and reduce the toxicity of water and its applications as shown in Fig. 16.1 [10, 11]. MFCs have a lot of advantages as compared to other sources of renewable energy such as solar and wind energy. MFC is comparatively inexpensive and can be used for domestic level as well, although it is recommended to scale it up, in order to gain the most effectiveness of the system, as with any other technology [12].

For MFCs, to be quickly assimilated into the mass production, thereby decreasing the costs even further and also spreading it into various fields and industries, it needs to be immediately applied into an area/field of great demand. All these conditions are met by using MFCs in the field of wastewater treatment [4]. Wastewater is a readily available substrate from the various industries [13]. Biomass and wastewater for creating renewable sources of hydrogen production and electricity generation is a great interest to fulfill the energy needs and simultaneously promote the sustainability of the environment. Additionally, the MFCs are able to achieve high levels of conversion efficiencies and degree of cleanness [4]. Newer design strategies were also proposed for the MFC where heavy metal ion recovery is also possible from the same system, along with efficient color and COD removal [5]. Some of the appreciable results show color removal up to 93.0% and COD removal of 75%. All these factors make the MFCs as novel and upcoming solution to the ever-increasing demand of energy, but more importantly, clean and green forms of energy production. Since the time of its inception, publications of experiments using pure and

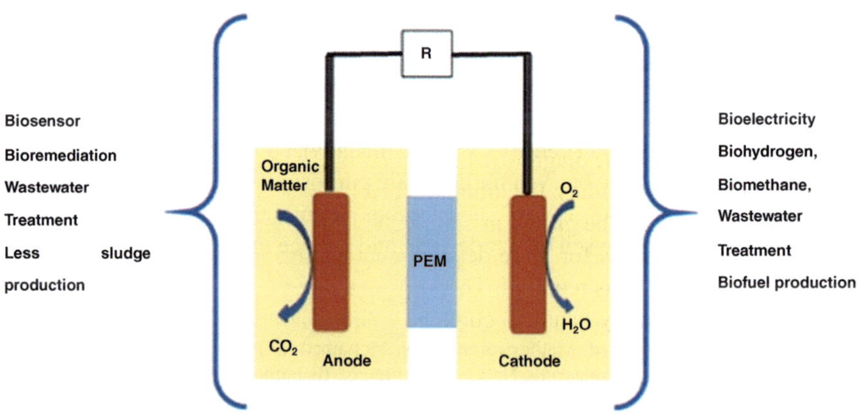

Fig. 16.1 Schematic representation of microbial fuel cell (MFC)and its applications

mixed species MFC have been growing in an exponential, demonstrating increasing interest by researchers globally. Also, MFCs exhibit a lot of versatility in the type of wastewater and substrates which can be used to produce either hydrogen or direct electricity, such as distillery wastewater [6], wood industry wastewater [14, 15], and synthetic wastewater [16]. Thus, MFC systems are proving to be an excellent alternative as the upcoming source of clean and green energies, with minimal disadvantages and various advantages over other forms of renewable sources, while also being capable fulfilling the simultaneous goals of achieving both cleaner environments, together with meeting the energy demands of the world.

16.3 Factors Affecting the Performance of MFC

In MFCs, substrate is regarded as one of the most important biological factors affecting electricity generation. Besides power generation, the other objective is to remove the pollutants, such as COD, TDS, color, nitrates, sulfide, and sulfates, from the wastewater. In most of the MFC studies, acetate has been preferred as a substrate for the electricity production because of the simple substrate and also end product of several metabolic pathways for higher order carbon sources [17]. Glucose is commonly used substrates in MFC next to acetate [18]. The abundance and renewability of lignocellulosic materials from agricultural residues renders them a promising feed-stock for cost-effective energy production. However, lignocellulosic biomass cannot be directly utilized by microorganisms in MFCs for electricity generation. When cellulose is used as the substrate, the electricity generation requires a microbial community with both cellulolytic and exoelectrogenic activities [19]. Several researchers have used the synthetic wastewater because of it has a well-defined composition and easy to control in terms of loading strength, pH etc. [20, 21]. Several researchers have also evaluated the MFC performance with real (industrial) wastewater, such as agricultural, domestic, paper, and food/dairy as a substrate [22, 23].

Bacteria oxidizes substrates and electrons are transferred to respiratory enzymes by NADH (reduced from of nicotinamide adenine dinucleotide). The microbes transfer electrons from electrolyte to an anode electrode surface via two mechanisms; direct electron transfer and indirect electron transfer [24–26]. The microorganism either in pure or mixed state should have high exoelectrogenic activity than the methanogenic activity which leads to enhance the power production. But the mechanism and role of the individual microorganism contribution to the power generation and treatment efficiency capability becomes difficult to understand when mixed microbial community is used as a biocatalyst [27]. Recently, the MFC populated by mixed microbial communities have gained much attention owing to their stability, robustness due to nutrient adaptability, stress resistance, and general tendency to produce higher current densities than those obtained using pure cultures. Several researchers have evaluated the performance of MFC in terms of electricity production and treatment of wastewater by inoculating with the mixed microbial communities [28]. Besides exoelectrogens, there are also metal reducing bacteria, denitrifying bacteria, hydrogen-scavenging microorganisms, methanogens, etc. [29]. Though a variety of microorganisms such as isolated or mixed culture are used in MFCs, it requires different pH environment for its exoelectrogenic activity and growth. The electrochemical activity and substrate degradation varied with respect to pH conditions involved which influences the power productions in MFCs. The bacteria require a pH close to neutral for their optimal growth; while oxygen reduction in the cathode results in an alkaline pH [30]. According to Nernst equation, the organic matter oxidation potential would shift −59 mV at each pH unit in an alkaline direction in the anolyte chamber and +59 mV at each pH unit in the acidic direction in the catholyte chamber in MFC [31, 32].

Catholyte is also an important factor that affects the performance of MFC. The current production in MFCs is largely dependent on the reduction kinetics at the cathode. The maximum potential of achieved in the MFC is 0.805 V when oxygen is used in the cathode. The maximum potential produced from the air-cathode MFC using acetate as a substrate is 1.105 V. However, the slow rate of oxygen reduction on the surface of graphite/carbon electrodes in the absence of catalyst on the electrode leads to a higher reduction over potential which is among the most limiting factors in the performance of MFCs. In the aqueous cathode chamber MFC, the most commonly used chemical catholyte next to oxygen is ferricyanide. It has faster reduction kinetics than that of oxygen in the cathode and a relatively higher redox potential [33]. Potassium permanganate, potassium persulfate dichromate aerated water is another most commonly used as electron acceptors by virtue of its high oxidization capacity as well as its environmental safety [34].

16.4 Major Significance of the Anode

The anode side plays a major role in affecting the overall performance of the MFC and has various parameters which can be tweaked in order to achieve the maximum potential of the system. It is also evident that the anodic side is highly significant as it imparts more impact on the system, when its various parameters are modified. For instance, the potential exhibited by the anode side is more or less stable except for the experiments with $KMnO_4$ (pH 10) and $KMnO_4$ (pH 8) [35], in which case a pH gradient is created between the two chambers. In MFC, the anode is designated area for microbial species proliferation, and hence it must possess the superior qualities, including large surface area, high conductivity, anaerobic conditions, good corrosion resistance, biocompatible for the enhanced growth of the desired microbial species [36]. Several materials have been tested for anode side over the past decade in the MFC research, and it was observed that the carbon-based electrodes provided better results (more energy efficient and also biocompatible as compared to other materials) [36]. Further researches have been concluded that anodes with nanocomposites give flexibility when the fore mentioned anode needs to be modified. Thus, in the MFC system, anode side parameters are significant in the process of power generation [37]. Taking into account the challenges, faced in the development of

MFC, the main aim boiled down to studying the different design parameters specially of the anode region, anode electrode size and surface area, and also the anode design, all of which have significant impact on MFC [38]. Recent developments also showed that for fixed anode surface area, the power density normalized to the anode surface area increased with Proton Exchange Membrane (PEM) size, and thus the PEM surface area was the limiting factor for power output due to increase in the internal resistance [33]. It is also found that the removal of the PEM in the system increased the power output, mainly due to the enhancement of the proton flux from the anode to the cathode [33]. Newer research also shows that the substrate-utilization rate exhibited by the anode-respiring bacteria (also known as the ARB) directly correlates to the current density, which forms one of the main factors in a microbial electrolysis/fuel cell [39].

16.4.1 Effect of Electrode Material

Electrode materials play a vital role in the performance of MFC as a whole, as they are extremely influential in the process of producing bioelectricity [40]. The extent to which the MFC is viable and efficiency/yield is ultimately determined by the electrode material. Even though, a wide variety of materials have been tested in various MFC, few special characteristics are desired for optimal performance namely, good conductivity of anode and cathode electrode, electrochemical stability, and microbial compatibility. It is studied that different types of electrode affect the extent of conductivity observed and also the level of pollutant degradation. The main reason is because the primary enrichment site for the microorganisms in the MFC is the surface of the electrodes. The major types of materials for the electrodes include carbonaceous materials, such as carbon fiber, carbon brush, carbon felt, activated carbon, graphite, graphite granules, graphite felt, graphite plates, and graphite rods. The main reason for the excessive usage of carbonaceous and graphite materials as electrode materials are the fact that they exhibit strong chemical stability at relatively low costs. There is also a requirement of highly conductive, at the same time, non-corrosive materials, which are also expected to exhibit high specific surface area (i.e., surface area per volume). One of such materials, which seem to fulfill all these conditions, and also have an open structure which helps to avoid biofouling [41].

According to the level of microbial proliferation, the expected trend was seen as graphite rod > carbon felt fiber > stainless-steel mesh, with the species *Proteobacteria* showed the highest growth rate (50.7–71.3%). As researches are leaning more towards the nanomaterials and nanocomposites domain, it has seen good advances in the fuel cell-specific materials as well. Carbon nanofibers, and their activated versions made by both chemical and physical method, have been used as catalyst in MFCs by Ghasemi et al. [42]. These results show that the nanocomposite enhanced materials are more superior to the plain materials as is the case of plain carbons and non-catalyzed graphite plates, and even plain carbon felts, carbon clothes, and stainless-steel mesh. Another innovative idea is the coating of carbon nanotubes with polypyrole which is synthesized by an in-situ polymerization method, extensively used specially for double chambered MFC, which is reported to have better efficiencies on the overall conversion rates. It was also noted, that electrons were preferentially harvested on the graphene anode, making it the ideal electron acceptor [43]. As advances in the materials production continues, graphene has now started to be produced from the reduction process of GO [43]. In similar research, the nanomaterial polyaniline (PANI) was electro-deposited on graphene-nanoribbons to fabricate a composite anode, and it was reported to form porous nanostructures with ease. With the intent of combining direct and mediated electron transfer of a few specific substrates like yeast, materials like carbon felt and carbon cloth are modified with PANI, the advantages were that the electron transfer chain was far more efficient, and it also minimizes the side reactions.

16.4.2 Carbonaceous Materials

Carbonaceous materials are widely considered as most suitable candidates to be used as the anode electrodes in MFC, mainly due to their desirable characteristics, such as good biocompatibility, excellent conductivity, chemical, and thermal stability (Table 16.1). These materials generally consist of CC, CP, CF, CB, GP, graphite sheet (GS), graphite rod (GR), granular activated carbon (GAC), reticulated glass

Table 16.1 Carbonaceous materials as anode electrodes in MFC

SR NO	Anode materials	Configuration	COD removal	Pmax (mW m − 2)	References
1.	Carbon paper	Flat MFC	79	43	[45]
2.	Carbon cloth	SCMFC	40–50	464	[46]
3.	Graphite rod	SCMFC	180	26	[47]
4.	Graphite cylinder	Tubular MFC	50	25	[48]
5.	Graphite plates	SCMFC	72.84	124.352	[49]
6.	Graphite granules with graphite rod	SCMFC	93	2981	[50]
7.	Carbon fiber	SCMFC	40	264	[51]
8.	Carbon felt	DCMFC	99.5	1600	[52]
9.	Carbon fiber brush	DCMFC	~100	51.2	[53]
10.	Graphite fiber brush	SCMFC	29	672	[54]
11.	Carbon cloth with CNTs	SCMFC	NA	65	[55]
12.	Carbon cloth with rGO	DCMFC	NA	1062 ± 58	[56]
13.	Carbon felt with PAni	DCMFC	67.34 ± 1.89	216	[57]
14.	Carbon cloth with PDA-rGO	DCMFC	NA	2047 ± 58	[56]
15.	Carbon felt with WO3	DCMFC	87 ± 3	321 ± 0.22	[58]
16.	Carbon paper with Cu doped FeO	DCMFC	75	161.5	[59]

carbon (RGC), and stainless-steel mesh (SSM). Of these, the SSM electrodes are very promising solutions, although the enriched biofilm was prone to falling off, especially after prolonged usage. A novel concept stating that the layered corrugated carbon (LCC) structure of anode electrodes greatly increased the power generation was put forward by Chen et al. (2012) [44]. The protogenic material was derived from abundant packing materials like corrugated cardboard. It was recorded that the power output was up to 390 A/m^2 at six corrugated layers, thus, researchers are more inclined to choose them for the novel anode electrodes. Various factors are evaluated comprehensively when choosing appropriate materials as MFCs anode. Carbonaceous materials are not only cost-effective anodes but are also capable of achieving higher performance, and also involves easy maintenance in the long-term operations,

16.4.3 Metal/Metal Oxides-Based Anode Electrodes

Majority of the metals and their oxides cannot be used directly in the MFC anodes due to their corrosion prone characteristics, but over the past decade, the selective metals and their alloys include, Pd, Au, NiO, Co_xO_y, TiO_2, WO_3, NiTi, Co/NiWO$_4$, and CoeMoO$_2$, have been used extensively. These precious metals are highly desirable in electrochemistry owing to their high conductivity, wide potential window, and high catalytic activity but it is not economically feasible manufacture such MFCs anode on a large scale due their high cost. The catalytic activity of the non-noble metal is almost comparable to that of the precious metals. As a result, it is growing as a highly potential research prospect. One of the novel ideas was a porous carbon anode electrode with 3D ordered honeycomb structure consisting of TiO_2 nanoparticles. After some modifications, the uniform macroporous distribution and high-specific surface area was obtained, which not only enhanced the anode roughness but also simultaneously increased biocompatibility. The obtained results showed that the maximum power density 973 mW/m^2 was 2.3 times higher than that of traditional carbon cloth anode. In summary, the anode electrodes which were modified via various synthesis methods, such as electrodeposition, using the metallic and metal oxide materials were shown to greatly improve the conductivity, reduce the internal resistance thus increasing the electrode roughness and specific surface area to increase bacteria attachment and reproduction.

16.4.4 Nanocomposites Supported Electrodes

In order to increase the performance of the anodes, increasing surface area will be a better method. In recent times, surface area of the anode have been increased using nanocomposite polymers made up of carbon which further resulted in increased current densities. Using nanocomposite polymers, such as polyaniline-carbon nanotube (CNT) composite, both electrical conductivity and surface area of the anode can be increased which ultimately results in better performance. CNT composites are known to be toxic to micro-organisms which grow in MFCs and thus to increase biocompatible nature of the anode, along with nanocomposites microorganisms like *Escherichia coli* are used as biocatalysts. Moreover, new developments have shown that the usage of iron oxide as a nanocomposite has increased the biocompatibility of the anode. When graphite electrode was coated with gold-based nanoparticles, current density recorded was 20 times higher. By utilizing CNT composite which is made up of one microscale porous layer and one macroscale porous textile, the results were able to find improved mechanical contact and electron transfer rate (ETR). Additionally, CNTs were made into sponges, films, fiber mats, etc., which provided high porosity and specific surface area. The preparation of these types of electrodes was simple and easy to scale up to a larger extent. Due to all its positive benefits and its economic factor, these types of electrodes have now gained increased interest and lots of research are progressing on this topic. Thus, soon the dominance of this electrode in the field of MFC will be well known.

16.5 Challenges and Opportunities

In recent years, MFC technology has been evolving as a popular technology, but it also has its fair share of challenges [60]. One of the main challenges is the poor oxygen reduction kinetics, which have been well documented as one of the limiting factors for optimal MFC operation [61, 62]. Results from various sources showed that anodic catalytic activity cannot be sustained due to limited proton transfer and electron consumption rate at the cathode. Another challenge related to the proton mass transfer in MFCs is the use of an aqueous solution as the electrolyte. The cations thus generated can interfere with proton transfer through the membrane, reducing the overall efficiencies and finally the yield generated by the MFC [63, 64]. This can also result in various electrochemical and biological problems obscuring the efficient operation of the MFC system. Also, in the case of aerobic processes, the microorganisms use up all the energy carried by the organic contaminants, while only a small part of the energy is available to microorganisms in MFCs for their growth, because a large part is converted to electricity, forcing the active microorganisms to use the free energy available from the oxidation of the contaminants which reduce the electrode efficiency [65, 66]. The electrode materials giving the most beneficial results, oftentimes not very suitable for scale-up because of their inherent lack of durability or structural strength (e.g., carbon paper), or cost (e.g., graphite rods) [41, 67]. As a result, we will have to consider conductive coatings on structurally strong supporting materials, which are also proving as a huge challenge for the researchers [68].

16.6 Conclusion

Experimental data from various MFC-related researches over the past decade have all pointed to the fact that the anode side plays a significant role in the overall per formance of system. When the various factors constituted by the anode side are maintained at their optimal levels, a huge increase in the performance of the system is reported in many cases. Apart from good COD removal efficiencies and electricity generation, anode side parameters can be regulated to provide excellent color removal (up to 89% in some cases), and also simultaneous removal of turbidity and meladonin content from substrates like distillery wastewater. Anode side also allows the use of bio-mass-derived materials with various metal oxides/polymers, to further decrease the overall cost of the system, and boosting its eco-friendly nature. Additionally, the combination of several factors, such as increased surface area, electrode materials, and also the types of substrate used had a positive influence on the efficiency of the overall process. Thus, the appropriate selection of the anode electrode, its parameters, especially the electrode materials has proved to be one of the most critical factors affecting the performance of the MFC systems. It is observed that by varying the anode side parameters to find the ideal combination, many challenging issues faced by MFC research can be addressed successfully.

References

1. Crist, E., Ripple, W. J., Ehrlich, P. R., Rees, W. E., & Wolf, C. (2022). Science of the Total Environment Scientists' warning on population. *Science of the Total Environment, 845*, 157166. https://doi.org/10.1016/j.scitotenv.2022.157166
2. Lowe, I., O'Sullivan, J., & Cook, P. (2019). *Population and climate change* (Vol. 15). Sustainable Population Australia.
3. Logan, B. E. (2007). *Microbial fuel cell*. Wiley.
4. Sun, D., Gao, Y., Hou, D., Zuo, K., Chen, X., Liang, P., et al. (2018). Energy-neutral sustainable nutrient recovery incorporated with the wastewater purification process in an enlarged microbial nutrient recovery cell. *Journal of Power Sources, 384*, 160–164. https://doi.org/10.1016/J.JPOWSOUR.2018.02.049
5. Abdirahman Elmi, H. S. (2015). Colour and COD removal from palm oil mill effluent (POME) using pseudomonas aeruginosa strain NCIM 5223 in microbial fuel cell. *International Journal of Waste Resources, 05*. https://doi.org/10.4172/2252-5211.1000181
6. Samsudeen, N., Radhakrishnan, T. K., & Matheswaran, M. (2016). Effect of isolated bacterial strains from distillery wastewater on power generation in microbial fuel cell. *Process Biochemistry, 51*, 1876–1884. https://doi.org/10.1016/j.procbio.2016.06.007
7. Samsudeen, N., Radhakrishnan, T. K., & Matheswaran, M. (2015). Bioelectricity production from microbial fuel cell using mixed bacterial culture isolated from distillery wastewater. *Bioresource Technology, 195*, 242–247. https://doi.org/10.1016/j.biortech.2015.07.023
8. Papaharalabos, G., Greenman, J., Melhuish, C., & Ieropoulos, I. (2015). A novel small scale microbial fuel cell design for increased electricity generation and waste water treatment. *International Journal of Hydrogen Energy, 40*, 4263–4268. https://doi.org/10.1016/j.ijhydene.2015.01.117

9. Gude, V. G. (2016). Wastewater treatment in microbial fuel cells – An overview. *Journal of Cleaner Production, 122*, 287–307. https://doi.org/10.1016/j.jclepro.2016.02.022
10. Chakraborty, I., Sathe, S. M., Dubey, B. K., & Ghangrekar, M. M. (2020). Waste-derived biochar: Applications and future perspective in microbial fuel cells. *Bioresource Technology, 312*, 123587. https://doi.org/10.1016/j.biortech.2020.123587
11. Urmetzer, S., Lask, J., Vargas-Carpintero, R., & Pyka, A. (2020). Learning to change: Transformative knowledge for building a sustainable bioeconomy. *Ecological Economics, 167*, 106435. https://doi.org/10.1016/J.ECOLECON.2019.106435
12. Jakhongir Turakul Ugli, T. (2019). The importance of alternative solar energy sources and the advantages and disadvantages of using solar panels in this process. *International Journal of Engineering and Information Systems (IJEAIS), 3*, 60–64.
13. Pant, D., Van Bogaert, G., Diels, L., & Vanbroekhoven, K. (2010). A review of the substrates used in microbial fuel cells (MFCs) for sustainable energy production. *Bioresource Technology, 101*, 1533–1543. https://doi.org/10.1016/j.biortech.2009.10.017
14. Toczyłowska-Mamińska, R., Szymona, K., & Kloch, M. (2018). Bioelectricity production from wood hydrothermal-treatment wastewater: Enhanced power generation in MFC-fed mixed wastewaters. *Science of the Total Environment, 634*, 586–594. https://doi.org/10.1016/J.SCITOTENV.2018.04.002
15. Kloch, M., & Toczylowska-Maminska, R. (2020). Toward optimization of wood industry wastewater treatment in microbial fuel cells-mixed wastewaters approach. *Energies, 13*, 263. https://doi.org/Artn26310.3390/En13010263
16. Saz, C., Ture, C., Turker, O. C., & Yakar, A. (2018). Effect of vegetation type on treatment performance and bioelectric production of constructed wetland modules combined with microbial fuel cell (CW-MFC) treating synthetic wastewater. *Environmental Science and Pollution Research, 25*, 8777–8792. https://doi.org/10.1007/s11356-018-1208-y
17. Liu, H., Cheng, S., & Logan, B. E. (2005). Power generation in fed-batch microbial fuel cells as a function of ionic strength, temperature, and reactor configuration. *Environmental Science & Technology, 39*(14), 5488–5493. https://doi.org/10.1021/es050316c
18. Chae, K. J., Choi, M. J., Lee, J. W., Kim, K. Y., & Kim, I. S. (2009). Effect of different substrates on the performance, bacterial diversity, and bacterial viability in microbial fuel cells. *Bioresource Technology, 1009*(14), 3518–3525. https://doi.org/10.1016/j.biortech.2009.02.065
19. Rezaei, F., Xing, D., Wagner, R., Regan, J. M., Richard, T. L., & Logan, B. E. (2009). Simultaneous cellulose degradation and electricity production by Enterobacter cloacae in a microbial fuel cell. *Applied and Environmental Microbiology, 75*, 3673–3678. https://doi.org/10.1128/AEM.02600-08
20. Venkata Mohan, S., Mohanakrishna, G., Reddy, B. P., Saravanan, R., & Sarma, P. N. (2008). Bioelectricity generation from chemical wastewater treatment in mediatorless (anode) microbial fuel cell (MFC) using selectively enriched hydrogen producing mixed culture under acidophilic microenvironment. *Biochemical Engineering Journal, 39*, 121–130. https://doi.org/10.1016/j.bej.2007.08.023
21. Rodrigo, M. A., Cañizares, P., García, H., Linares, J. J., & Lobato, J. (2009). Study of the acclimation stage and of the effect of the biodegradability on the performance of a microbial fuel cell. *Bioresource Technology, 100*, 4704–4710. https://doi.org/10.1016/j.biortech.2009.04.073
22. Nimje, V. R., Chen, C. Y., Chen, H. R., Chen, C. C., Huang, Y. M., Tseng, M. J., et al. (2012). Comparative bioelectricity production from various wastewaters in microbial fuel cells using mixed cultures and a pure strain of Shewanella oneidensis. *Bioresource Technology, 104*, 315–323. https://doi.org/10.1016/j.biortech.2011.09.129
23. Velasquez-Orta, S. B., Head, I. M., Curtis, T. P., & Scott, K. (2011). Factors affecting current production in microbial fuel cells using different industrial wastewaters. *Bioresource Technology, 102*, 5105–5112. https://doi.org/10.1016/j.biortech.2011.01.059
24. Huang, L., Regan, J. M., & Quan, X. (2011). Electron transfer mechanisms, new applications, and performance of biocathode microbial fuel cells. *Bioresource Technology, 102*, 316–323. https://doi.org/10.1016/j.biortech.2010.06.096

25. Rabaey, K., Boon, N., Siciliano, S. D., Verhaege, M., & Verstraete, W. (2004). Biofuel cells select for microbial consortia that self-mediate electron transfer. *Applied and Environmental Microbiology, 70*(9), 5373–5382. https://doi.org/10.1128/AEM.70.9.5373-5382.2004
26. Park, D. H., Laivenieks, M., Guettler, M. V., Jain, M. K., & Zeikus, J. G. (1999). Microbial utilization of electrically reduced neutral red as the sole electron donor for growth and metabolite production. *Applied and Environmental Microbiology, 65*, 2912–2917. https://doi.org/10.1128/aem.65.7.2912-2917.1999
27. Watson, V. J., Nieto Delgado, C., & Logan, B. E. (2013). Influence of chemical and physical properties of activated carbon powders on oxygen reduction and microbial fuel cell performance. *Environmental Science & Technology, 47*, 6704–6710. https://doi.org/10.1021/es401722j
28. Mohan, S. V., Raghavulu, S. V., Peri, D., & Sarma, P. N. (2009). Integrated function of microbial fuel cell (MFC) as bio-electrochemical treatment system associated with bioelectricity generation under higher substrate load. *Biosensors & Bioelectronics, 24*(7), 2021–2027. https://doi.org/10.1016/j.bios.2008.10.011
29. Zuo, Y., Xing, D., Regan, J. M., & Logan, B. E. (2008). Isolation of the exoelectrogenic bacterium Ochrobactrum anthropi YZ-1 by using a U-tube microbial fuel cell. *Applied and Environmental Microbiology, 74*, 3130–3137. https://doi.org/10.1128/AEM.02732-07
30. Raghavulu, S. V., Mohan, S. V., Goud, R. K., & Sarma, P. N. (2009). Effect of anodic pH microenvironment on microbial fuel cell (MFC) performance in concurrence with aerated and ferricyanide catholytes. *Electrochemistry Communications, 11*(2), 371–375. https://doi.org/10.1016/j.elecom.2008.11.038
31. Zhuang, L., Zhou, S., Li, Y., & Yuan, Y. (2010). Enhanced performance of air-cathode two-chamber microbial fuel cells with high-pH anode and low-pH cathode. *Bioresource Technology, 101*, 3514–3519. https://doi.org/10.1016/j.biortech.2009.12.105
32. Jadhav, G. S., & Ghangrekar, M. M. (2009). Performance of microbial fuel cell subjected to variation in pH, temperature, external load and substrate concentration. *Bioresource Technology, 100*, 717–723. https://doi.org/10.1016/j.biortech.2008.07.041
33. Oh, S. E., & Logan, B. E. (2006). Proton exchange membrane and electrode surface areas as factors that affect power generation in microbial fuel cells. *Applied Microbiology and Biotechnology, 70*, 162–169. https://doi.org/10.1007/s00253-005-0066-y
34. Pandit, S., Sengupta, A., Kale, S., & Das, D. (2011). Performance of electron acceptors in catholyte of a two-chambered microbial fuel cell using anion exchange membrane. *Bioresource Technology, 102*, 2736–2744. https://doi.org/10.1016/j.biortech.2010.11.038
35. Kumar, S. S., Basu, S., & Bishnoi, N. R. (2017). Effect of cathode environment on bioelectricity generation using a novel consortium in anode side of a microbial fuel cell. *Biochemical Engineering Journal, 121*, 17–24. https://doi.org/10.1016/j.bej.2017.01.014
36. Shahid, K., Ramasamy, D. L., Kaur, P., Sillanpää, M., & Pihlajamäki, A. (2021). Effect of modified anode on bioenergy harvesting and nutrients removal in a microbial nutrient recovery cell. *Bioresource Technology, 332*, 0–8. https://doi.org/10.1016/j.biortech.2021.125077
37. Hindatu, Y., Annuar, M. S. M., & Gumel, A. M. (2017). Mini-review: Anode modi fi cation for improved performance of microbial fuel cell. *Renewable and Sustainable Energy Reviews, 73*, 236–248. https://doi.org/10.1016/j.rser.2017.01.138
38. Boas, J. V., Oliveira, V. B., Marcon, L. R. C., Pinto, D. P., Simoes, M., & Pinto, A. M. F. R. (2015). Effect of operating and design parameters on the performance of a microbial fuel cell with *Lactobacillus pentosus*. *Biochemical Engineering Journal, 104*, 34–40. https://doi.org/10.1016/j.bej.2015.05.009
39. Lee, H.-S., Torres, C. I., & Rittmann, B. E. (2009). Effects of substrate diffusion and anode potential on kinetic parameters for anode-respiring bacteria. *Environmental Science & Technology, 43*, 7571–7577. https://doi.org/10.1021/es9015519
40. Huang, X., Duan, C., Duan, W., Sun, F., Cui, H., Zhang, S., et al. (2021). Role of electrode materials on performance and microbial characteristics in the constructed wetland coupled microbial fuel cell (CW-MFC): A review. *Journal of Cleaner Production, 301*, 126951. https://doi.org/10.1016/j.jclepro.2021.126951

41. Logan, B., Cheng, S., Watson, V., & Estadt, G. (2007). Graphite fiber brush anodes for increased power production in air-cathode microbial fuel cells. *Environmental Science & Technology, 41*, 3341–3346. https://doi.org/10.1021/es062644y
42. Ghasemi, M., Daud, W. R. W., Hassan, S. H. A., Oh, S.-E., Ismail, M., Rahimnejad, M., et al. (2013). Nano-structured carbon as electrode material in microbial fuel cells: A comprehensive review. *Journal of Alloys and Compounds, 580*, 245–255. https://doi.org/10.1016/j.jallcom.2013.05.094
43. Yuan, H., & He, Z. (2015). Graphene-modified electrodes for enhancing the performance of microbial fuel cells. *Nanoscale, 7*, 7022–7029. https://doi.org/10.1039/C4NR05637J
44. Chen, S., He, G., Liu, Q., Harnisch, F., Zhou, Y., Chen, Y., et al. (2012). Layered corrugated electrode macrostructures boost microbial bioelectrocatalysis. *Energy & Environmental Science, 5*, 9769–9772. https://doi.org/10.1039/C2EE23344D
45. Min, B., & Logan, B. E. (2004). Continuous electricity generation from domestic wastewater and organic substrates in a flat plate microbial fuel cell. *Environmental Science & Technology, 38*, 5809–5814. https://doi.org/10.1021/es0491026
46. Yuan, Y., Zhou, S. G., & Zhuang, L. (2010). Polypyrrole/carbon black composite as a novel oxygen reduction catalyst for microbial fuel cells. *Journal of Power Sources, 195*, 3490–3493. https://doi.org/10.1016/j.jpowsour.2009.12.026
47. Zuo, J., Cui, L., Fan, M., & Song, W. (2007). Production of electricity from artificial wastewater using a single chamber microbial fuel cell. *Taiyangneng Xuebao/Acta Energiae Solaris Sin, 28*, 320–323.
48. Rodrigo, M. A., Cañizares, P., Lobato, J., Paz, R., Sáez, C., & Linares, J. J. (2007). Production of electricity from the treatment of urban waste water using a microbial fuel cell. *Journal of Power Sources, 169*, 198–204. https://doi.org/10.1016/j.jpowsour.2007.01.054
49. Mohanakrishna, G., Venkata Mohan, S., & Sarma, P. N. (2010). Utilizing acid-rich effluents of fermentative hydrogen production process as substrate for harnessing bioelectricity: An integrative approach. *International Journal of Hydrogen Energy, 35*, 3440–3449. https://doi.org/10.1016/J.IJHYDENE.2010.01.084
50. Rabaey, K., Clauwaert, P., Aelterman, P., & Verstraete, W. (2005). Tubular microbial fuel cells for efficient electricity generation. *Environmental Science & Technology, 39*, 8077–8082. https://doi.org/10.1021/es050986i
51. Cheng, S., Liu, H., & Logan, B. E. (2006). Increased power generation in a continuous flow MFC with advective flow through the porous anode and reduced electrode spacing. *Environmental Science & Technology, 40*(7), 2426–2432. https://doi.org/10.1021/es051652w
52. Li, Z., Zhang, X., & Lei, L. (2008). Electricity production during the treatment of real electroplating wastewater containing Cr6+ using microbial fuel cell. *Process Biochemistry, 43*, 1352–1358. https://doi.org/10.1016/j.procbio.2008.08.005
53. Luo, Y., Zhang, R., Liu, G., Li, J., Li, M., & Zhang, C. (2010). Electricity generation from indole and microbial community analysis in the microbial fuel cell. *Journal of Hazardous Materials, 176*, 759–764. https://doi.org/10.1016/j.jhazmat.2009.11.100
54. Huang, L., & Logan, B. E. (2008). Electricity generation and treatment of paper recycling wastewater using a microbial fuel cell. *Applied Microbiology and Biotechnology, 80*, 349–355. https://doi.org/10.1007/s00253-008-1546-7
55. Tsai, H.-Y., Wu, C.-C., Lee, C.-Y., & Shih, E. P. (2009). Microbial fuel cell performance of multiwall carbon nanotubes on carbon cloth as electrodes. *Journal of Power Sources, 194*, 199–205. https://doi.org/10.1016/j.jpowsour.2009.05.018
56. Li, Y., Liu, J., Chen, X., Yuan, X., Li, N., He, W., et al. (2020). Enhanced electricity generation and extracellular electron transfer by polydopamine–reduced graphene oxide (PDA–rGO) modification for high-performance anode in microbial fuel cell. *Chemical Engineering Journal, 387*, 123408. https://doi.org/10.1016/j.cej.2019.123408
57. Rajesh, P. P., Noori, M. T., & Ghangrekar, M. M. (2020). Improving performance of microbial fuel cell by using polyaniline-coated carbon–felt anode. *Journal of Hazardous, Toxic, and Radioactive Waste, 24*, 1–7. https://doi.org/10.1061/(asce)hz.2153-5515.0000512

58. Das, S., & Ghangrekar, M. M. (2019). Tungsten oxide as electrocatalyst for improved power generation and wastewater treatment in microbial fuel cell. *Environmental Technology*, 1–8. https://doi.org/10.1080/09593330.2019.1575477
59. Aiswarya Devi, S., Jayabalan, T., Muthukumar, H., Chandrasekaran, N. I., Mohamed, S. N., & Matheswaran, M. (2019). Enhancing power generation and treatment of dairy waste water in microbial fuel cell using Cu-doped iron oxide nanoparticles decorated anode. *Energy, 172*, 173–180. https://doi.org/10.1016/J.ENERGY.2019.01.102
60. Zhang, Y., Liu, M., Zhou, M., Yang, H., Liang, L., & Gu, T. (2019). Microbial fuel cell hybrid systems for wastewater treatment and bioenergy production: Synergistic effects, mechanisms and challenges. *Renewable and Sustainable Energy Reviews, 103*, 13–29. https://doi.org/10.1016/j.rser.2018.12.027
61. Gil, G.-C., Chang, I.-S., Kim, B. H., Kim, M., Jang, J.-K., Park, H. S., et al. (2003). Operational parameters affecting the performannce of a mediator-less microbial fuel cell. *Biosensors & Bioelectronics, 18*, 327–334. https://doi.org/10.1016/S0956-5663(02)00110-0
62. Liu, H., & Logan, B. E. (2004). Electricity generation using an air-cathode single chamber microbial fuel cell in the presence and absence of a proton exchange membrane. *Environmental Science & Technology, 38*, 4040–4046. https://doi.org/10.1021/es0499344
63. Kim, B. H., Chang, I. S., & Gadd, G. M. (2007). Challenges in microbial fuel cell development and operation. *Applied Microbiology and Biotechnology, 76*, 485–494. https://doi.org/10.1007/s00253-007-1027-4
64. Zhao, F., Harnisch, F., Schröder, U., Scholz, F., Bogdanoff, P., & Herrmann, I. (2005). Application of pyrolysed iron(II) phthalocyanine and CoTMPP based oxygen reduction catalysts as cathode materials in microbial fuel cells. *Electrochemistry Communications, 7*, 1405–1410. https://doi.org/10.1016/j.elecom.2005.09.032
65. Kim, B. H., Kim, H. J., Hyun, M. S., & Park, D. H. (1999). Direct electrode reaction of Fe(III)-reducing bacterium, Shewanella putrefaciens. *Journal of Microbiolog and Biotechnology, 9*, 127–131.
66. Kim, G. T., Webster, G., Wimpenny, J. W. T., Kim, B. H., Kim, H. J., & Weightman, A. J. (2006). Bacterial community structure, compartmentalization and activity in a microbial fuel cell. *Journal of Applied Microbiology, 101*, 698–710. https://doi.org/10.1111/j.1365-2672.2006.02923.x
67. Cheng, S., Liu, H., & Logan, B. E. (2006). Increased performance of single-chamber microbial fuel cells using an improved cathode structure. *Electrochemistry Communications, 8*, 489–494. https://doi.org/10.1016/j.elecom.2006.01.010
68. Liu, H., Cheng, S., & Logan, B. E. (2006). Power densities using different cathode catalysts (Pt and CoTMPP) and polymer binders (nafion and PTFE) in single chamber microbial fuel cells. *Environmental Science & Technology, 40*(1), 364–369. https://doi.org/10.1021/es0512071

Chapter 17
Insights of Nanobiotechnology as Bio-adsorbents for Wastewater Remediation

Romana Naaz, Sameer Ahmad, and Weqar Ahmad Siddiqi

17.1 Introduction

Wastewater pollution has become a significant environmental concern, posing threats to both human health and ecosystems. Traditional wastewater treatment methods often fall short in efficiently removing various contaminants, including heavy metals, organic pollutants, and pathogens. In recent years, nanotechnology has emerged as a promising field with potential applications in wastewater remediation. Specifically, nanobiotechnology, which combines the principles of nanotechnology and biotechnology, offers innovative and effective solutions for the removal of contaminants from wastewater. Nanoparticles, which possess unique properties like small size, shape, inner structure, and a large specific surface area are commonly utilized for the removal of heavy metals due to their favorable affinity. Various metal oxides, in nanoparticle form, demonstrate a significant capability to uptake heavy metals. However, several adsorbents suffer from certain limitations, including high cost, the generation of by-products, low stability, and the inability to be recycled [1, 2]. This unique property enables them to effectively remove a wide range of contaminants, including heavy metals [2], organic pollutants [3], inorganic anions [4], and bacteria. Numerous studies have reported successful removal of these contaminants using diverse types of nanomaterials. Nanomaterials are characterized by their reduced dimensions at the nanoscale level, offer tremendous potential for environmental remediation due to their enhanced surface-to-volume ratios. Their small sizes result in significantly larger specific surface areas compared to

R. Naaz · S. Ahmad · W. A. Siddiqi (✉)
Department of Applied Sciences and Humanities, Faculty of Engineering and Technology, Jamia Millia Islamia, New Delhi, India
e-mail: wsiddiqui@jmi.ac.in

R. S. Chaughule, A. S. Lokur (eds.), *Applications of Nanotechnology in Microbiology*, https://doi.org/10.1007/978-3-031-49933-3_17

bulk materials, allowing for increased interaction with target pollutants. This increased surface area enables nanomaterials to exhibit exceptional adsorption capacities and reactivity, making them highly effective in the removal of a wide range of contaminants. The rapid expansion of industrial and agricultural activities has resulted in the widespread contamination of the environment with heavy metals. This pervasive heavy metal pollution has emerged as a significant environmental issue, posing a grave threat to human health. Heavy metals, such as cadmium (Cd(II)), lead (Pb(II)), and copper (Cu(II)), which are among the most dangerous and prevalent types, pose a significant challenge in terms of degradation compared to traditional organic compounds like ammonia nitrogen, phosphorus, and other common pollutants. These heavy metals exhibit persistence, resistance to breakdown, and the ability to accumulate through biological enrichment, resulting in severe harm to humans, animals, and plants. The primary sources of wastewater containing these heavy metals are electroplating water discharge, metal mining drainage, and the textile printing and dyeing industries [5]. Various techniques have been devised to eliminate heavy metal ions from industrial wastewater, including chemical precipitation, ion exchange, liquid–liquid extraction, electrodialysis, and the use of resins. However, each of these methods has its drawbacks in terms of cost, complexity, or efficiency. Conversely, adsorption is widely acknowledged as a cost-effective and efficient approach. Recently, nanoparticles have gained considerable attention for their unique properties and have become a subject of extensive research as potential adsorbents. Nanoparticles possess a high surface area to volume ratio, enabling a larger number of active sites for pollutant species to interact with, thereby enhancing the adsorption capacity. Pollutants in the environment are typically present in the form of molecules or ions dispersed in a medium (e.g., air or water). To remove these pollutants effectively, they need to come into contact with a reactive material capable of capturing or binding them. In the case of nanoparticles, their large surface area provides numerous active sites or binding sites that can interact with the pollutant species. Because of the increased number of active sites, nanoparticles offer a higher chance of adsorbing the pollutant molecules or ions. This leads to an enhanced adsorption capacity, as more pollutants can be trapped or immobilized on the nanoparticle surfaces. Additionally, the small size of nanoparticles allows them to reach and interact with pollutant molecules in areas where larger materials might not be as effective. The enhanced adsorption capacity of nanoparticles makes them valuable in various applications, including environmental remediation, water purification, and air filtration. They can effectively adsorb a wide range of pollutants, such as heavy metals, dyes, organic compounds, and even certain gases, contributing to cleaner and safer environments. Although nanoparticles have shown promise in wastewater treatment, further investigation is needed to fully understand their potential and limitations [6].

17.1.1 Nanomaterials for Wastewater Remediation

Nanoremediation for wastewater treatment involves the utilization of nanotechnology-based materials and processes to remediate and purify polluted water sources. This approach shows great promise in dealing with a wide range of pollutants and contaminants found in wastewater. Nanoparticles and nanocomposites are examples of nanomaterials that exhibit distinctive characteristics, including a large surface area, reactivity, and strong adsorption capabilities, making them highly efficient in eliminating pollutants from water. Water pollution is a pressing concern that requires immediate attention. Water pollution caused by the use of various dyes is a significant environmental issue. Nanomaterials possess unique physicochemical properties, such as high surface area, reactivity, and adsorption capacity, making them ideal candidates for wastewater treatment. Nanobiotechnology integrates these nanomaterials with biological entities, such as enzymes, cells, or biomolecules, to enhance their adsorption capabilities and selectivity. Various nanomaterials, including metal and metal oxide nanoparticles, carbon-based nanomaterials, and hybrid nanocomposites, have demonstrated remarkable potential in removing contaminants from wastewater. Nanoremediation refers to the utilization of appropriate nanomaterials (NMs) to remediate and remove environmental pollutants from soil, water, and air. This technology offers the advantage of conducting the cleanup process in situ, thereby eliminating the requirement for excavating and transporting contaminated soils. Additionally, various methods can be employed to regenerate and repurpose nanomaterials for contaminant treatment purposes, such as the magnetic separation of iron nanoparticles or the recovery of metals from spent nanosorbents [7]. The extensive surface modification capabilities and adjustable physical parameters of nanomaterials provide significant advantages over traditional approaches in addressing environmental contamination. Furthermore, methods that combine multiple materials (hybrids/composites) to leverage specific properties from each component offer potential enhancements in efficiency, selectivity, and stability compared to single nanoplatform-based methods. For example, incorporating nanoparticles onto a scaffold can enhance material stability compared to using nanoparticles alone. Additionally, functionalizing the material with specific chemicals designed to target desired contaminant molecules can enhance selectivity and efficiency [8]. One potential and promising solution to this severe problem is semiconductor photocatalysis. Several metal oxides, including ZnO, WO_3, In_2O_3, SnO_2, Cu_2O, and TiO_2, have been utilized in the photocatalytic degradation of contaminants [9]. The stability of quantum dots within the hydrogel network has garnered significant attention in the design of highly efficient photocatalysts for quantum dot nanocomposite hydrogels.

17.1.2 Bio-adsorbents in Nanobiotechnology

Bio-adsorbents, derived from living organisms or their components, have gained considerable attention due to their biocompatibility, renewable nature, and diverse functionalities. Nanobiotechnology exploits the unique properties of bio-adsorbents, coupled with the advantages of nanomaterials, to develop highly efficient and environmentally friendly approaches for wastewater remediation. These bio-adsorbents can be broadly categorized into three types: microbial bio-adsorbents, plant-based bio-adsorbents, and biomolecule-based bio-adsorbents.

17.1.2.1 Microbial Bio-adsorbents

A microbial-based nanobiosorbent is a mixture of microbial cells and nanomaterials that acts as an adsorbent to remove contaminants from diverse environmental matrices. It utilizes the unique features of both microbial cells and nanomaterials to improve the biosorbents adsorption capacity and efficiency. Microorganisms, such as bacteria, fungi, and algae, have inherent abilities to interact with various contaminants. Nanobiotechnology enhances the adsorption capacity of microbial bio-adsorbents by immobilizing them onto nanomaterials or modifying their surfaces with nanoparticles. Due to their versatile properties, including the ability to thrive in varying concentrations, pH levels, and temperatures, microorganisms are widely employed as biosorbents. They demonstrate remarkable efficiency in the uptake of metals, dyes, and pollutants. The composition of the cell wall in microorganisms plays a crucial role in the removal of pollutants [10]. The synergistic effect between microbial activity and nanomaterial properties improves the removal efficiency of pollutants. By combining microbial cells and nanomaterials, the resulting nanobiosorbent can harness the advantages of both components. The microbial cells provide specific binding sites for contaminants, while nanomaterials amplify the surface area and increase the availability of binding sites, leading to improved adsorption capacity and efficiency. Moreover, genetic engineering techniques can be employed to optimize the performance of microbial bio-adsorbents for targeted contaminants. The use of a nanobiosorbent that relies on microorganisms enables the effective elimination of various substances such as heavy metals, organic pollutants, dyes, pharmaceuticals, and other contaminants present in water and wastewater sources. The removal of pollutants from contaminated soils and sediments can be facilitated by utilizing the nanobiosorbent, which effectively adsorbs substances like pesticides, hydrocarbons, and heavy metals. In air pollution control, the nanobiosorbent has the capability to be incorporated into air filtration systems, aiding in the elimination of volatile organic compounds (VOCs), gases, and particulate matter. The inclusion of nanomaterials enhances the microbial activity and improves the efficiency of bioremediation processes, such as the degradation of organic pollutants or the reduction of heavy metal toxicity. This enhancement is particularly useful in bioremediation efforts. Microbial biosorbents offer a cost-effective and environmentally friendly solution for removing xenobiotic compounds from polluted environments. These synthetic compounds can pose risks to both human health and the environment. The microbial biosorbents efficiently adsorb or absorb xenobiotic compounds from water or soil. Moustafa studied the extracellular synthesis of silver nanoparticles using two filamentous fungi, *Penicillium Citreonigum Dierck* and *Scopulariopsis Brumptii Salvanet-Duval,* which were isolated from wastewater. Polyurethane foam was employed as a carrier for the silver nanoparticles. The resulting nanosilver solution was utilized for the removal of pathogenic bacteria from polluted water, showcasing its potential as an effective method for water purification [11].

17.1.2.2 Plant-Based Bio-adsorbents

Biosorbents derived from plant sources, also referred to as plant-based bio-adsorbents or biosorbents, exhibit the capacity to adsorb or attach specific substances in their vicinity. These biosorbents find extensive application in environmental scenarios, including wastewater treatment, air purification, and the elimination of contaminants from soil or water. Plants possess natural mechanisms to uptake and accumulate pollutants from the environment. By incorporating nanomaterials into plant tissues or utilizing nanoparticles for surface modification, nanobiotechnology facilitates the development of plant-based bio-adsorbents with enhanced adsorption capacity and selectivity. Additionally, the use of genetically modified plants can further enhance the remediation potential by optimizing pollutant-specific uptake pathways and accumulation. Plant-based adsorbents derived from agriculture possess economic advantages, as they are renewable and abundantly available. These adsorbents are primarily composed of cellulose and lignin, making them a viable choice for wastewater treatment [10]. Dubey et al. utilizes the waste biomass of the *Portulaca oleracea* plant as an environmentally friendly biosorbent to remove copper ions from an aqueous solution [12].

17.1.2.3 Biomolecule-Based Bio-adsorbents

Nanosorbents based on biomolecules refer to materials that employ biomolecules, such as proteins, enzymes, DNA, or other biological molecules, as the active components for sorption or adsorption of different substances. These nanosorbents have garnered considerable interest in recent times, thanks to their distinct properties and promising applications in areas such as environmental remediation, water purification, drug delivery, and bio separation. Biomolecules, including enzymes, proteins, and DNA, exhibit strong affinity towards specific contaminants. Nanobiotechnology exploits the inherent binding properties of biomolecules and combines them with nanomaterials to create highly efficient bio-adsorbents. The immobilization of biomolecules onto nanomaterial surfaces provides stability, reusability, and improved selectivity for targeted pollutants. Furthermore, the integration of nanosensors within the bio-adsorbent systems enables real-time monitoring of pollutant levels during the remediation process. Protein-based, enzyme-based, DNA-based, and polysaccharide-based nanosorbents offer unique properties and functionalities for a variety of applications. Proteins, such as albumin, can be utilized as nanosorbents for drug delivery due to their biocompatibility and ability to carry therapeutic agents. Antibodies or aptamers immobilized on nanoparticles enable selective binding for diagnostic or sensing purposes. Enzyme-functionalized nanoparticles, like those incorporating laccase or peroxidase enzymes, efficiently degrade organic pollutants. DNA-based nanosorbents can selectively bind specific molecules or ions, making them valuable for DNA-based sensors or the adsorption of heavy metals. Polysaccharide-based nanosorbents, such as chitosan or cellulose derivatives, can be chemically modified or combined with nanoparticles to enhance their adsorption capabilities, making them suitable for water treatment, metal ion removal, or drug delivery applications.

17.2 Nanomaterials as Effective Tools for Water Contamination and Remediating Pollutants

The implementation of green chemistry principles in nanomaterial preparation has introduced innovative techniques that reduce or completely eliminate the use of hazardous reagents in both research and industrial processes. This is achieved by leveraging nanomaterials that are readily available, non-toxic, exhibit superior adsorption efficiency, and are cost-effective. As a result, green chemistry has opened up new avenues for sustainable and environmentally friendly practices in the field of nanotechnology [13]. Industrial effluent streams are the primary pathways through which heavy metals and dyes become integrated into the ecological system. As a result, they present significant risks to human health. Nanotechnology is gaining increasing attention due to the widespread applications of nanoparticles (NPs) in various aspects of life. One of the most promising avenues for environmental remediation involves photocatalysis using nanostructured semiconductors, making it a focal point of research in nanoscience and nanotechnology. This approach has garnered significant interest for its remarkable capability to eliminate organic pollutants from the environment [14]. Nanomaterials can be classified into four main categories: (1) metal-based or inorganic nanomaterials, (2) carbonaceous nanomaterials, (3) polymer-based nanomaterials, and (4) composite nanomaterials. Figure 17.1 illustrates a diagrammatic representation showcasing different types of nanomaterials employed for the elimination of environmental pollutants. Metal-based nanomaterials such as iron-based nanoparticles, copper-based nanoparticles,

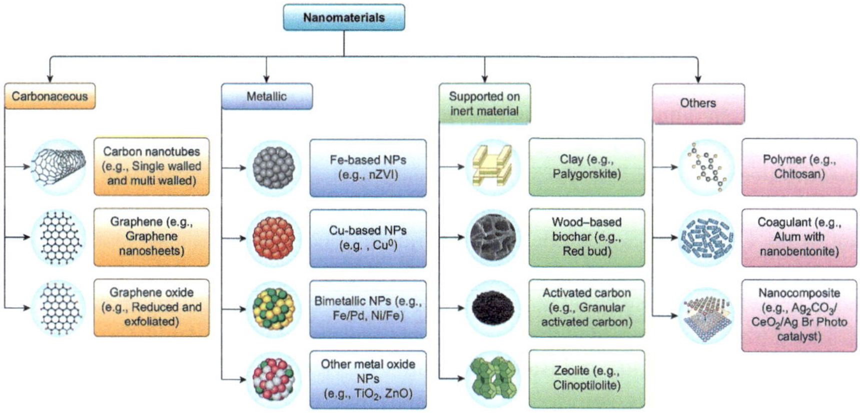

Fig. 17.1 Types of nanomaterials used for the removal of environmental contaminants

and zinc nanoparticles are extensively utilized in environmental remediation [7]. Industrial and agricultural development has resulted in widespread heavy metal pollution in the environment. Environmental pollution of this kind is a serious health concern that has become an important environmental problem. There is a serious threat to human health due to such pollution. Nanotechnology presents the promise of an effective removal of contaminants and microorganisms in the water treatment industry. These days, metals (such as cadmium, copper, lead, mercury, nickel, and zinc), microelements (such as phosphate, ammonia, and nitrite), cyanide, organics, algae (such as cyanobacterial toxins), viruses, bacteria, parasites, and antibiotics are all detected and eliminated using nanoparticles, nanomembranes, and nanostructured materials [15]. Streams and lakes are contaminated with these heavy metals because they are not biodegradable, but they can bioaccumulate in living organisms, causing health problems in animals, plants, and humans [16]. Chromium is among the heavy metals on which wastewater treatment has placed a lot of emphasis. Chromium is being used more and more in various sectors, which has resulted in a lot of polluted aqueous effluents with high chromium levels. Because of its solubility, it is extremely mobile in terrestrial and aquatic habitats and easily pierces the epidermis of plants and animals, harming tissues [17]. Currently, a variety of technologies, including filtration, adsorption, chemical precipitation, ion exchange, membrane separation techniques, and electro-remediation approaches, have been employed to remove heavy metals from an aqueous solution [18]. Adsorption provides various benefits over other techniques, including free operation, affordability, and compliance with rigorous effluent specifications. May be the majority of techniques could be very costly, and may not be efficient in removing heavy metal at low quantities. Due to their secondary effluent effects on the recipient ecosystem, these approaches are also ineffective. Because of this, the effluent wastewater treatment system needs to be fine-tuned using simple, efficient, affordable, and environmentally friendly ways [18]. There is an increasing concern over the contamination of water and soil by Cr (VI) due to its widespread distribution and high levels resulting from natural and anthropogenic causes. The oxidation of chromite by microbes in natural environments is caused by microbial interactions with mafic and ultramafic rocks. Some of these activities include mining and metalworking, making steel and metal alloys, producing paint, processing wood and paper, dyeing, and increasing wastewater chromium levels. Additional contributing factors to soil and water Cr (VI) levels include the incineration of coal or municipal waste, and the production of second-generation fertilizers. Global exposure to heavy metals has been known for a long time and continues to grow unabated. Heavy metal ions in drinking water are allowed within limits prescribed by the World Health Organization (WHO). Pakistan, India, and Bangladesh are experiencing an increasing amount of exposure. Therefore, cost-effective and environmentally friendly technologies are needed to remove heavy metal ions from wastewater. With its simplicity, recyclability, environmentally friendly characteristics, and high adsorption efficiency, adsorption holds immense potential to treat wastewater efficiently. As a result of the extensive research and progress on the adsorption process through nanomaterials to achieve environmental detoxification of toxic heavy metals and metalloids, this research work provides an analysis of the enduring research and progress activities.

Various types of nanomaterials are utilized for the purpose of removing environmental contaminants. These nanomaterials find application in a wide range of areas, including the following.

17.2.1 Metal-Based Nanomaterials

Nanoscale materials composed primarily of metal atoms or metal alloys are known as metal-based nanomaterials. These materials possess exceptional properties and display distinct behaviors in comparison to their larger counterparts, owing to their small size and significant surface-to-volume ratio. Metal-based nanomaterials are employed across diverse domains such as electronics, catalysis, energy, medicine, and environmental remediation due to their wide range of applications. Metal-based nanomaterials, such as iron-based nanoparticles, copper-based nanoparticles, and boron nanoparticles, are commonly employed for environmental remediation. They exhibit excellent reactivity and can effectively degrade or capture contaminants in water or soil. Pan et al. investigated the adsorption of Cr(VI) onto synthetic monodispersed magnetite nanoparticles with organic surface coatings, and the influence of water chemistry on the adsorption performance [19]. Yahya et al. fabricated Ag-NPs (silver nanoparticles) by utilizing *Thespesia populnea* bark extract. These Ag-NPs were intended for use as a photocatalyst to efficiently degrade MB dye (methylene blue) and also to test their antimicrobial activity [14]. Vanaja et al. investigated the synthesis of silver nanoparticles using *M. tinctoria* leaf extract and the photocatalytic activity of the synthesized silver nanoparticles was evaluated by studying the degradation of methylene blue under sunlight irradiation [20]. Afkhami et al. utilized maghemite nanoparticles as an efficient adsorbent for the removal of CR (Congo Red) from wastewater. The technique proved to be highly beneficial and cost-effective in enhancing dye removal during wastewater treatment. The maghemite nanoparticles displayed exceptionally high adsorption capacities for CR, surpassing many other adsorbents in their effectiveness [21].

17.2.2 Carbonaceous Nanomaterials

Carbonaceous nanomaterials are a group of nanoscale materials made predominantly of carbon atoms. Because of their large surface area, high aspect ratio, and great mechanical strength, these materials have unique physical, chemical, and electrical characteristics. Carbonaceous nanomaterials, including carbon nanotubes, graphene, and graphene oxides, have shown promise in pollutant removal. Their large surface area and unique properties enable adsorption, catalysis, and degradation of contaminants. Sharma et al. developed graphene oxide-calcium-zinc (GO@CZ) nanocomposite for the adsorption of Cr(III) and Cu(II) from aqueous solutions [22]. Sun et al. utilized multiwalled carbon nanotube immobilized ionic liquids for the adsorption of Cr(VI) from a water phase [23]. Djilani et al. evaluated the efficiency of activated carbons derived from agricultural waste, specifically coffee grounds (CCG), orange peels (COP), and melon seeds (CMS), for the purpose of removing organic micropollutants commonly present in agricultural residues and industrial effluents [24].

17.2.3 Polymer-Based Nanomaterials

Polymer nanomaterials, alternatively referred to as polymer nanocomposites or polymer-based nanomaterials, encompass materials that involve the amalgamation of polymer matrices with nanoscale fillers or additives. These additives, such as nanoparticles, nanofibers, nanoclays, or other nanoscale materials, serve to enhance the properties of the polymers and facilitate the creation of innovative materials possessing distinctive characteristics. Polymer-based nanomaterials, such as chitosan and alginate nanoparticles, offer advantages in terms of their biocompatibility and tunable properties. They can be functionalized to selectively adsorb and remove specific pollutants from water or air. Haddad et al. investigated the adsorption and tensile behavior of electro spun polyacrylonitrile (PAN) nanofiber mats that are loaded with varying quantities of ZnO [25]. Awang et al. utilized recycled newspapers as a raw material for the fabrication of a cellulose-based membrane and subsequently employed as the adsorbent for the adsorption of Cr(VI) [26].

17.2.4 Composite Nanomaterials

Composite nanomaterials combine different components to enhance their remediation capabilities. Examples include clay-polymer nanocomposites, zeolite-based nanocomposites, and biochar-supported nanocomposites. These materials leverage the synergistic effects of their constituents to effectively remove contaminants. Composite nanomaterials have emerged as promising tools for wastewater treatment, offering enhanced adsorption, catalytic, and photocatalytic properties. Graphene-based composites, such as graphene oxide (GO) and reduced graphene oxide (rGO), exhibit high surface area and strong adsorption capabilities, while metal-organic framework (MOF)-based composites combine porous MOFs with nanomaterials for improved adsorption and catalytic activities. Carbon nanotube-based composites leverage the mechanical strength and surface area of carbon nanotubes, while magnetic nanoparticle-based composites enable easy separation using an external magnetic field. TiO_2-based composites enhance photocatalytic efficiency for the degradation of organic pollutants. These composite nanomaterials hold great potential for the efficient removal of various contaminants from wastewater. These nanomaterials are applied in diverse environmental contexts, including water treatment, soil remediation, air purification, and wastewater management. Their use holds significant potential for mitigating environmental pollution and improving the overall quality of ecosystems. Dim et al. examined the adsorption characteristics of clay treated with acids for the purpose of removing Fe (III) and Cr (VI) [27]. Saranya et al. utilized a green and eco-friendly exfoliation method to prepare G-ZnO/Ag nanocomposites using *Tridax procumbens* leaves extract through a microwave-assisted hydrothermal process. The prepared G-ZnO/Ag nanocomposites were then tested for their ability to degrade methylene blue dye, as well as their antimicrobial and anticancer activities [13]. Abhilash et al. synthesized Fe_2O_3/Cu_2O nanocomposite for the photocatalytic degradation of Rhodamine-B and Janus green dye [28]. Javed et al. investigated the photocatalytic efficiency of Cu–ZnO/S-g-C_3N_4 (CZS) nanocomposites in degrading Methylene Blue dye. The CZS-25 nanocomposites demonstrated excellent photocatalytic stability. The researchers proposed a plausible mechanism for MB degradation over CZS-25 nanocomposites, based on observations from photoluminescence and reactive species scavenger tests [9]. Modwi et al. prepared and characterized Ag-doped Cu/ZnO photo composites and assessed their photocatalytic performance. The incorporation of Agx onto the surface of Cu/ZnO nanoparticles significantly improved their photocatalytic activity. Notably, the photo composite with 3% Ag doping in Cu/ZnO exhibited exceptionally high photocatalytic activity for the degradation of malachite green [29]. Hu et al. synthesized CuO-ZnO–Ag (CZA) photocatalyst through a simple sol–gel method. They investigated the photocatalytic activity of the samples using Methylene Blue, Methyl Orange, and rhodamine B aqueous solutions for degradation. The study reveals the impact of Ag introduction on the photocatalytic performance of CZA by analyzing the microstructures, optical properties, and charge transfer within the heterostructures of CZA in-depth [30].

17.2.5 ZnO-Based Nanoadsorbents

There are several adsorbents for organic adsorption, including zinc oxide nanoparticles (ZnO–NPs), a type of semiconductor metal oxide with excellent adsorption properties. Besides being non-toxic, it also has a high chemical and optical corrosion resistance, a high catalytic activity, and stable chemical properties. ZnO–NPs have garnered significant interest in research due to their biocompatibility, reasonable cost, and long-term stability. Moreover, studies have shown that ZnO–NPs possess the ability to absorb heavy metals through their hydroxyl groups, indicating their potential for effective removal of heavy metal ions from dental wastewater. Many researchers have explored the utilization of ZnO–NPs for this purpose, recognizing their promising capabilities [1]. Leiva et al. showed the application of ZnO–NPs as nanoadsorbents for the removal of Cu(II) ions from Acid Mine Drainage (AMD) waters [31]. Sanjay et al. demonstrated a straightforward method of chemical precipitation to synthesize ZnO nanostructures, which can be utilized for the removal of As(III) from wastewater [32]. Moustafa Ahmed and Yousef conducted the synthesis of zinc oxide nanoparticles for the purpose of their application in the removal of Cr^{+6} from wastewater. To investigate their effectiveness, batch adsorption experiments were conducted, considering variables such as contact time, solution pH, solution temperature, concentration of Cr^{+6} ions, and adsorbent dosage [18]. Faisal et al. developed bioaugmented zinc oxide nanoparticles (ZnO–NPs) using aqueous fruit extracts of *Myristica fragrans*. The biosynthesized ZnO nanoparticles exhibited excellent photocatalytic capabilities, leading to an impressive 88% degradation of methylene blue dye within 140 minutes. The outstanding antioxidant properties are proved to be biocompatible nanomaterials [33].

17.2.6 Composite-Based Nanoadsorbents

Sharma et al. investigated the cooperative interaction between a polymer and metal oxide for water bioremediation, a cost-effective ZnO–Ag/PPy nanocomposite with a core/shell structure is synthesized using a microwave irradiation technique. The idea behind this approach is to create a PPy shell on the core structure consisting of Ag and ZnO. The resulting ZnO–Ag/PPy nanocomposite exhibits effective removal of inorganic pollutants (Cd^{2+} and PO_4^{3-} ions) from wastewater, both in single and binary contaminant aqueous systems [22]. Mustapha et al. described the utilization of ZnO/kaolin nanocomposites for treating pollutants in tannery wastewater. This study concentrates on investigating the efficacy of kaolin and kaolin/ZnO nanocomposites in removing specific heavy metals (Cr and Fe) as well as other indicators of water quality (COD, BOD, and chloride) from the wastewater [34]. Xiao and colleagues focused their research on the preparation and assessment of 4-vinyl pyridine (4VP)-modified magnetic $Ni_3Si_2O_5(OH)_4$ nanotubes for the effective elimination of Cr(VI). Various characterization techniques were utilized to uncover the structure and properties of $Ni_3Si_2O_5(OH)_4$-g-P_4VPNTs. In order to investigate the adsorption behavior and efficiency of Cr(VI) removal, batch experiments were conducted, considering variables such as solution pH, contact time, and temperature. Additionally, the adsorption kinetics, isotherms, and mechanisms were evaluated [35].

17.3 Challenges and Future Perspectives

Nanobiotechnology holds significant potential for revolutionizing wastewater treatment, offering efficient contaminant removal, and enhanced treatment capabilities. However, its implementation faces several critical challenges that necessitate careful consideration and solutions. One key concern is the potential release of engineered nanomaterials used in the treatment process. While these nanomaterials effectively remove pollutants, their uncontrolled release into the environment may have adverse ecological and health impacts. To address this, researchers must design nanomaterials with reduced environmental persistence and understand their behavior during and after treatment.

Another challenge lies in ensuring the long-term stability and effectiveness of bio-adsorbents, such as Genetically Modified Organisms (GMOs), which can bind and remove pollutants from wastewater. Researchers need to develop bio-adsorbents that can withstand harsh conditions, maintain their activity over extended periods, and avoid unintended side effects. Additionally, the scalability of nanobiotechnological solutions for large-scale wastewater treatment is essential to make them practical and cost-effective. Improving production methods, standardization, and quality control are crucial steps in achieving this.

The use of genetically modified organisms in wastewater treatment also raises concerns about potential environmental impacts and risks. Rigorous risk assessments and thorough evaluation of GMOs' behavior and potential for unintended consequences are necessary before their deployment. Governments and regulatory bodies must establish stringent guidelines to govern the use and release of nanomaterials and GMOs in wastewater treatment, ensuring responsible practices and monitoring.

To address these challenges, interdisciplinary collaboration between experts in nanotechnology, biotechnology, environmental science, and risk assessment is vital. Continuous research and innovation are necessary to develop safer and more sustainable nanobiotechnological solutions. Conducting comprehensive life cycle assessments will help identify potential environmental impacts and guide improvements in design and implementation. Public awareness about the benefits and risks of nanobiotechnology can foster understanding and support for responsible implementation.

Future research in the field of nanobiotechnology for wastewater remediation should prioritize several key areas to overcome the challenges and improve the efficacy and safety of these technologies.

17.3.1 Optimizing Nanomaterials Synthesis and Functionalization

Researchers should focus on refining the methods of synthesizing nanomaterials used in wastewater treatment. This includes developing techniques that are more cost-effective, scalable, and environmentally friendly. Moreover, functionalization of nanomaterials can enhance their pollutant adsorption capacity and selectivity, making them more efficient in wastewater treatment. Investigating various functionalization approaches and understanding their impact on the behavior and fate of nanomaterials in the environment is essential.

17.3.2 Exploring New Bio-adsorbents from Untapped Biological Sources

While GMOs and other bio-adsorbents have shown promise in wastewater treatment, there is still much potential to be explored from untapped biological sources. Research efforts should be directed towards discovering and characterizing novel bio-adsorbents from diverse organisms, including microorganisms, plants, and natural materials. These bio-adsorbents may offer unique and efficient pollutant removal capabilities while potentially reducing concerns related to GMO use.

17.3.3 Comprehensive Risk Assessments

As nanobiotechnological approaches for wastewater treatment continue to develop, comprehensive risk assessments are essential to evaluate potential environmental and health impacts. These assessments should consider the behavior of nanomaterials and bio-adsorbents during and after treatment, their potential release into the environment, and any long-term effects on ecosystems and human health. Risk assessments will help identify potential hazards and guide the development of safer and more sustainable nanobiotechnological solutions.

17.3.4 Collaboration Between Experts and Stakeholders

Collaboration among scientists, engineers, policymakers, and environmental stakeholders is crucial for the successful and responsible implementation of nanobiotechnology in wastewater treatment. Interdisciplinary cooperation allows for a more holistic approach to addressing challenges, promoting innovation, and ensuring the technology's alignment with societal and environmental needs. Engaging with stakeholders and the public also fosters transparency, understanding, and acceptance of these new technologies.

17.3.5 Standardization and Regulation

To ensure the safe and consistent implementation of nanobiotechnological approaches, it is necessary to establish standardized protocols and regulations. These guidelines should cover the manufacturing, application, and monitoring of nanomaterials, bio-adsorbents, and GMOs in wastewater treatment. Adherence to strict regulations will minimize environmental risks and provide a framework for responsible and ethical use.

In summary, future research in nanobiotechnology for wastewater remediation should prioritize optimizing nanomaterials and bio-adsorbent synthesis, exploring novel bio-adsorbents, conducting thorough risk assessments, and fostering collaboration among different stakeholders. By addressing these aspects, scientists and policymakers can work together to ensure the safe and sustainable implementation of nanobiotechnology in wastewater treatment, contributing to cleaner and healthier water resources for the future. Overall, while nanobiotechnology holds great promise for wastewater remediation, a cautious and thoughtful approach is necessary to ensure its safe and successful implementation in the long run.

17.4 Conclusion

Nanobiotechnology offers innovative and sustainable solutions for wastewater remediation through the development of bio-adsorbents. By harnessing the unique properties of nanomaterials and biological entities, nanobiotechnology enhances the adsorption capacity, selectivity, and efficiency of wastewater treatment processes. While challenges exist, ongoing research and collaborations hold tremendous potential to address these issues and pave the way for the widespread application of nanobiotechnology in tackling wastewater pollution. The insights provided in this chapter underscore the importance of nanobiotechnology as a promising avenue for sustainable wastewater remediation.

References

1. Gu, M., Hao, L., Wang, Y., Li, X., Chen, Y., Li, W., & Jiang, L. (2020). The selective heavy metal ions adsorption of zinc oxide nanoparticles from dental wastewater. *Chemical Physics, 534*, 110750. https://doi.org/10.1016/j.chemphys.2020.110750

2. Sinha, V., Manikandan, N. A., Pakshirajan, K., & Chaturvedi, R. (2016). Continuous removal of Cr(VI) from wastewater by phytoextraction using *Tradescantia pallida* plant based vertical subsurface flow constructed wetland system. *International Biodeterioration and Biodegradation, 119*, 96–103. https://doi.org/10.1016/j.ibiod.2016.10.003
3. Fang, X., Wang, Y., Wang, Z., Jiang, Z., & Dong, M. (2019). Microorganism assisted synthesized nanoparticles for catalytic applications. *Energies, 12*, 190. https://doi.org/10.3390/en12010190
4. Applied, W., & Journal, S. (2012). Application of laser controlled nanoparticles in waste water treatment. *The Egyptian Journal of Environmental Change, 4*, 1–14. https://doi.org/10.21608/ejec.2012.96648
5. Li, A. Y., Deng, H., Jiang, Y. H., Ye, C. H., Yu, B. G., Zhou, X. L., & Ma, A. Y. (2020). Superefficient removal of heavy metals from wastewater by mg-loaded biochars: Adsorption characteristics and removal mechanisms. *Langmuir, 36*, 9160–9174. https://doi.org/10.1021/acs.langmuir.0c01454
6. Nalwa, K., Thakur, A., & Sharma, N. (2021). Synthesis of ZnO nanoparticles and its application in adsorption. *Advanced Materials Proceedings, 2*, 697–703. https://doi.org/10.5185/amp/2017/696
7. Mukhopadhyay, R., Sarkar, B., Khan, E., Alessi, D. S., Biswas, J. K., Manjaiah, K. M., Eguchi, M., Wu, K. C. W., Yamauchi, Y., & Ok, Y. S. (2022). Nanomaterials for sustainable remediation of chemical contaminants in water and soil. *Critical Reviews in Environmental Science and Technology, 52*, 2611–2660. https://doi.org/10.1080/10643389.2021.1886891
8. Guerra, F. D., Attia, M. F., Whitehead, D. C., & Alexis, F. (2018). Nanotechnology for environmental remediation: Materials and applications. *Molecules, 23*, 1760. https://doi.org/10.3390/molecules23071760
9. Javed, M., Qamar, M. A., Shahid, S., Alsaab, H. O., & Asif, S. (2021). Highly efficient visible light active Cu-ZnO/S-g-C3N4nanocomposites for efficient photocatalytic degradation of organic pollutants. *RSC Advances, 11*, 37254–37267. https://doi.org/10.1039/d1ra07203j
10. Jain, C. K., Malik, D. S., & Yadav, A. K. (2016). Applicability of plant based biosorbents in the removal of heavy metals: A review. *Environmental Processes, 3*, 495–523. https://doi.org/10.1007/s40710-016-0143-5
11. Moustafa, M. T. (2017). Removal of pathogenic bacteria from wastewater using silver nanoparticles synthesized by two fungal species. *Water Science, 31*, 164–176. https://doi.org/10.1016/j.wsj.2017.11.001
12. Dubey, A., & Mishra, A. (2017). A novel plant-based biosorbent for removal of copper (II) from aqueous solutions: Biosorption of copper (II) by dried plant biomass. *Journal of Renewable Materials, 5*, 54–61. https://doi.org/10.7569/JRM.2016.634127
13. Saranya, A., Murad, A., Thamer, A., Priyadharsan, A., & Maheshwaran, P. (2021). Preparation of reduced ZnO/Ag nanocomposites by a green microwave-assisted method and their applications in photodegradation of methylene blue dye, and as antimicrobial and anticancer agents. *Chemistry Select, 6*, 3995–4004. https://doi.org/10.1002/slct.202100413
14. Yahya Tahir, M., Ahmad, A. A., Alothman, A., Mushab, M. S. S., & Ali, S. (2022). Green synthesis of silver nanoparticles using *Thespesia populnea* bark extract for efficient removal of methylene blue (MB) degradation via photocatalysis with antimicrobial activity and for anticancer activity. *Bioinorganic Chemistry and Applications*, 1–12. https://doi.org/10.1155/2022/7268273
15. Singh, S., Kumar, V., Romero, R., Sharma, K., & Singh, J. (2019). Applications of nanoparticles in wastewater treatment. *Nanotechnology in the Life Sciences, 3*, 395–418. https://doi.org/10.1007/978-3-030-17061-5_17
16. Argun, M. E., & Dursun, S. (2008). A new approach to modification of natural adsorbent for heavy metal adsorption. *Bioresource Technology, 99*, 2516–2527. https://doi.org/10.1016/j.biortech.2007.04.037

17. Khalfa, L., Cervera, M. L., Bagane, M., & Souissi-Najar, S. (2015). Modeling of equilibrium isotherms and kinetic studies of Cr (VI) adsorption into natural and acid-activated clays. *Arabian Journal of Geosciences, 9*, 75. https://doi.org/10.1007/s12517-015-2104-0
18. Moustafa Ahmed, N., & Yousef, N. P. (2015). Synthesis and characterization of zinc oxide nanoparticles for the removal of CrVI. *International Journal of Scientific and Engineering Research, 6*, 1235–1243.
19. Pan, Z., Zhu, X., Satpathy, A., Li, W., Fortner, J. D., & Giammar, D. E. (2019). Cr(VI) adsorption on engineered iron oxide nanoparticles: Exploring complexation processes and water chemistry. *Environmental Science and Technology, 53*, 11913–11921. https://doi.org/10.1021/acs.est.9b03796
20. Vanaja, M., Paulkumar, K., Baburaja, M., Rajeshkumar, S., Gnanajobitha, G., Malarkodi, C., Sivakavinesan, M., & Annadurai, G. (2014). Degradation of methylene blue using biologically synthesized silver nanoparticles. *Bioinorganic Chemistry and Applications*, 1–8. https://doi.org/10.1155/2014/742346
21. Afkhami, A., & Moosavi, R. (2010). Adsorptive removal of Congo red, a carcinogenic textile dye, from aqueous solutions by maghemite nanoparticles. *Journal of Hazardous Materials, 174*, 398–403. https://doi.org/10.1016/j.jhazmat.2009.09.066
22. Sharma, R., & Kumar, D. (2018). Adsorption of Cr (III) and Cu (II) on hydrothermally synthesized graphene oxide–calcium–zinc nanocomposite. *Journal of Chemical & Engineering Data, 63*(12), 4560–4572.
23. Sun, L., Wang, M., Li, W., Luo, S., Wu, Y., Ma, C., & Liu, S. (2020). Adsorption separation of Cr(VI) from a water phase using multiwalled carbon nanotube-immobilized ionic liquids. *ACS Omega, 5*, 22827–22839. https://doi.org/10.1021/acsomega.0c02016
24. Djilani, C., Zaghdoudi, R., Modarressi, A., Rogalski, M., Djazi, F., & Lallam, A. (2012). SElimination of organic micropollutants by adsorption on activated carbon prepared from agricultural waste. *Chemical Engineering Journal, 189–190*, 203–212. https://doi.org/10.1016/j.cej.2012.02.059
25. Haddad, M. Y., & Alharbi, H. F. (2019). Enhancement of heavy metal ion adsorption using electrospun polyacrylonitrile nanofibers loaded with ZnO nanoparticles. *Journal of Applied Polymer Science, 136*, 1–11. https://doi.org/10.1002/app.47209
26. Awang, N. A., Wan Salleh, W. N., Ismail, A. F., Yusof, N., Aziz, F., & Jaafar, J. (2019). Adsorption behavior of chromium(VI) onto regenerated cellulose membrane. *Industrial and Engineering Chemistry Research, 58*, 720–728. https://doi.org/10.1021/acs.iecr.8b02366
27. Dim, P. E., Mustapha, L. S., Termtanun, M., & Okafor, J. O. (2021). Adsorption of chromium (VI) and iron (III) ions onto acid-modified kaolinite: Isotherm, kinetics and thermodynamics studies. *Arabian Journal of Chemistry, 14*, 103064. https://doi.org/10.1016/j.arabjc.2021.103064
28. Abhilash, M. R., Akshatha, G., & Srikantaswamy, S. (2019). Photocatalytic dye degradation and biological activities of the Fe2O3/Cu2O nanocomposite. *RSC Advances, 9*, 8557–8568. https://doi.org/10.1039/c8ra09929d
29. Modwi, A., Taha, K. K., Khezami, L., Bououdina, M., & Houas, A. (2019). Silver decorated Cu/ZnO photocomposite: Efficient green degradation of malachite. *Journal of Materials Science: Materials in Electronics, 30*, 3629–3638. https://doi.org/10.1007/s10854-018-00642-w
30. Hu, W., Zhang, Q., Luo, K., Yuan, H., Li, J., Xu, M., & Xu, S. (2020). Enhanced photocatalytic properties of CuO–ZnO nanocomposites by decoration with Ag nanoparticles. *Ceramics International, 46*, 24753–24757. https://doi.org/10.1016/j.ceramint.2020.06.235
31. Leiva, E., Tapia, C., & Rodríguez, C. (2021). Highly efficient removal of cu(Ii) ions from acidic aqueous solution using zno nanoparticles as nano-adsorbents. *Water (Switzerland), 13*, 1–21. https://doi.org/10.3390/w13212960
32. Sanjay, S., & Srivastav, Y. (1973). Removal of arsenic from waste water by using ZnO nanomaterials. *Journal of Materials Science and Engineering B, 89*, 101–106.

33. Faisal, S., Jan, H., Shah, S. A., Shah, S., Khan, A., Akbar, M. T., Rizwan, M., Jan, F., Wajidullah, Akhtar, N., Khattak, A., & Syed, S. (2021). Green synthesis of zinc oxide (ZnO) nanoparticles using aqueous fruit extracts of *Myristica fragrans*: Their characterizations and biological and environmental applications. *ACS Omega, 6*, 9709–9722. https://doi.org/10.1021/acsomega.1c00310
34. Mustapha, S., Tijani, J. O., Ndamitso, M. M., Abdulkareem, S. A., Shuaib, D. T., Mohammed, A. K., & Sumaila, A. (2020). The role of kaolin and kaolin/ZnO nanoadsorbents in adsorption studies for tannery wastewater treatment. *Scientific Reports, 10*, 1–22. https://doi.org/10.1038/s41598-020-69808-z
35. Xiao, C., & Lin, J. (2020). Efficient removal of Cr(VI) ions by a novel magnetic 4-vinyl pyridine grafted Ni3Si2O5(OH)4multiwalled nanotube. *ACS Omega, 5*, 23099–23110. https://doi.org/10.1021/acsomega.0c02874

Index

A

Adsorption, 15, 26, 44, 98, 111, 422–434
Advanced nanoparticles, 286
Anode, 407–415
Antibacterial effect, 60, 102, 137, 298, 302
Antifungals, 8, 13, 15, 25, 61, 105, 181, 240, 260, 271–286, 333, 334
Antimicrobial, 7, 59, 80, 97, 136, 173, 203, 271, 294, 365, 400, 428
Antimicrobial agents, 12–15, 20, 64–69, 99, 100, 104–106, 114, 118, 138, 140, 203, 230, 243, 244, 246–250, 253, 260, 278
Antiviral, 13, 59–70, 80, 333, 335, 350

B

Bacteria, 1, 41, 59, 89, 99, 136, 164, 214, 291, 316, 334, 376, 400, 409, 421
Bio-adsorbent, 426, 434
Biofilm, 17, 59, 105, 143, 201, 271, 291, 317, 413
Bioreceptors, xi, 193, 194, 200, 208–215, 218, 220, 369
Biosensors, 10, 18–19, 141–143, 193, 194, 196, 201, 206–212, 215–220, 310, 327–329, 357, 358, 363, 364, 368–376, 379, 380
Biosynthesis, 1, 2, 5, 10–11, 16, 44, 47, 48, 50, 169, 170, 334, 342, 343, 345

C

Candida albicans, 20, 61, 89, 118, 242, 271–286, 295
Caries, 116, 231, 232, 237, 239, 241, 242, 244, 255, 291–304
Copper oxide nanoparticles (CuO NPs), 44, 99–101, 117, 281, 282, 285, 294, 345

D

Dental, 116, 231, 232, 237–239, 241–244, 248, 251, 252, 255, 257, 260, 279, 282, 291–304, 431
Dental plaque, 230, 232, 237–244, 246, 257, 292, 299
Deoxyribonucleic acid (DNA), 16, 59, 79, 100, 135, 157, 194, 273, 302, 323, 334, 367, 387, 425
Detection, 11, 70, 111, 141, 182, 194, 310, 357, 364, 388
Diagnosis, 194, 203, 204, 208, 212, 220, 230, 310, 315, 327, 363–367, 377, 378, 380, 398–399, 403
Diseases prevention, 77–89
Drug delivery, 10–12, 17–18, 63, 83, 110, 147, 273, 280, 281, 333–359, 425, 426
Drug resistance, 60, 65, 105, 106, 248–251, 272, 274, 278, 286, 337, 367, 400

E

Electricity, 320, 407–409, 414, 415
Electrode, 200, 212, 215–218, 321, 322, 328, 370–373, 375, 376, 378, 393, 409–415

R. S. Chaughule, A. S. Lokur (eds.), *Applications of Nanotechnology in Microbiology*, https://doi.org/10.1007/978-3-031-49933-3

Encapsulation, 10, 62–64, 67, 84, 106, 107, 124, 136, 138, 140, 141, 201, 203, 282, 346, 350, 352
Endodontic treatment, 247, 294, 298

F
Food borne pathogens, 193–220
Food microbial, 136, 139, 143
Food safety issues, 21
Food shelf life, 144

G
Gene-delivery, 63, 70, 85, 158, 178, 179, 182, 280, 348
Genome editing, 158, 162, 164, 166, 168–176, 178, 181, 183
Genomics, 103, 163, 166, 173–176, 323, 324, 390–392, 399–402
Green chemistry, 426

I
Immobilization, 24, 27, 98, 111–112, 138, 213, 218, 373, 378, 426
Infectious diseases, 23, 59, 70, 77, 80, 89, 97, 208, 232, 327, 364, 376–378

L
Lab on chip, 315

M
Metabolites, 1–28, 46, 170, 233, 236, 333–335, 337, 342, 349, 354, 358
Microbe-based cosmetics, 351–354
Microbe-based nanosystems, 354–357
Microbes, 5, 23, 28, 42, 48–51, 59, 77, 100–103, 106, 113–116, 118, 121, 136, 137, 141, 143, 183, 194, 231, 232, 271, 277, 314, 315, 318, 322–324, 333–336, 342, 343, 352–354, 358, 401, 409, 428
Microbes mediated synthesis, 41–51, 343
Microbial fuel cell (MFC), 310, 320–322, 407–415
Microbiology, 2, 11, 135–149, 242, 309, 310, 314, 315, 323, 334, 398
Microfluidics, 194, 212, 213, 217–219, 309–330, 377–379
Microneedles, 355–357
Molecular dynamics (MD), 395–398

N
Nanocarriers, 18, 82, 83, 86, 104, 141, 146, 281, 292, 293, 296, 333, 337, 338, 340, 353
Nanomaterials, 2, 41, 82, 97, 135, 193, 272, 292, 333, 368, 411, 421
Nanoparticles (NPs), 1, 41, 59, 79, 97, 136, 179, 196, 278, 293, 333, 368, 413, 421
Nanopore, 196, 388–390, 392–398, 403
Nano SDF, 301–304
Nanosensors, 22, 114, 141, 143, 144, 149, 194, 364, 369, 426
Nanotechnology, 1, 60, 79, 98, 180, 200, 272, 293, 333, 368, 388, 421
Nanovaccines, 63, 81, 83–89, 338, 346
Nanoviricide, 64–68, 70
Next-generation sequencing (NGS), 367, 387–403

O
Oral diseases, 231, 237, 291

P
Periodontal diseases, 231, 237, 238, 241–243
Plant-microbe, xi

S
SARS-CoV-2, 64, 66, 68, 80, 116, 339, 348, 349, 371
Sequencing, 175, 214, 244, 367, 387–395, 397, 399, 400, 403
Silver diamine fluoride (SDF), 294, 296–302
Silver nanoparticles (AgNPs), 7, 14, 44, 60, 99, 136, 272, 278, 279, 284, 285, 293, 343, 425
Sustainable approach, 50

T
Traits, 12, 14, 15, 157, 158, 160, 161, 163, 166, 168, 169, 172, 179, 277, 353, 389, 401

V
Vaccine, 18, 65, 77, 138, 333
Virus, 1, 60, 81, 99, 174, 302, 329, 338, 363, 394, 427

W
Wastewater, 26, 27, 117, 118, 120, 320, 407–409, 415, 421–434

Z
Zinc oxide nanoparticles (ZnO NPs), 8, 44, 49, 98, 99, 102, 114, 115, 118, 280, 281, 294, 345, 430, 431

GPSR Compliance

The European Union's (EU) General Product Safety Regulation (GPSR) is a set of rules that requires consumer products to be safe and our obligations to ensure this.

If you have any concerns about our products, you can contact us on ProductSafety@springernature.com

In case Publisher is established outside the EU, the EU authorized representative is:

Springer Nature Customer Service Center GmbH
Europaplatz 3
69115 Heidelberg, Germany

Batch number: 10614857

Printed by Printforce, the Netherlands